Lecture Notes in Mechanical Engineering

Lecture Notes in Mechanical Engineering (LNME) publishes the latest developments in Mechanical Engineering—quickly, informally and with high quality. Original research reported in proceedings and post-proceedings represents the core of LNME. Volumes published in LNME embrace all aspects, subfields and new challenges of mechanical engineering. Topics in the series include:

- Engineering Design
- Machinery and Machine Elements
- Mechanical Structures and Stress Analysis
- Automotive Engineering
- Engine Technology
- Aerospace Technology and Astronautics
- Nanotechnology and Microengineering
- Control, Robotics, Mechatronics
- MEMS
- Theoretical and Applied Mechanics
- Dynamical Systems, Control
- Fluid Mechanics
- Engineering Thermodynamics, Heat and Mass Transfer
- Manufacturing
- Precision Engineering, Instrumentation, Measurement
- Materials Engineering
- Tribology and Surface Technology

To submit a proposal or request further information, please contact the Springer Editor of your location:

China: Ms. Ella Zhang at ella.zhang@springer.com
India: Priya Vyas at priya.vyas@springer.com
Rest of Asia, Australia, New Zealand: Swati Meherishi at swati.meherishi@springer.com
All other countries: Dr. Leontina Di Cecco at Leontina.dicecco@springer.com

To submit a proposal for a monograph, please check our Springer Tracts in Mechanical Engineering at http://www.springer.com/series/11693 or contact Leontina.dicecco@springer.com

Indexed by SCOPUS. All books published in the series are submitted for consideration in Web of Science.

More information about this series at https://link.springer.com/bookseries/11236

Francisco Cavas Martínez ·
Guillermo Peris-Fajarnes ·
Paz Morer Camo · Ismael Lengua Lengua ·
Beatriz Defez García
Editors

Advances in Design Engineering II

Proceedings of the XXX International Congress INGEGRAF, 24–25 June, 2021, Valencia, Spain

Springer

Editors
Francisco Cavas Martínez
Departamento de Estructuras, Construcción y Expresión Gráfica
Universidad Politécnica de Cartagena
Cartagena, Spain

Guillermo Peris-Fajarnes
CITG. Centro de Investigación en Tecnologías Gráficas
Universitat Politècnica de València
Valencia, Spain

Paz Morer Camo
Departamento de Mecánica
Tecnun - Universidad de Navarra
San Sebastián - Donostia, Guipúzcoa, Spain

Ismael Lengua Lengua
Departamento de Ingeniería Gráfica
Universitat Politècnica de València
Valencia, Spain

Beatriz Defez García
Departamento de Ingeniería Gráfica
Universitat Politècnica de València
Valencia, Spain

ISSN 2195-4356 ISSN 2195-4364 (electronic)
Lecture Notes in Mechanical Engineering
ISBN 978-3-030-92425-6 ISBN 978-3-030-92426-3 (eBook)
https://doi.org/10.1007/978-3-030-92426-3

This Springer imprint is published by the registered company Springer Nature Switzerland AG
The registered company address is: Gewerbestrasse 11, 6330 Cham, Switzerland

Preface and Acknowledgements

The INGEGRAF 2021 Conference originates as the 30th International Conference on GRAPHICS ENGINEERING "Digital Engineering, its application in Research, Development and Innovation. Engineering and digital transformation in education".

INGEGRAF 2021 has been organized by the Department of Graphical Expression and Graphics Technologies Research Center of the Universitat Politècnica de València. Cutting-edge topics in Product Design and Manufacturing, Innovative Design; and Computer-Aided Design were especially encouraged.

The list of topics (and subtopics) covered in the present edition is the following:

- Product design & development: Green engineering & Eco-design; User-centred design; Product lifecycle-based design; Robust design, reliability & maintenance; Modelling and simulation-based design; Ergonomics & human factors; Global product development.
- Computer-aided design and interactive design: Virtual Approaches for Interactive Design; Virtual prototyping-based design; CAD, CAE, IFC and BIM; Image processing and analysis; Geometric modelling and analysis; Reverse Engineering; Virtual and Augmented Reality.
- Manufacturing and industrial process design: Integrated/Advanced Manufacturing; Manufacturing Process and Production Management; Rapid Prototyping; Additive Manufacturing; Flexible assemblies; Remanufacturing; Industry 4.0.
- Graphical bioengineering: Biomechanics; 3D Modelling of biological structures; Computer-aided methods for pathologic diagnosis; Emotional engineering; Biomimicry for product design; Simulation and visualization of biological systems.
- Innovation in Design: Creativity & innovation methods; Collaborative engineering; Intellectual and Industrial Property Management; Design & Research Methods.
- Teaching – learning in Graphic Engineering: Teaching on Graphic Expression; Theoretical & applied geometry; Graphic Design; New approaches in teaching/learning process; Project-based Learning; Interactive 3D Modelling.

- Engineering and Construction: Sustainable Building. Sustainable Construction; Building Information Modelling; Photogrammetry and Remote Sensing; Geo-information. Data Capture; Virtual environments. Augmented Reality in AEC; Urban regeneration; Heritage and territory. Industrial Heritage Conservation.

We would like to thank our main organizer/institutions and the rest of the sponsoring/collaborating companies and institutions for their support and grants.

We would also like to express our gratitude to the members of the different committees for their support, collaboration and good work. Thanks to all reviewers for their selfless effort reviewing contributions, which positively influenced the quality of the final papers presented at the conference.

Last but not least, thanks to all the participants of INGEGRAF 2021.

October 2021

Francisco Cavas-Martínez
Guillermo Peris-Fajarnes
Paz Morer Camo
Ismael Lengua Lengua
Beatriz Defez García

Organization Committee

Conference Chair

Francisco Cavas Martínez	Universidad Politécnica de Cartagena, Spain

Conference Program Chair

Beatriz Defez Garcia	Universidad Politécnica de Valencia, Spain
Ismael Lengua Lengua	Universidad Politécnica de Valencia, Spain

Conference Advisory Chairman

Paz Morer	Universidad de Navarra – Tecnun, Spain

Scientific Committee

Fernando Aguilar Torres	Universidad de Almería, Spain
Manuel Alcala	Universitat de Girona, Spain
Silvia Aparisi	Universidad Politécnica de Valencia, Spain
Loris Barberi	Università della Calabria, Italy
Michele Bici	Sapienza Università di Roma, Italy
Eladia Beatriz Blázquez-Parra	Universidad de Málaga, Spain
Yuri Borgianni	Università di Bolzano, Italy
Itziar Castaño Goicoechea	Universidad de Vigo, Spain
Jesús Chacón	Universidad de Castilla la Mancha, Spain
Vicent Cheutet	INSA Lyon, France
Marina Corral	Universidad de la Rioja, Spain
Filippo Cucinotta	Università degli Studi di Messina, Italy
Alain Daidié	INSA Toulouse, France
Luigi De Napoli	Università di Cagliari, Italy
Óscar De Cózar Macías	Universidad de Málaga, Spain

Maria Gloria Del Río	Universidad de Sevilla, Spain
Lucía Díaz-Vilariño	Universidad de Vigo, Spain
Francisco Javier Espinach	Universitat de Girona, Spain
Claudio Favi	Università di Parma, Italy
Francesco Ferrise	Politecnico di Milano, Italy
Francesco Gherardini	Università degli studi di Modena e Reggio Emilia, Italy
Valentín Gómez Jáuregui	Universidad de Cantabria, Spain
Rafael Hidalgo	Universidad de Córdoba, Spain
Tommaso Ingrassia	Università degli Studi di Palermo, Italy
Julien Le Duigou	Université de technologie de Compiègne, France
Francesco Leali	Università degli Studi di Modena e Reggio Emilia, Italy
Rubén Lostado	Universidad de la Rioja, Spain
Muriel Lombard	Université de Lorraine, France
Cristina Manchado del Val	Universidad de Cantabria, Spain
Marco Mandolini	Università Politecnica delle Marche, Italy
Guiseppe Marannano	Università degli Studi di Palermo, Italy
Marco Marconi	Università della Tuscia, Italy
Cristina Martín Doñate	Univesidad de Jaén, Spain
Dominico Marzullo	L'Università di Trieste, Italy
Massimo Martorelli	Università degli Studi di Napoli "Federico II", Italy
Rikardo Minguez Gabiña	Universidad del País Vasco, Spain
Ramón Miralbes Buil	Universidad de Zaragoza, Spain
Maria Moncho	Universidad Politécnica de Valencia, Spain
Manuel Morato-Moreno	Universidad de Sevilla, Spain
Diego Padermo	Università degli Studi di Brescia, Italy
Manuel Paredes	INSA Toulouse, France
Dolores Parras	Universidad Politécnica de Cartagena, Spain
Marcello Pellicciari	Università degli Studi di Modena e Reggio Emilia, Italy
Nicolas Perry	Arts et Métiers ParisTech - ENSAM, France
Eugenio Pezzuti	Università di Roma "Tor Vergata", France
Alvaro Ramírez	Universidad Politécnica de Madrid, Spain
David Ranz	Universidad de Zaragoza, Spain
Roberto Razzoli	Università degli Studi di Genova, Italy
José Ignacio Rojas Sola	Universidad de Jaén, Spain
Bertrand Rose	Université de Strasbourg, France
Jacinto Santamaría	Universidad de la Rioja, Spain
Eneko Solaberrieta	Universidad del País Vasco, Spain
Félix Sanz-Adan	Universidad de La Rioja, Spain
Irene Sentana Gadea	Universidad de Alicante, Spain
Gaetano Sequenzia	Università di Catania, Italy
Fátima Somovilla	Universidad de la Rioja, Spain

Domenico Speranza	Università degli Studi di Cassino e del Lazio Meridionale, Italy
Miguel Suffo Pino	Universidad de Cádiz, Spain
Guillaume Thomann	Grenoble INP - UGA Institut d'ingénierie et de management, France
José Sebastián Velázquez Blázquez	Universidad Politécnica de Cartagena, Spain
Elisabetta Zanetti	Università degli Studi di Perugia, Italy

Organizing Committee

President

Guillermo Peris Fajarnés	Universidad Politécnica de Valencia, Spain

Secretary

Beatriz Defez Garcia	Universidad Politécnica de Valencia, Spain

Members

María Moncho Santonja	Universidad Politécnica de Valencia, Spain
Silvia Aparici	Universidad Politécnica de Valencia, Spain
Bernardo Pajares Moreno	Universidad Politécnica de Valencia, Spain
Ismael Lengua Lengua	Universidad Politécnica de Valencia, Spain
Fernando Garrigós Simón	Universidad Politécnica de Valencia, Spain

Contents

Product Design and Development

Manufacturing and Industrial Process Design

Graphical Bioengineering

Engineering and Construction. New Methodologies BIM

Digital Representation of the Terrain Associated with an Archaeological Site: Case Study of the 'Baker's House' in Torreparedones

Paula Triviño-Tarradas[1](✉), Diego Francisco García Molina[2], Rafael Hidalgo Fernández[1], and Irene Cáceres Criado[3]

[1] Department of Graphic and Geomatics Engineering, University of Cordoba, Córdoba, Spain
{ig2trtap,ig1hifer}@uco.es

[2] Department of Graphic Engineering, Design and Engineering Projects of the University of Jaen, Jaen, Spain
dfgarcia@ujaen.es

[3] Doctoral Program in Agricultural, Food, Forestry and Rural Development Engineering, University of Córdoba, Córdoba, Spain
z92cacri@uco.es

Abstract. Nowadays, it is essential to provide 3D digital reconstructions with the highest possible reliability and veracity. Archaeological sites within the framework of the building to be rebuilt are important. The data provided by Geographic Information Systems (GIS) can be extracted from different sources, and are a crucial tool to carry out the topographical survey in the context of Virtual Archaeology. Therefore, engineering and archaeology work hand in hand to form multidisciplinary teams with the aim of addressing the design of heritage cultural elements.

The aim of this study was to represent the terrain associated with the "the Baker's House", at the archaeological site of Torreparedones (Cordoba, Spain). The interest in this representation increased after the virtual reconstruction of the domus, which was performed to geographically place the house in its exact location. Two methodologies were used through different software and data to obtain the topographical survey. The first methodology was the one applied in QGIS 2.3.15 with data from the CNIG, and the second methodology was the one applied in Blender 2.90.0 using the Google Satellite database. Both methodologies were compared, analysing the time invested in the two processes, as well as the orography of the terrain, which was captured with a 360 panoramic picture taken on the site.

Keywords: GIS · Recreation · Virtual archaeology · Domus · Orography · Blender · QGIS

1 Introduction

Geographic Information Systems (GIS) are sets of related data with georeferenced information, that is, they are basic elements for the management of everything that presents a usable geographic component [1].

F. Cavas Martínez et al. (Eds.): INGEGRAF 2021, LNME, pp. 3–10, 2022.
https://doi.org/10.1007/978-3-030-92426-3_1

In every discipline, 70% of the information used is georeferenced [1]. In archaeology, this percentage is greater, as it is very important to place each element in space, in order to know the exact position of each heritage asset. These systems, in the archaeological scope, are key for the management of datasets that, combined in a single pack of information at a large scale, act as analytical systems [2]. In addition to the archaeological analysis, the combination of different tools and techniques allows documenting the heritage for subsequent valuation, and even solving complex research problems [3].

2 Objectives

The aim of this study was to compare the precision of two sources of terrain description data in order to integrate them into the visual representation of the archaeological site of Torreparedones (Cordoba, Spain), where the "Baker's House" is located. Both methodologies were described and we selected the one that required less investment time based on the quality of the work obtained. Then, qualitative results with a 360 panoramic picture taken at the study site were compared to determine the degree of accuracy of both methodologies.

3 Materials and Software

The geographic framework of a project is fundamental for data extraction, and it is not always obtained from a single, coherent and quality source [4]. This project had a single administration unit (Baena, Spain), which provided information from central and administrative organisations.

To create the terrain of this heritage asset and, for the first study case, the data were extracted from the Spanish Geographic Information Centre (CNIG) [5], specifically, the 1st coverage digital model of the terrain with a 5-m mesh step (DTM05). Regarding the second study case, the source used for data extraction was Google Satellite, which allows visualising maps of the entire planet based on satellite images [6].

It is worth mentioning the different qualities of the data to be used. On the one hand, digital terrain elevation data from the Shuttle Radar Topography Mission (SRTM) has a mean square error of around 10 m worldwide, at best, and even worse on steep terrain [6]. It also has a poor ground sampling distance (GSD) of 90 m. On the other hand, the vertical accuracies of the LIDAR DTM data can be less than one meter, and with a GSD of 5 m [5]. It is also necessary to mention the updating of said data. While the PNOA LIDAR data is relatively up-to-date, the SRTM data was taken in February 2000. The differences between the two data are notable, and it might be impractical to make a comparison between them. From the point of view of virtual archaeology, it is comparable for the visual integration of a virtual archaeological environment.

Two open-code programmes were used in this study: 1) QGIS 2.3.15, and 2) Blender 2.90.0. The former is a GIS, whereas the latter is a programme designed for the modelling, animation and creation of three-dimensional graphics.

4 Methodology

4.1 Creation of the Terrain Using LIDAR Data in QGIS

Firstly, to create the terrain, it is fundamental to be sure about the geographic framework of interest for the study. In our case, the geographic framework is the countryside of Cordoba, specifically, the archaeological site of Torreparedones, at the northern end of the municipalities of Castro del Río and Baena [7, 8].

The data were obtained from the Download Centre of CNIG, "Digital Models of Elevations", which allows searching by list and selecting the province of municipality of interest. The Digital Terrain Model (DTM) downloaded for this study was of 1st coverage, with a 5-m mesh step. It has two raster layers: one for the relief at the national level and the other one for elevations.

To obtain a DTM in QGIS using LIDAR data, the raster layers were entered. The study area was located at the archaeological site of Torreparedones. To export the 3D DTM files, we used the QGIS2threejs tool. This complement can be downloaded from Google by typing its name [9]; once downloaded, it has to be loaded in QGIS. This complement allows visualising the DTM.

The resolution was improved by changing the properties of the relief layer, increasing the geometry resampling level to 6 and the resolution of the material to 400% (5500 × 3456 px). This improvement allowed the DTM to be saved and used for the desired purpose. The model can be saved through two options: as an image (.png) or as a gITF file (.gltf, .glb); the latter is the most commonly used to import this model to other software.

The last step of this methodology consists in removing the rear walls and the base of the three-dimensional model, with the aim of reducing the size of the file. To this end, the Blender 2.90.0 software was used, importing the model obtained in QGIS. Each side of the terrain obtained in QGIS is an independent model. Therefore, it was only necessary to select and delete the unwanted sides.

The result was a 3D model with good resolution (Fig. 1), which, after being imported to other software, such as Blender, can be further manipulated to optimise the resolution in areas of interest; additionally, the model can also be optimised by modifying the terrain manually.

Furthermore, textures can also be incorporated into the model. Uniform textures can be applied to the entire terrain or specific areas. Lastly, other models can be added to this 3D model, to make the terrain more realistic. In this case, the domus of this archaeological site, and even vegetation, can be added to the environment.

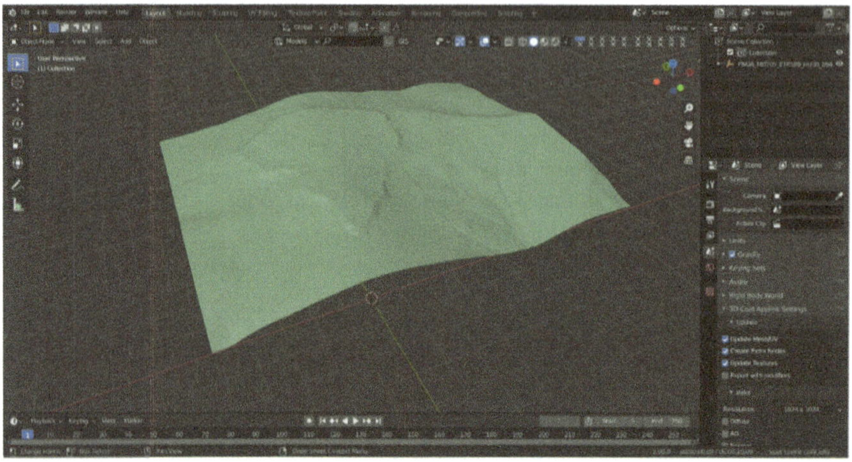

Fig. 1. Result of the digital terrain model of the obtained terrain (Developed by authors).

4.2 Creating the Terrain with Blender GIS

The second tool used in this study, i.e., Blender GIS, allowed adding maps of real locations of the entire world directly from this software. Although the detail of these maps is not very high, they can provide a good basis for the elevation of the terrain.

The first step for the creation of a terrain using Blender GIS was to install Add-ons 3D View: BlenderGis [10]. With the installation of this necessary complement, a work tab called "GIS" appears in the screen, where the terrain is developed ("Web geodata" – "Basemap"). To create the terrain of the archaeological site of Torreparedones with Blender GIS, we used Google as a source and Google Satellite as a layer. Once we had the desired view for the 3D map, we obtained the view in an image without relief. This was achieved through "Get SRTM", which, from the data of Open Street Maps, created the 3D area of the current view. To improve the visualisation of the relief, with better resolution, the model was subdivided, obtaining a mesh with smaller polygons.

To increase the horizon, the same steps must be followed, adding a new layer of Google GIS and selecting a broader view of the environment. The overlapping of the two objects can be prevented by adding a cube whose dimensions allow for a slight overlapping at the edges. Then, this cube can be subtracted from the new horizon terrain (Fig. 2). Thus, greater detail was achieved for the archaeological site of Torreparedones, where the domus was later placed, obtaining smaller polygons, while the horizon had lower detail and a broader view.

As in the first methodology, these models can be further manipulated to attain new results in them, or even add new models to them.

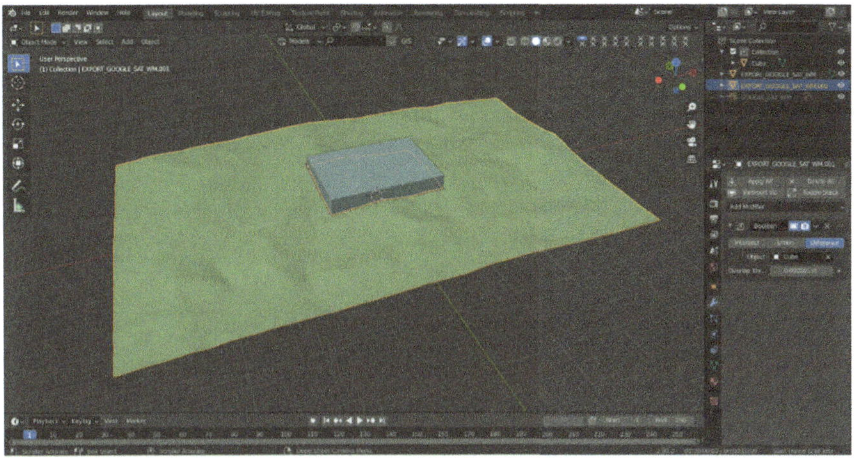

Fig. 2. Difference of the cube applied to the horizon using the "Boolean" modifier (Developed by authors).

5 Results and Discussion

The main results of this study were achieved from the data obtained from the download centre of CNIG [5] and Google Satellite [6]. Likewise, the use of the described programmes was of great help for the realisation of this work.

The digital representation of the terrain was successful with both methodologies. However, at this point, we must determine which methodology best suits the desired results. It is fundamental to know the resolution of the model; i.e., the greater the detail, the greater the quality of the representation. Therefore, the qualitative comparison of the two methodologies suggests associating greater detail to the creation of the terrain using LIDAR data in QGIS. The data extracted for the creation of the terrain are key for such resolution, thus the data from CNIG can be considered to be more precise than those from Google Satellite. As was previously mentioned, Blender GIS does not stand out for its detail level.

Another important factor to take into account is the time invested in each methodology. The programmes used are not intuitive, thus handling them for the first time can be tedious. For the creation of the terrain, QGIS requires fewer steps than Blender; i.e., the latter requires more time. Therefore, QGIS generates greater resolution in less time.

Lastly, the two methodologies were compared with a 360 panoramic picture taken *in situ*. It is worth mentioning that the 360 panoramic picture was divided into four sections (the cardinal points), as the landscape of the archaeological site of Torreparedones was too small due to the large size of the image. Four sub-images (north, south, east and west) of each methodology were obtained, with the same camera parameters used in the *in situ* panoramic picture. This accurately shows the common points of the landscape in both methodologies and the four cardinal points with respect to the 360 panoramic picture.

To calculate the approximation percentage of the points marked in the images obtained from the programmes with respect to the 360 panoramic picture (Table 1), we used the scale presented in Fig. 3. Supposing the 117 squares of each image to be 100% of the approximation, each square would be 0.85%. We subtracted half of a square (0.42%) from each point, since, although the points of the data from LIDAR and Google Satellite are in the same square with respect to the 360 panoramic picture, they are not in the exact position. The level of similarity for both methodologies is shown between brackets.

Table 1. Points and their locations associated with Fig. 3.

Points	On-site panoramic	LIDAR data	Google satellite data
1	N2-A4	N2-A4 (99.58%)	N1-A5 (97.88%)
2	N1-A7	N1-A7 (99.58%)	N1-A8 (98.73%)
3	N2-A12	N2-A13 (98.73%)	N2-A11 (98.73%)
4	E3-B2	E3-B2 (99.58%)	E3-B2 (99.58%)
5	E2-B6	E2-B6 (99.58%)	E3-B6 (98.73%)
6	E5-B8	E5-B8 (99.58%)	E4-B9 (97.88%)
7	S3-C3	S3-C4 (98.73%)	S3-C2 (98.73%)
8	S4-C4	S4-C4 (99.58%)	S3-C4 (98.73%)
9	S3-C9	S3-C9 (99.58%)	S3-C10 (98.73%)
10	O9-D4	O8-D4 (98.73%)	O8-D4 (98.73%)
11	O2-D8	O2-D7 (98.73%)	O3-D8 (98.73%)
12	O2-D13	O2-D13 (99.58%)	O2-E13 (99.58%)
		99.29%	98.73%

Thus, we can assert that the recreation of the terrain is more accurate using LIDAR data with respect to Google Satellite data, since the points marked in the first methodology are 99.29% similar to those of the 360 panoramic picture, whereas the points in the second methodology showed 98.73% similarity (Table 1).

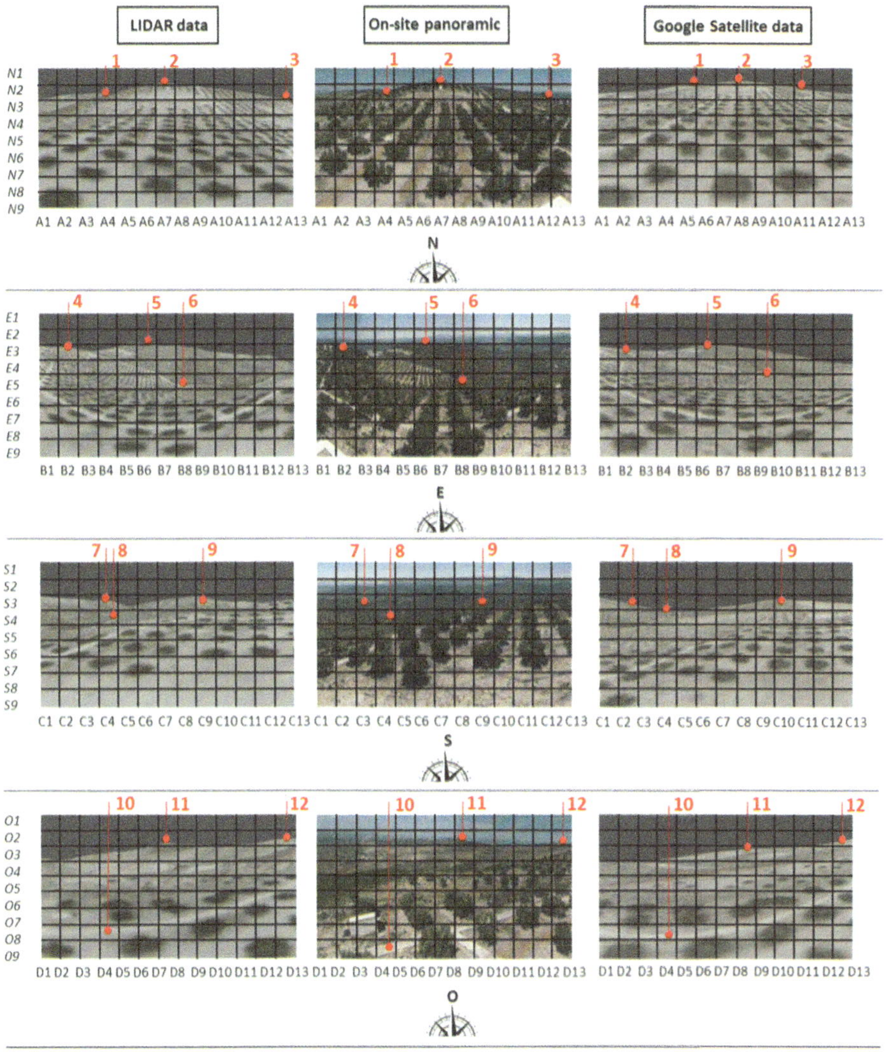

Fig. 3. Comparison of LIDAR data and Google Satellite data with the 360 panoramic picture.

6 Conclusions

Geographic Information Systems are indispensable in archaeological research. Virtual reconstructions guarantee the graphic recording and conservation of heritage, thereby playing an important role in the recreation of the landscape of heritage assets.

The data used to carry out the recreations of the terrain using the two methodologies were determinant. The LIDAR data showed greater detail and accuracy in the similarity of common points. Despite the fact that the methodology performed with Google Satellite produced lower detail and resolution, it provided a good basis for the elevation of the terrain, although the latter required manual modifications to obtain better results in some areas.

It can be concluded that the results of the present study were successful, since the two proposed methodologies were adequately compared. The recreation with QGIS using LIDAR data was more favourable, since, in addition to the above mentioned, it requires less time to produce a 3D terrain.

References

1. Olaya, V.: Sistemas de información geográfica. Cuadernos internacionales de tecnología para el desarrollo humano **8**, 1–6 (2009)
2. Patiño, D.: Sistemas de información geográfica y su aplicación en la Arqueología. Rev. Colomb. Antropología **34**, 194–205 (1998)
3. García, L., Wheatley, D., Murrieta, P., Márquez, J.: Los SIG y el análisis espacial en arqueología: aplicaciones en la prehistoria reciente del sur de España. In: Cau Ontiveros, M.A., Nieto Prieto, F.X. (coords.) Arqueología náutica mediterrània. Centre d'Arqueología Subacuàtica de Catalunya, pp.163–180. Girona (2009)
4. Domínguez, J.: Breve Introducción a la Cartografía y a los Sistemas de Información Geográfica (SIG). Informes Técnicos Ciemat **943**, 1–30 (2000)
5. CNIG Homepage. https://centrodedescargas.cnig.es/CentroDescargas/index.jsp. Accessed 16 April 2021
6. USGS EROS Archive – Digital Elevation - Shuttle Radar Topography Mission (SRTM) Homepage. https://www.usgs.gov/centers/eros/science/usgs-eros-archive-digital-elevation-shuttle-radar-topography-mission-srtm?qt-science_center_objects=0#qt-science_center_objects. Accessed 17 July 2021
7. Morena, J.A.: El Parque Arqueológico de Torreparedones (Baena, Córdoba). Arte, Arqueología e Historia (2012)
8. Morena, J.A., Sánchez, I.M.: Recientes metodologías para la caracterización arqueológica del paisaje de la colonia Itvci Virtvs Ivlia (Torreparedones, España), durante las épocas romana y tardoantigua. Romula **15**, 87–128 (2016)
9. QGIS Python Plugins Repository Homepage. https://plugins.qgis.org/plugins/Qgis2threejs/. Accessed 23 April 2021
10. GitHub Domlysz BlenderGIS Homepage. https://github.com/domlysz/BlenderGIS. Accessed 23 April 2021

Monitoring Industrial Plants from BIM Models with Extended Reality

Ana Carrera-Monterde[1], Valentin Gomez-Jauregui[2(✉)], Cristina Manchado[2], and César Otero[2]

[1] School of Industrial and Telecomunications Engineering, Universidad de Cantabria, Avda. Los Castros 46, 39005 Santander, Spain
[2] EgiCAD Research Group, School of Civil Engineering, Universidad de Cantabria, Avda. Los Castros 44, 39005 Santander, Spain
valen.gomez.jauregui@unican.es

Abstract. This project consists on creating a mobile application that, by means of Extended Reality (XR), allows the user to interact with any industrial plant, in our case a Wastewater Treatment Plant (WWTP). In order to make this possible, we have generated a digital twin of the facility from the BIM model. The user will be able to wander around the model immersed on it. The camera can be moved to explore the space by using, in VR, the mobile device's gyroscope (to rotate) and a joystick (to move around) or by freely walking inside the building in AR.

Its main purpose is to monitor the live status of the WWTP whether it is from a remote location using VR or inside the building using AR. In almost every equipment of the WWTP there is information from the computerized maintenance management systems (CMMS) and sensors uploading measurements to a server feeding the SCADA. The application will display this information in an intuitive 3D panel on top of each equipment once it is selected by the user. It will also be possible to interact with the model by moving equipment around or by taking measurements of the building on the device's screen.

Keywords: Extended Reality · Augmented Reality · Virtual Reality · SCADA · Mobile application · BIM · Digital twin

1 State of the Art

1.1 Brief Introduction to BIM

BIM (Building Information Modelling) is a collaborative work methodology that helps managing construction projects throughout their life cycle [1]. It consists of creating a digital model that integrates an extensive database, accessible for all the stakeholders [2]. Combining all the project information in the database allows more efficient management and helps optimizing and saving financial and human resources. It improves traceability and coordination between the involved parties in the lifecycle of the project. This methodology increases the collection of data which improves cost management, reduces time and helps with decision making [2]. This collection of data happens throughout the

F. Cavas Martínez et al. (Eds.): INGEGRAF 2021, LNME, pp. 11–19, 2022.
https://doi.org/10.1007/978-3-030-92426-3_2

whole project and allows generating updated reports and therefore monitoring the state of the project easier and enhancing its quality.

Nowadays, BIM technologies are widely spread throughout the world; some studies forecast an annual growth well above 20% of BIM technologies [3]. Some countries such as Denmark have set BIM technologies compulsory while developing local and regional projects as well as government buildings since 2011.

In Spain, the government created a national body called es.BIM to analyse how to implement this technology in public tenders [1]. This is interesting as it helps to implant public construction politics, sustainable building, public spend efficiency and national competitiveness [2].

BIM methodology integrates a digital model and a database assigning the BIM model dimensions depending on the nature of the data collected. These dimensions vary from 3D to 7D and are not sequential; it may be BIM 6D without a previous 5D [3]. BIM 3D is the three-dimensional model of the project. It serves as a foundation for the rest of the dimensions by assigning all the geometric information of the project. BIM 4D allows to establish completion times essential for developing an implementating schedules, that will allow to simulate the different phases of the life cycle. BIM 5D consists of cost control, managing expenses and predicting the operational cost of the use and maintenance stages. BIM 6D or Green BIM consists of looking for the optimal alternative, usually focused on eco-efficiency by making simulations about energy efficacy. BIM 7D will be implemented once the project is built and marks the appropriate optimized practices to follow for its maintenance [3]. It helps to manage the project throughout its whole life cycle.

The present project and paper will focus on this last dimension, because the application developed by the authors is used during the exploitation stage of the project. In order to implement BIM, we have used Autodesk Revit, which allowed us to model the buildings and access the building information from the database [4].

1.2 Use of BIM in Wastewater Treatment Plants (WWTP)

Due to the rapid growth of cities and industries, water pollution has become a concern producing high demand for sewage treatment technology with reduced process time and improved efficiency. This type of installations counts with complex special- shaped structures that hinder the design and arrangement of the parts. Also, due to its multidisciplinary nature, WWTP facilities requires engineers from different disciplines to collaborate, calling for a unified platform for them to work on a very useful tool [5]. Other projects, e.g. reprovisioning of Shek Wu Hui Sewage Treatment Works [6], show the potential of implementing BIM technologies to save time and cost in the building of the Sewage Treatment Plants. BIM technologies not only solved these problems (in the previous example) but also improved the design by including the opinions from the staff to make maintenance easier [6].

2 maXRii: The XR App

The app developed in this project, available in both English and Spanish, is called maXRii, "Mobile Application of eXtended Reality for Industrial Installations". The objective of this project was to create a mobile app that shows intuitively and simply a BIM7D of the WWTP of León y su Alfoz, using technology of Extended Reality (XR), a combination of Virtual Reality (VR), Augmented Reality (AR) and Mixed Reality (MR). The staff of the WTTP could easily download it on their mobile devices allowing access to the virtual model.

The WWTP is located in southern León (Spain). The plant treats 107.000 m3 per day of wastewaters of León and its surroundings, a population of 300.000 inhabitants [7].

The plant consists of 10 buildings and for the app, however, we have implemented it for 2 buildings (pressurization and digesting/drying of sludge), although the replication to other parts of the plant is quite straightforward.

For the making of the app, we have used the software Unity, a cross-platform game engine that has been utilized in many other projects due to its versatility while creating three-dimensional, two-dimensional, virtual reality games and simulations [8]. Even tough this program supports up to 25 platforms, we have developed it on Android for testing its performace.

2.1 Tracking the User

For programming Augmented Reality (AR) we used Vuforia, a widely spread AR software development kit (ARSDK) that supports phones, tablets and eyewear on any operating system [9].

Our greatest challenge in using AR was to ensure that the virtual model is projected aligned with the real-world objects when viewed through the mobile device´s camera. Vuforia uses image targets to position and orient the virtual objects. We have an image target located in the same position of the building in the real world, as well as in the virtual model allowing the alignment [9]. By tracking the distance and variations from the image target Vuforia positions to the user, allows them to explore the virtual model (as well as the real world) by moving and turning around.The image targets we are using are a combination of images, texts and QR codes for the different building of the WWTP, (Fig. 1).

Fig. 1. Image target

In VR the app will only show the virtual model, not the user's surroundings. To explore the building, we programmed a joystick to allow the user to move. Additionally, there are two options to rotate the camera, by means of the mobile device's gyroscope or by means of another joystick, both positioned in the lower half of the screen (Fig. 2, left).

In addition to VR and AR, maXRii offers a third visualization option: bird's-eye-view. The user will be able to see the different buildings form a top-down, third-person view perpendicular to the ground. The building will be explored by the user by dragging the finger on the screen (for changing the cameras position) and pinching (for changing the cameras height). For better control, one of the options in the user interface is a slider to regulate the camera height more precisely (Fig. 2, right).

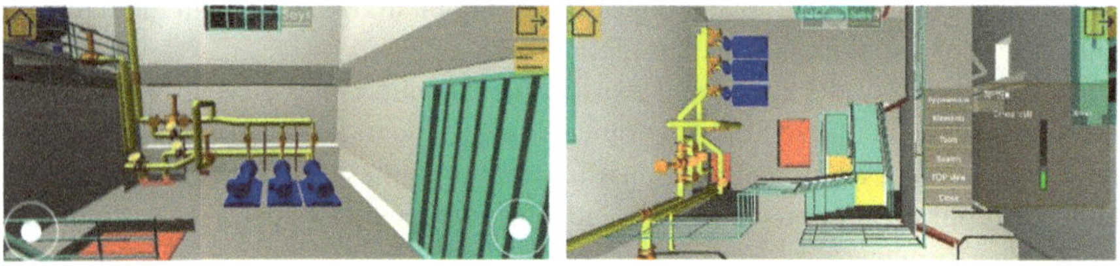

Fig. 2. VR option (left) and bird's-eye-view with slider (right)

2.2 Showing Object Properties

Knowing the properties and availability of the different equipment in the plant is essential in the exploitation stage. The attributes of the objects will be obtained from different sources such as the BIM database, CMMS, and data for the SCADA.

Regarding the SCADA database, the information of the sensors (PLC) are uploaded to a server. The app accesses the server to read the sensor values using a specific url and the server returns a JSON string that contains a parameter "v", with the actual measurement registered by the PLC on real time. A sample of a returned string is {"readResults":[{"id":"CANAL PLC3.FT11GB","s":true, "r":" ", "v":189, "t": 1616407681856}]}.

The app reads the databases' information and finds relations between the different items by crossing the common ID numbers and shows all the information on screen panels (Fig. 3).

Fig. 3. CMMS (GMAO in Spanish), BIM and SCADA panels for a pump

If the users want to check the properties of another object, they just must click on the tick (✓) symbol of the screen, in the upper left corner.

This option is very helpful for all the workers in the facility, because by a simple tap on the screen, they can check very useful information: the model and brand of the equipment, the state (busy, stopped, error…), the name of its PLC, the last or next planned revision, number of replacements in inventory, etc.

Another visualitation method used for SCADA information is by having a panel over each object of the scene that has a PLC, displaying the measurement of its sensor; the object to which the panel belongs is also highlighted in green colour. To ensure that the user can check the information from any position, the panel is programmed to face the user always by rotating around the vertical axis (Fig. 4, left).

As there are many sensors, the app implements a drop-down menu to clarify the visualization, from which the users can select the different equipment units they want to filter (Fig. 4, right).

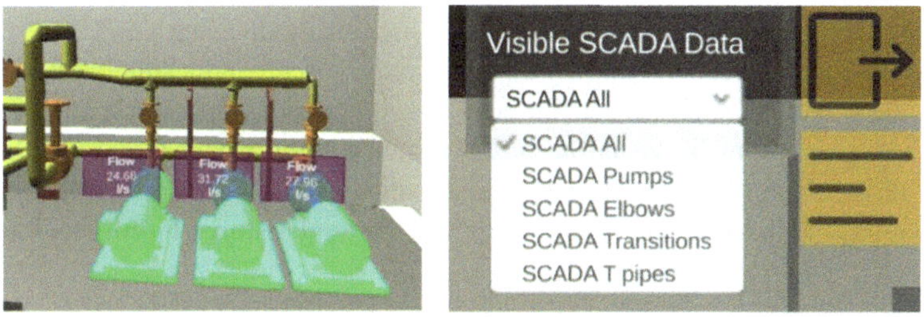

Fig. 4. SCADA panels (left) and dropdown menu (right)

2.3 Moving Equipment

Some of the equipment in industrial facilities, and in WWTP in particular, require delicate handling; any movement of the equipment for reposition, reparation, maintenance or replacement require a detailed planning as well as special machinery.

It is essential to avoid future problems, such as equipment not fitting in the new arrangement or interferences with the elements of the buildings when doing the transit of the equipment inside the building. maXRii includes an option to roughly simulate the movement of objects inside the building, previewing and identifying posible problems or clashes before the actual operation.

The user receives instructions on how to proceed (lower status bar), as well as 2 sets of buttons: the set on the right, for moving the object along the horizontal plane and turning around the vertical axis; and the set on the left, for changing the object elevation, up and down (Fig. 5). Once the user taps on an object, this object will be highlighted in yellow and any object colliding with it will be highlighted in red, showing the user where the clashes are happening (Fig. 5).

Fig. 5. Move option, set up and highlights

2.4 Measuring Distances

Sharing the purpose of helping the tasks of planning and simulating equipment movements, the maXRii implements a tool for measuring distances.

Firstly, the user selects a point by tapping on the screen and a small sphere appears marking the spot. The app instructs the user to select another spot, then another sphere defines the second point. To visualize the 3D distance in the model, a green line is drawn the two selected spots. Finally, a results panel shows the coordinates of the origin and end, as well as the distance in the x, y,z axis and the total distance highlighted in orange (Fig. 6).

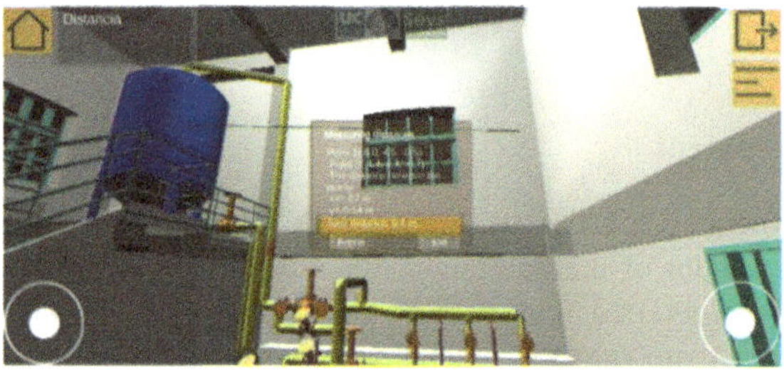

Fig. 6. Results panel for measuring

2.5 Searching Objects from ID Numbers

Another problem suffered by the staff of these industrial facilities is finding the exact equipments from their identification number or name. Due to the vast amount of equipment and different small devices in a WWTP, it is hard to locate a specific part when it needs to be changed or checked. Due to its relevance, we have defined a tool "Search by ID" in the options menu.

We have implemented two different methods to specify the equipment being looked for. For the first option, maXRii shows an input field in which the user can write the BIM ID of the object. Additionally, we have created a drop-down menu listing the equipment of the plant (Fig. 7). This menu can be edited to integrate the most searched equipment and has shortcut for finding them.

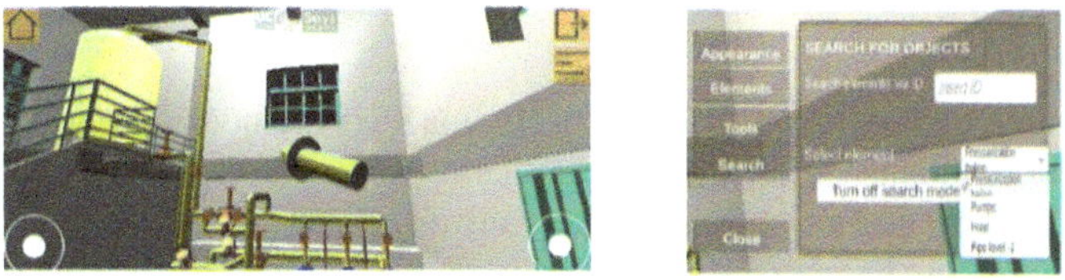

Fig. 7. "Search by ID", highlight and pointing arrow (left), input field and dropdown menu (right)

In order to help the user, the equipment to be found is highlighted in yellow and an arrow appears in front of the camera guiding the user towards the desired object. The arrow changes direction and position depending on the camera's attitude (Fig. 7).

2.6 Filtering Visualization of Element Groups

For the user to have a better understanding of the WWTP and for making it more intuitive, we have implemented a set of options that hide and show different elements, with the possibility of changing their textures.

Firstly, we have implemented an option to hide or show some of the element groups of the plant: walls and floors, pipes and valves, just pipes or just valves, equipments, etc. By hiding and showing the elements, the user can have a better feeling of the building space (Fig. 8). It is possible to disclose elements that are hidden behind construction elements helping with remodelling or installing new equipment.

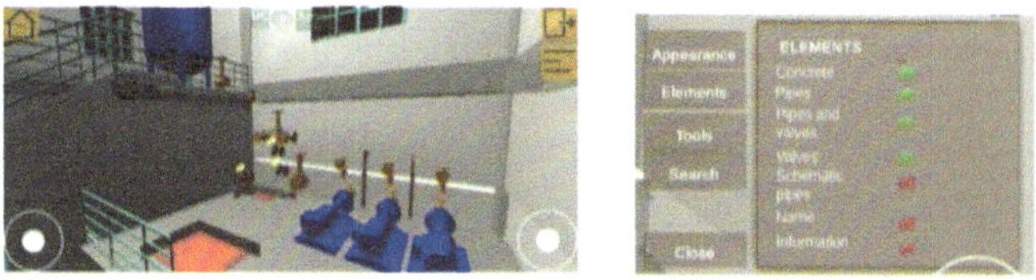

Fig. 8. Hide/show elements option

Secondly, to ease with visualisation, maXRii allows the user to change the textures of the different elements to make it more realistic and intuitive. These textures can be activated in the "Aspect" tab in the options menu (Fig. 9).

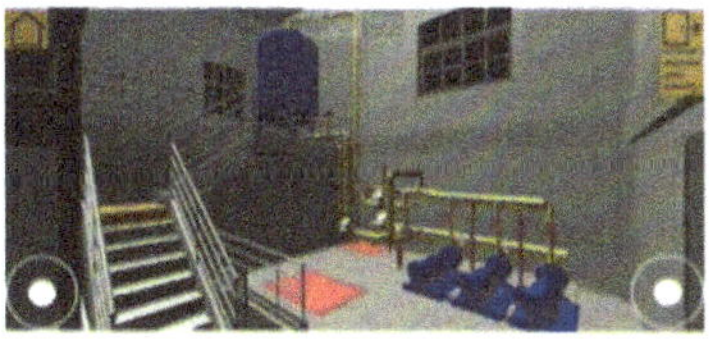

Fig. 9. Realistic texture option

2.7 Adding Notes and Comments

In order to help collaboration between users, maXRii includes an option that allows users to add comments on different parts of the virtual model. These comments are uploaded to a server and are accessible from any device connected to it. The object with an attached comment has a post-it sign marking it. Once selected, maXRii asks the user whether they want to edit or visualize the comment (Fig. 10).

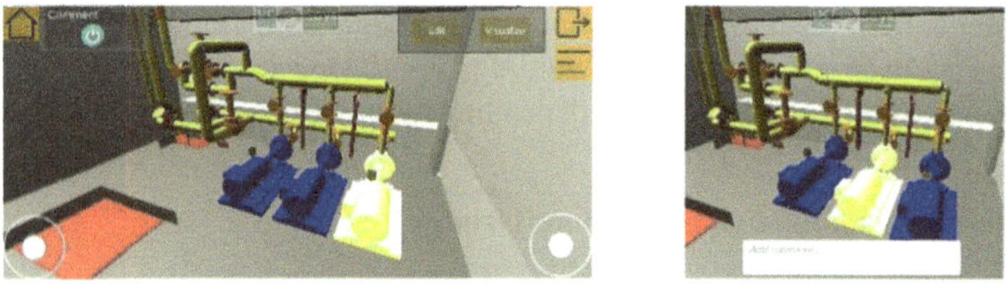

Fig. 10. Editing existing comments (left) and add comments option (right)

3 Conclusions

Through the different options we have implemented in the app, users can explore this WWTP, or any other industrial facility, and familiarize themselves with the building without safety risks. In addition to this, the plant worker is able to check in real time the PLC sensorss, the data from the CMMS, simulate any machine moving as well as take measurements and find equipment. We believe that this app is helpful in many situations, as it allows a wide range of activities, through a simple layout, without the need for any additional devices that could hinder the operations taking place.

References

1. Building SMART International: What We Do? 2017. [Online]. Available: https://www.buildingsmart.org/. Accessed 6 April 2021
2. BIMes: Qué es BIM, 2020. [Online]. Available: https://cbim.mitma.es/. Accessed 6 April 2021
3. BIMnD Building new dimensions: Las 7 Dimensiones BIM, 2019. [Online]. Available: https://www.bimnd.es/7dimensionesbim/. Accessed 08 April 2021
4. Autodesk: Autodesk Revit, 2021. [Online]. Available: https://www.autodesk.es/products/revit/overview?term=1-YEAR. Accessed 12 April 2021
5. Sobke, H., Theiler, M., Tauscher, E., Smarsly, K.: BIM-based description of wastewater treatment plants. In: 17th International Conference on Computing in Civil and Building Engineering. Tampere, Finland (2018)
6. H. G. S. T. Drainage Services Department: BIM Success in Construction of Seage Treatment Works Project with lasting benefits. In: Autodesk Hong Kong BIM Awards 2017. Shek Wu Hui, Hong Kong (2017)
7. García Fernández, O., et al.: Modelling As-Built MEP Facilities in a BIM Environment. In: Adnvances in Design Engineering, p. 498. INGEGRAF 2019 (2019)

8. Unity: Unity Platform, 2021. [Online]. Available: https://unity.com/products/unity-platform. Accessed 16 March 2021s
9. Vuforia engine developer portal: Home, 2021. [Online]. Available: https://developer.vuforia.com/. Accessed 19 April 2021

Development of a System for Using Transcranial Doppler Monitoring with Virtual Reality Head Mounted Displays

Beatriz Rey[1](✉), Almudena Palacios-Ibáñez[1], Jose M. Monzo[2], and José Tembl[3]

[1] Departamento de Ingeniería Gráfica, Universitat Politècnica de València, Camino Vera s/n, 46022 Valencia, Spain
beareyso@dig.upv.es
[2] Instituto de Instrumentación para Imagen Molecular (I3M), Centro Mixto CSIC – Universitat Politècnica de València, Camino Vera, s/n, 46022 Valencia, Spain
[3] Servicio de Neurología, Hospital Universitari i Politècnic La Fe, Valencia, Spain

Abstract. New head mounted display devices for virtual reality (VR) have appeared in the last years and have made this technology closer to the population, allowing its wide diffusion in the mass market. Currently, VR is a tool with applications in many areas, including motor training and rehabilitation. In many of these applications, it is mandatory to evaluate participants' brain responses during the performance of tasks in VR. One of the techniques that has been applied in psychophysiological research for this purpose is Transcranial Doppler (TCD), a non-invasive technique based on ultrasound that monitors blood flow velocity in main cerebral vessels. However, up to now, VR systems based on head mounted displays are not prepared for their simultaneous use with TCD monitoring. In the present work, a VR system based on Oculus Rift that allows a simultaneous recording of the TCD signal has been designed and developed. Special adapters have been proposed to allow the correct holding of the ultrasound probes to the VR head mounted display. It has been checked that the system allows a correct monitoring of the TCD signal during VR exposure.

Keywords: Virtual reality · Head mounted displays · Transcranial Doppler · Design · 3D printing

1 Introduction

Virtual Reality (VR) is a technology that is currently been applied in many fields, including motor training and rehabilitation for people with motor disabilities, pain or stroke patients. Exercises can be performed immersed in a virtual environment that can help to increase the therapeutic adherence. However, that is not its only advantage. With an adequate programming of the virtual environment, it is possible to obtain data about the performance of the exercise and the therapist can use this information to evaluate the patients' evolution during the training and propose modifications in the protocol if

F. Cavas Martínez et al. (Eds.): INGEGRAF 2021, LNME, pp. 20–29, 2022.
https://doi.org/10.1007/978-3-030-92426-3_3

needed. Furthermore, VR can be combined with neurophysiological monitoring techniques that can provide additional information about the evolution of the brain activations associated to the performance of the motor tasks.

1.1 Transcranial Doppler Monitoring

One of the techniques that can be useful to monitor brain activation in this context is Transcranial Doppler (TCD). It is a non-invasive diagnosis technique used to monitor blood flow velocity (BFV) in the main cerebral vessels. With an appropriate neurovascular coupling, BFV changes in a vessel reflect changes in regional cerebral blood flow generated by brain activation in the regions supplied by that vessel [1]. Consequently, one of the main applications of the technique is to evaluate changes in brain activation during the performance of cognitive or motor tasks [2, 3].

The main limitation of TCD monitoring is its low spatial resolution, as it cannot give information related to brain regions smaller than those supplied by the monitored vessel. Usually, functional studies with TCD include at least one of the following vessels, in both cerebral hemispheres (right and left): middle cerebral arteries (MCA), anterior cerebral arteries (ACA) and posterior cerebral arteries (PCA). The vessels for a specific study are selected according to the brain regions they supply [4].

In spite of these limitations, the non-invasiveness and high temporal resolution of TCD make it highly suitable for functional studies, as an alternative or complement to other techniques such as functional magnetic resonance.

1.2 Transcranial Doppler and Virtual Reality

TCD has been applied in combination with VR to evaluate brain activation associated to presence in virtual environments [5–7]. However, in all these previous studies the hardware settings that were used for the VR exposure were not based in head mounted displays. Either simple configurations where the virtual environment was visualized in a computer screen or complex configurations based on projection systems were used. The users did not have to put any special hardware on their heads to visualize the virtual environments, so it was possible to correctly place the ultrasound probes required to monitor BFV of brain vessels during the exposure.

In the present work, the goal is to design and develop a system to allow TCD monitoring during the exposure to a virtual environment with a head mounted display. With this system, the possibility of monitoring brain activation during the exposure to a virtual environment will be easily extended to the new generation of VR devices that have arrived in recent years to the mass-market, making easier the adoption of these technological solutions in hospitals and rehabilitation centers.

2 Materials and Methods

In this section, the system that has been designed and developed to allow TCD monitoring during the exposure to a virtual environment with head mounted displays is described.

2.1 Virtual Reality Head Mounted Display

The head mounted display that has been used to present the virtual environment to the participant is the Oculus Rift. It has a resolution per eye of 1080 × 1200 pixels, a refresh rate of 90 Hz and a field of view of 110°. It includes two controllers (Oculus Touch) to interact with the virtual environment using the hands. An infrared outside-in tracking system is used to track head and hands movements with 6 degrees of freedom. The head mounted display is connected to a PC with an Intel® Core™ i5-8600K CPU @ 3.60 GHz processor, 16 GB RAM, an NVIDIA GeForce GTX 1080 graphic card and Windows 10 Pro Operating System. A volunteer navigating in a virtual environment using this device is shown in Fig. 1.

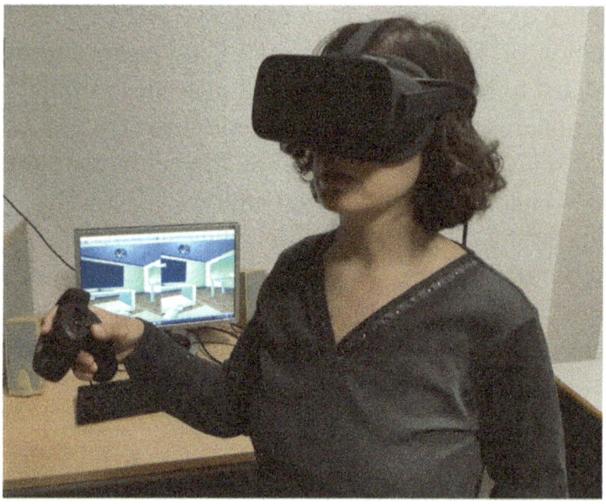

Fig. 1. Oculus Rift head mounted display.

2.2 TCD Monitoring

The TCD device that is used to monitor BFV is a 2-MHz pulsed-wave TCD unit: Multi-Dop T (DWL® Compumedics Germany GmbH, Singer, Germany). In order to conduct the monitoring, two 2-MHz probes should be placed and oriented correctly in the temporal bone window. This procedure should be performed by a neurologist.

If the technique is used while the volunteer is performing motor or cognitive tasks in the context of functional studies, it is necessary that the probe remains fixed during the total length of the experiment. DWL includes in the TCD system a head-set that allows the adjustment and fixation of the probes, which can be observed in Fig. 2. This system acts as a probe holder that can be regulated to the user's head dimensions and can be used to maintain the two probes in their adjusted location to correctly monitor the vessels that are relevant for the study.

The insonation depth of each probe for each study is indicated in the QL software from DWL (installed in the TCD Device). Details about the insonation procedure are described in previous works [8]. The QL software registers the BFV signals with a

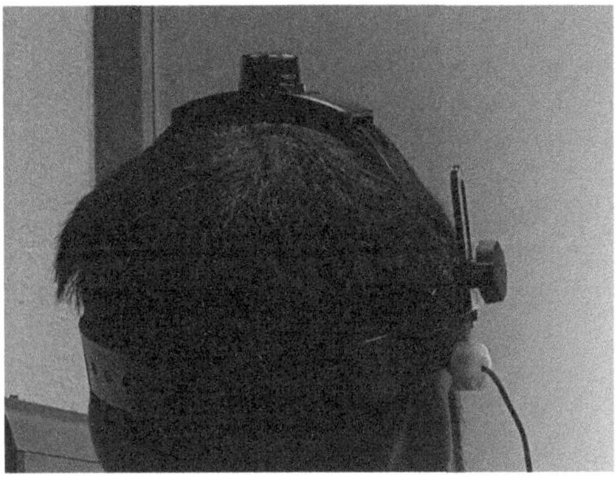

Fig. 2. Probes placed in their correct location using the DWL probe holder.

sampling frequency of 100 Hz. These signals can be saved and exported from the QL software for subsequent analyses.

The head-set that is provided to adjust the probes does not allow the simultaneous use of the Oculus Rift head mounted display. Consequently, it is necessary to find an alternative way of fixing the probes during the exposure to VR with this system.

2.3 Adapters to Adjust the Ultrasound Probes

In order to hold the probes in their correct location and orientation, an adapter has been designed and developed. Instead of using the default head-set provided by DWL to place the probes, the current proposal assumes that the probes will be attached to the Oculus Rift head mounted display structure.

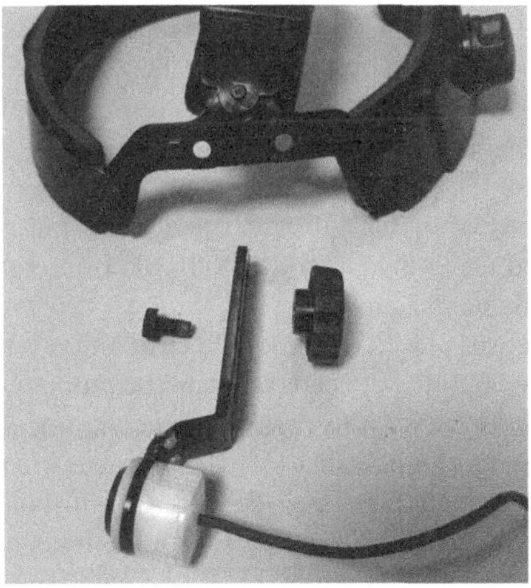

Fig. 3. Different parts of the system provided by DWL to hold the probe in place: probe holder (head-set), probe with an arm with a central hole, bolt and nut.

The ultrasound probes already have an arm with a central hole in order to join them to the probe holder using a bolted joint. The bolt is placed through the central hole of the arm, and the nut is twisted at the other end. All the parts can be observed in Fig. 3.

The adapter has been developed according to the following requirements:

- The adapter should provide a way to get attached to the lateral part of the structure of the Oculus Rift head mounted display.
- Another region of the adapter should allow the use of the bolted joint to attach the probe.

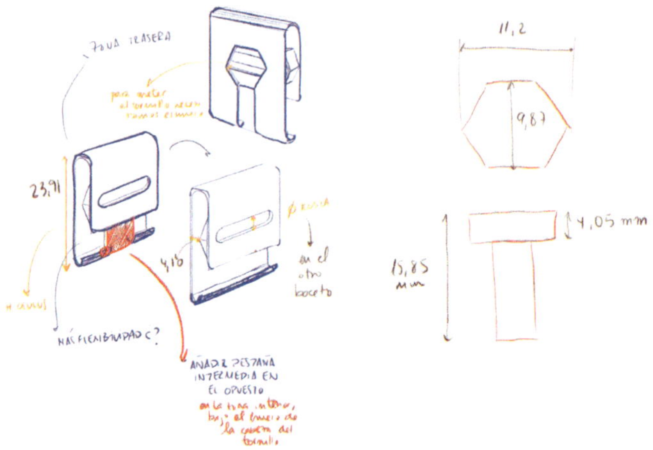

Fig. 4. First drafts of the adapter design.

The first drafts of the initially proposed design can be observed in Fig. 4.

In this first proposal (represented in the left part of Fig. 5), the height of the background face has the same height as the Oculus Rift lateral part to allow joining them. On the other hand, the foreground face has a slot that allowed the placement of the twisted joint. The probe with its arm will be placed in the outer part of the foreground face and fixed with the twisted joint.

The different versions of the design were modeled using Autodesk Inventor Professional and 3D printed using the Creality CR-10 3D printer with 1.75 mm diameter polylactic acid (PLA) filaments.

The initial prototype was placed on the Oculus Rift and it was evaluated by an experienced neurologist. The neurologist needs to adapt the probes comfortably to correctly acquire the BFV signals. However, this first prototype did not allow it. Some problems were observed when the neurologist tried to adjust the probes to a volunteer. Firstly, it was difficult to move the probe laterally because the access to the bolt was hidden. Secondly, the range of lateral movement that was allowed was limited and not enough to adjust for different head sizes. Finally, the probe remained parallel to the lateral part of the head, and it was not possible to achieve a good positioning for the probe. Consequently, additional versions of the adapter were designed.

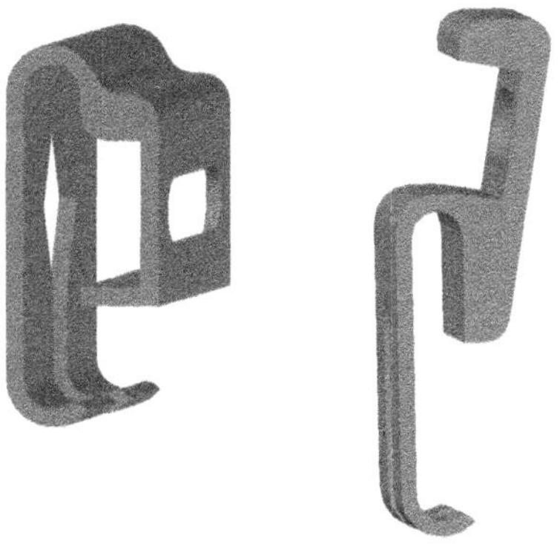

Fig. 5. Adapter design. First version (left) and final version (right).

Subsequent versions of the design were developed increasing the distance between the background face and foreground face to allow an easier manipulation of the bolted joint, increasing the horizontal length of the faces to allow a greater range of lateral adjustment for the probe and adding an additional tab to reinforce the adjustment to the Oculus Rift lateral part. However, these modifications were not enough.

Finally, a change in the adapter structure and shape was proposed. The part of the adapter used to adjust the probe using the jointed bolt was placed externally to the part of the adapter used to attach it to the Oculus Rift lateral part. Furthermore, the orientation of the external part was slightly modified, so it was not exactly vertical, allowing the probe to get closer to the head. With these two modifications, the final version of the adapter (see Fig. 5, right part) was designed. Figure 6 shows its orthographic views and the main dimensions.

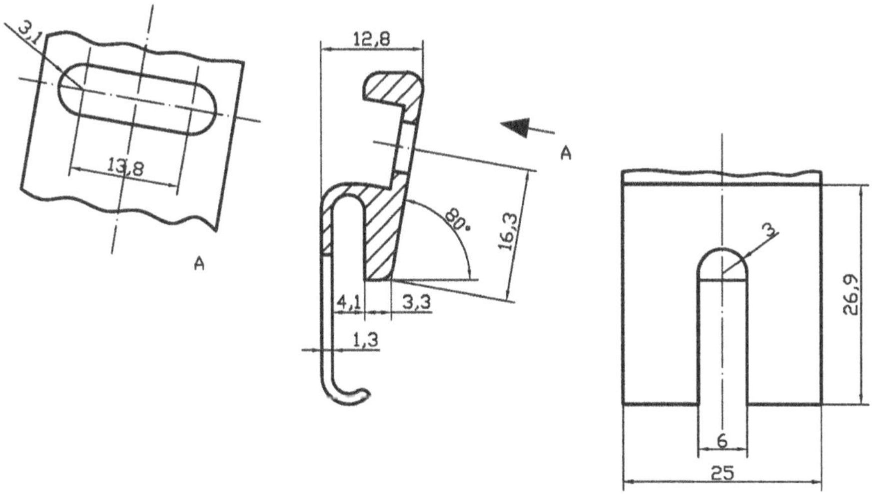

Fig. 6. Adapter orthographic views with its main dimensions

2.4 Software

A virtual environment was used to validate the final design. The environment allowed the participants to perform a simple motor task with their hands (moving a cube from an initial position to a final position) in a repetitive way, while their BFV signal was monitored with the previous system. This environment can be observed in Fig. 7.

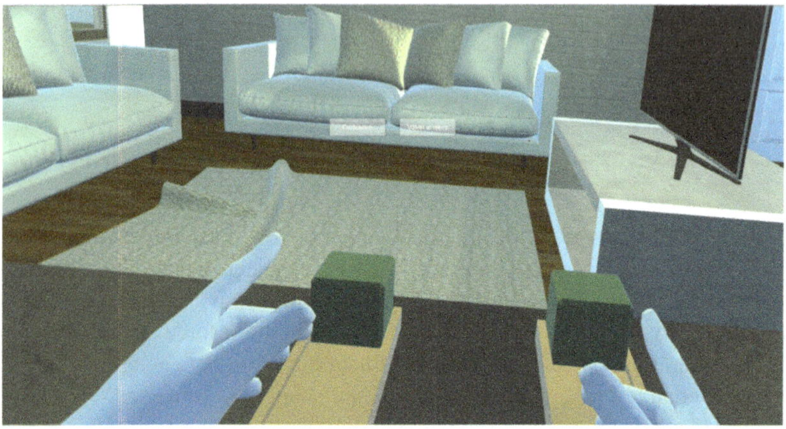

Fig. 7. Virtual environment for motor training.

This environment was developed using Unity, version 2018.2.15f1, a leading game development engine that allows the design, creation and operation of interactive environments. Oculus integration package was used to enable the use of virtual reality.

2.5 Validation of the Developed System

The designed system was evaluated by an experienced neurologist. A volunteer was instructed to remain still while the neurologist placed the ultrasound probes in their correct location using the developed system. Once fixed, the volunteer had to perform in the virtual environment, in a repetitive way, the previously described motor task, while TCD in the middle cerebral arteries was monitored using the ultrasound probes by means of the developed system.

3 Results and Discussion

With the developed system, the probes can be located in the temporal window of the participant head with the correct orientation to monitor the relevant vessels for the study, while the participant is immersed in a virtual world using the Oculus Rift. Figure 8 shows how the adapters are placed in the Oculus Rift head mounted display with the probes fixed by means of the bolted joint.

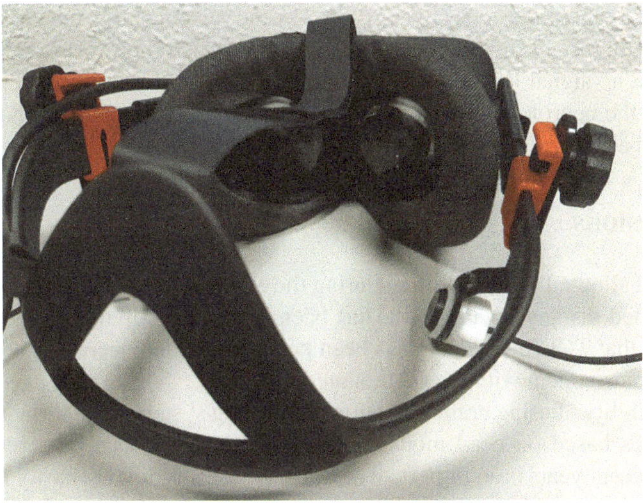

Fig. 8. Probes attached to the Oculus Rift using the designed adapters.

The experienced neurologist evaluated the procedure to position the probes, the quality of the registered signals and the stability of the fixation during the exposure to the developed virtual environment.

The developed system was adequate to place the probes in their correct location and the adjustment procedure could be performed in a comfortable way and reduced amount of time, similarly as happens with the probe holder provided by the DWL company. Figure 9 shows an image of a volunteer with the Oculus Rift head mounted display and the TCD monitoring system.

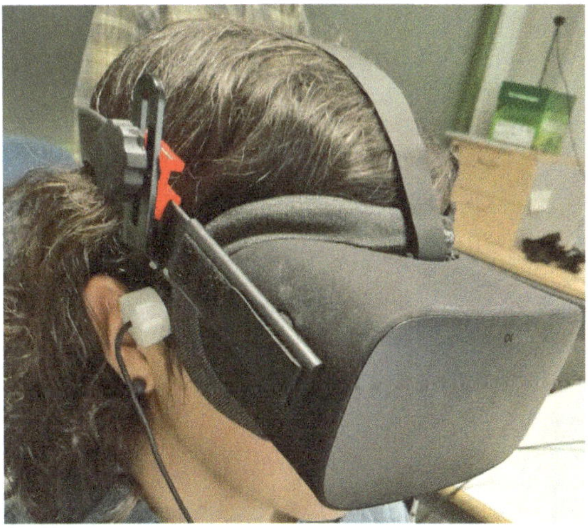

Fig. 9. Volunteer visualizing a virtual environment with the Oculus Rift while the BFV is monitored using TCD by means of the designed system.

The BFV signals were correctly registered by the system during the virtual exposure, showing the stability of the fixation procedure. The registration was made without any incidence and no signal losses were observed during the monitoring. The signals were evaluated by the neurologist who found them morphologically correct and suitable for subsequent analyses.

4 Conclusions

A system that allows TCD monitoring during the exposure to a virtual environment with a head mounted display (Oculus Rift) has been developed. Up to our knowledge, it is the first time that TCD monitoring has been performed simultaneously to the exposure to a virtual environment with a head mounted display.

The possibility of using brain activation monitoring systems with the new generation of VR devices based on head mounted displays that have been made commercially available in recent years is an important advance. These systems are easily available and can be adopted as a technological help to conduct motor training procedures in hospitals and rehabilitation facilities.

The system is also compatible with other technological advances that can be easily integrated in the set-up. A non VR-based TCD neurofeedback system has already been developed and validated in a previous study from out group [9]. This neurofeedback system can be adapted so the feedback is given by means of a virtual environment. The same system can also be used to modify a motor training procedure in real time depending on the observed BFV responses. This opens many possibilities in the field of motor training and rehabilitation.

Acknowledgments. The research work reported here was made possible by Conselleria de Innovación, Universidades, Ciencia y Sociedad Digital, Comunitat Valenciana, Spain, grant number AICO/2019/029.

References

1. Kontos, H.: Validity of cerebral arterial blood flow calculations from velocity measurements. Stroke **20**(1), 1–3 (1989)
2. Stroobant, N., Vingerhoets, G.: Transcranial Doppler ultrasonography monitoring of cerebral hemodynamics during performance of cognitive tasks: a review. Neuropsychol. Rev. **10**(4), 213–231 (2000)
3. Duschek, S., Schandry, R.: Functional Transcranial Doppler sonography as a tool in psychophysiological research. Psychophysiology **40**(3), 436–454 (2003)
4. Angevine, J.B., Cotman, C.W.: Principles of Neuroanatomy. Oxford University Press, England (1981)
5. Rey, B., Alcañiz, M., Tembl, J., Parkhutik, V.: Brain activity and presence: a preliminary study in different immersive conditions using Transcranial Doppler monitoring. Virtual Reality **14**(1), 55–65 (2010)
6. Alcañiz, M., Rey, B., Tembl, J., Parkhutick, V.: A neuroscience approach to virtual reality experience using Transcranial Doppler monitoring. Presence Teleoperators Virtual Environ. **18**(2), 97–111 (2009)

7. Rey, B., Parkhutik, V., Alcañiz, M.: Breaks in presence in virtual environments: an analysis of blood flow velocity responses. Presence Teleoperators Virtual Environ. **20**(3), 273–286 (2011)
8. Ringelstein, E.B., Kahlscheuer, B., Niggemeyer, E., Otis, S.M.: Transcranial Doppler sonography: anatomical landmarks and normal velocity values. Ultrasound Med. Biol. **16**(8), 745–761 (1990)
9. Rey, B., et al.: Design and validation of an FPGA-based configurable Transcranial Doppler neurofeedback system for chronic pain patients. Sensors **18**(7), 2278 (2018)

Requirements of a Common Data Environment (CDE). Study Case of VIRCORE

Silvia Odriozola[1], Cristina Manchado[2(✉)], Valentin Gomez-Jauregui[2], and César Otero[2]

[1] School of Civil Engineering, Universidad de Cantabria, Avda. de los Castros 44, 39005 Santander, Cantabria, Spain
[2] R&D EgiCAD, School of Civil Engineering, Universidad de Cantabria, Avda. de los Castros 44, 39005 Santander, Cantabria, Spain
manchadoc@unican.es

Abstract. Nowadays, it is unquestionable to consider BIM when dealing with a project in the AEC (Architecture, Engineering and Construction) sector. BIM is more and more present in the construction industry due to the technological evolution and the challenges that appear every day. The use of BIM provides advantages such as: reliable and accurate information, simple and instant communication, updated register of actions, error reduction, teams connection, etc. BIM methodology involves collaboration between different disciplines and people, allowing them to have controlled access to the project data depending on their role.

From this collaboration between stakeholders arises the need for a Common Data Environment (CDE). This collaborative platform makes it possible to work on the different parts and disciplines of a project on a distinct federated model at the same time along its entire life cycle.

The main goal of this article is to establish the minimum requirements for a CDE, as well as to describe the functionalities of the collaborative platform VIRCORE. This article discusses its potential, advantages and versatility.

Keywords: BIM · CDE · Common Data Environment · ISO 19650

1 Introduction

When talking about the degree of BIM implementation and its maturity levels, explicit reference is made to the existing collaboration level, and therefore to the implemented collaborative environment. Collaboration is based on the generation and interchange of data between all the agents involved in a project; therefore, there exist three implementation levels.

Collaboration concept, initially understood as a central repository that allows the storage and digital interchange of all the information of a construction project without a common data format, arises at level 1. These can be either files created in a BIM environment, or any type of documentation or file, and it basically focuses on the transition from

F. Cavas Martínez et al. (Eds.): INGEGRAF 2021, LNME, pp. 30–37, 2022.
https://doi.org/10.1007/978-3-030-92426-3_4

2D CAD drawings to 3D models. At this level, the concepts of roles and responsibilities, naming conventions, project codes, etc. appear. At level 2, common working formats as IFC (Industry Foundation Class) or COBie (Construction Operations Building Information Exchange) are implemented, allowing a collaborative flow, whereas level 3 implies a maximum degree of collaboration, with all participants working on a single information model, ensuring consistency, open and collaborative working [1].

From the first level, the implementation of a Common Data Environment (CDE) is required. It allows the definition of a common workspace, an information exchange strategy, the assignment of roles and permissions and collaborative work. This brings a number of benefits, such as cost reduction, quality improvement and time reduction during the construction process [2]. Furthermore, the improvement of coordination and interference detection processes is enhanced, as well as the design change management in infrastructure projects [3], and the improvement of the organisation of information from existing resources, both structured and unstructured information at the exploitation stage [4].

This paper describes the minimum requirements that a CDE must meet for a correct BIM implementation, showing some of these requirements on VIRCORE, an existing software that allows to fulfil several of the BIM Level 2 requirements.

2 Regulations and Minimum Requirements

It is not possible to talk about BIM without paying attention to the guidelines established by current regulations. Nowadays, the management of information along the life cycle of an asset is governed by ISO 19650 [5], closely linked to PAS 1192-2:2013 (withdrawn in 2018). It sets the standards for the generation, verification and exchange of information between the different stakeholders involved in the project, as well as its organisation and digitisation.

ISO 19650 divides the information management process into two phases. In the first development phase, the asset is designed, built and delivered; in the second phase, the focus is on the maintenance and operation of the asset, which may have different requests. All the information requirements can be defined by the client, during the tender process, or in the BIM Execution Plan (BEP). In any case, the new Collaborative Modelling Systems (CMS) [6] makes it possible to host all project information allowing collaborative operations through a common environment. Multidisciplinary collaboration requires at least a common working scenario, specified in ISO 19650 as a CDE, which gives the possibility to complement model collaboration according to established protocols. The CDE is in fact a combination of a technical solution and a set of operational processes. [7].

The CDE must be provided by the CMS in such a way that users are able to perform data collaboration and management, meet requirements and avoid deficiencies. Defining processes and procedures, roles and responsibilities and storing information in a secure and accessible way ensures an "auditable, transparent and controllable process" [8].

The requirements to be fulfilled by the CDE, according to ISO 19650, are the following:

1. Each information container must have a unique identification and its own authorship. This identification should be based on a pre-agreed model, allowing easy identification and quick access to the documents in the repository.
2. Rate each file according to its **status.**

- Work in progress. Contains all types of documents being developed by a particular team in a particular discipline that are not visible or accessible to other work teams.
- Shared work. As its name indicates, it is only shared information and it is not editable, in which case the "Work in progress" phase is resumed and those responsible for the file will be in charge of editing it.
- Published work. Stage where the BIM Project Manager authorises releasing the information.
- Archived work. Used to be able to keep a record of all the information that has been shared and approved.

The CDE should establish check/review/approval procedures to move information from "work in progress" state to the next states and allow the different disciplines to share this information, until common agreement is reached to publish the final information.

3. Rate each file according to its **revision**. Keep an updated record of modifications made to the files, indicating both the name of the person responsible and the time they were made. This classification provides information on the revisions that have been made to the file and the person responsible for them, ensuring the traceability of the file.
4. Rate each file according to its **classification**. Identify each file with all types of information linkable to it.

These classifications are contained in clauses 12.1 of ISO 19650-1 and 5.1.7 of ISO 19650-2. The former requires a code for revision and status and the latter for classification, both of which can be extended to include other information.

5. Having a working directory properly structured in folders with a secure and comprehensive maintenance that facilitates the order and classification of files according to the dependence degree between each one of them.
6. Being able to specify the Delivery Plan reflecting important dates or milestones agreed with the client and define the BIM Execution Plan with model update strategies.
7. Having different compatible information exchange formats (proprietary or not, depending on the project needs) with the model such as: .rvt, .nwc, .ifc, .dwg, .dwf, .pdf, .dgn, .alg, using distributed information containers for structured (i.e. graphical and non-graphical data) and unstructured (i.e. documents) data.
8. Identifying each one of the roles involved in the project according to which the corresponding permissions will be granted, thus the access security to the information is guaranteed.

3 VIRCORE

This section shows the application of the requirements of ISO 19650 for the CDE, by means of an existing software in a BIM project. The software used is VIRCORE [9], a CDE that implements the BIM methodology and collaborative development, reaching level 2 of BIM development, as it allows linking all the project information (unstructured data) with the 3D graphic models and elements.

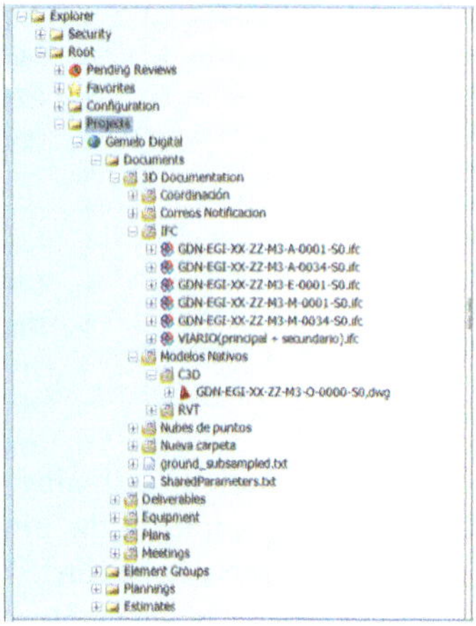

Fig. 1. Folders hierarchy

The software supports the management of 3D objects in Navisworks files. Its central system allows the creation of hierarchical folder structures (Fig. 1), with a bidirectional linking system that allows interlinking unstructured information with existing 3D elements in the models. In this way, it is possible to associate information concerning these 3D elements, either drawings, constructive elements data, maintenance history, user instructions, etc. From this initial structure. the hierarchy explained in the BEP project can be created.

Within this software, all the projects being carried out can be viewed. Each of them has an initial information structure classified in three main folders: "documents", "planning" and "estimates". These folders can be related to the BIM dimension [10] they correspond to. "Documents" folder is intended for storing all types of files that contain key information for the project, from reference data to 3D; "Planning" allows time control (BIM 4D) with the scheduling of workflows, all types of milestones and tasks that are established in the project; and "estimates" in which the cost estimation data of the work units are stored (BIM 5D). A fourth folder, "elements group", contains all the 3D elements imported into the Vircore database. These elements are selected on the basis of certain chosen properties of the model itself (Fig. 2). This allows accelerating the selection and linking of structured and unstructured data.

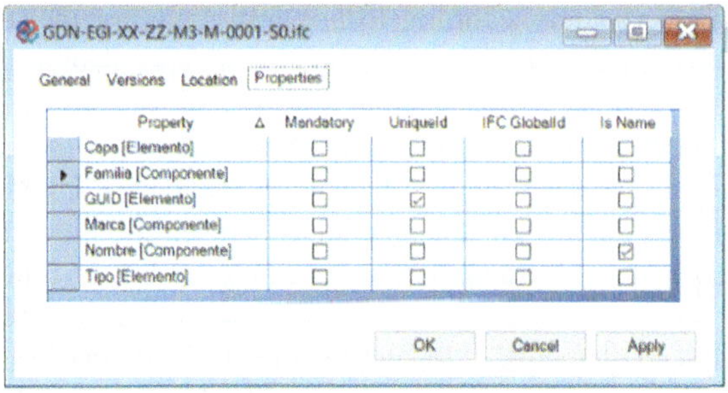

Fig. 2. Sample properties for importing 3D elements into Vircore

Fig. 3. Roles and users.

The platform works with a SQL database capable of referencing all available files and performs the function of a search engine, which, in addition to identifying the file being searched, provides all the properties of the file. This file search system is very efficient and simple, its results depend on the user's role type (see below) and it has a filtering option that makes the search easier, whatever the type of file in question: projects, folders, documents, people, roles, elements, groups, logical operations, schedules, tasks or types of tasks.

Role generation, assigning people to them, and applying read, edit, review and approval permissions to the different files stored is one of the main requirements for a CDE. All of this can be carried out to develop quality and effective collaborative work. The role of each person involved in the project (Fig. 3) should be stated from the beginning. In each part of the project, both in the different disciplines and at a general level, the permissions that each worker has are established, so the control is maintained over each one of the decisions taken in the project, such as modifying elements, approving files, discarding others, etc., and favouring good work management. VIRCORE differentiates between three possible permissions at user level:

- *Reading Permission.* Workers can access all the information loaded in the system only to review, obtain data and make queries in the system. Users with this permission will not be able to make any decisions about the files or have access to modify them.
- *Editing Permission.* It involves a greater responsibility in the handling of the files, being able to exchange versions, discard information, upload new information, create modifications, etc.
- *Validation Permission.* The user is able to make revisions, approve or notify aspects of the uploaded information.

This distinction favours good information management by allowing or disallowing access to specific project folders according to the permissions assigned to their role within the project's organisational structure. A person can have as many roles as specified in the project.

As previously mentioned, one of the main fonctionalities is the ability to generate relationships between files according to their interdependence. Users can link all files relating to a given document. All kinds of additional information can be associated, regardless of whether the information is structured or not (e.g. time schedules, construction aspects, economic aspects, etc.) All of them are summarized in the element children folders "linked elements", "linked documents" and "linked tasks", depending on whether there are interrelationships between 3D objects, with documents or with specific tasks created in the project planning. Figure 4 shows the information linking capability of the CDE. It is possible to associate any 3D element (a compressor in the figure) with the necessary unstructured information (e.g., a maintenance history).

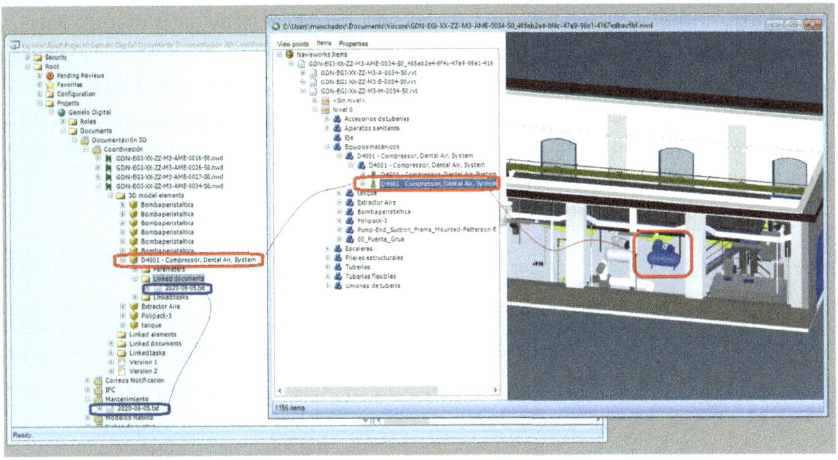

Fig. 4. File interdependence

From the "planning" folder the user is able to create as many plans as wanted or import them from Project (project management software) together with their assigned tasks. In a newly created schedule, it is also possible to create as many tasks as necessary, all of them with their specific information, such as: start and end time, duration, predecessor tasks and to link all the necessary information to the task. Its planning viewer allows the user to see the organisation of the project by means of a Gantt chart (Fig. 5).

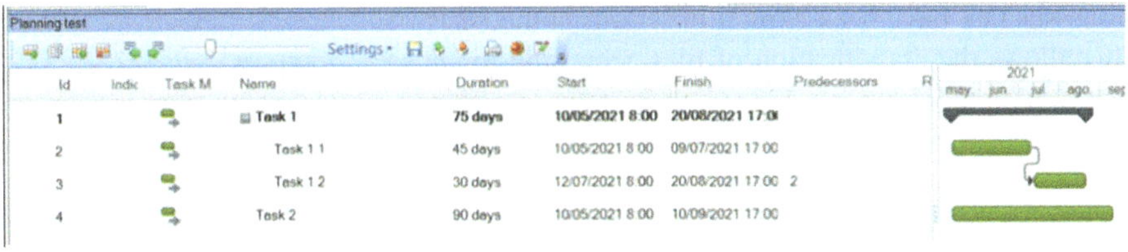

Fig. 5. Gantt chart

A highlight of this module is that it can be linked to Vircore's 3D viewer to visualise the planning in a three-dimensional way through a simulation.

The third module, cost estimation, works in a similar way as previous one. BC3 type files, also known as FIE-BDC format, can be imported, thus easily exchanging budgeting information and construction databases (Fig. 6) or created manually from Vircore with their detailed information: performance, quantity, machinery, labour, unit price, total, etc. All this information is collected in the folder "estimates" and has another independent viewer.

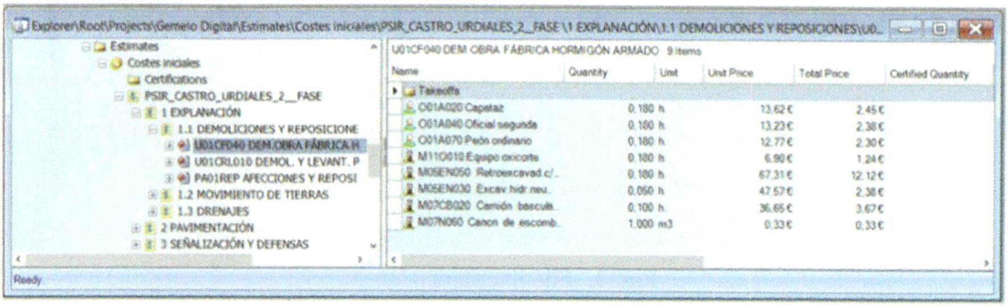

Fig. 6. Budget data imported into Vircore

The other great power of this CDE is the possibility of making requests for review and approval as many times as necessary, as well as versions management, according to the different roles assigned. This is done by ensuring traceability and transparency in every process, in accordance with the review and approval processes that have been explicitly set out in the BEP. This allows the classification of files according to, as discussed in requirements 2, 3 y 4, their status, revision and classification.

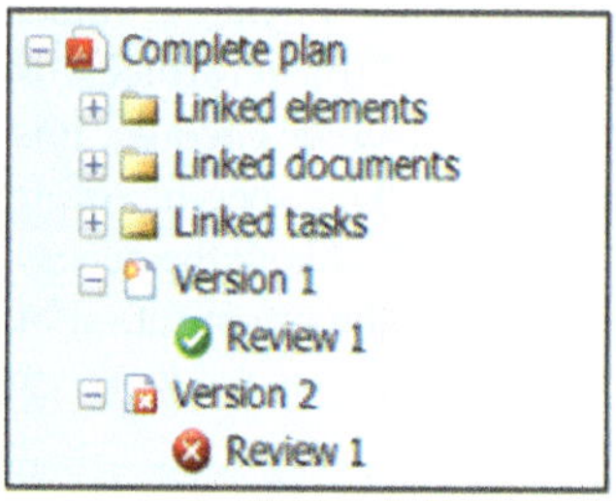

Fig. 7. File rating

This can be done in all existing versions of a file (Fig. 7).

Requests generate notices, via email, to notify the users involved, informing about all kinds of decisions (both rejection and acceptance of versions), revisions, modifications or simply comments on the file in real time.

4 Conclusions

Once covered the CDE requirements and analysed the VIRCORE, we can affirm that it is feasible comply satisfactorily not only with the BIM methodology but also with the main functionalities that ISO 19650 recommends to a CDE. Some requirements, as file naming or folders structure can be easily achieved for many systems, but other complex tasks, as automatizing review/approval processes, assigning roles and responsibilities, or even more, linking structured and unstructured data and queries are essential for a suitable BIM implementation.

As we know, one of the pre-conditions of BIM is collaborative work; after having analyze the potential of the VIRCORE, we are able to fulfil BIM level 1 and to reach the requirements of level 2. Achieving this level of collaboration between users brings great advantages in work management and risk reduction, directly affecting improved performance, reduction of unnecessary time and costs and achievement of reliable and quality information.

References

1. McPartland, R.: BIM Levels explained. Definitions for levels of BIM maturity from Level 0, through Level 1, Level 2 and Level 3 and beyond. 2014/11/01. NBS: https://www.thenbs.com/knowledge/bim-levels-explained. Accessed 14 May 2021
2. Radl, J., Kaiser, J.: Benefits of implementation of common data environment (CDE) into construction projects. IOP Conf. Ser. Mater. Sci. Eng. **471**, 022021 (2019)
3. Kumar, B., Cai, H., Hastak, M.: An Assessment of benefits of using BIM on an infrastructure project. In: International Conference on Sustainable Infrastructure (2017)
4. Patacas, J., Dawood, N., Kassemb, M.: BIM for facilities management: a framework and a common data environment using open standards. Autom. Constr. **120**, 103366 (2020)
5. UK BIM Alliance. UK BIM FRAMEWORK: https://www.ukbimframework.org/standards-guidance/. Accessed 14 May 2021
6. Tariq Shafiq, M., Matthews, J., Stephen, R.: A study of BIM collaboration requirements and available features in existing model collaboration systems. ITcon **18** (2013)
7. Information management according to BS EN ISO 19650 Guidance Part C Facilitating the common data environment (workflow and technical solutions), 1st edn. Published by UK BIM FRAMEWORK (2020)
8. Bouzas, M.: What is a CDE? 2017/04/01 Building Smart. https://www.buildingsmart.es/2017/04/01/qu%C3%A9-es-un-cde/. Accessed 14 May 2021
9. INGECID Research and development of projects, VIRCORE: https://vircore.es/. Accessed 21 May 2021
10. UNITED-BIM INC. What are BIM Dimensions – 3D, 4D, 5D, 6D, and 7D BIM Explained UNITED BIM: https://www.united-bim.com/what-are-bim-dimensions-3d-4d-5d-6d-7d-bim-explained-definition-benefits/. Accessed 14 May 2021

IFC for Infrastructures: New Open Standards for Intelligent Data

Ayoub El-Amraoui-Farssi[1], Valentin Gomez-Jauregui[2](✉), Cristina Manchado[2], and César Otero[2]

[1] School of Civil Engineering, Universidad de Cantabria, Avda. Los Castros 44, 39005 Santander, Spain

[2] EgiCAD Research Group, School of Civil Engineering, Universidad de Cantabria, Avda. Los Castros 44, 39005 Santander, Spain

valen.gomez.jauregui@unican.es

Abstract. This paper presents an analysis and evaluation of the last Industry Foundation Classes (IFC) version, including minor releases and addenda developed by BuildingSMART. This document focuses on the definition of the new elements or entities that are directly related to the infrastructure projects included in the current IFC Release Candidate (IFC4.3 RC2). Up to now, IFC's exchange format in infrastructure and civil works have undergone much less development than those in buildings, in terms of exchanging BIM data. Currently, the official IFC version available (IFC4 ADD2 TC1) has a set of limitations or loopholes in the infrastructure sector; for this reason, there is a constant effort in updating and extending this format specification to manage particular entities related to civil works, as alignment, profiles, etc. The evolution of this exchange data standard relies on integrating data structures to better understand complex geometries. Therefore, to conclude this study, the characteristics of the intermediate IFC versions between the above-mentioned versions (the two withdrawals IFC4.1 and IFC4.2, as the archived IFC4.3 RC1) will be cited and compared.

Keywords: BIM · IFC · Standard · Infrastructures · BuildingSMART · Civil engineering

1 Introduction

Over the last few years, the evolution of the different agents involved in the creation and management of BIM models has potentially improved [1]. The capacity of the whole working system integrates more and more information about different elements that define a BIM project and facilitates knowledge between them.

Currently, infrastructure information modeling refers to an activity of drawing and identifying physical and spatial components in a three-dimensional space. So far, IFC is not well suited to share the data model of infrastructure building products, such as roads, tunnels, or bridges [2]. The need to implement software capable of exporting detailed information from these structures becomes a necessity.

F. Cavas Martínez et al. (Eds.): INGEGRAF 2021, LNME, pp. 38–45, 2022.
https://doi.org/10.1007/978-3-030-92426-3_5

The BIM model in IFC format is one of the requirements increasingly requested by public administrations [3].

Industry Foundation Classes (IFC) was developed by BuildingSMART and set out in UNE-EN ISO 16739-1: 2020 [4]. It is defined in the construction industry as a standardized and open data format that allows information about construction projects to be described, exchanged, and shared between software applications. Until version IFC4, the IFC standard focused on building projects, but due to the high demand for BIM models in IFC format in other sectors, an extension of the standard is underway to make it compatible with civil engineering installations infrastructures.

This study performs an analysis of the new items integrated in version IFC 4.3 RC2 (Release Candidate 2) and which, unlike previous versions (IFC4 ADD2 TC1, IFC4.1, IFC 4.2 and IFC 4.3 RC1), includes a greater number of functions related to infrastructure projects.

Much of the information collected in this article comes from the documentation reflected on the BuildingSMART website [5].

2 IFC Format and Its Hierarchical Structure

The IFC format is characterized by being an object-oriented format, based on class definitions that represent objects and contain information generated throughout the life cycle of the project (from conception and design to demolition); being structured in layers with their entities, types, and relationships between them; and having a Model View Definition (MVD) subset that covers a specific area of the project.

This format allows smoother communication between heterogeneous programs, being able to carry enough information to read and analyze the spaces in the reference model. With these files, an engineer, for example, can calculate the energy efficiency of a building using the virtual model provided by an architect, without having to reconstruct the entire building in 3D.

The structural schema of IFC data is configured in four layers (Fig. 1). Each individual schema stores a series of definitions and is assigned to a conceptual layer. In this way, the definitions included in each layer are as follows: Domain layer: stores definitions for exchanging and sharing information in the domain; Interoperability layer: includes definitions for the exchange between different domains and for sharing construction information; Main layer: contains more general entity definitions; Resource layer: comprehends resource definitions and does not include an identifier like the rest; therefore, it will be necessary to declare the prior definition of the preceding layers.

In parallel, there is an IFC tree structure (IfcProyect => IfcSite => IfcBuilding => IfcBuildingStorey => ifcElements) which is significant to ensure a certain order. This second structure is always the same regardless of the IFC version; for this reason, it is not discussed thoroughly in this document.

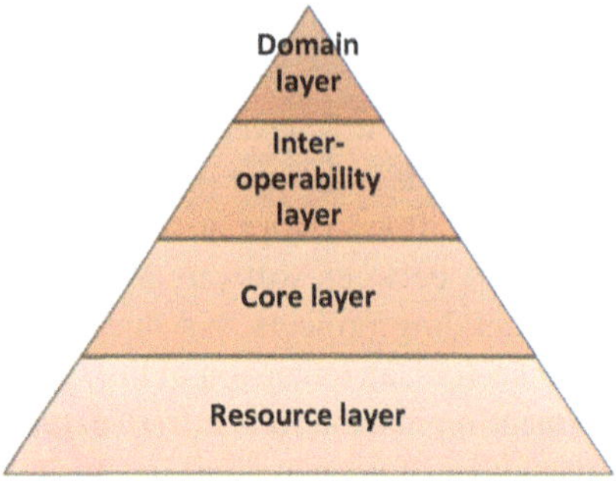

Fig. 1. Data schema architecture of IFC [5]

The schema specification aims to describe the use of the project, as well as how it works. Then, the different entities that make up each of these layers are collected and used as a basis for comparing versions of IFC prior to IFC 4.3 RC2 with this one:

I. **Domain layer**

It contains the final specifications of the entities. These stated entities reflected in this layer are autonomous and cannot be referenced by the rest of the layers. The organization of information is carried out corresponding to the discipline of the sector (Table 1):

Table 1. Entities of domain specific data schemas

IfcArchitectureDomain	IfcPortsAndWaterways
IfcBuildingControlsDomain	IfcRail
IfcConstructionMgmtDomain	IfcRoad
IfcElectricalDomain	IfcStructuralAnalysisDomain
IfcHvacDomain	IfcStructuralElementsDomain
IfcPlumbingFireProtectionDomain	

II. **Interoperability layer**

These schemas contain intermediate specifications of the entities. Likewise, these entities can be referenced and specified by other entities located higher in the hierarchy.

This shared element layer provides more specific objects and relationships shared by multiple domains (Table 2).

Table 2. Entities of shared element data schemas

IfcSharedBldgElements	IfcSharedFacilitiesElements
IfcSharedBldgServiceElements	IfcSharedInfrastructureElements
IfcSharedComponentElements	IfcSharedMgtElements

III. **Core layer**

The most general layer is integrated into the hierarchy of IFC schemas. The entities reflected in this layer can be referenced and specified by entities superior to this one in the IFC schema architecture. It provides a basic structure and all entities defined in the middle and upper layers are derived from IfcRoot and consist of a unique identifier, name, description, and change control information (Table 3).

Table 3. Entities of core data schemas.

IfcControlExtension	IfcProcessExtension
IfcKernel	IfcProductExtension

IV. **Resource layer**

These data schemas for defining resources consist of supporting data structures. The entities in this layer can be referenced by all entities in upper layers. The difference with the rest of the layers is that the resource definition data structures cannot exist independently but must be referenced by one of the entities that are derived from IfcRoot (Table 4).

Table 4. Entities of resource definition data schemas

IfcActorResource	IfcPresentationAppearanceResource
IfcApprovalResource	IfcPresentationDefinitionResource
IfcConstraintResource	IfcPresentationOrganizationResource
IfcCostResource	IfcProfileResource
IfcDateTimeResource	IfcPropertyResource
IfcExternalReferenceResource	IfcQuantityResource
IfcGeometricConstraintResource	IfcRepresentationResource
IfcGeometricModelResource	IfcStructuralLoadResource
IfcGeometryResource	IfcTopologyResource
IfcMaterialResource	IfcUtilityResource
IfcMeasureResource	

3 Analysis and Evaluation of IFC Version

The IFC format, although operational for more than 20 years, is still a format under development. Figure 2 shows the evolution of the different versions of this format, where those highlighted in green are the official versions and the latest version shown in yellow (IFC 4.3 RC2) is the current candidate. The other versions have been withdrawn.

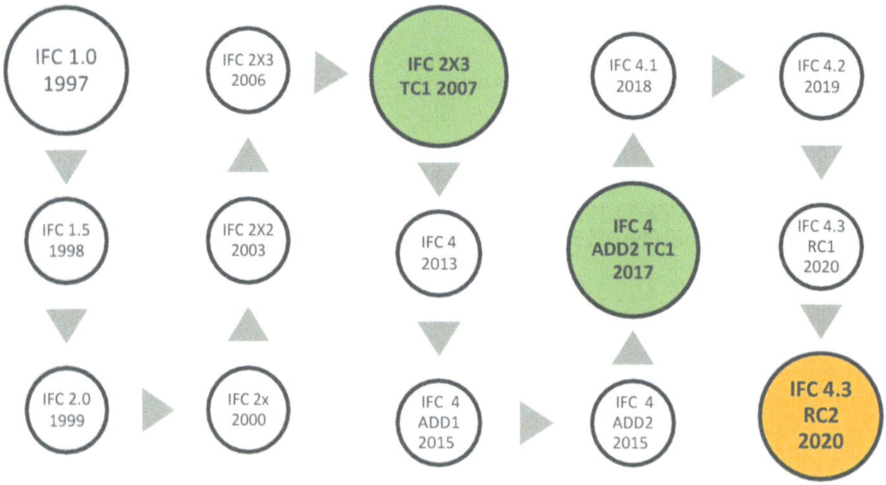

Fig. 2. Evolution of the IFC versions

I. IFC4 ADD2 TC1 to IFC4.1

Before commenting on the improvements reflected in IFC 4.1 compared to the current official version (IFC4 ADD2 TC1), it is important to explain the changes that this version has incorporated compared to previous versions:

The official version of IFC resolves a set of issues and makes improvements to geometric problems detected in the implementation of one of the versions prior to this one. Among the main entities that have been incorporated into the schema to improve the tessellated geometry and thus ensure more complex tessellations, the following are worth highlighting: IfcPolygonalFaceSet, IfcIndexedPolygonalFace, IfcIndexedPolygonalFaceWithVoids. Another potentially improved aspect was the representation of the usage limits of elementary surfaces and intersection curves, for which the following items have been incorporated: IfcSphericalSurface, IfcToroidalSurface, IfcSurfaceCurve, IfcInter-sectionCurve, IfcSeamCurve.

One of the essential elements of infrastructure projects is the alignment. The definition of bridges, tunnels, roads, railroads ... depends largely on the alignment. The first proposal of IFC 4.1 integrated the alignment scheme as a common resource for exchanging and sharing information in infrastructure works. The alignment-related elements included in this version are alignments as a valid combination of horizontal, vertical, and 3D alignments; horizontal and vertical alignment segments; the existing terrain surface model; transforming the local coordinate system into a geographic coordinate system; spatial zones; and geographic and civil physical features. For example, alignments in Autodesk Civil 3D and their associated longitudinal profile data for the vertical alignment can be exported using this schematic version 4.1 (for each alignment, only one longitudinal profile of the vertical alignment can be exported) [6].

The objective of adding alignment information to the IFC schematic is mainly based on exchanging information in the most efficient and secure way from the planning phase to the construction phase and later from the planning phase to the asset management phase. Furthermore, through the implementation of the alignment, an improvement is sought in the linkage of this element with the rest of the project information, such as elevations, cross sections, and 3D geometry. But despite the inclusion of this element, whose importance in any infrastructure project is relevant, the version is far from being effective, because in infrastructure projects the capacity of exporting and importing an isolated element without any link, in this case the alignment brings almost no added value to the IFC format. These are the reasons why BuildingSMART declared its withdrawal and worked on a new IFC proposal (IFC4.2) with the aim of improving the above-mentioned problems.

II. IFC4.1 to IFC4.2

The focus of the IFC4.2 release was on specific types and entities for bridges in the IFC Bridge extension. Model views for bridges were added, in addition to improving the entities related to the alignment element. In the base data schema, the ifcProductExtension (specialized in physical product concepts, i.e. a component with a shape and a location in the project context) and directly linked to the Bridge element, the following functions should be highlighted: Types: IfcBridge-PartTypeEnum and IfcBridgeTypeEnum; Entities: IfcBridge and IfcBridgePart; Property Sets: Pset_BridgeCommon.

The problem with this version is that it was able to represent some parts of the bridge like the "ifcBeam", "ifcColumn" or "ifcSlab", but nevertheless the definition of the IFC specification was not sufficient to represent the concept of structural element in terms of civil works [7].

In summary, despite all these improvements, this IFC version continued to have deficiencies in the field of infrastructure as in previous versions, because it maintained the problem of not being able to represent the structural element itself. For this reason, and like the previous version (IFC4.1), this version was withdrawn before its approval and the proposal of a new version began to solve the problems detected.

III. **IFC4.2 to IFC4.3 RC1**

IFC4.3 RC1 version was published in April 2020, the focus of which is on researching and incorporating new elements to improve the export and import of infrastructure information. BuildingSMART has been successful in improving the hierarchy of the IFC system through the implementation of new entities and improvements to existing ones.

Figure 3 shows, highlighted in red, the new domain and interoperability layer entity specifications (all of them are infrastructure-related). Within the domain layer: ports and waterways, rail and road, while the tunnel element has not yet been developed; the interoperability layer includes shared infrastructure elements, linked to the shared infrastructure data schema.

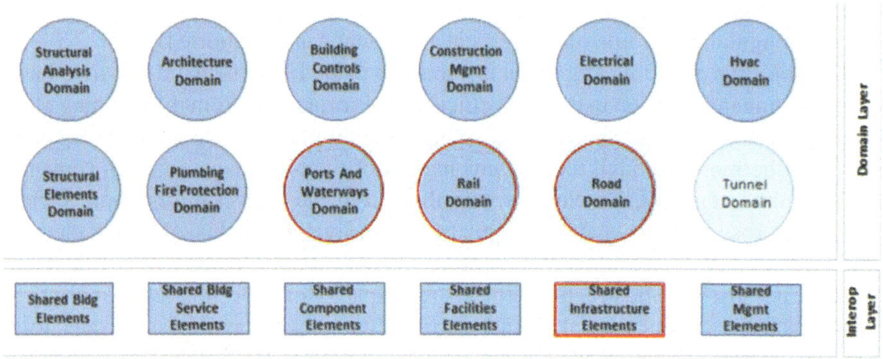

Fig. 3. Data schema with conceptual domain and interoperability layers. [5]

In addition, significant improvements are included with respect to the elements embedded in the data schema. A large part of these improvements is the ability to reference structures from already installed elements, such as IfcSpace, IfcSite, IfcGrid, IfcBridge... whose purpose is to ensure that more information about the building blocks can be read. Also, other details have been removed or replaced for better software configuration.

Finally, after an exhaustive analysis of this first version of IFC incorporating more infrastructure-related components, BuildingSMART filed, that is, withdrew the version due to deficiencies presented in the new integrated elements and in November 2020 proposed a new improved version, currently with a candidate status.

IV. **IFC4.3 RC1 to IFC4.3 RC2**

In this version, new rules related to infrastructure have been added and corrected compared to the previous IFC version. The time between the publication of these two versions is only 7 months, so the changes are not very relevant. Here are some of the main changes observed in this release candidate related to infrastructure: IfcRail: removal of components such as IfcLinearAxisSelect, ifcDirectrixDistanceSweptAreaSolid..., the name of the element ifcAlignment2DCant changes to ifcAlignmentCant, types such as ifcDirectrixCureaveSweptAreaSolid IfcRail to IfcGeometryResource; IfcRoad: remove the ifcLinearSpanPlacement; types such as ifcAlignmentCantSegmentTypeEnum, ifcAlignmentHorizontalSegmentTypeEnum, etc. were included in ifcGeometricConstraintResource.

4 Conclusions

The IFC format has been operational for 20 years and, to date, there is still no official version capable of representing the structural element concept in terms of civil works. The latest versions have focused on including new elements to ensure the exchange of information on infrastructure projects. The IFC4.3 RC1 version, which is currently archived, i.e., discarded, configures a structure similar to the current candidate (IFC4.3 RC2). The latter allows it to improve the IFC schema with the objective of having on the market updated software capable of exporting and importing all kinds of information on construction projects.

References

1. Pastor García, R., Collado Mariscal, D., Cuevas Murillo, A.: El Papel de BIM en la Industria 4.0. Cimbra **417**, 6–14 (2020)
2. Industry Foundation Classes (IFC) for data sharing in the construction and facility management industries — Part 1: Data schema: International Organization for Standardization (ISO), Novembre 2018. [Online]. https://www.iso.org/standard/70303.html. Accessed 15 April 2021
3. Comisión BIM del Ministerio de Transportes, Movilidad y Agenda Urbana: Análisis de la Inclusión de Requisitos BIM en la Licitación Pública Española, Observatorio de Licitaciones Públicas BIM, 2019. [Online]. https://cbim.mitma.es/observatorio/informes-trimestrales-del-observatorio. Accessed 12 April 2021
4. Liebich, T.: Unveiling IFC2x4 – the next generation of openbim. AEC3 Deutschland, Germany (2010)
5. buildingSMART: Industry Fundation Classes (IFC), [Online]. https://technical.buildingsmart.org/. Accessed 10 March 2021
6. AUTODESK: Civil 3D, 2019. [En línea]. https://www.autodesk.es/products/civil-3d/overview?term=1-YEAR. Último acceso 12 Abril 2021
7. Amann, J., Singer, D., Borrmann, A.: Extension of the upcoming IFC alignment standard with cross. In: Proceedings of the ICCBEI. Tokio, Japan (2015)s

The Digital Simulation of the Visual Impact of Power Lines as a Tool to Optimize the Design of the Power Line Network

Jacinto Santamaría-Peña(✉), Eduardo Martínez-Cámara, Félix Sanz-Adan, and Efrén Tarancón-Andrés

Department of Mechanical Engineering, University of La Rioja, 26004 Logroño, La Rioja, Spain
jacinto.santamaria@unirioja.es

Abstract. The current thrust of the ecological transition implies a significant increase in the number of electric generation installations of renewable origin and, as a consequence, the increase in the power lines necessary to connect these installationsis with the evacuation points. In addition, the population is increasingly aware of the visual impact generated by these lines in their localities. The environmental impact assessments of the different administrations specifically assess the visual impact of power lines, as well as their possible cumulative and synergistic effects with other power lines.

For all these reasons, it is becoming increasingly relevant to provide tools that allow simulating and comparing the visual impact of the technically and economically viable alternatives of the power lines to be projected. In such a way that the design of the different lines that make up the power grid can be optimized to reduce or minimize the visual impact that they may cause in the region.

This article proposes a digital simulation system of the visual impact of power lines for the optimization of the design of power networks.

The case of visual optimization of the drawing of electric lines is studied.

Keywords: Visual impact · Landscape simulation · Environmental impact assessment · Power network · Electric line

1 Introduction

The economic structure and welfare state of Spanish and European society makes it unthinkable, today, not to have electric energy. Its discovery is at the top of the list of technological advances that have contributed most to the improvement of the quality of life of humanity. Its great lack is not being able to be stored on a large scale; which conditions the activities of generation, transport, distribution and distribution from the generation plants to the points of consumption. As a consequence, to meet the imp demand of society, there are more than 600,000 km of power lines in Spain.

The network of electric lines makes it possible for the electric energy required by industries, utilities and homes to be transported, at the time and with the necessary

F. Cavas Martínez et al. (Eds.): INGEGRAF 2021, LNME, pp. 46–53, 2022.
https://doi.org/10.1007/978-3-030-92426-3_6

quantity and quality, from the generation plants, generally remote, in optimal conditions of safety and economy.

A well-developed network allows generation in sites and developments far from the place of consumption, allowing to search for the most suitable sites.

Electric energy consumption in Spain went through a decline in 2020 due to the COVID-19 pandemic. Specifically, the demand for electric energy in Spain in 2020 was 249,819 GWh, 5.6% lower than in 2019 [1]. But it is presumable that it will rise again as activity reaches more normal levels.

In addition, the generation used to meet this demand is increasingly renewable. In 2020, renewable energy has accounted for 44% of power generation and the objective set by the government in its long-term decarbonization strategy is to reach values close to 100% renewables by 2050 [2].

Therefore, in the coming years the number of electrical installations of renewable origin and the kilometers of power lines necessary to transport the generated energy to the distribution points will continue to increase. This increase in transmission power lines is becoming increasingly visible [3] and makes considerations regarding the visual impact of the same [4] more critical when granting administrative licenses for these new facilities.

1.1 Visual Impact of Overhead Power Lines

Among the circumstances that contribute to a power line and his towers going unnoticed is the complexity of the landscape that demands the attention of the observer.

It is necessary to look for the best solutions to integrate power lines into the landscape without destroying it. Power lines are usually simple, slender, lightweight, porous and repetitive structures.

Taking in account that the most difficult impact to reduce is that produced by the difference in character between what is intended to be built and the character of the place, power generation is usually difficult to reconcile with the rural environment.

To minimize this impact, different non-exclusive strategies can be used:

- Looking for the most appropriate route.
- Reducing the visual basins of the supports and conductors taking advantage of the relief and vegetation that prevent their perception from the most visited places.
- Mimicking their appearance so that they blend in with the vegetation or the sky, playing with colors.
- Designing the towers as sculptures related to the characteristics that define the country, in our case with viticulture and enology.
- Placement of birds' nests on their summits to attract storks and birds of prey.
- Burying the power lines, so that the visual basins are null.

Three of the fundamental aspects to take in account when evaluating the visual impact of an installation are:

- The technical problem of how to visualize future landscape changes.
- The social problem of how to evaluate scenic beauty.

- The administrative problem of how to integrate visual aspects into the Environmental Impact Assessment (EIA) process.

With the idea of facilitating solutions for these problems, this article proposes a methodology that allows simulating and comparing the visual impact of the alternatives of the power lines to be projected.

1.2 Methodology

Simulation techniques as a working tool allow the designer, the promoter and the competent Administration to get close enough to reality to predict the visual impact of a power line.

The integration of cartographic techniques, Geographic Information Systems (GIS) and landscape simulation techniques, supported by three-dimensional infographics, models or photomontage serve to simulate the built work.

In the phase of generation of alternatives, the anticipation of future landscape and visual impacts allows the detailed study of the site, making modifications and adjustments to adapt the project to the optimal solution.

At the preliminary project level, geographic information systems (GIS) are a valuable design tool to test various alternatives to build on the same land that cannot be carried out with traditional and laborious methods.

For the modelling of overhead power lines in order to study their visual impact, a methodology based on the following steps is proposed:

1. Digital terrain modelling and digital surface modelling.
2. Modelling of supports and circuits.
3. Definition of static and dynamic observation places from which visual impact occurs (tourist viewpoint, see-sight, villages and roads)
4. Three-dimensional views of landscape.
5. Integration of previous graphic information in the Geographic Information Systems.
6. Virtual dynamic simulation.

The provision throughout the national territory of high-resolution LIDAR Data allows to obtain digital terrain (MDT) and surface (MDS) models of great precision. To this, we should add the possibility of having digital orthophotographs with resolutions of around 0.5 m/pixel or greater, which allows to provide the whole with a realism difficult to achieve with other media.

In order to evaluate the visual impact of a power line, it is necessary to know in detail the types of support that can be used in each situation, its dimensions, its location (beginning and end); the number of circuits, their positioning (horizontal or vertical), the type of crosshead, the insulators, the color of the support, the bird-saving measures, the access roads to the supports (layout, profiles), the type of construction, the preventive measures to be adopted during construction to mitigate temporary and permanent impacts and the corrective measures to be adopted during its useful life.

In the sizing of the bars of each of the supports, it is necessary to try to re-strip the reality to the maximum, starting from the current existing commercial models. Support geometries built with solid bodies must include thickness and elements of metallic texture, which gives the models the maximum possible realism.

For the definition of static observation points (population enclaves, observatories, viewpoints, see-sights) and dynamic observation routes (walks, roads, roads, railways), from which the visual impact is evaluated, aspects such as the greatest influx of tourist or cultural public in the visual domain space must be considered.

Photo-composition has been relegated in a certain way to the development of three-dimensional reconstruction tools of the terrain and construction models. It is no longer necessary to see the effect of a new support within an image, but the support is fully integrated as such into the terrain model. Once this is achieved, infinite points of view of the new model can be obtained in its environment and determine its greater or lesser adequacy to the proposed objectives.

That is why, in this methodology, photocomposition tools have not been used, but the functionalities of three-dimensional vision have been exploited to the maximum.

Geographic Information Systems are georeferenced graphic data management systems with associated alphanumeric information and powerful spatial analysis tools, both vector and raster.

The data obtained from the 3D simulation of the environment can be quickly integrated into the Geographic Information Systems, with other information that may be of interest in the future, such as, for example, the boundaries of cadastral parcels to address the process of land expropriation and new electrical infrastructures.

2 Case Study: Results and Discussion

The proposed methodology is applied to Iberdrola's 13.2 kV Sajazarra-Villalba-San Felices Distribution Network, which crosses the A-68 highway and is located in the European environmental protection area "SPA Obarenes" in the west of La Rioja (Spain).

2.1 Terrain Modelling

In Fig. 1 you can see the Special Protection Area for Birds Obarenes (SPA), as well as the route of the line that crosses it. It consists of a main line of about 3.7 km in length and a secondary line of 1.8 km.

In addition, the SPA protection zone, the power line itself and the Vasco-Aragonesa A-68 highway coincide in this area, which has always generated a certain conflict of interest and environmental controversy.

Fig. 1. Power lines (red) and SPA (yellow)

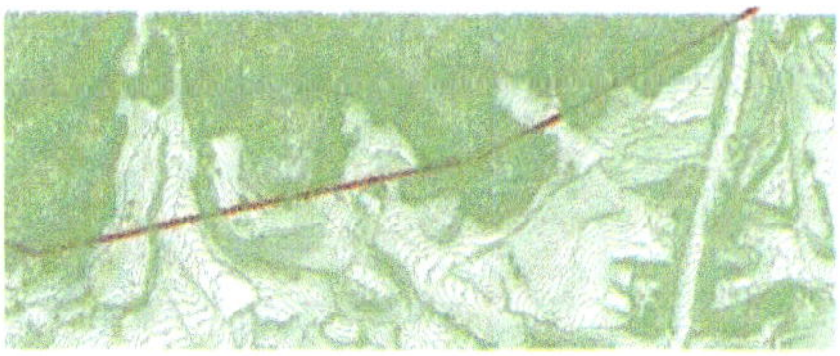

Fig. 2. DSM with Contour lines at 1 m MDT

In Fig. 2 you can see the contour lines of the Digital Surface Model (DSM) with equidistance 1 m between contour lines, which have been used as support for the generation of the Digital Terrain Model (DTM). The original data belong to the 2016-PNOA LIDAR flight, of 2 points/m^2, promoted by the National Geographic Institute and Government of La Rioja, within the framework of the National Cartographic System (NCS). In order to work the surface more easily, two simplification processes and a cut have been carried out, reducing it to 1,470,941 million points and 2,934,416 million triangles. Originally the mesh had about 4 million points.

As a texture for the realization of the three-dimensional views, the ECW orthophotos of 0.25 m/px of the year 2020 have been used (Fig. 3).

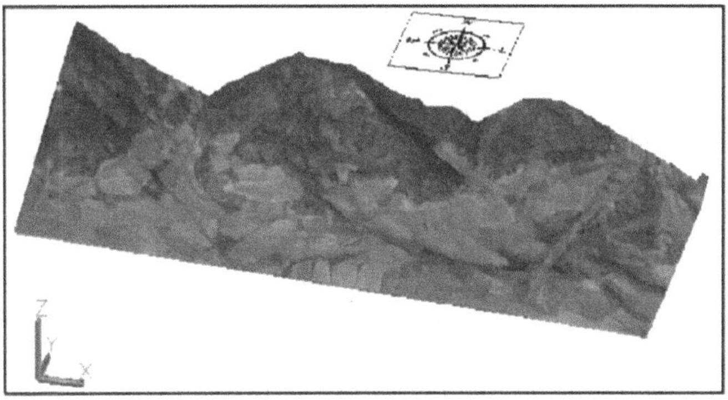

Fig. 3. Digital Terrain Model (DTM), with orthoimage coverage year 2020.

2.2 Modelling of Towers

The next step is to model the towers to be able to introduce them into the simulation. In Fig. 4 you can see real images of the environment and the existing towers and a 3D model of these supports.

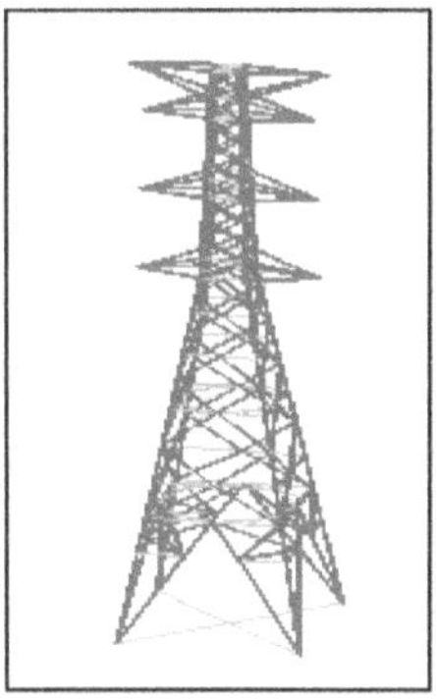

Fig. 4. Photographs of the power line towers (left). 3D Model of a tower (right).

2.3 Definition of Static Observation Points from Which Visual Impact Occurs

For the drawing of this line, 5 significant points of observation have been considered from which it is necessary to study the visual impact of the same (see Fig. 5).

- Visibility from an interior point, in the cultivation area. Position 0, green color.
- Visibility from an interior point, in the cultivation area. Position 1, red color.
- Visibility from a route on the A-68 motorway. Static positions 2,3,4.

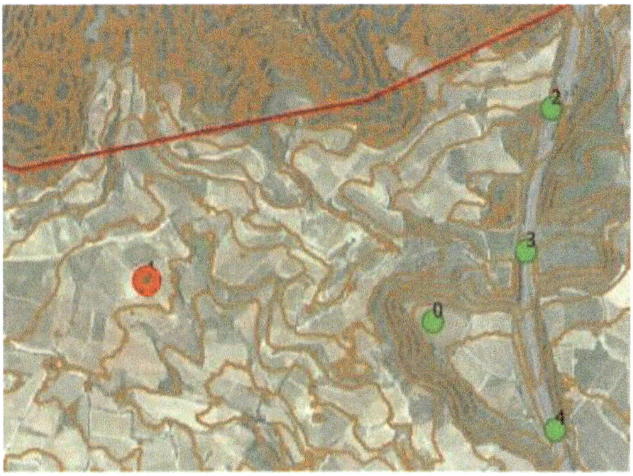

Fig. 5. Static observation points (0, 1). Dynamic observation route (A-68 highway: 2,3,4,...)

2.4 Three-Dimensional Views

Once the Digital Terrain Model and the line towers are included in Autocad® Civil 3D, it is possible to obtain three-dimensional views from the different defined observation points.

Figure 6 shows the resulting view from the point 0 of the cultivation area, and Fig. 7 shows the view of the path from a point that varies from frame to frame.

Fig. 6. View from a static point (position 1)

Fig. 7. A frame from highway (position 2)

2.5 Incorporation into Geographic Information Systems

With this data you can check what the visibility would be from each point and incorporate these results into any geographic information system for later analysis.

In Figs. 8–10 you can see the visibility maps from three observation points (QGIS environment, using Visibility Analysis plugin). Where the red zone indicates a non-visible area, the yellow zone a visible area with difficulty and the green zone a clearly visible area. A similar study has also been conducted in a CIVIL3D environment, with similar results.

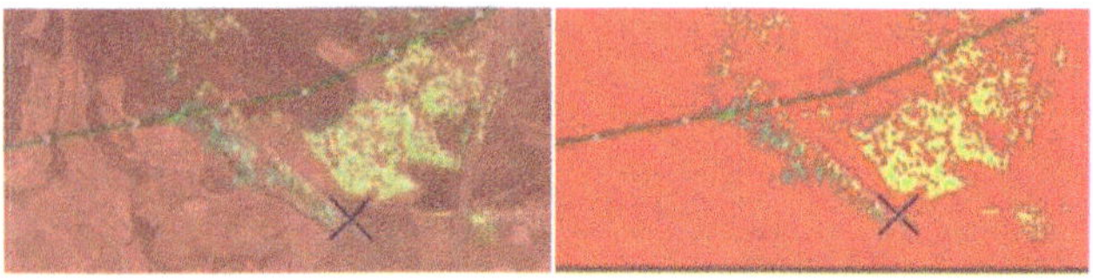

Fig. 8. Visibility from an interior point, in the cultivation area. Position 0.

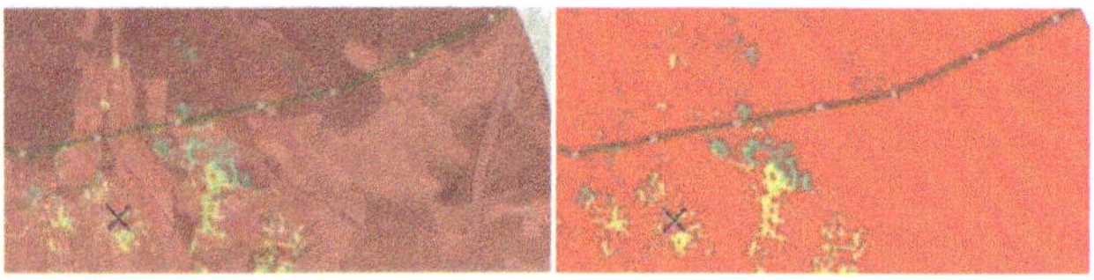

Fig. 9. Visibility from an interior point, in the cultivation area. Position 1.

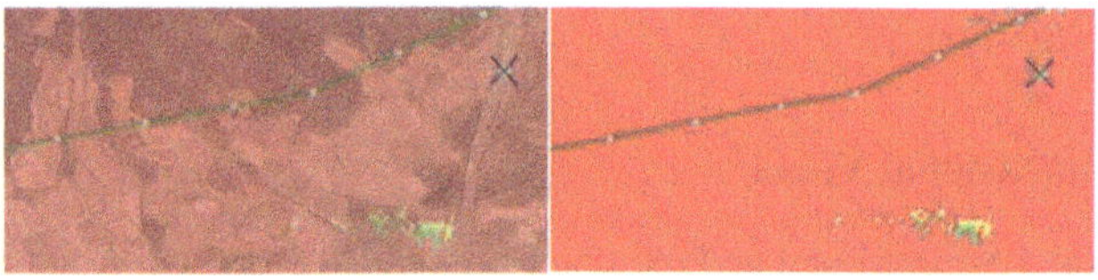

Fig. 10. Visibility A frame from A-68 highway. Position 2.

2.6 Incorporation into Terrain Viewer Software

Georeferenced data can be incorporated into a terrain viewer specific to a geographical region such as the one developed by the authors [5] or a universal one such as Google Earth© and simulate the view of the landscape from as many routes as you want (dynamic impact points). This visualization will be all the more real the more objects existing on the ground are represented in 3D (buildings, trees) in addition to the power line (Fig. 11).

Fig. 11. Virtual simulation

3 Conclusions

The foreseeable increase in new renewable energy installations in the coming years will make it necessary to significantly increase the number of power lines to evacuate the energy produced in a generation plant.

This increase of lines has the risk of producing an important visual impact at a global level that will generate a strong rejection by the affected population, either for seeing the vision of "their landscape" altered or for fear of resenting the influx of tourists.

Environmental impact studies must be pedagogical, easy to understand and reliable. With photomontage techniques, reality is falsified and causes mistrust.

With the proposed methodology, it is possible to faithfully reflect the effect of a power line within the landscape, seen from any point of view and from any route.

The virtual dynamic simulation allows any interested person to visualize the effect of the different alternatives of the route of the power line to be built without the need for technical knowledge.

References

1. Red Eléctrica de España: The Spanish Power System. Avance 2020, Spain (2021)
2. Ministry for the Ecological Transition and Demographic Challenge: Long-Term Strategy for a Modern, Competitive and Climate-Neutral Spanish Economy in 2050 (2020)
3. Spielhofer, R., et al.: Does rated visual landscape quality match visual features? An analysis for renewable energy landscapes. Landsc. Urban Plan. **209** (2021)
4. Azraff Bin Rozmi, M.D., et al.: Role of immersive visualization tools in renewable energy system development. Renewable Sustainable Energy Rev. **115**, 109363 (2019)
5. Martínez, E., Jiménez, E., Sanz, F., Pérez, M., Blanco, J., Santamaría, J.: Virtual representation of terrain through the web with VRML-Web3D and graphic libraries. Int. J. Interact. Des. Manuf. **4**(2) (2010)

Product Design and Development

Study of the Interlayer Behaviour of 3D Printing Materials and Optimisation of an Application

David Ranz(✉), Ramón Miralbes, Diego Bernal, and José Antonio Gómez

Departamento de Ingeniería de Diseño y Fabricación, EINA, Universidad de Zaragoza, C/ María de Luna S/N, 50018 Zaragoza, Spain
dranz@unizar.es

Abstract. This study addresses one of the major weaknesses of the FDM (Fused Deposition Modelling) additive manufacturing process, which is the low mechanical strength between the deposited layers. For this purpose, the characterisation of the interlayer behaviour of PETG and its carbon fibre reinforced version, a material commonly used in applications with structural requirements, is carried out. During this experimental characterisation, the influence of different manufacturing parameters and post-processing treatments on the interlayer strength is analysed. This initial study is completed with further experimental tests, other manufacturing tests and a more experiential approach to the material, which allows the application of the MDD (Material Driven Design) design methodology for the development of an application. MDD allows the competitive advantages offered by both the material and the FDM 3D printing process itself to be put to good use, in this case for the development of a face protection mask.

Keywords: 3D printing · FDM · Interlayer · PETG · Methodology design

1 Introduction

Additive manufacturing using FDM (Fused Deposition Modelling) is a manufacturing technology that was initiated to the 1980s (Sánchez Restrepo 2017), but has only become popular in recent years. This is mainly due to the liberalization of many of the patents for this type of manufacturing. This has opened up the world of 3D printing to many people and companies, who are developing different solutions for this technology, from new materials, new machines or post-processing for this industry. Although FDM has many advantages over other traditional manufacturing processes, such as: speed, customization, low cost and manufacturing of complex geometries, however, it shows a great weakness which is the dependence of the mechanical properties with the printing direction. As it is a layered manufacturing process, the level of adhesion between layers can be a problem for certain types of materials and/or applications, especially if they are going to be subjected to stresses outside the printing plane. Some authors have investigated the effect of material type and printing parameters on interlayer behaviour. In most cases, ASTM D 5528-01 is used to perform the DCB (Double Cantervile Beam) test which determines the Mode I interlaminar fracture toughness. There are studies using

F. Cavas Martínez et al. (Eds.): INGEGRAF 2021, LNME, pp. 57–63, 2022.
https://doi.org/10.1007/978-3-030-92426-3_7

ABS (Aliheidari et al. 2018) and PLA (Spoerk et al. 2017) in which the influence of varying printing temperatures, filling patterns or layer height on the strength is studied. Other authors have worked to address this weakness from two perspectives. Some include short fibre reinforcement materials (Young et al. 2018) and (Van de Werkena et al. 2019), or by incorporating continuous carbon fibre into the PLA filament, (Heidari-Rarani et al. 2019) and (Mohammadizadeh et al. 2019). However, others (Dong Goh et al. 2018) and (Meng et al. 2019) have enhanced interlayer fusion by using a laser incident on the deposited layer of material to improve interlayer adhesion.

In this study, PETG material, which can also be found reinforced with short carbon fibre, is selected. It is a relatively new material in 3D printing, for which there are no previous studies on its interlayer behaviour. PETG as a material used in 3D printing is characterised by slightly superior properties to ABS (good mechanical strength, good temperature resistance, weather resistant) and the ease of being printed like PLA (Mohammadizadeh et al. 2019). In addition, it shows good initial adhesion between layers, little deformation during printing, resistance to low temperature environments for prolonged intervals of time, chemical resistance (bases and acids) and the absence of odour during printing, characteristics that give it the attribute by a large majority of users of "the new PLA".

The study is completed with further mechanical tests (tensile, short beam bending), observation of breakage modes, analysis of manufacturing processes, observation of surface finishes and characteristics of both processed materials. This additional information allows the application of the MDD (Material Driven Design) design methodology for the development of an application (Karana et al. 2015). MDD allows the competitive advantages offered by both the material and the FDM 3D printing process itself to be put to good use, in this case for the development of a face protection mask.

2 Experimental Characterisation

One of the main objectives of this study is to increase the interlayer strength of the material and to obtain a better understanding of it. To this end, the influence of the layer height, the presence of carbon fibre (PETG-Carbon) and the performance of a post-processing treatment are analysed. As for the analysis of the influence of the layer height, tests were carried out at 0.19, 0.29 and 0.39, the first of these being discarded as it presented problems in the case of the reinforced material. As mentioned above, the reinforced version of PETG is used, in this case with a percentage of short carbon fibre between 8 and 12%, to see if there is a significant difference in adhesion between layers. Finally, the post-processing treatment consists of applying a pressure of 1 bar and three temperature levels (95 °C, 110 °C, 130 °C) to PETG and PETG-Carbon specimens for 30 min. These temperatures exceed in all cases the glass transition temperature (Tg) of the material without and with reinforcement, which is 70 and 80 °C, respectively.

2.1 Manufacture of Test Samples

To manufacture the test specimens for the different mechanical tests, a Zortrax M200 printer was used, to which a 0.4 mm stainless steel nozzle was installed in order to be able to print with PETG-Carbon, due to the fact that the reinforcements are very abrasive. For both materials, samples have been made with a solid filler alternating layers at ±45°, an extruder nozzle temperature of 250 °C has been used, and samples have been made with a layer height of 0.29 and 0.39 mm.

2.2 DCB Testing

This test, which follows the requirements of ASTM D5528-01, is initially designed to evaluate the interlaminar strength of laminated composite materials, but existing research experience is used to adapt it to this type of layered deposition materials. The type B specimen indicated in the standard (Fig. 1) is selected for the generation of the initial crack, 50 mm long, the printing process is stopped and a thermoplastic film (blue) of 0.025 mm thickness is introduced.

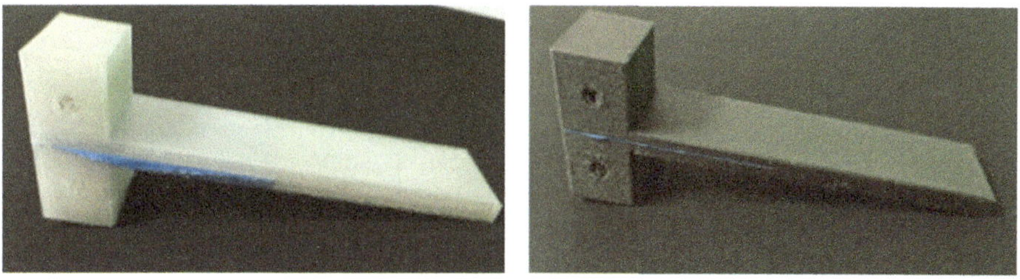

Fig. 1. PETG sample (left) and PETG-Carbon sample (right)

The tests were carried out at a speed of 10 mm/min, and it was observed that the PETG pieces had a greater bending of the arms and a greater delamination advance than the pieces made of PETG-Carbon (Fig. 2). These behaviours were observed with specimens of any layer height.

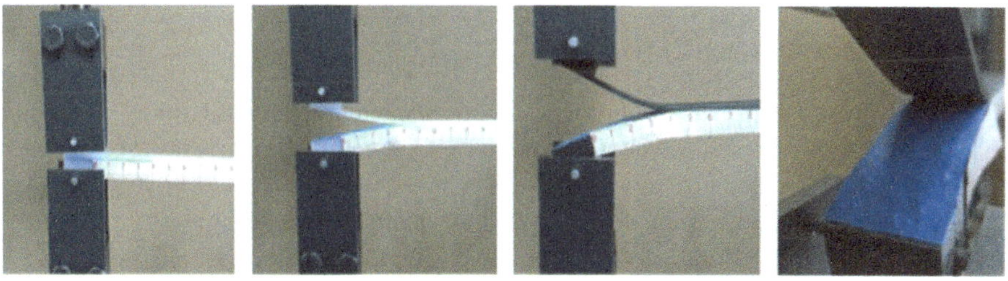

Fig. 2. Double Cantervile Beam (DCB) testing

2.3 Results

In the DCB test, the values obtained for the mode I interlaminar fracture toughness (GI) are plotted against the crack length (Fig. 3), which progresses throughout the test until the specimen rupture is reached. Looking at Fig. 3-left, which shows the graphs of the tests with different layer heights for both materials, it can be seen that the average value of GI is higher for PETG-Carbon in both layer heights. Being the layer height of 0.29 mm the one that presents better results in both materials.

The heat and pressure treated samples offer significant improvements in interlaminar fracture toughness when subjected to 95 and 110 °C cycles compared to the untreated specimens. Figure 3-right shows that for the 0.39 mm layer height PETG-Carbon specimens, average GI values of 1500 J/m^2 are achieved, compared to 1100 J/m^2 for the untreated material. On the contrary, when the specimens are subjected to cycles of 130 °C, the average GI value falls below that of the untreated specimen, with significant permanent deformations being observed that can affect the overall behaviour of the specimen, invalidating this post-treatment for a large number of applications.

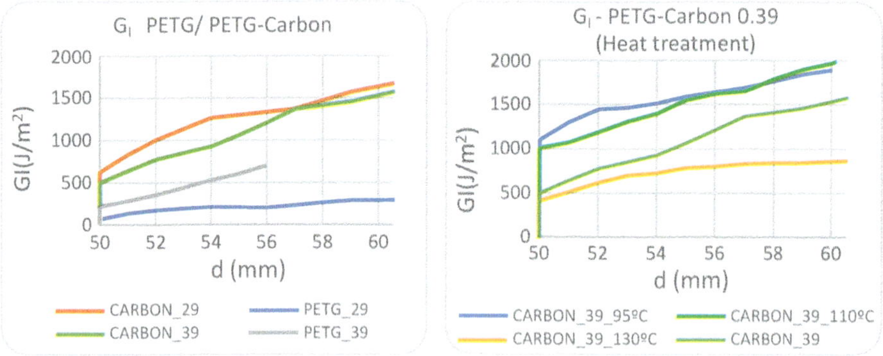

Fig. 3. PETG y PETG-Carbon DCB testing: Height layer (left); Heat treatment (right)

3 Design and Development of an Application

3.1 Application Search by MDD

To make appropriate use of the properties and characteristics of the study materials, PETG and PETG-Carbon, the Material Driven Design methodology (Karana et al. 2015) has been used. This methodology (Fig. 4) provides designers with tools to define and design meaningful experiences with and for an available material, not only for what it is, but also for what it offers.

In the first phase of compression of the material, a summary of the characteristics, techniques of the material and its experiential properties is made, allowing to visualise the competitive advantages that the materials to be studied have. In addition, the results obtained from the mechanical tests provide a better compression of the materials. The next phase works with the creation of the material experience vision. This compares the materials with others of similar characteristics or manufacturing processes to establish the advantages and disadvantages. In this case PETG and PETG-Carbon in a greater way, it is a material with a higher resistance than other 3D printing materials analysed,

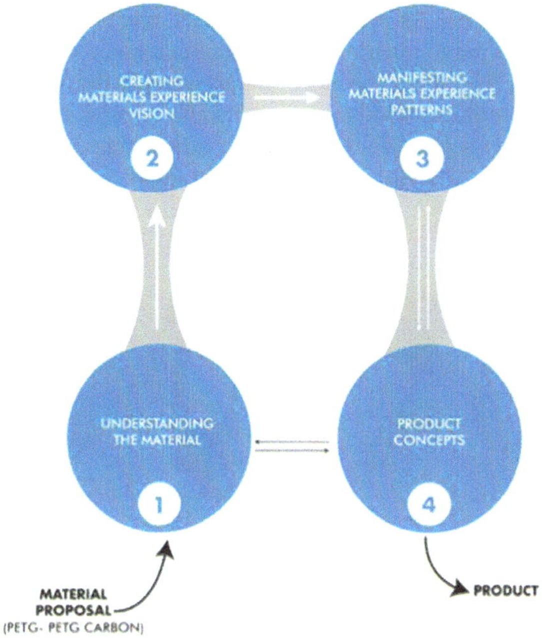

Fig. 4. Material Driven Design methodological scheme (Karana et al. 2015)

economical and easy to print. Furthermore, in the previous stages it has been seen that, for FDM printing, it is PETG-Carbon that has the best adhesion between layers. The third phase of the MDD is called "material experience patterns", the aim of which is to find patterns that evoke the use of the selected material, taking into account the user, material and products in which to apply it. The last phase is the creation of material concepts, with all the data obtained after applying the first three phases, possible applications for PETG-Carbon were proposed, where the characteristics of the process and the material lead to customised solutions that require complex geometries (support for bicycles, tools for manufacturing, structures for drones, dolly for hammer tools, splints, facial prostheses, skateboard protections, etc.). In all of them, the use of PETG-Carbon processed by FDM provides competitive advantages over other materials, due to its mechanical strength, ease of machining, production speed and low cost. Finally, a prosthetic splint has been chosen for development, in this case, a facial protection mask.

3.2 Prototype

Taking advantage of the possibility of customising and manufacturing highly complex geometries, the generation of a protective mask adapted as much as possible to the user's facial morphology was planned. To do this, a model of the user's face was created using the photogrammetry technique. This technique consists of taking a sequence of photographs from different positions and then processing them, in this case using Agisoft PhotoScan software, generating a model based on a point cloud (Fig. 5-left). This point cloud is used as a reference for the elaboration of a refined model of the mask (Fig. 5-middle), incorporating the geometric characteristics necessary to provide functionality to the final product. The STL file of the model is transferred to the manufacturing software, where the printing profile is introduced, in terms of temperatures, speed, layer height, which provides a better behaviour of the PETG-Carbon material. The final prototype of the protective mask (Fig. 5-rigth) needed a printing time of 4.5 h.

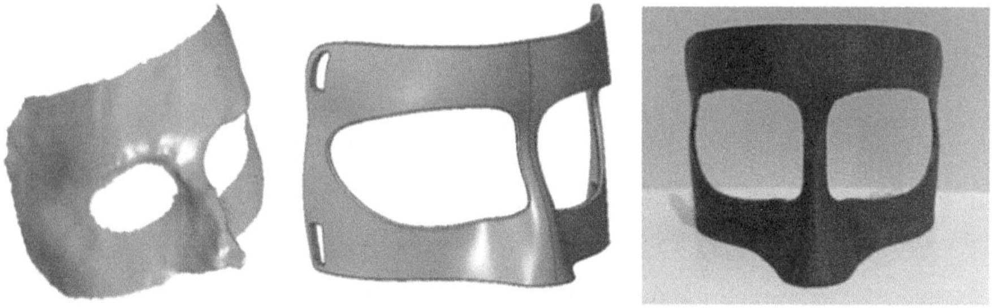

Fig. 5. Facial protective mask: Dot cloud (left); CAD design (middle); PETG-C prototype (right)

4 Conclusions

The results of the PETG specimens with and without carbon reinforcement show that for a layer height of 0.29 mm, the best results for interlaminar fracture toughness in mode I are achieved. Moreover, in the case of PETG-Carbon, the results for this same parameter are superior to those of the unreinforced material.

The heat treatments performed on the manufactured specimens have provided very limited improvements in the interlaminar fracture toughness values. However, small deformations occur, due to pressure and high temperatures, which can make the parts unsuitable for applications with high dimensional tolerances.

MDD can offer both end products and high-end industrial applications. This methodology can serve as a flexible guide to enhance the use of new 3D printing materials, find a suitable application and fully characterize them from different points of view.

References

Aliheidari, N., Triouranemi, R., Christ, J., Nadimpalli, S.: Interlayer adhesion and fracture resistance of polymers printed through melt extrusion additive manufacturing process. Mater. Des. **156**, 351–361 (2018). https://doi.org/10.1016/j.matdes.2018.07.001

Dong Goh, G., Ling Yap, Y., Agarwala, S., Yee Yeong, W.: Recent progress in additive manufacturing of fiber reinforced polymer composite. Adv. Mater. Technol. **4**(1), 1800271 (2018). https://doi.org/10.1002/admt.201800271

Heidari-Rarani, M., Rafiee-Afarani, M., Zahedi, A.M.: Mechanical characterization of FDM 3D printing of continuous carbon fiber reinforced PLA composites. Composites B: Eng. **175**, 107147 (2019). https://doi.org/10.1016/j.compositesb.2019.107147

Karana, E., Barati, B., Rognoli, V., Zeeuw Van der Laan, A.: Material driven design (MDD): a method to design for material experiences. Int. J. Des. **9**(2), 2015 (2015)

Luo, M., Tian, X., Shang, J., Zhu, W., Li, D., Qin, Y.: Impregnation and interlayer bonding behaviours of 3D-printed continuous carbon-fiber-reinforced poly-ether-ether-ketone composites. Composites A: Appl. Sci. Manuf. **121**, 130–138 (2019). https://doi.org/10.1016/j.compositesa.2019.03.020

Mohammadizadeh, M., Imeri, A., Fidan, I., Elkelany, M.: 3D printed fiber reinforced polymer composites – structural analysis. Compos. B: Eng. **175**, 107112 (2019). https://doi.org/10.1016/j.compositesb.2019.107112

Sánchez Restrepo, S.: FDM o modelado por deposición fundida, te explicamos todo! (2017). https://www.3dnatives.com/es/modelado-por-deposicion-fundida29072015/. (consulted at May 2021)

Spoerk, M., Arbeiter, F., Cajner, H., Sapkota, J., Holzer, C.: Parametric optimization of intra- and inter-layer strengths in parts produced by extrusion-based additive manufacturing of poly(lactic acid). Appl. Polym. Sci. (2017). https://doi.org/10.1002/app.45401

Van de Werkena, N., Hurleyb, J., Khanboloukia, P., Sarvestania, A., Tamijanib, A., Tehrania, M.: Design considerations and modeling of fiber reinforced 3D printed parts. Composites B: Eng. **160**, 684–692 (2019)

Young, D., Wetmore, N., Czabaj, M.: Interlayer fracture toughness of additively manufactured unreinforced and carbon-fiber-reinforced acrylonitrile butadiene styrene. Addit. Manuf. **22**, 508–515 (2018). https://doi.org/10.1016/j.addma.2018.02.023

Study of the Application of Gyroid Structures in Cyclist Helmets

Ramón Miralbés(✉), David Ranz, and Saul Higuera

University of Zaragoza, 50018 Zaragoza, ZA, Spain
miralbes@unizar.es

Abstract. Gyroid structures are triple periodic minimum surface (TPMS) structures that maximize specific properties and the energy absorption capacity of the materials used to manufacture them. In addition, they are generated by additive manufacturing (AM), which disrupts traditional manufacturing methods. These structures can also be customized by varying their main parameters. Consequently, they could be an effective substitute for expanded polystyrene (EPS) foam, which is traditionally used in applications such as protective helmets. In addition, the use of TPMS structures and AM could lead to the combined generation of the helmet's main parts from the same material. This would improve the item's recyclability, reduce assembly operations, and make possible the use of renewable-origin materials such as poly-lactic acid (PLA). On the other hand, additive manufacturing also allows helmet designers to optimize the protection of each area of the head by configuring the parameters of the gyroid structure. Therefore, this article examines the applicability of this type of structure for the production of protective cyclist helmets.

Keywords: Gyroid · Energy absorption · Helmet · Additive manufacturing · Protection

1 Introduction

Cellular solids have always been present in nature. They possess extraordinarily efficient structures, but not until the generation of the mathematical models that compose them and the appearance of additive manufacturing (AM) technology have people been able to use them.

The latest research on these structures [1] highlights the potential of the gyroid structure, a particular type of the triple periodic minimum surface (TPMS) structures; these structures are defined by a mathematical function (see Eq. (1) for the gyroid and Fig. 1) and a series of constitutive variables (mainly the volume fraction "t" and the unit cell size "a"). The volume fraction is understood as the ratio between the volume/weight of the printed material and the volume/weight of a solid specimen. The unit cell size is the size of the basic repeating unit of the structure.

$$F(x,y,z) = sin\left(\frac{2\pi x}{a}\right)cos\left(\frac{2\pi y}{a}\right) + sin\left(\frac{2\pi y}{a}\right)cos\left(\frac{2\pi z}{a}\right) + sin\left(\frac{2\pi z}{a}\right)cos\left(\frac{2\pi x}{a}\right) - t \quad (1)$$

F. Cavas Martínez et al. (Eds.): INGEGRAF 2021, LNME, pp. 64–70, 2022.
https://doi.org/10.1007/978-3-030-92426-3_8

In addition, they are characterized by optimizing the mechanical properties of the base material due to their minimal surface area [2], so they can be a more efficient substitute for various traditional materials such as polystyrene foam (EPS), which is used in applications such as hulls of all types and protective packaging for goods [3]. EPS is generated from petroleum, so it has a non-renewable origin and, although it is easily recyclable by itself, in the case of helmets that are composed of various materials – EPS for the inner shell, ABS for the outer shell and for various elements and other types of foam – recyclability is significantly reduced due to the need to separate the various materials.

However, there is still a lack of knowledge on how the variables that model its equation and materials affect its performance in shock absorption [4]. TPMS structures can be configured so that their constitutive variables vary with the geometry, modifying the local stiffness and energy absorption capacity. This is why, compared to EPS, which has constant mechanical properties throughout the helmet determined by its manufacturing density, TPMS structures can be used to adapt these mechanical properties depending on the area of the head to be protected. This aspect is particularly important, as studies [5] indicate the need to protect certain areas of the head due to the lower stiffness of the certain zones of the skull or due to the adjacent brain area.

In the case of oblique impacts during a cycling accident, rotational moments generated in the head cause of certain traumas; moreover, as indicated by some authors [6], it depends directly on the mass of the helmet. Therefore, by reducing the mass of the helmet, it is possible to reduce the lesivity.

2 Objectives

The main objective of this article is to study the applicability of gyroid structures in bicycle helmets so that this structure can replace EPS. In addition, the possibility of using renewable materials such as poly-lactic acid (PLA) and the possible use of AM to manufacture all the parts of the helmet with the same material to increase recyclability will be studied. Another objective is to analyse the possibility offered by AM to modify the local mechanical properties of the inner liner of the helmet depending on the area of the brain to be protected by varying the constitutive variables, and even the possibility of varying them using gradients.

3 Materials and Methods

This innovation was developed based on a study of the possible materials that could be used to manufacture the hull (ABS, PLA, TPU, PA12 and PA66 were selected). These materials were characterised using a quasi-static pure compression test according to the ASTM D695-15 standard. An INSTRON 8032 uniaxial testing machine was used for this purpose. In addition, several specimens of EPS with different densities (60, 70, 80, 100 and 120 kg/m^3) were characterized to compare them with the gyroid structures developed. The next step was the generation of CAD models using n-Topology software for various cubic specimens of the gyroid structures, varying both the cell size and the

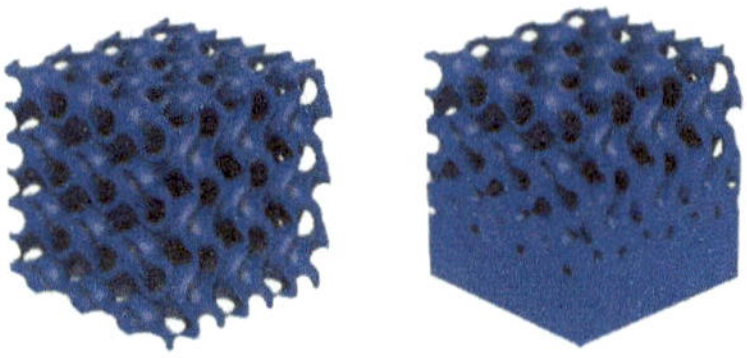

Fig. 1. CAD models of gyroid structure with constant (left) and variable volume fraction (right).

volume fraction. Models were also generated with a variable volume fraction by means of a gradient (Fig. 1).

These structures (Table 1) were printed using a Zortrax 300 fused deposition printer (FDM). They were then tested according to ASTM D695-15, and various material curves were obtained, including stress-strain, specific stress-strain, energy absorbed per unit mass and per unit volume, efficiency and ideality. These were compared with the base material and with the various EPS densities.

Table 1. Tested specimens.

	Continuous												Gradiant	
		10mm			15mm					20mm			15mm	
	100%	15%	20%	25%	10%	15%	20%	25%	30%	15%	20%	25%	0,74-1,24	0,6-3
PLA														
ABS														
TPU														
PA12														
PA66														
Resin														
FPU														

Subsequently, some specimens were tested using a drop tower to analyse the influence of strain rate on their mechanical properties. All were recorded using photographic cameras, video cameras and high-speed cameras to analyse the local failure modes and the deformation process of each gyroid structure (Fig. 2). Moreover, using a pattern previously printed on the specimens, and using the photo- and video-extensometry technique with the GOM-Correlate software, the internal local stress and strain maps of the specimens and the failure modes were obtained.

Finally, a CAD prototype helmet was developed in which the parameters of the gyroid structure were optimized in each area of the head to be protected to optimize the helmet's protection capacity; in addition, the helmet's weight was reduced to decrease the rotational moments in the event of an oblique impact.

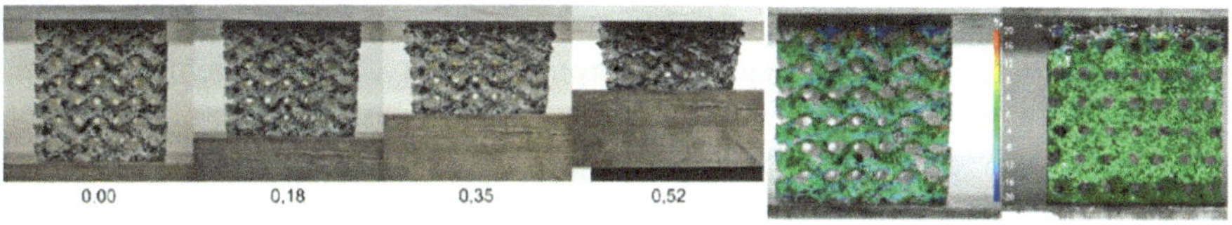

Fig. 2. Compression process of a 15-mm unit cell and 20% volume fraction PLA specimen with the standard for video and photo extensometry (left) and video extensometry results (right).

4 Results

After analysing the mechanical properties of the various gyroid structures, it was found that they follow the Ashley-Gibson model [7], which is characterized by three distinct zones: an elastic zone with a linear stiffness, a plateau zone with a more or less constant stiffness and a final zone of densification with an exponential stiffness. This exponential zone should be avoided so the hull's stiffness is not excessive, as this would cause brain damage. This type of behavioural model is the same that governs foams such as EPS, so the behaviour of gyroid structures is similar, indicating that they can be a suitable substitute.

However, the analysis of the results of the quasi-static compression teste indicated that the specimen's volume fraction determines the final mechanical properties, whereas the unit cell size has no special relevance except for generating a plate zone with more or less oscillation (Fig. 3).

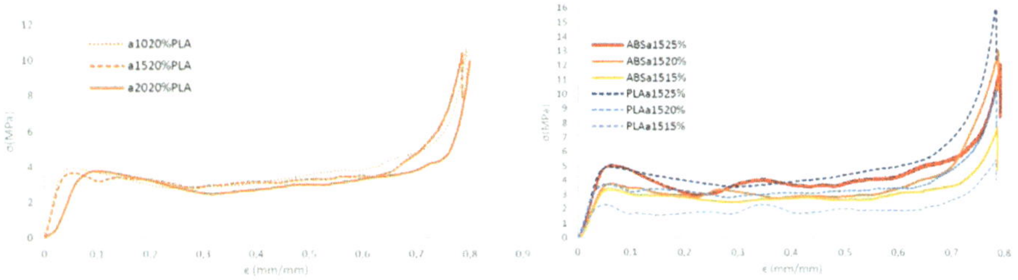

Fig. 3. Stress-strain curves. First two numbers: cell size, third and fourth numbers: volume fraction and letters: material. Influence of cell size (left) and material and volume fraction (right).

On the other hand, a study of these properties that used gradients for the volume fraction showed that, in the case of a high gradient of 5 (Fig. 4. right), the material stops following the Ashley-Gibson model and has an approximately linear stiffness until densification is reached. However, the amount of energy absorbed (i.e. the area under the curve) is identical to that in the model with constant volume fraction equivalent. In the case of a low gradient of 2, the material and its constant volume equivalent have the same behaviour (Fig. 4 left).

Subsequently, the comparative study of materials (Table 2) revealed that TPU has low energy absorption capacity, PLA and PA12 have a medium capacity and ABS has the highest energy absorption capacity, both in terms of energy per unit mass and per unit volume. Furthermore, compared to EPS100, the aforementioned materials, with the exception of TPU, have a higher energy absorption capacity. This is much higher in the case of ABS. In addition, TPU, which is an elastomer resin, is capable of returning to its initial geometry once deformed. Thus, in the case of multiple impacts, it could be a suitable material if not for its low mechanical properties.

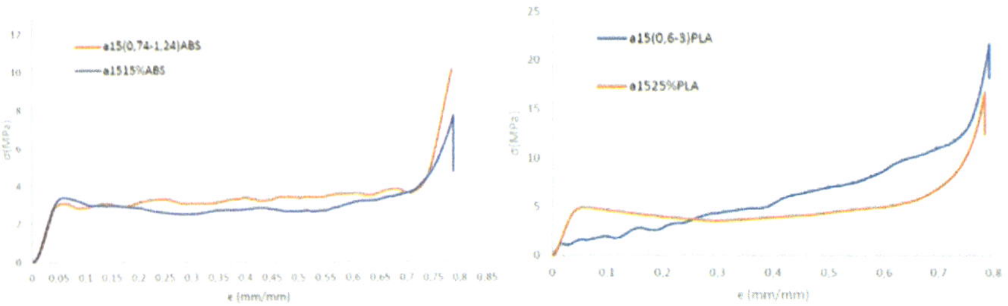

Fig. 4. Stress-strain curves. The first two numbers indicate cell size, and the third and fourth numbers indicate volume fraction. The letters indicate the material used with low gradient (left) and high gradient (right).

Table 2. Comparison of absorbed energies for the various materials for a 15 mm unit cell size and 15% volume fraction specimen.

Material	Energy absorbed per unit of mass (kJ/kg)	Energy absorbed per unit of vol. (kJ/m^3)
PLA	6.39	1165.56
ABS	10.74	1949.68
PA12	9.64	1028.60
TPU	0.2	36.28
EPS100	6.65	498.48

Finally, based on data obtained from various studies [8] analysing the type of impacts to which bicyclists are subjected and the areas of the brain that are damaged, helmets should be designed using materials with high energy absorption and, therefore, high weight in the frontal, parietal and occipital lobe areas. The temporal and frontotemporal lobe areas can use materials with lower energy absorption capacities (Fig. 5).

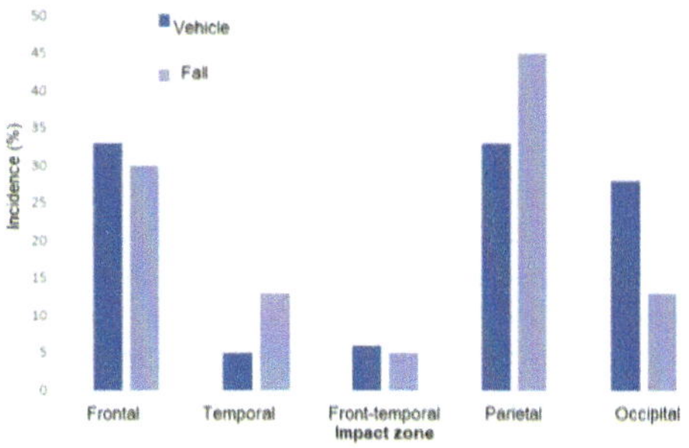

Fig. 5. Percentage of impacts resulting from collisions with vehicles or falls in each area of the head.

By scanning an existing helmet with an HP 3D scanner, the geometry of a new helmet was generated, including various gyroid structures depending on the area of the head to be protected (Fig. 6).

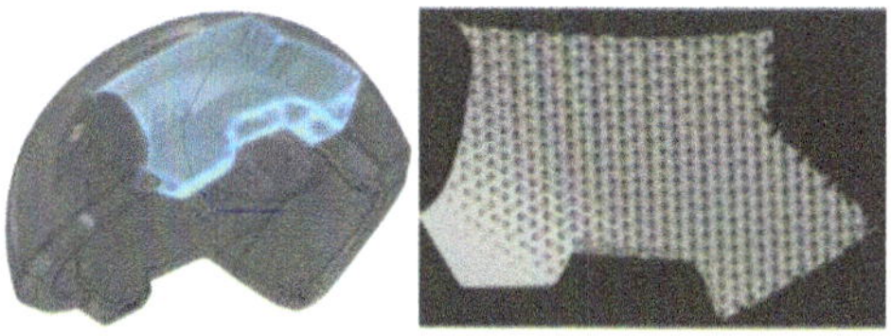

Fig. 6. Final design of the helmet in the occipital area.

5 Conclusions

The main conclusion of the study is that it is possible, by means of AM and gyroid structures, to generate helmets with a higher energy absorption capacity than traditional EPS helmets, allowing greater protection for the user. For this purpose, ABS has been postulated as the ideal material, despite its low recyclability, although PLA's properties are also superior to those of EPS. In addition, AM facilitates the generation of different structures depending on the area of the brain to be protected and reducing the mass of the helmet, which has a significant influence on injury in the case of oblique impacts. An in-depth study using advanced numerical models and dummies such as THUMS would be necessary to evaluate and determine the optimal parameters of the gyroid structure in each area of the head to be protected. In addition, although tests were carried out using a drop tower, these analyses are still processing.

References

1. Miralbes, R., Ranz, D., Pascual, F.J., Zouzias, D., Maza, M.: Characterization of additively manufactured triply periodic minimal surface structures under compressive loading. Mech. Adv. Mater. Struct. (2020)

2. Al-Ketan, O., Al-Rub, R.K.A.: Multifunctional mechanical metamaterials based on triply periodic minimal surface lattices. Adv. Eng. Mater. **21**(10), 1900524 (2019)
3. Chen, W., Hao, H., Hughes, D., Shi, Y., Cui, J., Li, Z.X.: Static and dynamic mechanical properties of expanded polystyrene. Mater. Des. **69**, 170–180 (2015)
4. Zhang, L., et al.: Energy absorption characteristics of metallic triply periodic minimal surface sheet structures under compressive loading. Addit. Manuf. **23**, 505–515 (2018)
5. Mustafa, H., Pang, T.Y., Ellen, T., Nasir, S.H.: Impact attenuation of user-centred bicycle helmet design with different foam densities. J. Phys. Conf. Ser. **1150**, 012043 (2018)
6. Pang, T.Y., Thai, K.T., Mcintosh, A.S., Grzebieta, R., Schilter, E., et al.: Head and neck responses in oblique motorcycle helmet impacts: a novel laboratory test method. Int. J. Crashworthiness **18**, 297–307 (2019)
7. Gibson, L.J., Ashby, M.F.: Cellular Solids: Structures and Properties. Cambridge University Press, Cambridge (1997)
8. Depreitere, B., et al.: Bicycle-related head injury: a study of 86 cases. Accid. Anal. Prev. **36**(4), 561–567 (2004)

Comparison of Tools for Simplified Life Cycle Assessment in Mechanical Engineering

Mylène Pongérard, Flavien San Augustin, and Manuel Paredes(✉)

ICA, Université de Toulouse, UPS, INSA, ISAE-SUPAERO, MINES-ALBI, CNRS, 3 rue Caroline Aigle, 31400 Toulouse, France
paredes@insa-toulouse.fr

Abstract. Current trends indicate that Life Cycle Assessment (LCA) tools are going to be extremely important for companies in the forthcoming years. New laws and standards are increasingly taking environmental aspects into account but, for non-experts in LCA, it can be difficult to find software suited to their studies. This paper aims to analyze the ability of simplified LCA software to combine a simple and convenient interface with relevant and exploitable results. This work tries to show which LCA tools are usable in preliminary analysis performed by non-experts like students in mechanical engineering or designers. Various assessments were studied with different tools: ArtoACV, Ecodesign Studio, Bilan Produit, and Umberto LCA+. An evaluation grid was drawn up to compare the tools according to objective and non-objective criteria. The case studies were carried out separately in order to remain as objective as possible. In this paper, the studies were made by two students in the first year of a Master's degree in mechanical engineering at INSA Toulouse. This paper shows that the four software packages enable a preliminary LCA to be finalized and propose graphs and tables for a better understanding of results. Even though tools generally allow the users to reach the same conclusions, some parameters persist that can strongly influence the result. Designers should keep in mind that results may not be consistent for their projects and may depend strongly on the hypothesis of the study.

Keywords: Life Cycle Assessment (LCA) · Tools · Mechanical engineering · Simplified LCA · Comparison

1 Introduction

The Life Cycle Analysis method has emerged in a global context of awareness of pollution, environmental degradation, and resource management issues. This awareness began in the 1970s and 1980s because of environmental damage that appeared progressively. A major responsibility has been found to be linked to consumer products [1–3]. Considering the entire life cycle of a product has become crucial, and the LCA method can achieve this.

F. Cavas Martínez et al. (Eds.): INGEGRAF 2021, LNME, pp. 71–80, 2022.
https://doi.org/10.1007/978-3-030-92426-3_9

1.1 Literature Review

LCA means both the eco-design methodology and the tools associated with it. Ecodesign considers environmental aspects in the design of the product without excluding the other essential parameters (cost, quality, etc.) [1].

Some studies thus propose to link LCA and LCC (Life Cycle Cost) on mechanical case studies like those concerning diesel engines [4] or Steyr engines [5].

The LCA method assesses the potential environmental impact of a product, a service, or a system for all steps of its life cycle: "extraction of raw materials – production – distribution – consumption – recovery – elimination" [6]. It relies on a database that can provide an inventory of information related to the life cycle of a product.

Life cycle assessment is standardized by ISO 14040 and ISO 14044. The first provides general guidelines for good practices for performing LCA. The second offers technical assistance for carrying out LCA correctly [7]. The standardization of the method ensures "the robustness and the validity of the results obtained" [1]. Nevertheless, LCA studies appear to be very sensitive [8, 9].

LCA aims to help designers choose the best improvement targets for the product and compare several design solutions. LCA tools have been created to simplify this method by integrating databases and calculation methods directly into the software to extract environmental impacts. Some other works have been developed to link LCA with CAD software [10–12].

LCA software packages have their own characteristics because they do not all integrate the same databases, calculation methods, or even indicators. They comply with all the standards mentioned above, which justify their qualification as LCA software. There are no "rules" for choosing a tool and users are completely free to select the one they want. Making documentation available would help to guide users in their choices but there is still a lack of information and comparison of tools in the literature.

1.2 Purpose of the Study

This paper aims to highlight Life Cycle Analysis software that is suitable for use by non-experts to carry out simplified studies. We have chosen to analyze the capacity of software to lead a user to a consistent conclusion whatever the subject of his study in the field of mechanical engineering. To fully compare the users' experiences, numerous aspects must be considered, such as the explanation given by the software for successfully performing a study or the time required to carry out the simplified LCA.

2 Materials and Methods

This paper compares four LCA software tools through three different case studies. This type of comparison can be altered by the lived experiences and judgement capacity of those carrying out the case studies. To enable them to remain objective, a methodology was set up as objectivity was crucial to our study.

The reader should know that Artogreen (ArtoACV) and Altermaker (Ecodesign) propose a Saas (Software as a service) solution and can adapt their software with ease.

2.1 Software and Studies

The four software tools studied as part of this research are briefly presented below.

- **ArtoACV** [13]**:** Online software used at INSA Toulouse in Mechanical Engineering and created by ArtoGreen.
- **Bilan Produit** [14]**:** Developed by ADEME, this LCA software is free and therefore simplified in terms of database and modeling. Two accounts have been created on the website.
- **Ecodesign Studio** [15]**:** Created by Altermaker, Ecodesign Studio is an online platform allowing collaborative environmental analysis. Two licenses were obtained free of charge thanks to Altermaker.
- **Umberto LCA+** [16]**:** Compared with the other software, Umberto is not an online tool and is more suitable for experts in LCA. Access to it was possible because Umberto is used in the department of Process Engineering at INSA Toulouse.

After a study of all the software, it was decided not to concentrate on Umberto LCA+ because it is intended for more expert users.

This paper concerns three case studies. A wind turbine was studied first with the four software tools mentioned above, thanks to a paper comparing several life cycle impact assessment (LCIA) software tools [17]. The second case studied was an ecological solution named Sylcat, created by SOFRINOV [18]. It allows wooden frameworks to be made from pallets connected by pieces of wood also named Sylcat (Sylcat A,B,C,D or E). The last study was of a hoverboard that was entirely disassembled to note and weigh all pieces. Umberto LCA+ was only used on the wind turbine study. Although its high level of flexibility is convenient for making detailed analyses, it requires considerable personal investment from a non-expert in LCA.

2.2 Methods and Hypothesis

The difficulty of this research was to compare LCA software through case studies while remaining objective and neutral. To reduce influence throughout the work on the case studies, it was decided to perform two separate autonomous analyses in parallel. In this paper, the studies were carried out by two students in their first year of a Master's degree in mechanical engineering at INSA Toulouse. The evaluation grids drawn up aimed to list the criteria used to compare the tools. The objective of these grids was to keep a written record of anything done during the study and to have the same criteria at the beginning of the research work.

The criteria chosen to compare the LCA tools were various and could be objective or non-objective. They were chosen for their possible impact on the results or the conclusion about the software. The criteria or information used were:

- online/offline accessibility
- possibility import/export the results/data
- form of the results
- available languages

- databases
- relative cost of the tools
- possibility of collaboration
- information given by the software to help the user
- handling time
- average time to do a study

They are detailed in Table 1 and Table 2.

Making a simplified LCA requires making some hypotheses at the beginning of and during the study. It was decided to use only the information available in the case study documentation in order to avoid making very different assumptions. That is why there is no inflow and outflow for some points, such as transport, packaging, or the end of the product's life. Making hypotheses during the case study depends on the databases of the software and the user. A database cannot have all inflows available. Hence, the user could make some choices and therefore put forward a hypothesis. The results show that these choices could considerably change the conclusion of the user about his study.

3 Results and Discussion

3.1 Comparison of Software from Criteria

While using the software, we separately completed a grid of criteria useful to compare the tools with each other. The objective of this part is to summarize those criteria and compare software as clearly as possible. Table 1 presents criteria that depend only on the characteristics of the tools.

Table 1. Relevant information about LCA software

Criterion	Bilan Produit	ArtoACV	Ecodesign Studio	Umberto LCA+
Software location	Online	Online	Online	Local
Safeguards	Local	Online	Online	Local
LCA data exportation	Yes (.xml)	Yes (.xls)	Yes (.csv)	Yes (.umberto)
LCA data importation	Yes (.xml)	Yes (.xlsm)	Yes (.csv)	Yes (.umberto)
LCA results exportation	Yes (.xls)	Yes (.pdf or.xls)	Yes (.pdf or.xlsx)	Yes (.xls)
Form of the results	Graphs and percentages	Graphs, tables, labels and global impacts	Graphs detailed by impact factor	Excel with graphs and tables
Language databases	En, Fr Base IMPACTS	Fr ELCD APME Other French databases	En, Fr Base IMPACTS FD E01–008	En, De Ecoinvent 2 Ecoinvent 3
Relative cost	0	+	+	++

The accessibility of software is generally online (Online) except for Umberto LCA+, which is usable only on the computer on which it is installed (Local). Each tool provides the possibility to import and export data or results and the associated format is detailed in Table 1.

It is particularly important for software to be in English (En) to make it more accessible. At the time of our studies, ArtoACV was only available in French (Fr), which could be a problem for international users.

Other criteria were also studied (Table 2) to compare software tools. Collaborative work was an important feature because it simplifies remote work by making it possible for users to access the same project and modify it simultaneously. Ecodesign Studio is particularly useful for group projects because it makes feedback available to the other members of the project throughout the study.

Table 2. Features useful for the user

Criteria	Bilan Produit	ArtoACV	Ecodesign studio	Umberto LCA+
Collaboration	No	Yes	Yes	No
User guidance	Yes	Yes	Yes	Yes
Handling time	30 mn	1 h	1 h 30	10 h
Average time to do a study	45 mn	1 h	1 h	2 h

Each software tool helps the user during his/her study (Table 2). ArtoACV gives some information and advice directly on the software, which makes the tool easier to handle. The same is true for Bilan Produit, which gives complete information about the Base IMPACTS. Ecodesign Studio does not have information to guide the user within the tool because the tool itself is intuitive. Nevertheless, information about LCA is available on the Altermaker website, which could help non-expert users. Umberto LCA+ puts complete tutorials at the user's disposal for the LCA method and other features. In our case, for Umberto LCA+, only the tutorials concerning LCA were completed, which represent the half of the tutorials available.

By using software, we noticed some advantages and problems with each simplified LCA tool (excluding Umberto because he needs more time to handle). In addition to the previous comments, ArtoACV has the particularity of presenting its results in labels. This way of presenting the results includes an overall costing of the impact and consumption of water, electricity, and hazardous wastes. This can be useful for pedagogical purposes. ArtoACV has a large database composed, partly, of ELCD (European Life Cycle Database) and the APME (Association of Plastics Manufacturers in Europe) database. These databases made ArtoACV the most complete tool for inflow for the moment. However, this software presents a problem for studying an object when using the multiplication factor feature in the nomenclature part. That led to problems in graphics and expressions of mass in percentages in the results part.

Bilan Produit is a fast software tool which is free of charge and easy to use, and is a way to learn the LCA method. The editor does not guarantee full respect of ISO14040 and ISO14044 and orients the user toward other compliant LCA software for their models. It is not possible to use a database other than Base IMPACTS on Bilan Produit but it is easy to signal your need if a flow required for a given study is missing in Base IMPACTS.

Ecodesign Studio has the particularly interesting possibility to give very complete results with numerous graphs showing the impact of each component. Aside from using Base IMPACTS, Ecodesign Studio has another, specific database for mechanical studies (FD E01–008) (Table 1).

The following part aims to synthesize the results of the case studies made for this research project. They are then used to highlight what could influence the interpretation of the results by a non-expert user.

3.2 Results of Case Studies on Four Software

The objective of an LCA tool is to make the most impactful areas (component, phase of the life cycle, etc.) of its product clear to the user. Important parts that will then have to be redesigned or carefully developed are thus highlighted. These results aim to check the consistency of the possible conclusion drawn by the user with each tool. They are not necessarily easy to compare because the software packages studied do not always present the results in the same way (different impact factors, detailed information on the impact of the components). Let us now check that the results given by the different software tools converge towards the same conclusions or, if not, try to explain the differences.

It was decided that only the information already available would be used and that hypotheses would be avoided. This was to prevent the use of assumptions that were very different between the two students carrying out the study. That is why some parts, such as transportation and packages, did not have inflow and outflow.

The first case studied was a hoverboard, which is the only product that was disassembled to analyze the components. There were no difficulties in conducting this study and the results given by the various software tools were very close to each other.

The environmental balance (Fig. 1.a., Fig. 1.b. and Fig. 1.c.) shows that "raw materials" and "use" are the two most impactful phases of the product. The designer should conclude that he needs to concentrate on the raw materials if he wants to reduce the environmental impact of his product. The global impact of the components indicates that the ferromagnetic core made with Nickel has the greatest environmental impact (Fig. 2). The user should conclude that he must concentrate specifically on this part of the product. Note that ArtoACV is the only software capable of giving the global impact of each component or material. It contains a dedicated part for sensitivity evaluation of materials or parts on each impact factor. The other software tools (Bilan Produit and Ecodesign Studio) give the impact of each component according to impact factors.

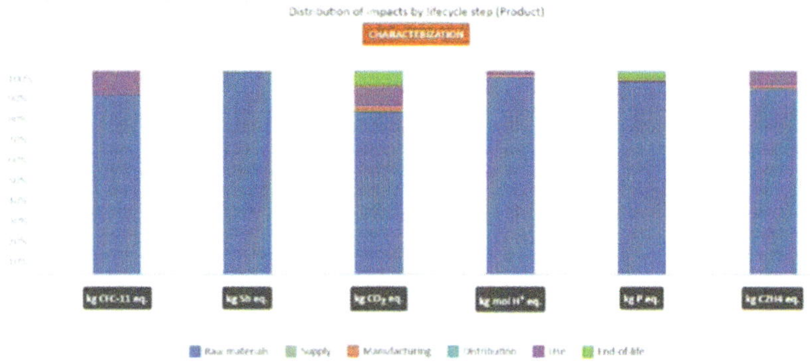

(a)

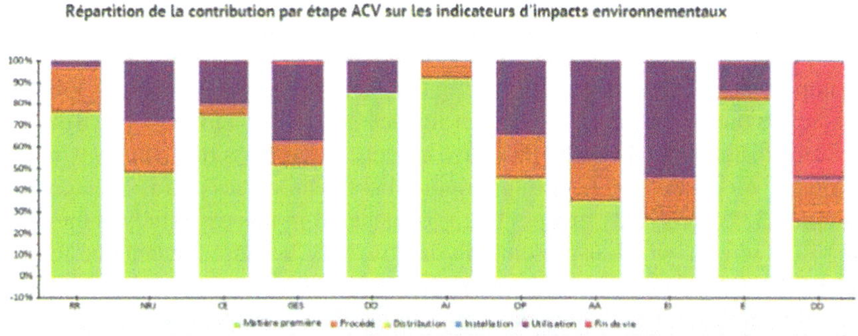

(b)

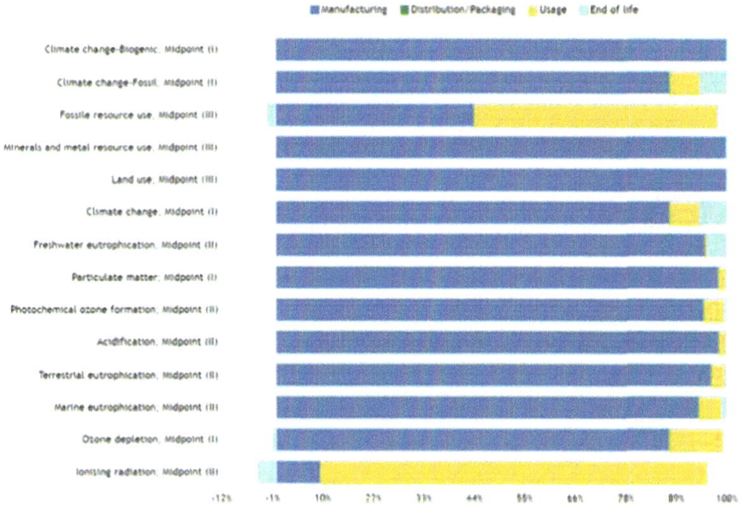

(c)

Fig. 1. **a** Environmental balance of the hoverboard on Ecodesign Studio. **b.** Environmental balance of the hoverboard on ArtoACV. **c.** Environmental balance of the hoverboard on Bilan Produit

	% en masse	masse en g	% impacts
Ferromagnetic core	19.38	1938	33.17
Electronic card	6.75	675	14.83
Plastic crankcase	8.19	819	14.14
Rotor	14.72	1472	12.09
Metallic crankcase	6.65	665	5.54
Motor copper wire	4.5	450	4.81
Magnets (x15)	4.48	448	3.53

Fig. 2. Global impact of the hoverboard components on ArtoACV.

The second case study was about wood pieces and the results were more dispersed. This study is very interesting because few parts are involved, making it easier to visualize the importance of the choice of incoming flows on the result. The ArtoACV balance sheet indicates that processes have a higher impact than raw materials. Since Bilan Produit does not differentiate between these two aspects, it is impossible to know which has more impact. We observe on Ecodesign Studio that the processes part is not significant (maximum 0.33%). The only process that is considered in this case study is the manufacture of the screws, which is present only in ArtoACV because it cannot be found in the Base IMPACTS database. Cold rolling was selected in the case of Ecodesign and Bilan Produit because they did not offer a greater choice of processes. However, it was not the manufacturing of the screws that generated the difference of impact but the calculation method employed in the software. It seems that ArtoACV considers not only the processes entered by the user but also those upstream at the level of the extraction of raw materials. This can be misleading as the user may think that the manufacturing of the screws has a huge impact. This problem showed that the calculation method can impact the understanding of the results.

Another problem with this study was the conclusion on the impact of the components. ArtoACV indicated that the pieces named Sylcat D (36.67% of the global impact) had a larger impact than the pallets (22.52%). Bilan Produit and Ecodesign Studio indicated the contrary (Fig. 3) in the acidification factor (chosen for this comparison).

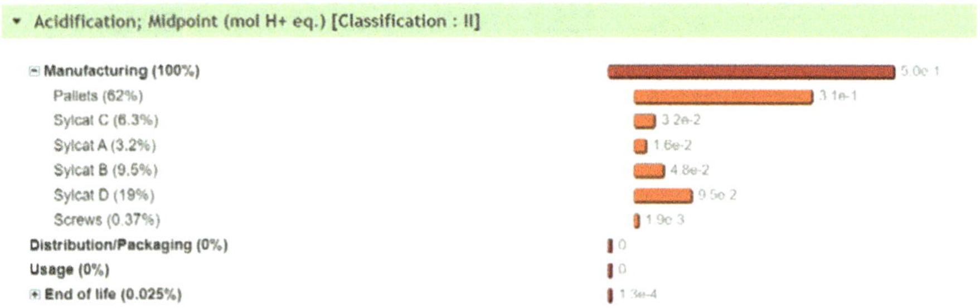

Fig. 3. Impact on acidification factor calculated by Bilan Produit.

This difference was due to a choice made for the inflows. On ArtoACV, the veneer wood used to make the Sylcat pieces was not a durable one, unlike the wood used for the pallets. On Bilan Produit and Ecodesign Studio, we found durable veneer woods and the results were greatly impacted (Fig. 3). This problem shows that a hypothesis made by the user about an inflow can change the results and so the conclusion. In the worst case, it can change the user's conclusions and eco-design decisions.

For the third study on the wind turbine, Umberto allowed the LCA to be carried out and even showed the flow of energy converted as described in the functional unit. The other tools can help in an LCA within the limits of their databases, which are evolutive. Concrete was available on ArtoACV and Umberto LCA+ but not on Ecodesign Studio and Bilan Produit because they use the Base IMPACTS database. This lack of inflow shows the importance of an exhaustive database if a study is to be conducted and consistent results obtained.

4 Conclusions and Perspectives

The purpose of the study described in this paper was to compare some LCA (Life Cycle Assessment) software packages usable for simplified analysis. Through various case studies and criteria, we analyzed the capacity of LCA tools to combine ease of use and consistency of results. These data make a comparison of the possible tools and enable some advice to be proposed for non-expert users of LCA. The reader should be aware that ArtoACV and Ecodesign studio are Saas (Software as a service) and adaptable.

The results of this paper show that a "perfect" LCA tool, suitable for all studies, does not exist. The software choice depends strongly on the type of study and the needs of the designer. Through the cases studied, it appears that the quality of the results depends strongly on the hypotheses made by the user, the available incoming flow (databases), and the calculation methods.

From this point of view, ArtoACV had a very complete database which needed fewer hypotheses to be made than the other software studied in this paper (except Umberto LCA+). Nevertheless, this tool has the drawback of still being available only in French.

Ecodesign Studio is a very intuitive software and is accompanied by a website with quality content about LCA, which allows users to become informed about the issue. Its management of collaborative work makes this tool interesting and convenient for companies.

Bilan Produit, is an accessible, fast, free software with lots of information about the Base IMPACT database.

These three tools are particularly suitable for non-experts because they propose a simple form of results and the possibility to perform preliminary LCA. Even though Umberto LCA+ allows for some preliminary LCA too, it is more suitable for experts in the field.

The last step of this research project aims to have the tools tested by a larger number of people. It would be interesting to integrate the tools into an industrial context in companies and also to ask students to test them. The results and feedback will be collected and analyzed to know whether our preliminary conclusions can be generalized.

References

1. Clusel, F.: L'ACV comme outil de reconception et de conduite du changement. Technique de l'ingénieur, Ref. 0240, pp. 1–3 (2014). https://www-techniques-ingenieur-fr.gorgone.univ-toulouse.fr/fiche-pratique/innovation-th10/deployer-l-innovation-dt30/download/0240/l-acv-comme-outil-de-reconception-et-de-conduite-du-changement.html
2. Klöpffer, W.: Life cycle assessment. Environmen. Sci. Pollut. Res. **4**(4), 223–228 (1997)
3. Guinee, J.B., Heijungs, R., Huppes, G., et al.: Life Cycle Assessment: Past, Present, and Future (2011)
4. Deng, C., et al.: Integration and optimization of LCA and LCC to eco-balance for mechanical product design. In: 2008 7th World Congress on Intelligent Control and Automation, pp. 1085-1090. IEEE (2008)
5. Shi, J., Wang, Y., Fan, S., Ma, Q., Jin, H.: An integrated environment and cost assessment method based on LCA and LCC for mechanical product manufacturing. Int. J. Life Cycle Assess. **24**(1), 64–77 (2018). https://doi.org/10.1007/s11367-018-1497-x
6. Rousseaux, P.: Analyse du Cycle de Vie (ACV). Technique de l'ingénieur. Ref. G5500 V1, pp.2–4 (2005). https://www-techniques-ingenieur-fr.gorgone.univ-toulouse.fr/base-documentaire/42552210-environnement-et-construction/download/g5500/analyse-du-cycle-de-vie-acv.html
7. Finkbeiner, M., Inaba, A., Tan, R., Christiansen, K., Klüppel, H.-J.: The New International Standards for Life Cycle Assessment: ISO 14040 and ISO 14044. Int. J. Life Cycle Assess. **11**(2), 80–85 (2006). https://doi.org/10.1065/lca2006.02.002
8. Finnveden, G., et al.: Recent developments in life cycle assessment. J. Environ. Manage. **91**(1), 1–21 (2009). https://doi.org/10.1016/j.jenvman.2009.06.018
9. Guo, M., Murphy, R.J.: LCA data quality: sensitivity and uncertainty analysis. Sci. Total Environ. **435**, 230–243 (2012). https://doi.org/10.1016/j.scitotenv.2012.07.006
10. Cappelli, F., Delogu, M., Pierini, M.: Integration of LCA and EcoDesign guideline in a virtual cad framework. In: Proceedings of LCE, pp. 185–188 (2006)
11. Marosky, N., Dose, J., Fleischer, G., et al.: Challenges of data transfer between CAD and LCA software tools. In: Proceedings of the 3rd International Conference on Life Cycle Management. University of Zurich, Irchel, Switzerland (2007)
12. Chen, Z., Tao, J., Suiran, Y.: A feature-based CAD-LCA software integration approach for eco-design. Procedia CIRP **61**, 721–726 (2017). https://doi.org/10.1016/j.procir.2016.11.228
13. ARTOGREEN Homepage. http://www.artogreen.com/
14. Base IMPACT Homepage. http://www.base-impacts.ademe.fr/
15. ECOdesign Studio Homepage. https://www.ecodesign-studio.com/
16. Ifu Hamburg Homepage. https://www.ifu.com/en/umberto
17. Martínez, E., Blanco, J., Jiménez, E., Saenz-Díez, J.C., Sanz, S.F.J.: Comparative evaluation of life cycle impact assessment software tools through a wind turbine case study. Renew. Energy **74**, 237–246 (2015). https://doi.org/10.1016/j.renene.2014.08.004
18. Sylcat Homepage. https://sylcat.eu/

Conceptual Design Using Virtual Reality: Case Study with Portable Light

Lucía Rodríguez-Parada(✉), Miguel-Ángel Pardo-Vicente, Alejandro Sánchez-calle, and Pablo Pavón-Domínguez

Engineering Faculty, Av. Universidad de Cádiz, 10 Puerto Real, Cadiz, Spain
lucia.rodriguez@uca.es

Abstract. In the creation of new products, the conceptual design is the key to generating valid and competitive ideas. In this paper, which applies Virtual Reality (VR) to the generation of new concepts, it is explored the generation of previous ideas through the modelling of virtual sketches. For this purpose, Gravity Sketch software has been used as a tool for conceptual design. Gravity Sketch allows the functional evaluation of the product, as the concepts are generated as 3D models that can be tested in the immersive VR environment. The case study focuses on the design of portable light concepts made in VR, which had subsequently been analysed and tested with several users who have entered the virtual environment. Users had been able to visualise the concepts and interact with them to test the product's grip and operation. The results obtained in this immersive study had been valuable for the final development of the product. Furthermore, the conceptual design applying VR technology allows for intuitive designs derived from the three-dimensional environment, reducing design time and the costs associated with prototyping.

Keywords: Product design · Customization · Virtual reality VR · Lighting · User experience

1 Introduction

Nowadays, we are living in a digital era that encompasses all fields, among them is the Smart device in the home. These devices are gaining momentum and surround us in our daily lives [1]. Users routinely use devices such as smartphones, smartTVs and even smart appliances in addition to smart homes [2]. Accordingly, it can be said that technology is here to stay in society, progressively extending to other products of common use and also to work processes. These types of products are born from the idea of creating new experiences with the actions that are carried out on a daily basis, creating sensations, feelings and emotions never before experienced by users.

The development of immersive technology offered by virtual reality (VR) is changing the way engineers design, create and experience. Factories, buildings, cars and any consumer product can be developed using [3].

F. Cavas Martínez et al. (Eds.): INGEGRAF 2021, LNME, pp. 81–90, 2022.
https://doi.org/10.1007/978-3-030-92426-3_10

Thus, in the design of new products, designers spend much of their time looking for new forms, aesthetic and functional, through concept generation [4]. In this context, visualisation techniques help designers to carry out creative work. The most prominent technique is sketching, which assists in the development of design solutions and are an effective means to design solutions by recognising problems and solutions in projects [5–7]. In this sense, Hyun et al. [8] highlights that design teams often explore a wide range of possibilities that may contradict each other, thus asserting that with the help of computational design it is possible to discover combinations that otherwise could not be exploited. Furthermore, James [9] studied the influence of experience with stimuli, such as sketches, on the interpretation of designs by different stakeholder groups. In short, the application of VR-based technologies facilitates the realisation of rapid conceptual designs, observing the overall scale of the product, visualising how the design changes in real time and experimenting with the object as if it were already built [10].

In the digital age, reinventing the world of design is one of the great strengths of this new technology [11]. The use of VR offers the opportunity to transform 2D designs into interactive, immersive digital models, giving context to the original information [12]. Engineers, designers and architects can quickly and easily turn their CAD data into real-time interactive experiences. VR allows the user or engineer to navigate in immersive, data-rich environments to make better informed decisions and create new and more engaging experiences [13, 14]. Notably, with VR, users can train in scenarios complex and dangerous ones within the safety of a virtual world [15]. An example of this is the case studies presented by Mandic where virtual technologies are integrated for the design, analysis and validation of intelligent products and manufacturing within the concept of Industry 4.0 [16].

On the other hand, these tools favour the analysis of human and ergonomic factors during product design [17]. In fact, these tools are used to make virtual prototypes and evaluate product designs for both functionality and ergonomics [18]. It should be noted that in product design and development it is important that models can be created and modified easily and intuitively in the virtual environment. However, the VR design software currently available on the market is easy to use but limited in the working area. One such system is the one used in this work, Gravity scketch, which allows the creation of ready-to-use 3D objects [3]. In addition, virtual prototypes reduce the use of costly, time-consuming and resource-intensive physical prototypes. With VR, full-scale prototypes can be realised and modified while being tested [19]. Furthermore, in the later stages of the design process, the physical object could be built thanks to the generated CAD model.

In this work, a study of VR as a tool for product design has been carried out by conducting a case study focused on the design of a portable lighting system in which it was designed and evaluated entirely with VR techniques.

2 Materials and Methods

A case study has been carried out using VR goggles as a means for the design, development and evaluation of early concepts for a portable luminaire. First, the concepts were generated in the virtual environment and then the selected designs were analysed using the same technology. It should be added that the initial concepts designed entirely by means of VR were analysed and selected according to the design specifications. Finally, as part of the final validation of the virtual concept, a prototype was generated by additive manufacturing.

2.1 Materials

The VR glasses used for the study are the HTC VIVE, manufactured by HTC in cooperation with Valve. The device is designed to use the space of a room to create a virtual environment, allowing the user to walk and touch objects by detecting body movements through sensors. The device has two screens, one for each eye, each with a resolution of 1080 × 1200 pixels and uses more than 70 sensors, including gyroscopic, accelerometer and laser sensors. On the other hand, to carry out both the sketching and the VR evaluation, a working area of 16 m^2 was set up in a room, which is the minimum space necessary to be able to move around in the virtual environment. The movements of the hands and head are captured by means of motion sensors.

For the conceptual development of the VR designs, the Gravity sketch programme was used. This 3D design tool is intuitive and immersive. It also allows designers to visualise and communicate concepts quickly and in real time in a fully virtual environment. The 3D design was exported in two formats, OBJ for 3D printing and IGES for export to a parameter programme such as Solidworks. The dimensions used in the programme were in millimetres. For 3D printing, a FFF Ender pro desktop printer and PLA material were used.

2.2 Method of RV Sketching and Evaluation

The work began with the ideation process for the design of the proposals using the VR tool, Gravitiy scketch. Virtual quick sketches were made based on the design requirements of the product under study. An image of the working environment is shown in Fig. 1.

The designs were drawn and modelled in three-dimensional space and aspects of the actual size of the product proposal could be modified at the same time as it was conceptualised. An example is shown in Fig. 2.

The next step in the experimental methodology was the selection of 8 proposals by meeting with experts for the evaluation of the technical requirements. All proposals were redesigned in VR in order to better communicate the 8 ideas to the users, whose input led to the selection of 3 final proposals. The evaluation of the concepts and the selection of the 3 proposals involved 5 users of different age and gender. Each user interacted with the proposals, picking them up and visualising them within the VR environment. Their impressions were recorded and each user was given a questionnaire in which they were able to evaluate 8 key points of the proposals: cleanliness, assembly, weight, ignition, grip, stability, battery and charging points. The 5-point Likert scale was used for scoring, where 1 is not suitable at all and 5 is totally suitable.

a) b)

Fig. 1. Example of working environment and controls: a) sketching with lines, b) filling and shading.

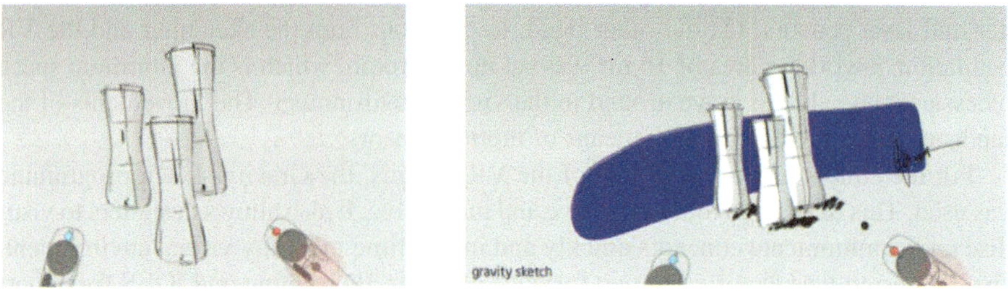

a) b)

Fig. 2. Example of a quick sketch: a) several unit sketches superimposed on each other, b) creation of a background simulating a conventional sketch.

In the third phase of the experiment, the 3 final proposals were evaluated to obtain a final proposal. For the evaluation, a virtual living room was created, which creates a feeling of total immersion for the user. By connecting the user to a familiar environment, it immerses the user in the experience making them feel comfortable and at ease, according to [12]. On the other hand, predefined hands were used to simulate the grip.

Five different users participated in the previous phase, also of different ages and sexes. The proposals were scored according to the 5-point Likert scale. In this case, 3 questions related to the context of use were asked: grip, assembly and transport. The users were able to sit and walk around the space, interacting with the luminaire concept and the environment in a free and intuitive way. Finally, the final validation of the procedure and the product was carried out by manufacturing the 3D model of the final proposal by additive manufacturing.

3 Results and Discussion

Using the VR tool has allowed for additional information that hand-drawn 2D sketches do not offer. Moreover, by being able to observe the sketches in their real dimension and interact with them, it has been possible to modify aspects of the concepts to improve them, thus reducing the time in the design process. It is worth noting that it has enabled early decisions to be made, thus increasing reliability.

As a result of the application of VR, in the first phase, more than 20 early sketches were obtained, Fig. 3, all of which were made without taking into account technical requirements. As can be seen, the sketches generated are 3D models.

Fig. 3. Early sketches obtained with Gravitiy scketch and VR using HTC glasses.

8 proposals were selected based on design specifications and technical requirements in a meeting with experts. These proposals were reworked in Gravity Sckech and VR to elaborate the concepts in detail and to better communicate the ideas to the users, Fig. 4.

From these, the 3 best performing ones have been selected, assessing the functional properties with respect to the specifications of use. Therefore, it has been possible to evaluate the concepts through VR. It should also be noted that the VR tool has facilitated decision-making with regard to other functional aspects by directly and unconsciously relating the context of use in the virtual simulation. However, other aspects such as weight, battery life and charging points cannot be evaluated in a timely manner with VR.

According to the results in Fig. 5, the proposals selected to continue with the virtual design process are proposals C1, C3 and C4. Specifically, the concept C1, Fig. 4 a), has stood out for the adaptability to the hand, the ease of switching on and the durability of the battery, in all of them the score has been 4.5 or higher, as a negative point the users consider it with a lot of weight. The design proposal C3, Fig. 4 c) has stood out in terms of ease of switching on and stability, although it has been rated positively in the rest of the requirements, although users considered that it does not fit well in the hand. On the other hand, C4, Fig. 4 d) was rated positively on all but intuitive use and battery life, which was rated less than 2.5.

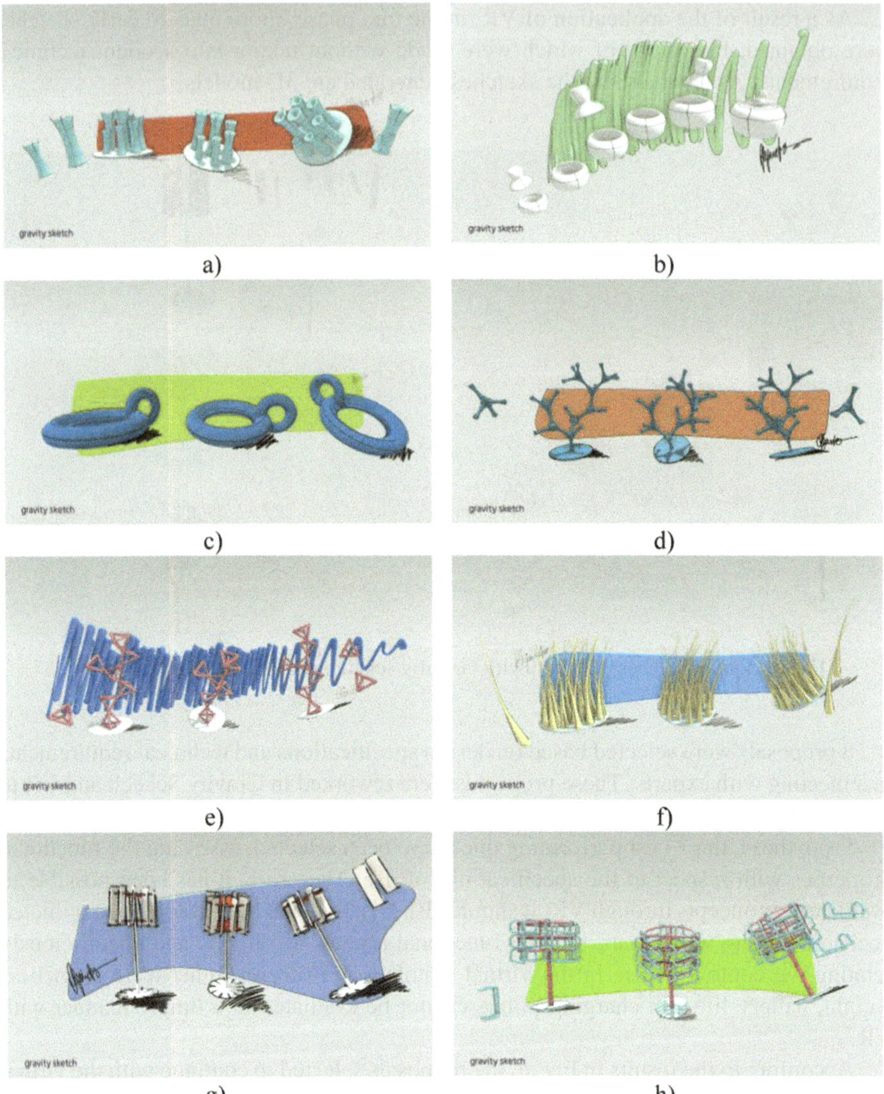

Fig. 4. Sketches developed with VR in the second phase: a) concept C1, b) concept C2, c) concept C3, d) concept C4, e) concept C5, f) concept C6, g) concept C7, h) concept C8.

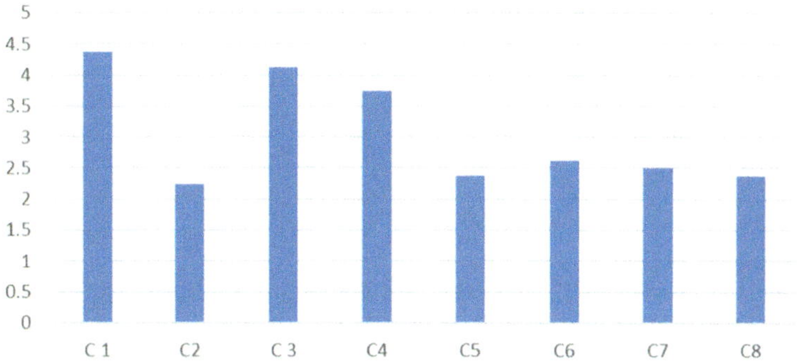

Fig. 5. Overall results of the evaluation of the properties where C is the ranking of the concepts presented.

As shown in Fig. 6, the 3 selected proposals have been evaluated by means of VR introduced in the environment. The proposal selected as the final one was Cf2, Table 1. This proposal has obtained the best rating although it has been observed that it needs to be improved in terms of its portability. On the other hand, proposal Cf1, Fig. 6 a) obtained the worst results in terms of assembly and grip, but the users considered that having a handle made it the best option for transporting it. For this reason, this solution has been taken into account in the final proposal for Cf2, Fig. 6 b). As for Cf3, Fig. 6 c), it has been discarded by the users as it did not meet any of the requirements analysed.

The use of VR has been instrumental in the conclusion of the final design of this work. The feedback from the users has been fundamental in the design process and is considered reliable thanks to VR. On the other hand, the participants considered the user experience with the VR environment to be very satisfactory.

Putting the user in a virtual environment so that they can use the product and see it in real time has generated a lot of information and feedback for the final design. This information has allowed for redesign and assessment of factors that had not been considered before.

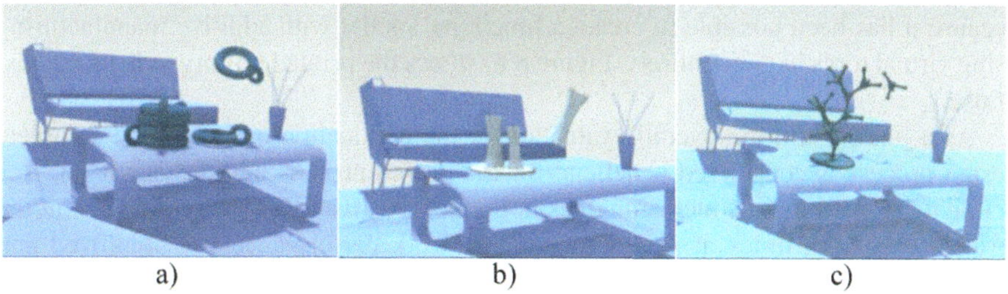

Fig. 6. Propuestas finales donde a) es el concepto Cf1, b) es el concepto Cf2 y c) es el concepto Cf3.

Table 1. Global Results of evaluation where c is concept

Concept	Cf 1	Cf 2	Cf 3
Hand grip	2.8	4.2	2.4
Transport	3.4	2.8	2.8
Assembly	2.6	4.6	2.2

The selected concept, Cf2, has been developed and improved to obtain the final design shown below and has also been evaluated with VR in the same virtual environment previously created. Figure 7 shows the product in the environment being used.

The final product has been improved in terms of the transport of the luminaires which is directly related to the grip mode. As shown in the image, this has been solved through the inclusion of grooves in the surfaces that allow the product to be more manageable.

Fig. 7. Virtual evaluation environment of the final proposal: a) overview of the room, b) interaction with the lamp design.

On the other hand, as a result of the application of the final virtual design, presentation images have been produced which, as shown in the image, are perfectly compatible with 3D models created with parametric programmes, Fig. 8 a).

The prototype realised has validated the design procedure by means of virtual reality because it has been possible to create a functional model with additive manufacturing using virtual tools in their entirety. Figure 8 b) shows the printed prototype in its context of use.

As a final result, the overall evaluation of the application of VR in the complete design process is considered beneficial for the design process. On the one hand, it is not demonstrated that its inclusion does not affect creativity. On the other hand, at the level of execution times, it was observed that work performance was improved and process times were reduced due to agile decision making. Training in the use of the tool is simple because the interface is similar to other programmes used in the field of product design. Also, the time spent on sketching is reduced compared to conventional representation techniques, as the drawing and colouring is done momentarily to the drawings. Furthermore, there is no need for subsequent 3D modelling with parametric modelling programs.

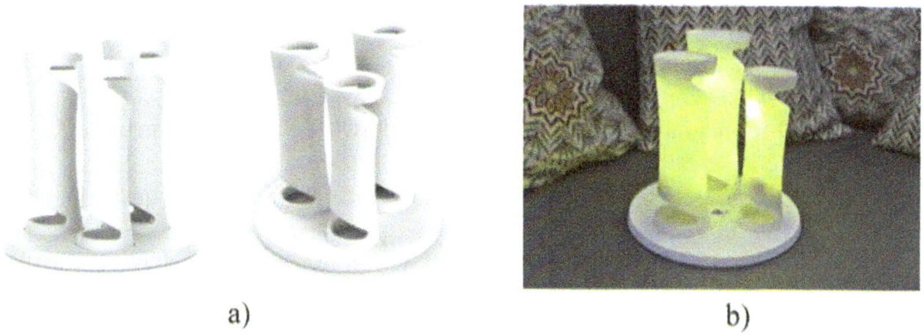

Fig. 8. Final validation results: a) infographics of the final product, b) photograph of the prototype generated with additive manufacturing.

4 Conclusions

In this study, where a project was carried out applying VR sketching using the Gravity Scketch programme, it was observed that the result of the concepts was good and that it was easy to create shapes freely. However, there is a limitation in the working environment.

Also, this VR technology has been used to efficiently evaluate design proposals with users of different ages and genders. Users were able to functionally evaluate the luminaire, interacting with the product in isolation and also in a virtual environment. This meant that they received a lot of information about the proposal for the selection of the final proposal.

The final result has been positive, obtaining a fully functional, intelligent lighting device. At the same time, thanks to digital sketching, it was possible to create a 3D printed prototype without the need for subsequent computational design using CAD techniques.

Finally, the ability of this technique as part of the design, evaluation and development process of new products is considered key to obtaining products adapted to a new generation of users, with changing and very specific needs. Furthermore, adapted to the new technologies that are becoming more and more integrated in society.

From the overall analysis of the research carried out, it can be seen that the inclusion of virtual tools in the design process does not limit creativity in the creation of new proposals and improves both the quality of decisions and the time taken to make them. Likewise, technical errors are reduced due to the possibility of evaluating the product in an environment similar to the real one.

References

1. Ng, P.C., She, J., Jeon, K.E., Baldauf, M.: When smart devices interact with pervasive screens: a survey. ACM Trans. Multimed. Comput. Commun. Appl. **13**(4), 1–23 (2017)
2. Liang, Y., Cai, Z., Yu, J., Han, Q., Li, Y.: Deep learning based inference of private information using embedded sensors in smart devices. IEEE Netw. **32**(4), 8–14 (2018)
3. Balzerkiewitz, H.-P., Stechert, C.: The evolution of virtual reality towards the usage in early design phases. Proc. Des. Soc. Des. Conf. **1**, 91–100 (2020)

4. Ban, S., Hyun, K.H.: 3D computational sketch synthesis framework: assisting design exploration through generating variations of user input sketch and interactive 3D model reconstruction. CAD Comput. Aided Des. **120**, 102789 (2020)
5. Ha, D., Eck, D.: A neural representation of sketch drawings. In: 6th International Conference on Learning Representations, pp. 1–16. Vancouver (2018)
6. Bilda, Z., Gero, J.S., Purcell, T.: To sketch or not to sketch? That is the question. Des. Stud. **27**(5), 587–613 (2006)
7. Arora, R., Darolia, I., Namboodiri, V.P., Singh, K., Bousseau, A.: SketchSoup: exploratory ideation using design sketches. Comput. Graph. Forum **36**(8), 302–312 (2017)
8. Hyun, K.H., Lee, J.H.: Balancing homogeneity and heterogeneity in design exploration by synthesizing novel design alternatives based on genetic algorithm and strategic styling decision. Adv. Eng. Informatics **38**, 113–128 (2018)
9. Self, J.A.: Communication through design sketches: Implications for stakeholder interpretation during concept design. Des. Stud. **63**, 1–36 (2019)
10. Caldas, L., Keshavarzi, M.: Design immersion and virtual presence. Technol. Archit. Des. **3**(2), 249–251 (2019)
11. Ng, L.X., Ong, S.K., Nee, A.Y.C.: Conceptual design using functional 3D models in augmented reality. Int. J. Interact. Des. Manuf. (IJIDeM) **9**(2), 115–133 (2014). https://doi.org/10.1007/s12008-014-0207-6
12. Rieuf, V., Bouchard, C., Meyrueis, V., Omhover, J.-F.: Emotional activity in early immersive design: Sketches and moodboards in virtual reality. Des. Stud. **48**, 43–75 (2017)
13. Obeid, S., Demirkan, H.: The influence of virtual reality on design process creativity in basic design studios. Interact. Learn. Environ. 1–19 (2020)
14. Huang, Y.C., Chen, Y.R.: The study of virtual reality product design in education learning. In: ACM International Conference Proceeding Series, pp. 9–11 (2019). https://doi.org/10.1145/3369199.3369207
15. Berg, L.P., Vance, J.M.: Industry use of virtual reality in product design and manufacturing: a survey. Virtual Real. **21**(1), 1–17 (2017)
16. Mandic, V.: Model-based manufacturing system supported by virtual technologies in an industry 4.0 context. Lect. Notes Mech. Eng. 215–226 (2020)
17. Aromaa, S., Väänänen, K.: Suitability of virtual prototypes to support human factors/ergonomics evaluation during the design. Appl. Ergon. **56**, 11–18 (2016)
18. Arrighi, P.A., Mougenot, C.: Towards user empowerment in product design: a mixed reality tool for interactive virtual prototyping. J. Intell. Manuf. **30**(2), 743–754 (2019)
19. Berni, A., Borgianni, Y.: Applications of virtual reality in engineering and product design: Why, what, how, when and where. Electronics **9**(7), 1–29 (2020)

Applying Sustainable Design in the Production Process of Cash Management Machines

José Ignacio Valero[1(✉)], Anna Biedermann[1], Natalia Muñoz[1], Enrique Lacasa[2], and José Luis Santolaya[1]

[1] Design and Manufacturing Engineering Department. EINA, University of Zaragoza. C/María de Luna, 3 (50018), Zaragoza, Spain
ivalero@unizar.es

[2] I + D Engineer at "Partner Company" for This Case Study, Zaragoza, Spain

Abstract. Product development based on sustainability principles is one of the poles of innovation in design activity. Public policies, consumer preferences and corporate social responsibility lead to the growing importance of applying sustainability criteria in the current design of products.

In this work, collaboration with a company which main activity is the manufacturing of cash management machines is carried out in order to project a more sustainable deposit unit. An approach that takes into account the initial specifications of the product, contemplates a life cycle perspective and simultaneously evaluates environmental, economic and social impacts is applied. Thus, product requirements are analyzed, flows entering and leaving the industrial system are studied and sustainability indicators are obtained. A product design with improved levels of indicators is finally proposed.

Keywords: Engineering design · Sustainable design · Sustainability indicators

1 Introduction

Nowadays, designers face the challenge of reaching the best possible combination of both technical and sustainable aspects in the product design process. A continuous effort has been made along these decades by integrating the environmental performances of industrial products together with their technical characteristics [1, 2]. The life cycle perspective has become the base of a multitude of concepts and tools to calculate and analyze environmental impacts in a system as well as to find solutions that can lead to reducing this impact. ISO standard [3] defines the product life cycle as a succession of interconnected stages, beginning from the stage of raw material obtaining in the product creation to end-of life stage, in which activities of recycling or waste processing are carried out. According to this life-cycle approach, the Life Cycle Assessment (LCA) technique [3, 4] was developed to evaluate environmental impacts and Eco-design methodology [5, 6] was applied to improve environmental behavior of product systems.

Nevertheless, applying a sustainable approach not only means taking into account environmental impacts, but also introducing socio-ethical principles and models economically feasible. Three dimensions, looking simultaneously at economic aspects

F. Cavas Martínez et al. (Eds.): INGEGRAF 2021, LNME, pp. 91–99, 2022.
https://doi.org/10.1007/978-3-030-92426-3_11

(profit), environmental issues (planet) and social concerns (people) should be considered [7]. According to this approach, the Life Cycle Sustainability Assessment (LCSA) methodology [8, 9] combines three techniques to effectively assess the sustainability of a system: the Environmental Life Cycle Assessment (E-LCA) to examine environmental issues, the Life Cycle Costing (LCC) to evaluate economic aspects and the Social Life Cycle Assessment (S-LCA) to get social impacts. LCSA is based on the same structure of four phases that is already applied in LCA: i) goal and scope definition; ii) Life cycle sustainability inventory; iii) Life cycle sustainability impact assessment; iv) interpreting LCSA results.

Different impact categories and indicators can be used to evaluate each sustainability dimension. In the case of the environmental dimension, the most common indicator used is the Global Warming Potential (GWP) (also referred as climate change or greenhouse emissions) but other indicators as acidification (A), impact of land use (LU), ozone depletion (OD) or eutrophication (E) can be also studied according to the CML method [10]. The economic impacts are usually measured by the costs analysis along the product life cycle [11]. Total costs can be calculated by compiling the execution costs in different stages and operations (manufacturing, transport, final disposition…). In addition, the difference between revenues and costs or net operating profit obtained from the development of an activity is also an economic indicator proposed in different works [12, 13]. While, social issues are usually analysed according to different stakeholders' categories, such as workers, local community, society, consumers and value chain actors [14]. A number of subcategories can be distinguished within these categories. Child labour, fair salary, working hours or health and safety are subcategories established within the workers group. Qualitative or semi-quantitative information is obtained in different studies [15], due to the social aspects are difficult to quantify. However, the use of quantifiable indicators is proposed in those works in which the ability to improve the product sustainability is analysed [13, 16].

Numerous studies have been carried out about sustainable product design applying different tools, methods and approaches. Nevertheless, sustainability is mainly addressed from an environmental perspective and studies, in which three dimensions are analysed, are focused on the product sustainability assessment and improvements are not projected. In this work, LCSA methodology is used to effectively achieve the sustainability evaluation of a product and engineering design principles are applied to project an improved design. Methods used and results obtained for the case of a cash management machine, are shown in the following sections.

2 Methods

As Fig. 1 shows, a more sustainable product design can be attained using methods to assess the sustainability of a product system as well as applying design methods to improve it. Sustainability analysis and improvement is focused on the production stage.

The following phases can be carried out: i) sustainability assessment of the initial product; ii) product design applying sustainability strategies; iii) comparing both initial and final designs.

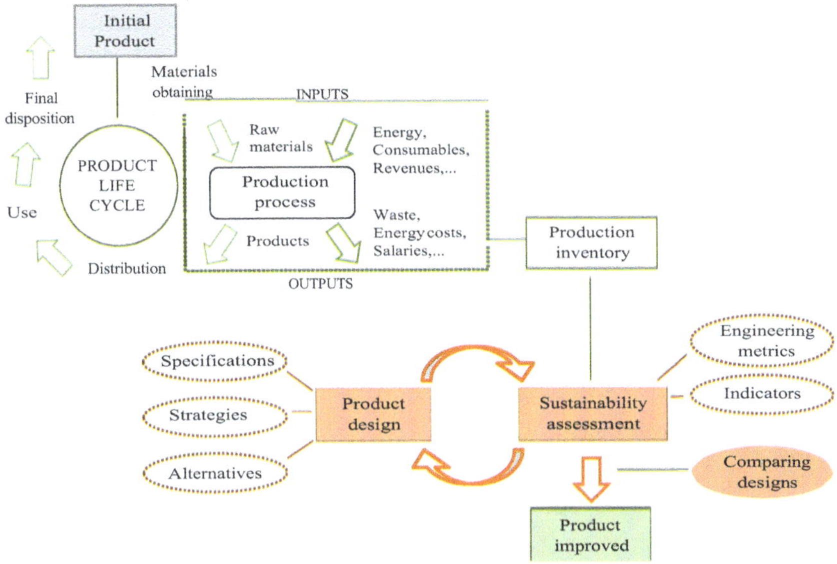

Fig. 1. Sustainable design methodological scheme.

Flows exchanged along the production process are valued to obtain detailed production inventory. With a nearer approach to the practice of engineering, the use of engineering metrics and quantitative indicators is, respectively, suggested to effectively express operating parameters and assess each sustainability dimension. The following engineering metrics have been used: product mass, waste percentage, raw material costs, production costs, machining time and annual production.

The number of indicators for measuring the three dimensions of sustainability can vary according to the difficulty of obtaining practical information and the scope of the study. In this study, quantitative indicators are selected to measure the impacts of each sustainability dimension. The global warming potential (GWP100) indicator is used to evaluate the environmental dimension. It represents greenhouse gas emissions. The value added (VA) or net operating profit is used to evaluate the economic dimension. Finally, the workers are selected as the main stakeholders' category and the social dimension of sustainability is valued by the quantifiable indicators of working time (TW) or average time required to carry out one product unit. A reduction of the working time can be associated to the productivity improvement (from the company perspective) and the reduction of the time developing an activity (from the worker point of view). The environmental unit indicators of different materials and products are found in the ProBas database [11]. In addition, emission factors associated to commercial electric companies [12] were applied.

3 Case Study

This approach was implemented in the sustainability improvement of a cash management machine. The main parts and characteristics of this product are shown in Fig. 2. The cash machine device has been designed with two elementary functions: counterfeit detection and money accounting and safe storage. Four different modules can be differentiated. Money notes are placed in feeding module where they are sequentially rolled on to the detection module and then to the safe deposit box through the transportation module. Closing module act as general cover for the machine (excepting safe box). Rejected banknotes from detection module are driven out from the machine.

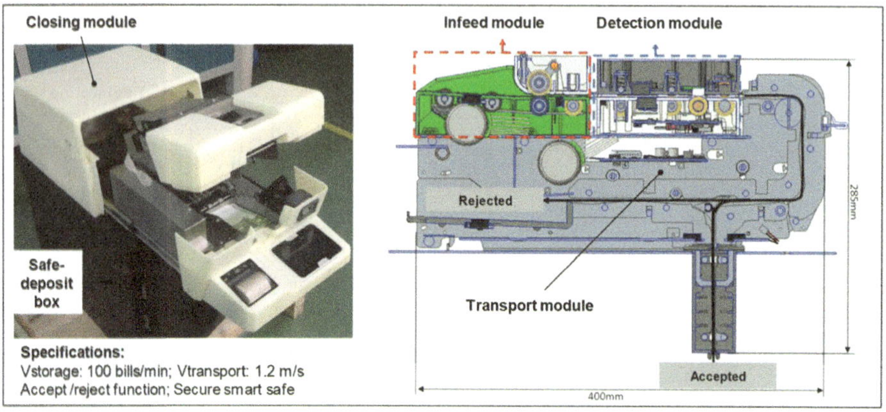

Fig. 2. Cash machine scheme.

Data of the production process of the cash management machine were obtained from a detailed study supported by the manufacturing company. This process is schematically shown in Fig. 3. Raw materials required in the manufacture of mechanism and structural parts are mainly steel (AISI 430, AISI 303 and ST 37–2) and aluminium alloy (Al 6061), which are mainly received in plant as sheets and bars. A wide number of components are manufactured in external facilities and are incorporated to the production process in the final assembly process. The manufacture of this product includes cutting, folding, machining as well as welding operations, components assembly and testing of the product. Times required by operation and energy consumptions are shown in Fig. 3, as well as money flows corresponding to one product unit.

Welding of the safe box parts is the operation with the highest energy consumption and machining of involved components in the transport module is the operation in which more time is required. Total mass of raw materials, including waste, is 50,4 kg. An important waste percentage is generated and the energy consumption for the whole manufacturing process is 917,9 kW h. Engineering metrics for the initial production process are summarized in Table 2. These can be comparatively examined with those obtained later from the application of sustainability strategies in the product design. Production costs are 1911,3 €. Commercial parts are about 36% of the total cost of the machine. The machining time is 4,15 h and the production is about 2000 units per year.

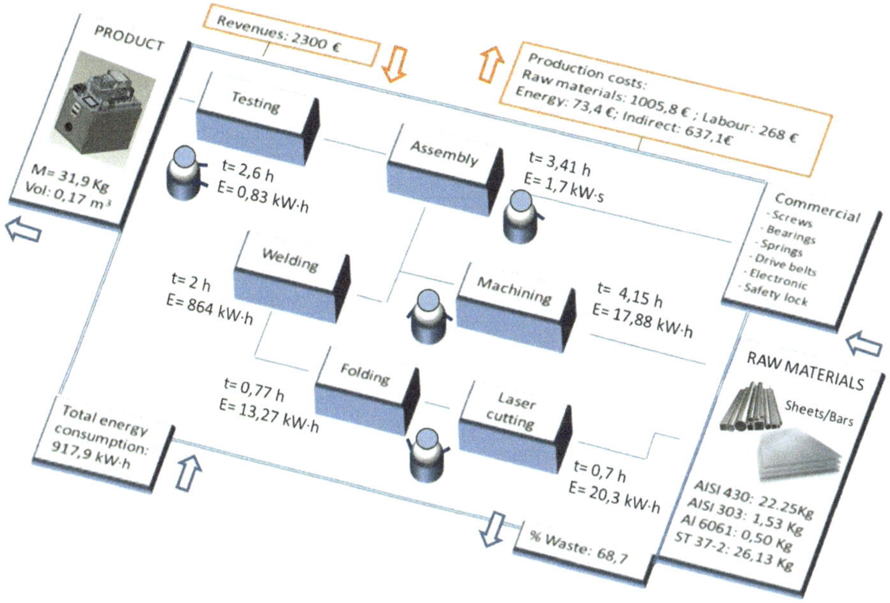

Fig. 3. Production process scheme.

Next, three different strategies were applied as supposed more sustainable design options of this product:

- Material change (Str1): Rollers on transportation module were redesigned from aluminium to POM (Polyoxymethylene) material.
- Alternative production techniques (Str2): the bending technique is replaced by die stamping in the manufacture of some components.
- Reuse of components (Str3): The reuse of the safe box from ancient models has been studied as an option.

These strategies are proposed taking into account that all previous product performance and specifications in terms of storage capacity, notes processing speed and detection level. Data analysis shows results on Table 1. Costs and energy consumptions along the manufacturing processes have been examined.

When POM is selected in Str1, machining operations related of aluminium are avoided, involving a cost saving and also energy consumption decrease of 42%. If process design includes switch some bending operations into stamping (Str2), also shows an improvement in terms of cost (21%) and energy consumption (6%). Moreover, it may seem obvious, when reusing safe box (Str3) material cost and energy consumption cuts down to minimum. Note that this case study does not include transport and additional operations to guarantee the assembly of the safe box from one to another. Also, being precise, the initial cost (materials, energy…) from the first time that the safe box was manufactured should be divided and considered.

Table 1. Material redesign comparison.

			Machining/ material cost (€)	Energy Consumption (kW·h)
Material change (Str1)	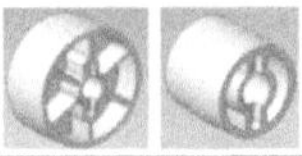	Al 6061	22,8	2,17
		POM	6,3	1,26
Alt. production technique (Str2)	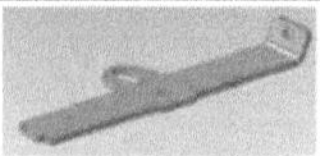	Bending	9,55	13,26
		Stamping	7,56	12,41
Reuse of components (Str3)		Safe box	560	873,6
		Reused safe box	0	12,41

As seen on Table 2, when rollers are made of POM instead of aluminum there is a slight reduction in the product mass and energy consumption metrics. Machining time decreases so production could be improved. If Str2 is applied (stamping instead of bending), waste reduces 4,3% and small reductions in manufacturing costs are also obtained. Finally, when reusing components (safe box), reduction of raw materials, energy consumptions and production costs are quite high due to welding operation in considerably simplified. At the same time, machining time reduce 4,5 % and production could be improved 37%.

Table 2. Engineering metrics comparison.

	Product mass (kg)	Waste (%)	Energy cons. (kW·h)	Raw mat. costs (€)	Production costs (€)	Machining time (h)	Annual production
Initial Product	31,97	69	915,48	301,09	1911,3	4,15	2000
(Str1) Material change	31,90	69	914,32	301,09	1874,8	3,86	2110
(Str2) Alt. techniques	31,97	66	914,04	295,12	1904,1	4,15	2000
(Str3) Reuse components	31,97	79	41,88	111,09	1071,3	3,96	2738
Final Product (Str1 + Str2 + Str3)	31,97	73	39,88	105,12	1027,7	3,61	2848

Some key sustainability indicators were calculated from rough data in order to make a better understanding and comparison between the initial product and the redesigned one. As shown on Table 3, significant improvements were achieved on environmental, economic and social dimensions when all sustainability strategies are applied. The GWP100 reduces from 146,6 to 94,2 kgCO2 eq, VA increases from 338,7 to 1272,3 € and Tw reduces from 13,04 to 9,02 h.

Table 3. Sustainability indicators.

Sustainability indicators	GWP_{100} ($kgCO_2eq$)	VA (€)	T_W (h)
Initial Product	146,8	388,7	13,04
(Str1) Material change	146,6	425,2	11,28
(Str2) Alt. production technique	133,5	395,9	11,57
(Str3) Reuse of components	107,5	1228,7	9,38
Final Product (Str1 + Str2 + Str3)	94	1272,3	6,15

If Str1 is applied significant improvements are achieved in both economic and social indicators, meanwhile the GWP100 indicator hardly modifies. In the case of Str2 application, GWP100 reduces 9,05%, VA increase 1,8% and Tw reduces 3,6%. On the other hand, Str3 is the sustainability strategy that causes a more substantial improvement in all indicators, particularly in environmental and economic dimensions. It was assumed that revenues and workforce of the manufacturing company are not modified.

Resulting values for each sustainability indicator are comparatively shown in Fig. 3. The effects of each strategy are clearly exposed. Since positive results are obtained a more sustainable product could be developed. In view of these results, Str1 and Str2 are currently being implemented by the manufacturing company and Str3 is under study. The reuse of the safe box is a very promising strategy although its implementation is complex. Recycling processes are increasingly in efficiency on recovered products but a better tracking of those products leaving the factory should be also necessary to facilitate the recovery process (Fig. 4).

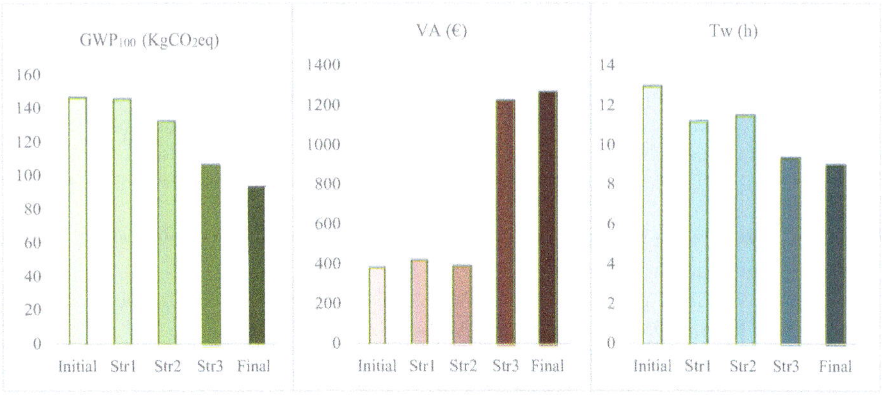

Fig. 4. Sustainability indicators comparison.

4 Conclusions

A wide number of research efforts have been carried to evaluate and reduce environmental impacts using Eco-design methodology. In this work sustainability, considering simultaneously environmental, economic and social aspects is integrated in the product design process. An engineering approach is applied using a set of metrics and quantifiable indicators that facilitate both results communication and decision-making process about product design.

The case of a cash management machine is addressed in this work. Study is focused on the production stage. A more sustainable product design in three phases: sustainability assessment of the initial product, application of sustainability strategies and comparing initial and new designs. The collaboration of the manufacturing company was essential to obtain production inventory. Data of materials and manufacturing operations (material costs, energy consumption, workforce…) have to be precisely valuated in each component and process to be able to make a right assessment. Specifications of every component or part of the machine were taken into account in order to propose adequate sustainability strategies to project a new machine.

The approach applied in this work shows evidence about how to make a comparison between a product design and some changes of itself based on sustainability indicators. Since data mining and analysis is based on same method, results can be compared to achieve improvements on indicators depending on the changes of the design.

Even the great challenge is to achieve a general tool for evaluating sustainability on product design stage, this case study shows how sustainability can be evaluated as a part of design process evaluating different options based on accurate data from manufacturing processes, materials and parts design.

References

1. Wimmer, W., Züst, R.: Ecodesign PILOT: Product Investigation Learning and Optimization Tool for Sustainable Product Development. Kluwer Academic Publishers, Dordrecht (2003)
2. Boothroyd, G., Dewhurst, P., Knight, W.A.: Product Design for Manufacture and Assembly, 3ed edn. CRC Press, Taylor and Francis Group, USA (2011)
3. ISO 14040 International Standard: Environmental Management–Life Cycle Assessment–Principles and Framework. International Organisation, Geneva, Switzerland (2006)
4. ISO 14040 International Standard: Environmental Management–Life cycle Assessment–Requirements and Guidelines. International Organisation, Geneva, Switzerland (2006)
5. Brezet, J.C., Van Hemel, C.G.: Ecodesign: A Promising Approach to Sustainable Production and Consumption. UNEP United Nations Publications, Paris (1997)
6. Van Hemel, C.G.: Eco-Design, Empirically Explored. Thesis, Delf University of Technology, Delf, ND (1998)
7. WCED (World Commission on Environment and Development): Our Common Future. Oxford University Press (1987)
8. Kloepffer, W.: Life Cycle Sustainability Assessment of Products (With Comments by Helias A. Udo de Haes, p. 95). Int J Life Cycle Assess 13:2, 89–95, (2008)
9. Finkbeiner, M., Schau, E.M., Lehmann, A., Traverso, M.: Towards life cycle sustainability assessment. Sustainability **2**, 3309–3322 (2010)

10. Guinée, J., et al.: Life Cycle Assessment–An Operational Guide to the ISO Standards. Centre of Environmental Sciences (CML), Leiden University (2001)
11. Hunkeler, D., Rebitzer, G., Lichtenvort, K., (eds.) Ciroth, A., Hunkeler, D., Huppes, G., Lichtenvort, K., Rebitzer, G., Rüdenauer, I., Steen, B. (Lead authors): Environmental Life Cycle Costing. SETAC Publications (2008)
12. Azapagic, A., Perdan, S.: Indicators of sustainable development for industry: a general framework. Trans IChemE, Process Safety Environmental Protection, Part B **78**(4), 243–261 (2000)
13. Lacasa, E., Santolaya, J.L., Biedermann, A.: Obtaining sustainable production from the product design analysis. J. Clean. Prod. **139**, 706–716 (2016)
14. UNEP/SETAC: Guidelines for Social Life Cycle Assessment of Products. United Nations Environment Programme, Paris (2009)
15. Ciroth, A., Franze, J.: LCA of an Ecolabeled Notebook. Consideration of Social and Environmental Impacts Along the Entire Life Cycle. GreenDeltaTC GmbH, Berlin (2011)
16. Santolaya, J.L., Lacasa, E., Biedermann, A., Muñoz, N.: A practical methodology to project the design of more sustainable products in the production stage. Res. Eng. Design **30**(4), 539–558 (2019). https://doi.org/10.1007/s00163-019-00320-w
17. Umweltbundesamt (UBA): German Environmental Protection Agency (2007)
18. Mapama. http://www.miteco.gob.es/es/cambio-climatico/temas/mitigacion-politicas-y-medidas/factores_emision_tcm30-446710.pdf, Last Accessed 16 Apr 2018

Geometrical Study of the Barrel Stave in a Procedure of Restitution of the Original Flat Shape

A. Conde Fernández(✉), E. Zurita de la Vega, P. Vila Lameiro, and P. Tato-Sánchez del Valle

Departamento de Ingeniería Agroforestal, Universidad de Santiago de Compostela (Escuela Politécnica Superior de Ingeniería), Benigno Ledo S/N, 27002 Lugo, Spain
adriana.conde@usc.es

Abstract. The cooperage is still an artisan trade almost entirely. The price of barrels, leaving to a side the cost of the raw material, is determined to a large extent based on the labour required in its manufacture. The process of internal al lateral lowering of the staves and their subsequent curving are decisive to obtain a useful and quality barrel. This communication is directed towards the parameterization of the stave as a fundamental element in the cooperage production process and its standardization using CAD tools. Through the digitization of the stave of this work, it will be possible to determine the possibilities of streamlining this artisan manufacturing process and thus lay the foundations for, in the future, study the ability to develop specialized machinery for sawing wooden parts with a specific shape and even robotization of the staves manufacturing process, as the bending and internal lowering of the wood, which would consequently reduce costs in the manufacture of the barrels.

Keywords: Cooperage · Stave · Model · Standardization

1 Introduction

In previous papers such as Conde et al. 2019, the equations of the inner surface and the volume of a barrel have been determined as a shape generated by the revolution of a circumference arc. From these expressions and the analysis of the barrel shape, its relationship with the production process of wine has been analyzed, differentiating its influence depending on the type of wine.

The present geometrical study allows us to address another fact in the production process: the cost of manufacturing the barrel. The barrel must be renewed as a consequence of aging, the clogging of its pores and, consequently, the degradation of the oxygenation process of the wine through the wood barrel.

The indicative prices of timber for cooperage are between €150 to €400 /m3 for standing timber, €350 to €550 /m3 for timber on site and €2,200 to €2,500 /m3 of stave [1]. This means that the average cost of a new barrel in France is €500. Other reports raise this price to €700.

F. Cavas Martínez et al. (Eds.): INGEGRAF 2021, LNME, pp. 100–106, 2022.
https://doi.org/10.1007/978-3-030-92426-3_12

The manufacture of a barrel is an artisan process, its cost is more related to the number of hours that the artisan spends in its manufacturing than to the cost of the materials. Thus, if the objective is to achieve greater competitiveness with the reduction of manufacturing costs without altering the quality, the aim is to reduce the manufacturing costs of the barrel. It is necessary to consider that the optimal exchange of aromatic compounds by the barrel takes place for 5–8 years, therefore using a barrel more than 8 years old would be acting to the detriment of the wine's quality.

If the barrel is the manufacturing element that contributes to the added value to the wine through flavor and odor compounds during the maturation process, we will understand the value of the barrel, which leads us to think about which industry must renovate the most part of its manufacturing process machinery every eight years or which type of industry has machinery with an amortization period of eight years. Being aware of this, the reduction of the barrel manufacturing costs becomes particularly important.

The unit of barrel manufacturing is the stave. If we observe the barrel staves, we detect that not all are equal, driving us to think that, in the manufacturing process, with the aim of making good use of the timber, the manufacturers choose the staves on the spot so that they adjust to nonstandardized sizes and achieve a barrel of an specific volume.

In this assembly process, where the constraining fact is the expertise of the artisan when choosing each stave, the time of the manufacturing process is increased as well as the cost of the labor. Therefore, it is necessary to address the automation of the manufacturing process and even go beyond because nowadays we have the knowledge that, with the help of robotization, we can create intelligent systems that detect the superficial shape and state and, according to this, act on the elaboration process of any material or even reduce the intervention of artisan labor.

2 Objectives

The main objective of this communication is to determine the shape of the stave and, through inverse engineering, determine the shape of the flat stave.

The secondary objective is the establishment of specific models of staves with the aim of standardizing the manufacturing process for barrels.

3 Methodology and Development

The starting point of the manufacturing process of a stave is a flat wooden board that will be curved. The curving process implies that the most external fibers will suffer tractive stress while the inner fibers will suffer compression stress.

As timber is an isotropic material, the capacity of this material to resist both stresses are variable and depends directly on whether it withstands compression and suffers shortening. This does not imply that the maximum yield strength would be equal unless it is within the recovery yield strength of the material. In any case, the curving on the stave is within the yield strength of the material. The fiber that suffers the transition from compressing area to traction area would be the one that does not suffer deformation and the one that can define the form of the board before the curving process.

The deformation, as presented in the Fig. 1, shows that in A the same elongation in the symmetric fibers was produced. In diagram *B*, the fiber suffers less shortening (C) than elongation and in *C*, the contrary, a greater shortening and a minor elongation regarding the central fiber.

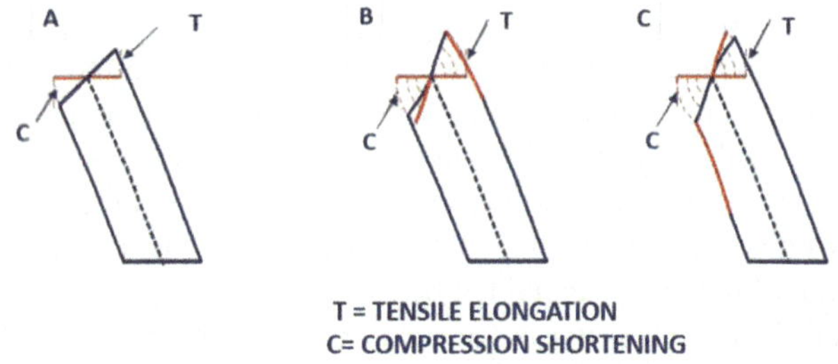

Fig. 1. Behavior of the stave fibers regarding the inner fibers.

Observing the cross section in the head of the staves, we can conclude that in artisan manufacturing, the curving is carried out within the recovery yield strength. Maintaining the cross section, without resulting in an evolution of concavity or convexity in the cross section, constituting the inner fibers, the reference for the restitution of the shape of the stave from the deformed stave.

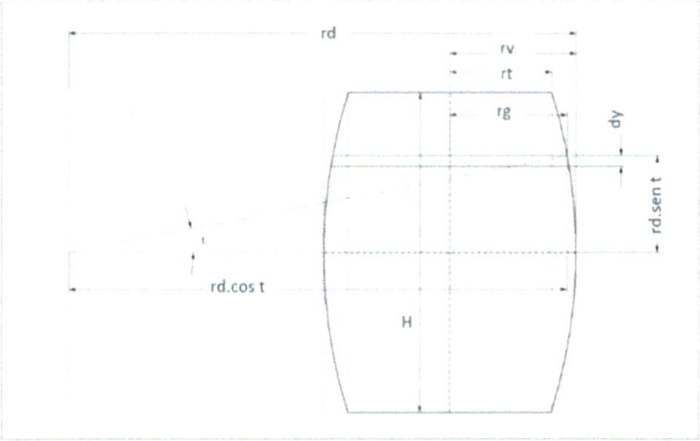

Fig. 2. Parts of the barrel

In previous studies [2], the shape of the barrel was determined, as well as the equations that defined the geometry of the barrel in relation to its volume and inner surface, from the parameters defined in the scheme of Fig. 2.

The restitution of the shape of the stave that is represented in Fig. 3 draws from the rectification of the inner fiber of the stave, on which we associated in each point the end of the variable polygon associated to the circumferences and the variable distance from the plan of the bilge of the barrel.

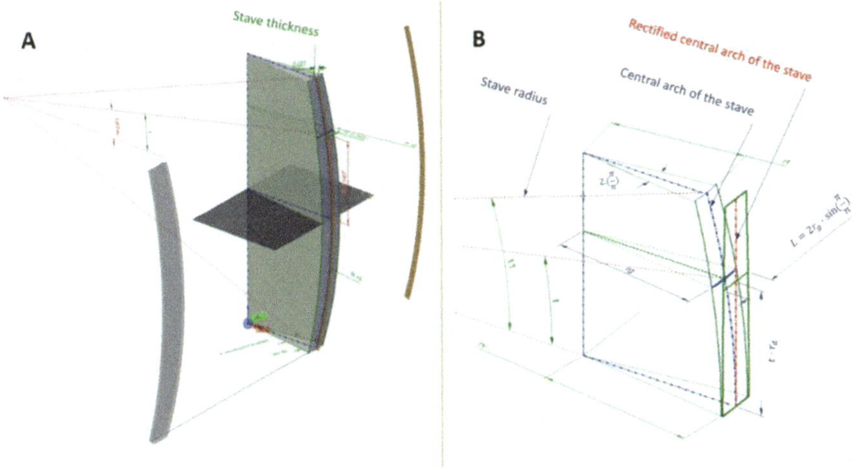

Fig. 3. Restitution of the stave flat shape.

The following expressions where we define r_g (1), that is the radius of the variable rotation associated to the point of the circumference that, through a revolution generates the surface of the barrel, where H is the height of the barrel, r_t is the radius of the head hoop, r_v is the radius of the bilge and r_d (2) is the radius of the arc associated to the stave, elements represented in the Fig. 3.

It must be considered that all the expressions would be valid if, when working with the inner fiber, all the ratios increase in half of the thickness of the stave.

$$r_g = r_d(\cos t - 1) + r_v \tag{1}$$

$$r_d = \frac{\frac{H^2}{4} + (r_t - r_v)^2}{2(r_t - r_v)} \tag{2}$$

$$t_1 = arcsen\frac{H}{2r_d} \tag{3}$$

From the expressions gathered in Conde et al. 2020, we can conclude that the parametric Eqs. (4) and (5) of the rectified stave curve, associated with the flat shape of the stave (Fig. 4) before the deformation process of the same, where n is the number of staves.

$$x = \frac{\frac{H^2}{4} + (r_t - r_v)^?}{2(r_t - r_v)}.t \tag{4}$$

$$y = \left[\frac{H^2/4^+(r_t - r_v)^2}{2(r_t - r_v)} \cdot (\cos t - 1) + r_v\right] \cdot \operatorname{sen}\left(\frac{\pi}{n}\right) \tag{5}$$

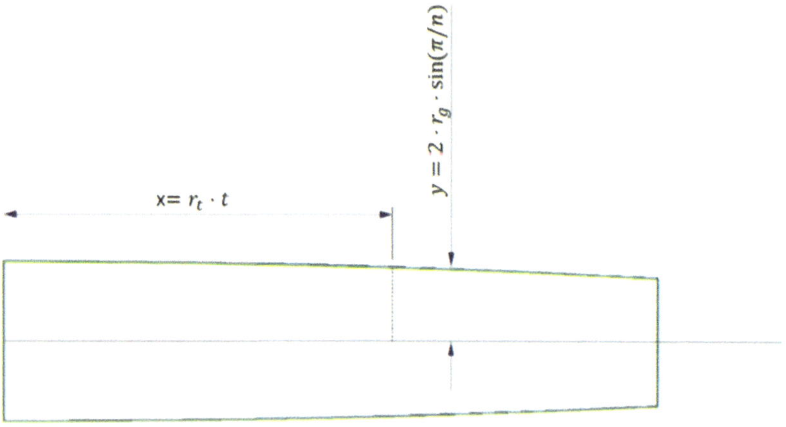

Fig. 4. Parameters of the rectified stave curve.

Table 1. Commercial dimensions of barrels

Diameter (mm)					
Type of barrel	Weight (kg)	Height (mm)	Head hoop	Bilge diameter	Thickness (mm)
300l	58	1008	610	780	27
350l	63	1020	670	830	27
400l	67	1050	690	850	30
500l	82	1092	780	950	32

Table 1 shows the dimensions of the most common types of casks, which have been extracted from a current commercial brochure [3] belonging to a Spanish cooperage company.

The previous data define the external geometry of the barrel, nevertheless, in order to define the geometry of the stave, they must be corrected to place that data on the medium fiber, subtracting half of the thickness of the stave.

Table 2 shows the adjusted dimensions in millimeters, the diameter as well as the radius for the commercial barrels described above.

For example, for a 300l cask, for which it would be necessary to have 24 staves for its manufacturing, the values of these would be given by the equations that define the geometry of the curve that is described. Working with the values of the dimensions in mm is:

$$\mathrm{x} = \frac{1008^2/4{+}^{(298.25-383.25)^2}}{2 \cdot (298.25 - 383.25)} \cdot \mathrm{t}$$

$$y = \left[\frac{1008^2/4 + (298.25 - 383.25)^2}{2(298.25 - 38325)} \cdot (\cos t - 1) + 383.25\right] \cdot \mathrm{sen}\left(\frac{\pi}{24}\right)$$

Table 2. Adjusted measurements for commercial barrels.

Diameter (mm)					Radius (mm)		
Cask volume (l)	Weight (kg)	Height (mm)	Head hoop	Bilge diammeter	Head hoop	Bilge diameter	Thickness (mm)
300l	58	1008	596.5	766.5	298.25	383.25	27
350l	63	1020	656.5	816.5	328.25	428.25	27
400l	67	1050	675	835	337.5	417.5	30
500l	82	1092	764	934	382	467	32

With the values of t and r_d given by the following expressions.

$$t_l = \arcsin(H/2 \cdot r_d)$$

$$r_d = \left(H^2/\left(4 + (r_t - r_v)^2\right)/(2(r_t - r_v))\right.$$

$$r_d = \left(1008^2/\left(4 + (298.25 - 383.25)^2\right)\right)/(2 \cdot (298,25 - 383,25)) = 1536,7117$$

$$t_l = \text{arcsen}\frac{1008}{2 \cdot 1536.7117} = 0.334157\,\text{rad}$$

The importance of having equations that accurately define the shape of the stave is that this is the first step to carry out an automation in the manufacturing process. Once we know the shape of the stave, the CAD program can be implemented so that it can be associated and synchronized with the CAD_CAM systems, permitting automated serration of the stave in the first stage of the manufacturing process.

The bending of the stave is currently carried out by artisans through the simultaneous application of temperature and humidity during a process that, traditionally, is named assembly. This could be carried out partially, thus reducing the manufacturing times of the casks with the use of, for example, hydraulic systems adjusted to the size of the stave.

4 Conclusions

- The CAD systems provide accuracy to the decisionmaking process regarding the manufacturing process of staves.
- This system complements the development in innovation with CAD-CAM systems aimed at the manufacturing of casks, which is a first step towards the automation of the manufacturing process.
- The standardization of the length of the staves and manufacturing the casks with a preestablished number of the same would make it possible to reduce the manufacturing times.
- The modeling and subsequent standardization of the staves permits automation in the production process, which would permit the reduction of the manufacturing costs.

References

1. BNPParibasWealthManagement(2016). https://wealthmanagement.bnpparibas/es/es/news/agrifrance.html. Accessed 21 Jan 2021
2. Conde-Fernández, A., La Vega, E.-D., Vila-Lameiro, P., del Valle, P.-S.: Geometrical Study of the Barrel as an Element of the Wine-Making Productive Process. Parameterization and Modelling. In: Cavas-Martínez, F., Sanz-Adan, F., Morer Camo, P., Lostado Lorza, R., Santamaría Peña, J. (eds.) INGEGRAF 2019. LNME, pp. 115–122. Springer, Cham (2020). https://doi.org/10.1007/978-3-030-41200-5_13
3. Ibérica de Tonelería. http://www.ibericadetoneleria.com/es/dargaud-jaegle-es/. Accessed 21 Jan 2021

Geometric Inspection in Surfboard Manufacturing by Using Reverse Engineering and 3D Inspection Tools

Xabier Amezua[1(✉)], Eneko Solaberrieta[1], Xabier Garikano[1], Angel Perez[1], Florencio Fernandez[2], and Mikel Iturrate[1]

[1] Department of Graphic Design and Engineering Projects, Faculty of Engineering Gipuzkoa, University of the Basque Country UPV/EHU, 20018 San Sebastian, Spain
xabier.amezua@ehu.eus

[2] Department of Chemical and Environmental Engineering, Faculty of Engineering Gipuzkoa, University of the Basque Country UPV/EHU, 20018 San Sebastian, Spain

Abstract. Although computer-aided design (CAD) and computer-aided manufacturing (CAM) tools have been integrated into the surfboard production process in recent years, this process still relies on the expertise of the shapers, as several stages of the process are still done by hand. This can lead to manufacturing surfboards with geometric errors that can worsen their performance. Therefore, it is important to perform geometric inspections of surfboards. In view of the lack of methodologies and standards to perform geometric inspections of surfboards, this paper presents a methodology to perform this type of inspections using reverse engineering and 3D inspection tools.

Keywords: Reverse engineering · Inspection · 3D scanner · Design · Symmetry · Surf

1 Introduction

Traditionally, surfboards have been handmade by shapers, who have continuously improved their design based on the intuitive feedback they received from surfers, especially those at the advanced or professional level. Therefore, surfboard production has been considered more as an art than an engineering process [1].

However, with technological advances in recent years, computer-aided design (CAD) and computer-aided manufacturing (CAM) tools have been integrated into the surfboard production process. These tools have simplified to shapers both the design and manufacturing of surfboards, allowing them, among others, to save time and to systematically modify or reproduce similar surfboards [1–3]. These tools also offer them the possibility to design surfboards with the support of Computational Fluid Dynamics (CFD) software programs, which allow them to study and evaluate the performance of their designs [1–7].

F. Cavas Martínez et al. (Eds.): INGEGRAF 2021, LNME, pp. 107–113, 2022.
https://doi.org/10.1007/978-3-030-92426-3_13

Despite the integration of CAD and CAM tools, the surfboard production process still depends on the expertise of the shapers, as several stages of the process are still done by hand [5, 8]. This can lead to the production of surfboards with geometric errors that worsen their performance. Therefore, it is important to carry out geometric inspections of surfboards to evaluate their conformity to the intended design. Currently, in most cases, surfboards are only evaluated visually and by surfers' sensations, as the surf industry still lacks methodologies and standards to perform geometric inspections of surfboards.

In this sense, this work aims to provide tools to perform geometric inspections of surfboards. For this purpose, a geometric inspection of a surfboard was presented.

2 Material and Methods

Geometric inspection of a surfboard designed and manufactured by shapers Johnny Cabianca and Mikel Agote of Basque Country Surf Company (BCSC) [9] for professional surfer Gabriel Medina [10] was performed. The area of the surface that is in contact with the water when surfing (the so-called wetted surface area) was inspected (Fig. 1), as its geometry influences the performance of the surfboard [2].

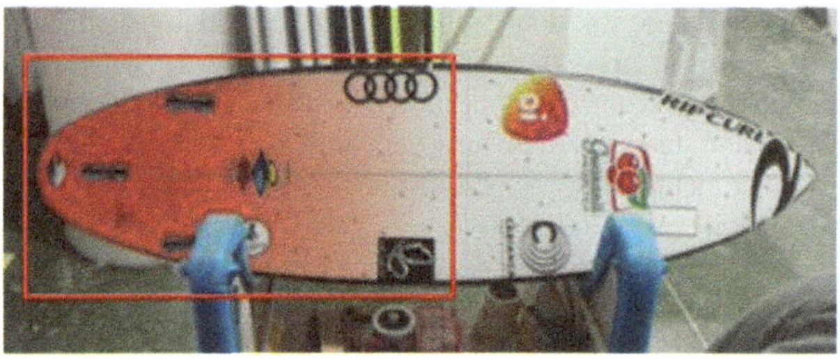

Fig. 1. Wetted surface area of surfboard.

The wetted surface area of the surfboard was scanned with a handheld 3D scanner based on structured white light scanning technology (Go!SCAN 20 scanner with VX element 6.3 SR1 software; Creaform, Inc) (Fig. 2). For that, targets used by the 3D scanner for positioning were stuck on the wetted surface area of the surfboard. As a result, a 3D triangle mesh of the wetted surface area of the surfboard was obtained in the object (OBJ) file format (Fig. 2B).

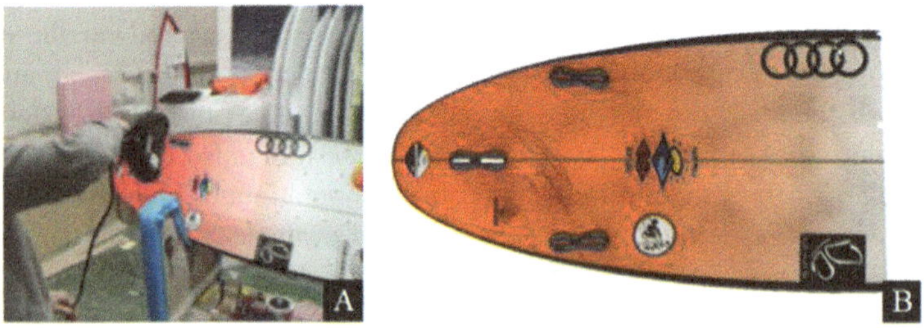

Fig. 2. Acquisition of 3D triangle mesh of wetted surface area of surfboard. A, Scanning of wetted surface area of surfboard. B, 3D triangle mesh of wetted surface area of surfboard.

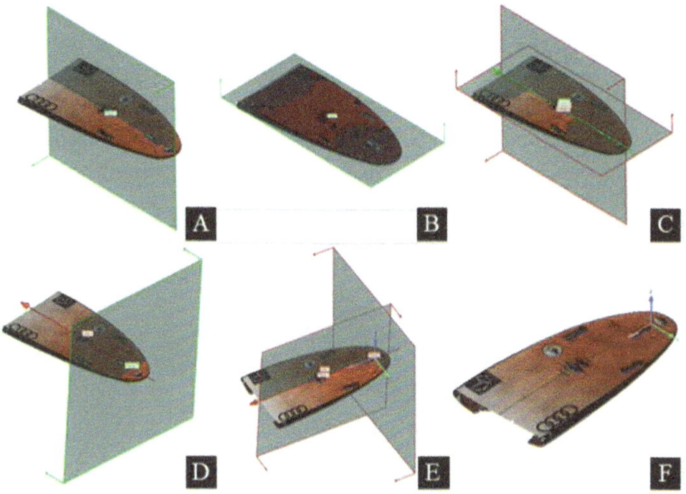

Fig. 3. Alignment of 3D triangle mesh of wetted surface area to global coordinate system through geometric elements. A, Creation of Plane 1. B, Creation of Plane 2. C, Creation of Line 1. D, Creation of Plane 3. E, Alignment of 3D triangle mesh of wetted surface area to global coordinate system. F, Aligned 3D triangle mesh of wetted surface area.

The 3D triangle mesh of wetted surface area was imported into a reverse engineering software program (Geomagic Studio 2013; Geomagic, Inc) to align it with the global coordinate system. For that, the following geometric elements were created: symmetry plane of the 3D triangle mesh of the wetted surface area (Plane 1) (Fig. 3A); best-fit plane of the 3D triangle mesh of the wetted surface area (Plane 2) (Fig. 3B); intersection line between Plane 1 and Plane 2 (Line 1) (Fig. 3C); and a plane normal to Line 1 (Plane 3) (Fig. 3D). Once the elements were created, the 3D triangle mesh of the wetted surface area was aligned to the global coordinate system by aligning Plane 1 with the XZ plane, Line 1 with the X axis, and Plane 3 with the YZ plane (Fig. 3E). Finally, the aligned 3D triangle mesh of the wetted surface area (Fig. 3F) was exported in the standard tessellation language (STL) file format.

Aligned 3D triangle mesh of the wetted surface area was then imported into a 3D inspection software program (GOM Inspect 2019; GOM GmbH). There, both the symmetry of the wetted surface area and the positioning and symmetry of the fin attachment systems were inspected.

To inspect the symmetry of the wetted surface area, its 3D triangle mesh was divided into two parts through the XZ plane (Fig. 4). Once the 3D triangle mesh of the wetted surface area was divided into two parts, transverse sections parallel to the YZ plane (Fig. 5) and longitudinal sections parallel to the XZ plane (Fig. 6) were created on both parts. Then, corresponding sections in both parts were superimposed to inspect their symmetry (Figs. 5 and 6). As a result, the deviation between superimposed sections was obtained (Figs. 5 and 6).

Fig. 4. Aligned 3D triangle mesh of wetted surface area divided into two parts through the XZ plane.

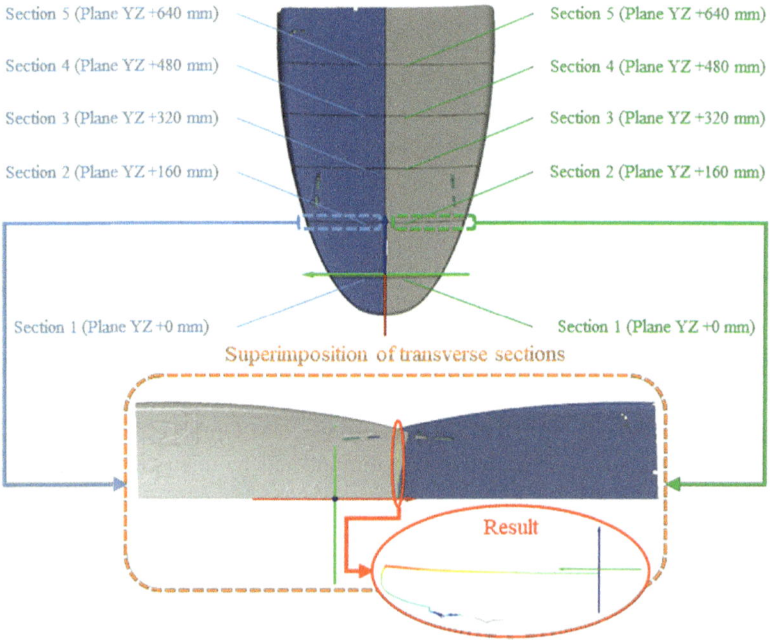

Fig. 5. Inspection of symmetry of transverse sections of wetted surface area.

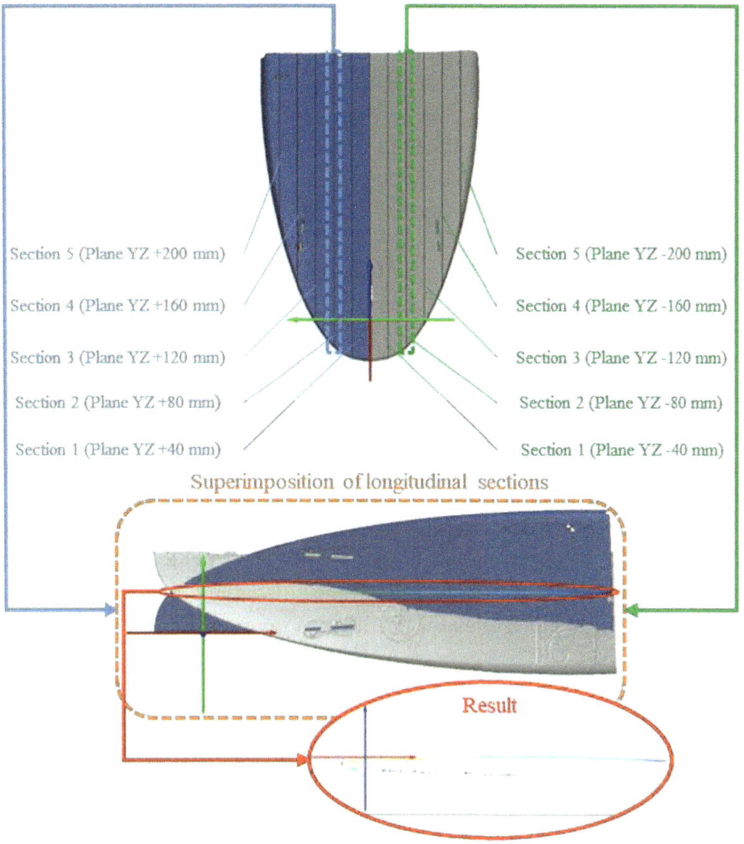

Fig. 6. Inspection of symmetry of longitudinal sections of wetted surface area.

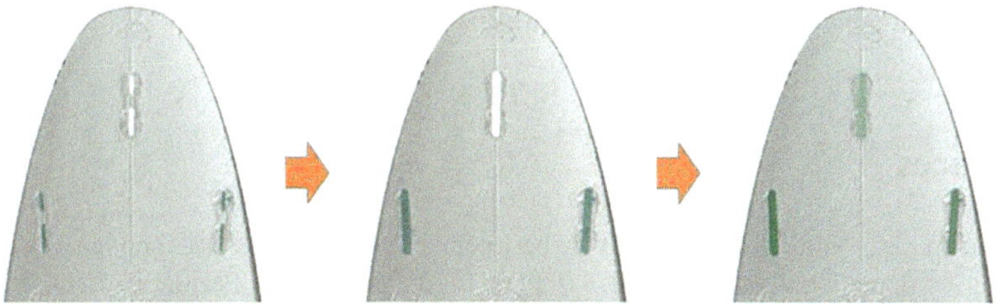

Fig. 7. Creation of fitting slotted holes.

Finally, to inspect the positioning and symmetry of the fin attachment systems, a fitting slotted hole was created in each of them (Fig. 7). Then, the positioning of the axes of the 3 slotted holes and the symmetry with respect to the XZ plane of the axes of the two side slotted holes were inspected.

3 Results and Discussion

For confidentiality issues, the numerical results of the inspections could not be published. However, it can be confirmed that these results provided valuable information to the shapers. Thus, inspection of the symmetry of the wetted surface area of the surfboard revealed a lack of symmetry and indicated that the deviations between corresponding transverse sections increased with distance from the symmetry plane (XZ plane) and that the deviations between corresponding longitudinal sections were larger at the tail of the surfboard and in the sections more distant from the symmetry plane. In turn, inspection of the positioning and symmetry of the fin attachment systems revealed that both the positioning of the attachment systems and the symmetry with respect to the symmetry plane between the lateral attachment systems were correct.

The inspection presented in this paper demonstrates that with available technology it is possible to perform geometric inspections of surfboards that provide valuable information to shapers. The results of this type of inspections allow shapers to correct errors made during the manufacturing process. In this work, geometric inspection of the finished surfboard was performed, but this type of inspection could also be performed at different stages of the manufacturing process to detect the origin of the errors and thus be able to correct them. In this regard, it would be convenient to develop methodologies or standards that provide guidelines for performing geometric inspections of this type.

4 Conclusion

By using reverse engineering and 3D inspection tools, geometric inspections of surfboards that provide valuable information to shapers can be performed.

Acknowledgments. The authors thank the Faculty of Engineering Gipuzkoa of the University of the Basque Country UPV/EHU for locating the DEHI laboratory (www.ehu.eus/dehi) in their facilities, and the Spanish Ministry of Economy and Competitiveness (MINECO) for financing this work (grant number PID2019-108975RA-I00).

References

1. Oggiano, L.: Numerical comparison between a modern surfboard and an alaia board using computational fluid dynamics (CFD). In: Vilas-Boas, J., Pezarat-Correia, P., Cabri, J. (eds.) The 5th International Congress on Sport Sciences Research and Technology Support (icSPORTS 2017), pp. 75–82. SCITEPRESS, Funchal (2017). https://doi.org/10.5220/0006488400750082
2. Oggiano, L., Pierella, F.: CFD for surfboards: comparison between three different designs in static and maneuvering conditions. Proceedings **2**(6), 309 (2018). https://doi.org/10.3390/proceedings2060309
3. Oggiano, L., in het Panhuis, M.: Modern surfboards and their structural characterization: towards an engineering approach. Proceedings **49**(1), 65 (2020). https://doi.org/10.3390/proceedings2020049065

4. Gudimetla, P., Kelson, N., El-Atm, B.: Analysis of the hydrodynamic performance of three- and four-fin surfboards using computational fluid dynamics. Aust. J. Mech. Eng. **7**(1), 61–67 (2015). https://doi.org/10.1080/14484846.2009.11464579
5. Falk, S., et al.: Computational hydrodynamics of a typical 3-fin surfboard setup. J. Fluids Struct. **90**, 297–314 (2019). https://doi.org/10.1016/j.jfluidstructs.2019.07.006
6. Falk, S., Kniesburges, S., Janka, R., O'Keefe, T., Grosso, R., Döllinger, M.: Numerical investigation of the hydrodynamics of changing fin positions within a 4-fin surfboard configuration. Appl. Sci. **10**(3), 816 (2020). https://doi.org/10.3390/app10030816
7. Carswell, D., Lavery, N., Brown, S.: Computational modelling of surfboard fins for enhanced performance. In: Moritz, E.F., Haake, S. (eds.) The Engineering of Sport 6: Volume 1: Developments for Sports, pp. 425–430. Springer New York, New York, NY (2006). https://doi.org/10.1007/978-0-387-46050-5_75
8. Zapelloni, G., García Rellán, A., Bello Bugallo, P.M.: Sustainable production of marine equipment in a circular economy: deepening in material and energy flows, best available techniques and toxicological impacts. Sci. Total Environ. **687**, 991–1010 (2019). https://doi.org/10.1016/j.scitotenv.2019.06.058
9. Bio of the athlete Gabriel Medina on the website of World Surf League: https://www.worldsurfleague.com/athletes/1085/gabriel-medina. Accessed 15 May 2021
10. Basque Country Surf Company: https://www.basquecountrysurf.com/. Accessed 15 May 2021

Custom Fit Overgrips for Bats Used in Different Basque Pelota Modalities

Xabier Amezua[1(✉)], Eneko Solaberrieta[1], Xabier Garikano[1], Mikel Iturrate[1], Jose-Antonio Oriozabala[1], and Iñaki Martin[2]

[1] Department of Graphic Design and Engineering Projects, Faculty of Engineering Gipuzkoa, University of the Basque Country UPV/EHU, 20018 San Sebastian, Spain
xabier.amezua@ehu.eus

[2] Department of Graphic Design and Engineering Projects, Faculty of Engineering Gipuzkoa, University of the Basque Country UPV/EHU, 20600 Eibar, Spain

Abstract. Adapting a sport practiced by people without disabilities to a group of people with certain disabilities may require the development of adapted equipment. Available technology makes it possible to manufacture simple adaptations that can open the door to people with disabilities in existing sports and competitions. In this sense, custom fit overgrips could improve the grip of racquets and bats to one-handed people, making it easier for them to practice some racquet and bat sports, including the Basque pelota modalities in which a bat is used. To illustrate how these custom fit overgrips would be manufactured, in this work, a prototype for a bat used in one of the Basque pelota modalities was manufactured.

Keywords: Adaptive sports · Custom fit · Reverse engineering · Computer-aided design · Additive manufacturing

1 Introduction

Participation in adaptive sports (also known as disability sports or parasports) is growing internationally [1]. Adaptive sports are those sports adapted to people with disabilities, including people who have long-term physical, mental, intellectual or sensory impairments which in interaction with various barriers may hinder full and effective participation in society on an equal basis with others [2]. Adaptive sports are beneficial for people with disabilities, especially because, according to several studies, physical activity has benefits on the health of most people with disabilities [1].

Most adaptive sports are adaptations of sports practiced by people without disabilities. The adaptation of a sport practiced by people without disabilities to a group of people with certain disabilities may require the development of adapted equipment [3–7]. Technological advances in recent years have led to progress in adapted equipment, especially those adapted for wheelchair users and amputees. For them, wheelchairs and prostheses that meet their individual needs and preferences as well as the functional requirements of the sport activity have been developed [5–7]. However, technological advances also make it possible to manufacture simple adaptations that can open the door to people with

F. Cavas Martínez et al. (Eds.): INGEGRAF 2021, LNME, pp. 114–119, 2022.
https://doi.org/10.1007/978-3-030-92426-3_14

disabilities in existing sports and competitions. In this sense, custom fit overgrips could improve the grip of racquets and bats to onehanded people, making it easier for them to practice some racquet and bat sports, including the Basque pelota modalities in which a bat is used [8–10].

Traditionally, to improve the grip of the bats used in the different Basque pelota modalities, overgrips have been elaborated by wrapping the handle of the bat with a tape with glue. In this work, it is proposed to replace for one-handed people these overgrips with custom fit overgrips that further improve the grip of the bat. Therefore, a prototype of a custom fit overgrip was manufactured to illustrate how they would be developed.

2 Material and Methods

The handle of a bat used in one of the Basque pelota modalities was scanned with a handheld 3D scanner based on structured white light scanning technology (Go!SCAN 20 scanner with VX element 6.3 SR1 software; Creaform, Inc) (Fig. 1). For that, targets used by the scanner for positioning were stuck on the surface of the handle (Fig. 1A). As a result, a 3D triangle mesh of the handle was obtained (Fig. 1B).

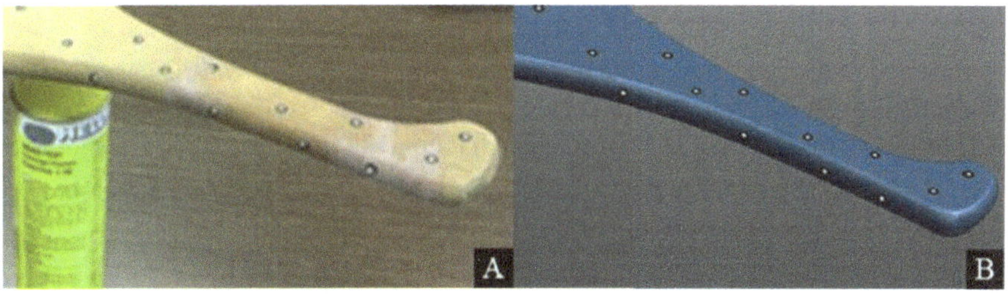

Fig. 1. Bat handle and its 3D triangle mesh. A, Bat handle. B, 3D triangle mesh of bat handle.

Then, the impression of a player's grip was obtained. For that, without removing the positioning targets, the handle of the bat was covered with a layer of vegetable-based modeling clay (Plastilina; Jovi, SA) and the player was asked to grip it.

The handle with the impression was then scanned with the 3D scanner (Fig. 2). To facilitate the positioning of the 3D scanner, different auxiliary objects with positioning

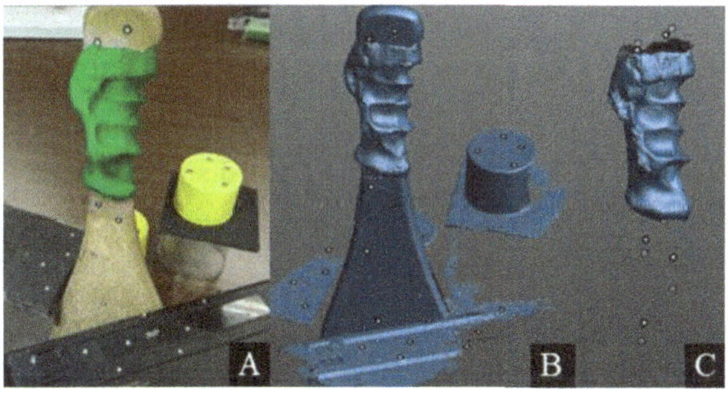

Fig. 2. Bat handle with grip impression and its 3D triangle mesh. A, Bat handle with grip impression. B, 3D triangle mesh obtained after scanning. C, 3D triangle mesh of grip impression.

targets were placed around the handle (Fig. 2A). After finishing the scan, from the obtained 3D triangle mesh (Fig. 2B), the regions that did not belong to the impression of the grip were eliminated, keeping the positioning targets (Fig. 2C).

Once both 3D triangle meshes were obtained, they were aligned. The alignment was performed by best-fit of common positioning targets (Fig. 3).

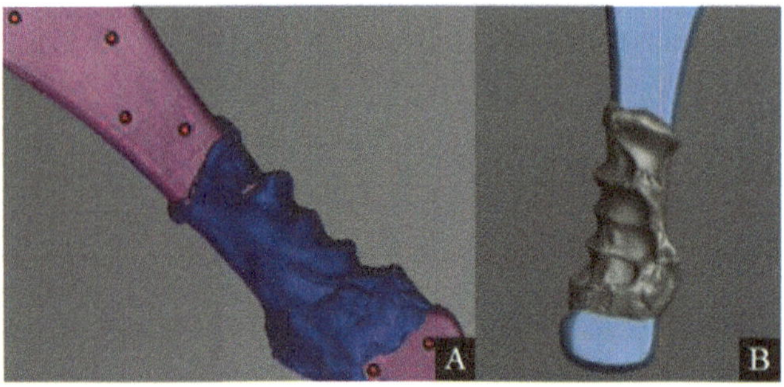

Fig. 3. Alignment of 3D triangle meshes. A, Alignment of 3D triangle meshes by best-fit of common positioning targets. B, 3D triangle meshes after completion of alignment.

After completion of the alignment, the 3D triangle meshes were cleaned and the holes in them were filled. Then, the extent of the overgrip was delimited by two planes and both 3D triangle meshes were cut through them (Fig. 4).

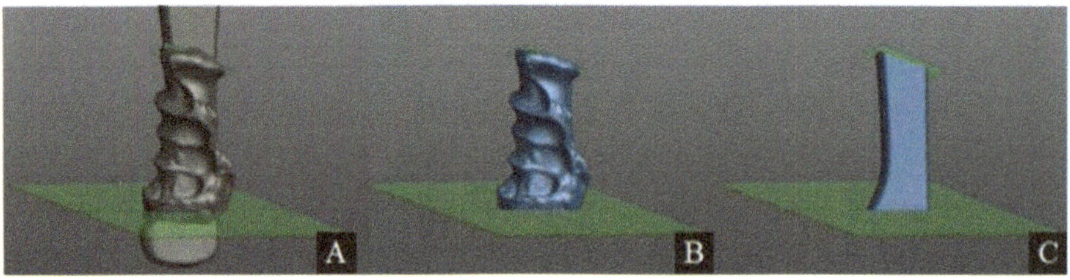

Fig. 4. Planes used to delimit overgrip extension and 3D triangle meshes cut through them. A, Planes used to delimit overgrip extension. B, 3D triangle mesh of bat handle cut through the planes. C, 3D triangle mesh of grip impression cut through the planes.

On each of the two resulting 3D triangular meshes, an organized patch layout surface was created (Fig. 5). For that, the automatic surface generation tool incorporated in the software of the 3D scanner was used. The created surfaces were transferred to an engineering CAD software program (Solid Edge ST10; Siemens).

In the CAD software program, the solid model of the overgrip was created from the two surfaces (Fig. 6). For that, the two surfaces were connected along the cutting sections with flat surfaces and then the obtained closed surface was transformed into a solid model.

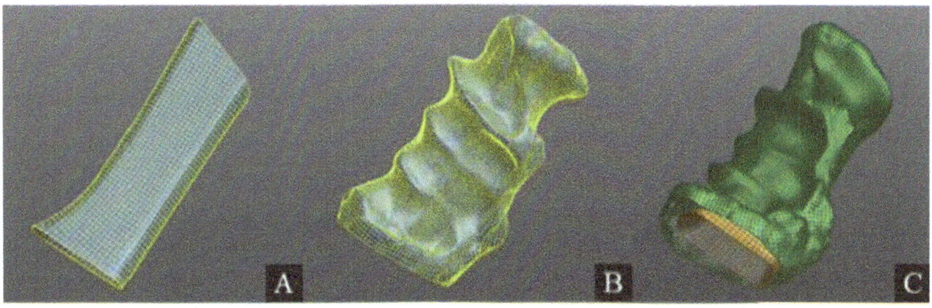

Fig. 5. Creation of organized patch layout surfaces. A, Creation of an organized patch layout surface on 3D triangle mesh of handle. B, Creation of an organized patch layout surface on 3D triangle mesh of grip impression. C, Organized patch layout surfaces created.

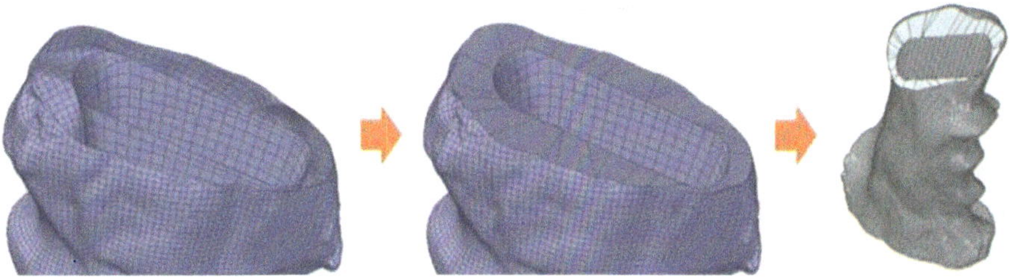

Fig. 6. Transformation of surfaces into a solid model.

Due to the geometry of the bat and its handle, the solid model of the overgrip was divided into two parts so that it could be mounted on the handle (Fig. 7A). Two lugs were also added to each part of the overgrip so that they could be attached by screws or nails to the handle (Fig. 7B). Each of the two parts of the overgrip were exported in the standard tessellation language (STL) file format.

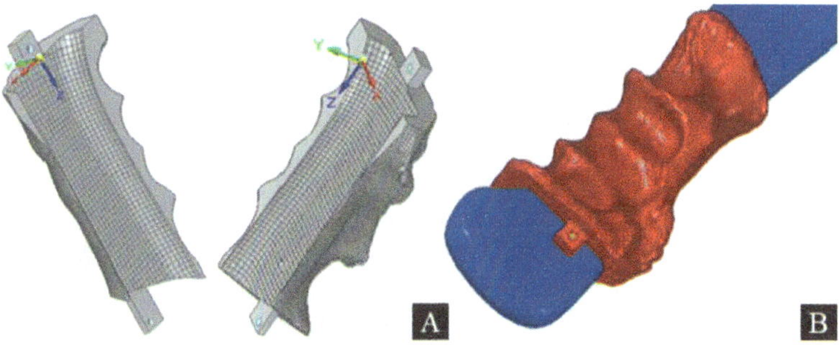

Fig. 7. Overgrip's two parts and its digital assembly. A, overgrip's two parts. B, virtual assembly of overgrip's two parts.

Finally, the STL files of the two parts of the overgrip were transferred to a 3D printer based on fused deposition modeling (FDM) technology (uPrint SE; Stratasys, Inc) to manufacture them in a white-colored, nontoxic thermoplastic material (AB-Splus-P430; Stratasys, Inc) (Fig. 8). It should be noted that, thanks to the wide variety of additive manufacturing machines available, it would be possible to manufacture this type of overgrip in a wide variety of materials [11].

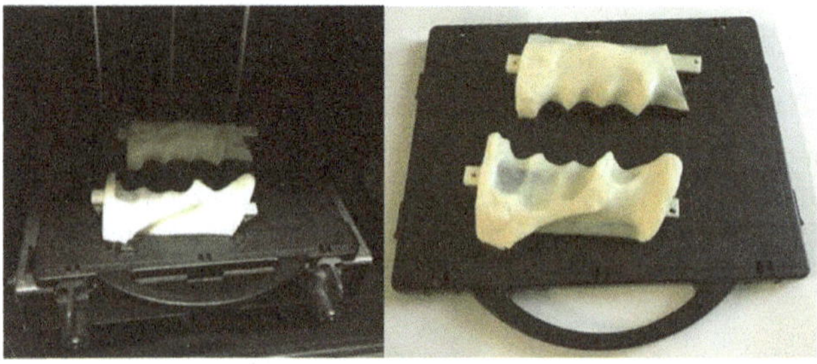

Fig. 8. Manufacturing of overgrip's two parts.

3 Results and Discussion

It was illustrated how it would be the manufacturing of custom fit overgrips for bats used in different Basque pelota modalities by manufacturing a prototype for a bat used in one of those modalities. It was found that the manufactured overgrip fitted the bat perfectly and improved the grip of the bat (Fig. 9).

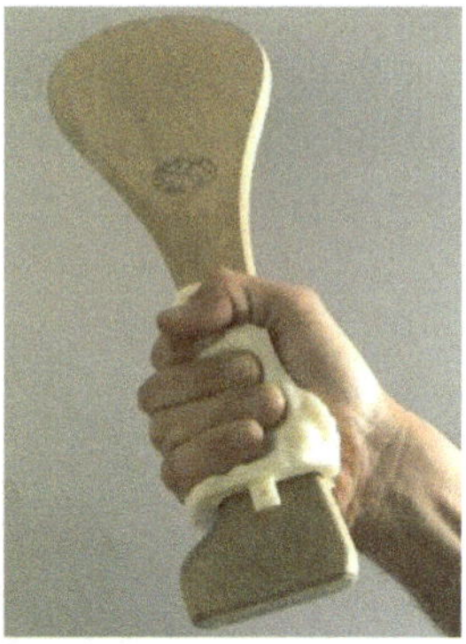

Fig. 9. Manufactured overgrip mounted on the bat.

Therefore, it is to be expected that this type of overgrips will improve the grip of the bats used in the different Basque pelota modalities to one-handed people, thus making it easier for them to practice these sports. However, more studies are needed to verify this.

It should be noted that the procedure followed in this work to manufacture custom fit overgrips for bats used in different Basque pelota modalities could be easily adapted to manufacture custom fit overgrips for other racquet and bat sports. It is also worth noting that these custom fit overgrips could also be useful for two-handed people in those racquet and bat sports where switching the racquet or bat from one hand to the other is not allowed.

4 Conclusion

Available technology makes it possible to manufacture custom fit overgrips that improve the grip of bats used in different Basque pelota modalities to one-handed people.

Acknowledgments. The authors thank the Faculty of Engineering Gipuzkoa of the University of the Basque Country UPV/EHU for locating the DEHI laboratory (www.ehu.eus/dehi) in their facilities, and the Spanish Ministry of Economy and Competitiveness (MINECO) for financing this work (grant number PID2019-108975RA-I00).

References

1. Martin Ginis, K.A., et al.: Participation of people living with disabilities in physical activity: a global perspective. The Lancet **398**(10298), 443–455 (2021). https://doi.org/10.1016/S0140-6736(21)01164-8
2. UN General Assembly, Convention on the Rights of Persons with Disabilities: Resolution adopted by the General Assembly on 13 December 2006, A/RES/61/106. https://undocs.org/en/A/RES/61/106. Accessed 15 May 2021
3. Matsuwaka, S.T., Latzka, E.W.: Summer adaptive sports technology, equipment, and injuries. Sports Med. Arthrosc. Rev. **27**(2), 48–55 (2019). https://doi.org/10.1097/JSA.0000000000000231
4. Oh, H., Johnson, W., Syrop, I.P.: Winter adaptive sports participation, injuries, and equipment. Sports Med. Arthrosc. Rev. **27**(2), 56–59 (2019). https://doi.org/10.1097/JSA.0000000000000236
5. Cooper, R.A., De Luigi, A.J.: Adaptive sports technology and biomechanics: wheelchairs. PM&R **6**, S31–S39 (2014). https://doi.org/10.1016/j.pmrj.2014.05.020
6. De Luigi, A.J., Cooper, R.A.: Adaptive sports technology and biomechanics: prosthetics. PM&R **6**, S40–S57 (2014). https://doi.org/10.1016/j.pmrj.2014.06.011
7. Duvall, J., Satpute, S., Cooper, R., Cooper, R.A.: A review of adaptive sport opportunities for power wheelchair users. Disabil. Rehabil. Assist. Technol. **16**(4), 407–413 (2020). https://doi.org/10.1080/17483107.2020.1767220
8. General Secretariat for Foreign Action of the Basque Government: Euskal Etxeak 74, Central Publications Service of the Basque Government, Vitoria-Gasteiz (2006)
9. Basque Pelota Federation EEPF. https://www.euskalpilota.org/pub/modalidades.asp?idioma=ca. Accessed 15 May 2021
10. International Federation of Basque Pelota GAISF. http://fipv.net/images/PRES_FIPV/FIPV-2020-interactivo.pdf. Accessed 15 May 2021
11. Tofail, S.A.M., Koumoulos, E.P., Bandyopadhyay, A., Bose, S., O'Donoghue, L., Charitidis, C.: Additive manufacturing: scientific and technological challenges, market uptake and opportunities. Mater. Today J. **21**(1), 22–37 (2018). https://doi.org/10.1016/j.mattod.2017.07.001

Adjustment of Friction Models in Elastomers Using the Finite Element Method and Model Updating Techniques

Saúl Íñiguez Macedo[1(✉)], Asier Rodríguez San Miguel[1], Enrique Fernández Martínez[1], Manuel Rubio Sampedro[1], Álvaro Pérez-Sala[2], Ignacio M. Larráyoz[2], Rafaél Peláez Cristóbal[2], Fátima Somovilla Gomez[3], María Ángeles Martínez Calvo[3], Marina Corral Bobadilla[3], and Rubén Lostado Lorza[3(✉)]

[1] Centro Tecnológico del Calzado de La Rioja (CTCR), Calle Raposal, 65, 26580 Arnedo, La Rioja, Spain
siniguez@ctcr.es

[2] Grupo de Biomarcadores y Señalización Molecular, Centro de Investigación Biomédica de La Rioja, Logroño, La Rioja, Spain

[3] Departamento de Ingeniería mecánica, Universidad de la Rioja, C/Luis de Ulloa, 4, 26004 Logroño, La Rioja, Spain
ruben.lostado@unirioja.es

Abstract. Defining the best friction model in elastomers for studies with the Finite Element Method (FEM) is a complex task. Hyper-elastic materials have a non-linear mechanical behavior that favors the storage and transfer of deformation energy in an unpredictable way when they are subjected to an external action. This abrupt change determines unpredictable friction coefficients and even with negative values. Furthermore, the coefficient of friction depends on the surface and the contact pressure. This article shows the process of adjusting a friction model based on the MEF according to the standardized test UNE-EN ISO 13287 for the characterization of friction in shoe soles. Using "model updating" techniques, a search was made for the most appropriate friction model for a specimen made of nitrile butadiene rubber (NBR). Five friction models based on Coulomb and Shear together with hexahedral and tetrahedral elements with different mesh sizes, contact pressure and sliding speed were studied according to a Design of Experiments (DoE) and validated experimentally. The best friction model obtained in this case was the Coulomb model with a hexahedral element size of 0.5 mm.

Keywords: Friction models · Elastomers · Hyper-elastic · FEM · Footwear

1 Introduction

When a new security shoe model is designed, its sole must accomplish UNE-EN ISO 13287 requirements [1]. These requirements demand that the footwear meet a certain friction value when testing on different surfaces such as steel and tile. The basis of the

F. Cavas Martínez et al. (Eds.): INGEGRAF 2021, LNME, pp. 120–127, 2022.
https://doi.org/10.1007/978-3-030-92426-3_15

test is to speed up the contact surface to 300 mm/min while a normal force is applied. That normal force varies depending on the sole size. Friction is then determined when the velocity has reached the target, being thus, a dynamic coefficient of friction.

It is very difficult to estimate the friction that a sole design will present after performing real tests. That is the reason of trying the adjust of a finite element simulation to solve this problem. If a rubber test specimen is modelled and its friction behaviour determined as a function of the normal force, the global model of a complete sole could be properly defined. From the previous hypothesis, a nitrile butadiene rubber (NBR) specimen was studied, trying to adjust the best friction model by means of a finite element simulation and model updating techniques. This adjusting methodology is capable of determine the best fitting between the real test and the simulation model.

The first step when simulating a rubber, like every other material, is to characterize its mechanical behaviour. That necessary step was accomplished in previous work [2], developing a novel technique of adjusting by means of simulation and machine learning methods. Once the material was properly characterized, and the best energy function model selected, the next step was the design of a new test based on UNE-EN ISO 1328 methodology. Specimens of NBR rubber with dimensions of 7 × 3 cm were tested. Having 21 cm^2 of effective flat surface, the normal force was varied within the range of 300 to 700 N.

Several models of friction were simulated according to a Design of Experiments (DoE) [3], changing the type and size of elements, searching the best fit between the models' output and real tests.

2 Materials and Methods

The present study is based in the selection of the best slip model to reproduce the behaviour of the real slip test. To achieve this objective, a battery of models was configured by modifying four variables, such as the type of element being hexahedral (HEX) or tetrahedral (TET), the size of the element, the friction model, and the normal force. A DoE was performed, to explore the whole range of possibilities.

Table 1. shows a small summary of the first five combinations from a total of 40 where force was modified between the range of 300 and 700 N.

Table 1. First five rows from the performed DoE

Combina*tion*	Element type	Element size	Slip model	Normal force
1	HEX	3,0	Coulomb_arctan	300–700 N
2	HEX	3,0	Coulomb_bilinear	300–700 N
3	HEX	3,0	Coulomb_stickslip	300–700 N
4	HEX	3,0	Shear_arctan	300–700 N
5	HEX	3,0	Shear_bilinear	300–700 N
...	...	...	...	...

In order to replicate the slip test, the specimen was modeled as a 70 × 30 × 4 mm 3D solid. This solid was meshed using two types of element, hexahedral and tetragonal in different sizes according to the defined DoE. Below the test specimen a surface was placed imposing the condition of non-deformability. That surface was attached to a velocity ramp, just like the one programmed in the test machine. A comparison between the real test and FEM model can be seen in Fig. 2.

Boundary conditions (Fig. 1) were shared between every model, no matter what kind of element had been used to the meshing step. Like the real test, all the nodes belonging to the solid upper surface were imposed a fixed displacement of 0, preventing thus any movement and simulating by this boundary condition the glued contact to the steel holder.

To improve the model and guarantee the correct equilibrium, every node in the symmetry plane along the displacement direction were prevented to move in the perpendicular direction. Finally, the normal force was applied as a nodal force in the upper surface. Like the velocity of the floor surface, this load was not applied suddenly, but gradually, according to another ramp definition.

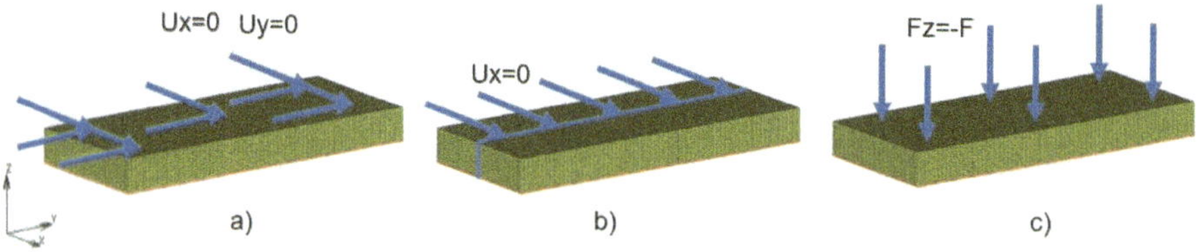

Fig. 1. a) Fixed displacement, b) equilibrium condition, c) normal force

Friction is very complex since it involves not only the characteristics of the surface but also other variables like normal stress, temperature and relative velocity. An example of this complexity is that quite often neutral lines develop [4]. This means that along the contact surface, the material flows in one direction in one side of the neutral line and in the opposite direction in the other side. This phenomenon make sometimes the appearance of negative coefficients of friction (COF) if the rubber releases its stored elastic energy.

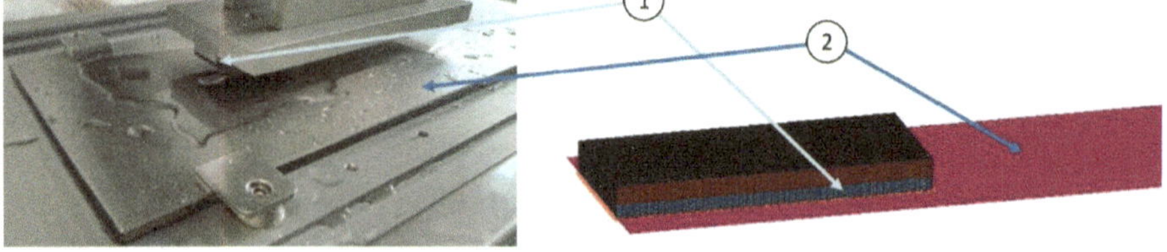

Fig. 2. 1) Test specimen, 2) Contact surface

2.1 Coulomb Model

MSC Marc software was used to develop the FEM model. In this program, the general friction model usually used for most applications is the Coulomb model [5]. The Coulomb model can be formulated according to (1).

$$\|\sigma\| < \mu\sigma_n(stick)\ and\ \sigma_t = -\mu\sigma_n \cdot \boldsymbol{t}(slip)\quad \boldsymbol{t} = \frac{v_r}{\|v_r\|} \tag{1}$$

Where σ_n is the normal stress, σ_t is the tangential stress (by friction), t is the tangential vector in the direction of the relative velocity, in which v_r is the relative sliding velocity and μ is the friction coefficient.

The friction stress or force has a step function behaviour, starting in the stick region and changing to the slip region. This transition is based on the relative sliding velocity v_r or the tangential relative incremental displacement Δ_{ut}. Since this discontinuity in the friction value may easily cause numerical difficulties, different approximations of the step function have been implemented by MSC Marc (Fig. 3).

Arctangent Model: This model is based on a continuously differentiable function that is formulated in terms of the relative sliding velocity (2). Physically, the value of C (Relative Velocity Constant) can be seen as the transition value of relative velocity below which sticking is present [6].

$$\sigma_t = -\mu\sigma_n\frac{2}{\pi}arctan\left(\frac{\|v_r\|}{c}\right)\cdot \boldsymbol{t} \tag{2}$$

Stick-slip Model: This model is based on a step function that has been modified, and it can be used to simulate a true stick-slip behavior [7]. Each node in the contact area gets a friction status, being possible stick or slip. Depending on the status, different constraints are applied and after each iteration, the correctness of the friction status is checked and changed if necessary. The parameters that Marc uses are: the friction force tolerance e, the slip to stick transition region β and the friction coefficient multiplier α.

Bilinear Model: This model is similar to the modified step function model, and it is based on relative tangential displacements [8]. It assumes that the friction conditions correspond to permanent (plastic) relative displacements and reversible (elastic) respectively, instead of defining special constraints to enforce sticking conditions.

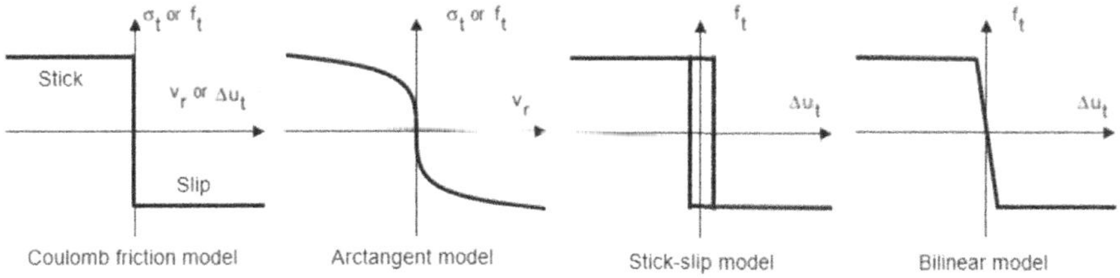

Fig. 3. Coulomb model variations behaviour.

When the normal force or stress is huge, the Coulomb friction model is not a good choice because its results could not fit well with experimental data. This is due to the Coulomb model predicts that the frictional shear stresses always increase. This situation can even cause the stresses to reach such a high level that they can even exceed the failure of the material. Since this is not possible, a different friction mode is applied. There exist three possible options, define a nonlinear coefficient of friction, introduce the friction stress limit in the bilinear model or use the shear-based friction model.

2.2 Shear Model

The shear-based model defines the frictional stress as a fraction of the equivalent stress $\overline{\sigma}$ in the material (3).

$$|\sigma_t| < m\frac{\sigma}{\sqrt{3}}(stick)\ and\ \sigma_t = -m\frac{\sigma}{\sqrt{3}} \cdot t(slip) \tag{3}$$

where m is the friction factor.

As in the Coulomb model, in this one there are also approximations of the theoretical step function (4), (5) [9]. One is the arctangent function that smooths the step function making a better transition.

$$\sigma_t = -m\frac{\overline{\sigma}}{\sqrt{3}}\frac{2}{\pi}arctan\left(\frac{v_r}{c}\right) \cdot t \tag{4}$$

The other model considered is the bilinear model. This model is based on the assumption that the shear stress in a node (generated by friction) is proportional to the applied shear force. The shear stress due to friction is limited by the minimum of two terms (5) (Similar to the friction stress limit).

$$\sigma_t = min\left(m\sigma_n, m\frac{\overline{\sigma}}{\sqrt{3}}\right) \tag{5}$$

3 Results and Discussion

After the simulation of every DoE case, results were post-processed and compared to the real data. The NBR rubber tests gave the results in Fig. 4 for different load values.

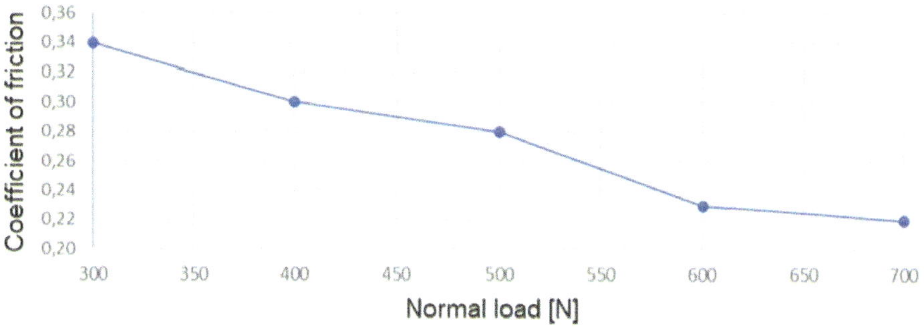

Fig. 4. NBR coefficient of friction (COF) for different load values

As an example of the process, from this point on, every data will be related to the maximum slip value (0.34) regarding 300 N of normal force. Having fixed the load, it is possible now to compare how a different model behave depending on the type of element and the size of the used mesh. Figure 5 shows the difference in behavior of the five studied models, besides each size of the image represents one element type. It is clear that the use of TET elements creates a strange peak just in the transition area when the model changes from the stick to the slip formulation. The appearance of these peaks favored the decision of using HEX elements for this simulation, discarding the TET elements.

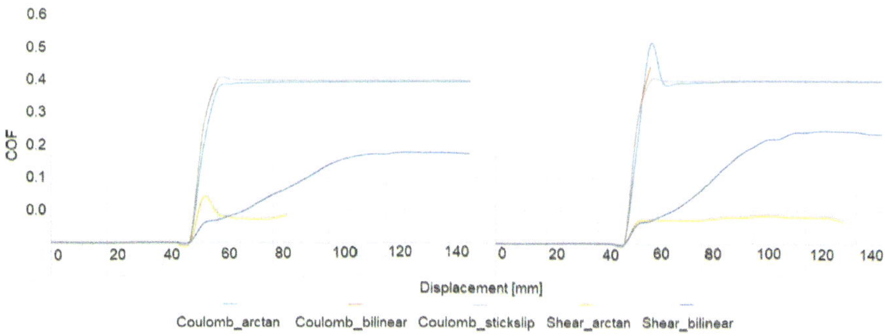

Fig. 5. Friction models for 1 mm HEX elements (Left) and for TET elements (Right)

Regarding the mesh sensibility, when using HEX elements, the evolution in each model behaviour can be seen in Fig. 6, while mesh size influence in stress fields is depicted in Fig. 7 for the Coulomb Arctangent model.

Coulomb was the best model in all cases, specifically in its arctangent and stick slip variants. In the case of the bilinear variant, good results were not achieved, since the model did not converge. The shear arctangent and bilinear models did not offer good results. The simulations did not reach the desired friction values and oscillations appeared after the transition region, something that did not fit with a real test.

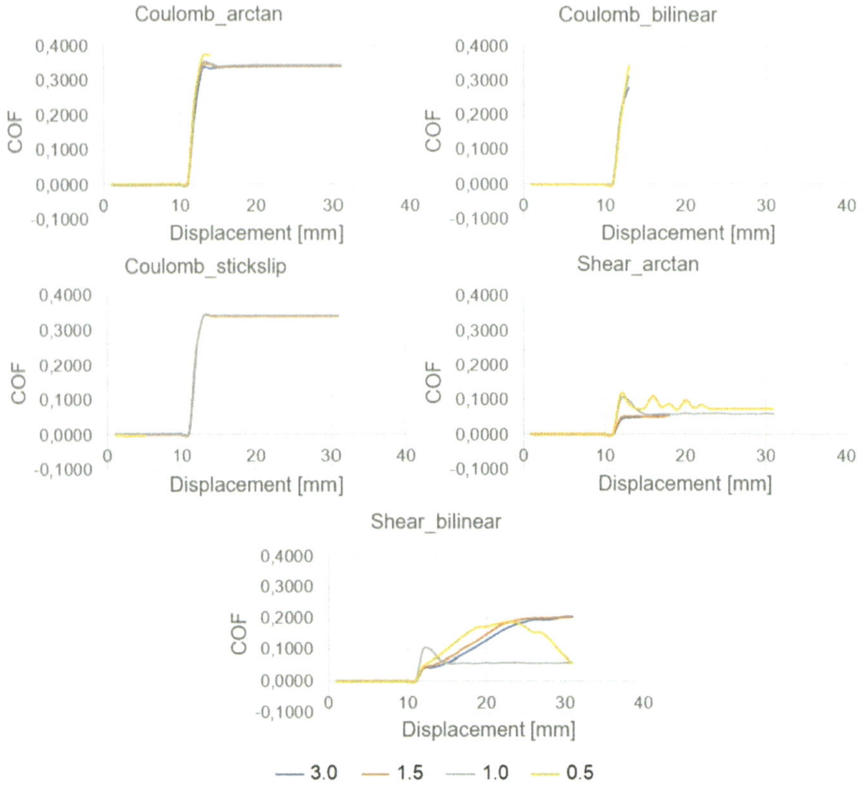

Fig. 6. Results for different friction models for mesh sizes of 3, 1.5, 1 and 0.5 mm

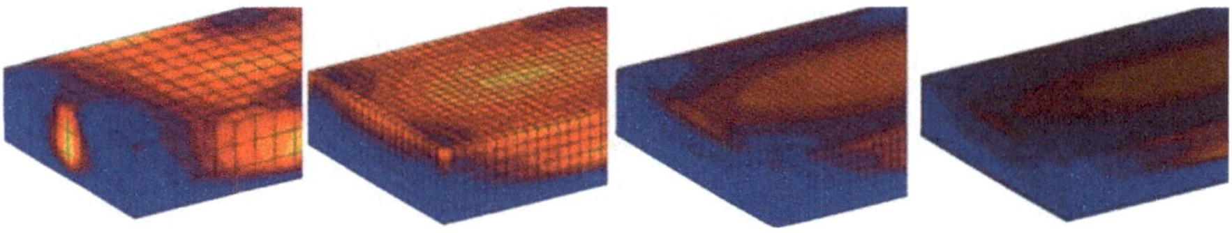

Fig. 7. Stress field differences between mesh sizes. Left to Right, 3, 1.5, 1, and 0.5 mm

These results made determine that the best model for the simulation of this type of test in rubbers is Coulomb in the two mentioned variants. The type of element must be hexaedric and the mesh size 0.5 mm. If the element size gets bigger, results start to be very different from the reality. In the other hand, smaller values do not add extra precision, so no extra computational resources should be used to have a good performance.

4 Conclusions

As has been seen, the simulation of a slip problem with rubber elements implies taking certain considerations. In addition to a previous study that properly characterizes its mechanical properties, a subsequent study is needed to adjust the model.

In this case, an experiment design was the mean to vary the parameters of the same model, thus finding the best possible combination. The best friction model and element size was determined to maximize precision and minimize computational cost.

Several improvements can be made to enhance the achieved results. The most important could be the experimentation with larger test specimens in order to detect possible changes in the stress distributions.

Future work may continue to develop complex elements that use the models trained here, thus achieving a better global fit.

This brief study is only a basis to continue developing more complex models of complete sole. Models that will be capable of determine the exact coefficient of friction based on the friction each point represent in the whole part depending on the applied pressure.

References

1. UNE-EN ISO 13287: Personal protective equipment – Footwear – Test method for slip resistance. Madrid: AENOR Marth (2013)
2. Íñiguez-Macedo, S., Lostado-Lorza, R., Escribano-García, R., Martínez-Calvo, M.Á.: Finite element model updating combined with multi-response optimization for hyper-elastic materials characterization. Materials **12**, 1019 (2019). https://doi.org/10.3390/ma12071019
3. Montgomery, D.C.: Design and Analysis of Experiments. John Wiley & Sons Inc., Hoboken, NY, USA (2008)
4. Marc, M.S.C.: Computer Software. MSC Software Corporation, Newport Beach, USA (2020)
5. Cull, S.J., Tucker, R.W.: On the modelling of Coulomb friction. J. Phys. A Math. Gen. **32**(11), 2103 (1999)
6. Bray, J.W., Goodman, R.E.: The theory of base friction models. Int. J. Rock Mech. Min. Sci. Geochem. Abstr. **18**(6), 453–468 (1981)
7. Marton, L., Lantos, B.: Modeling, identification, and compensation of stick-slip friction. IEEE Trans. Industr. Electron. **54**(1), 511–521 (2007)
8. Zhang, Q.Q., Li, L.P., Chen, Y.J.: Analysis of compression pile response using a softening model, a hyperbolic model of skin friction, and a bilinear model of end resistance. J. Eng. Mech. **140**(1), 102–111 (2014)
9. Santos, P.M., Júlio, E.N.: A state-of-the-art review on shear-friction. Eng. Struct. **45**, 435–448 (2012)

Design of an Ergonomic and Inclusive Hairdresser's Washroom

María Alonso-García[1,2](✉) and Almudena Palacios-Ibañez[3]

[1] Department of Mechanical Engineering and Industrial Design, University of Cádiz, Cádiz, Spain
maria.alonso@uca.es
[2] Graphic Engineering Research Group (TEP189), University of Malaga, Málaga, Spain
[3] I3B, Universitat Politècnica de València, Valencia, Spain

Abstract. Nowadays, beauty and aesthetics industry is an essential activity for all types of users, from children and adults to people with functional diversity. However, due to the lack of wheelchair-friendly equipment, this kind of users often rely on in-home services, facing an additional barrier for wheelchair users. Today's hairdressing washbasins can cause injuries, affecting customer and worker health. These include neck pain, headaches, dizziness or the "Beauty Parlor Stroke Syndrome", a rare disease that causes strokes due to the impingement of a vertebral artery during rotation or hyperextension of the neck over the sink basin. We suggest that designing a more ergonomic product and by changing the posture adopted by the customer and worker, health problems can be minimized or eliminated. This article presents the design of an inclusive washroom and the study of its perception based on "User-Centered Design" methodology (where the user is the main character of the design process). We concluded that the resulting product reduces and eliminates client and worker injuries by adopting a non-inclined head posture. Also, it can be used by any kind of user, respecting the posture of the worker at any time.

Keywords: Inclusive design · Industrial design · Product design · Product evaluation · Emotional design · Kansei engineering · Innovation

1 Introduction

Industrial design and development of new products can satisfy generally users and customers' needs and desires, including ergonomic improvements, increasing the usability of each product [1–3] and ensuring a good user experience [4]. This tool is therefore useful for companies to develop innovative products in any area. In particular, the beauty industry is one of the fastest growing and most profitable business sectors in the world [5], being essential for any user of any age range (including diverse minorities). Despite this and the increase of products and services adapted to disabled people, children and the elderly, this has not always been the case. There is still a certain lack in the market, as in the case of hairdressing services adapted to wheelchair users.

F. Cavas Martínez et al. (Eds.): INGEGRAF 2021, LNME, pp. 128–136, 2022.
https://doi.org/10.1007/978-3-030-92426-3_16

Due to the lack of appropriate equipment and environments for these people, these users often resort to in-home services, becoming an additional barrier for them. Methodologies such as "Design for All" try to eliminate these obstacles under the premise of designing products that can be used by any user regardless of their physical and/or intellectual condition. This should not be just a methodology, it should be included as a requirement when implementing solutions to allow the use of the product by minority users and fully meet their needs and desires. In this context, there are more and more working tools which generates valuable information to help achieve market success for each product, regardless of the characteristics of the target user. In particular, Emotional Design [6] and User-Centered Design (or UCD) understand the user and his/her opinion as a key factor in the design and development of new products, with the user being able to define his/her own needs [7].

UCD is a methodology for new product development that puts the user at the center of the design process [8, 9]. This methodology is able to identify user desires and needs, improving product specifications [10] by involving the user throughout the whole process (from the early stages to the validation of the final prototypes) [11]. On the other hand, Emotional Design seeks to establish strong and lasting relationships between user and product [9, 12], ensuring the success of a product in the market. To do so, it focuses on the analysis of the user's emotional response at three different levels [6, 13]: visceral (VL), behavioral (BL) and reflexive (RL). The emotions perceived by the user are taken into account in the context of the user's interaction with the object itself. The first level (VL) corresponds to the first impression, it creates a first expectation about what will come next [14]. The subject level (BL), on the other hand, corresponds to the experience (either positive or negative) that the user has during the use of a product [15]. The las level (RL) relates to the meaning of the product and affects the message and image that the product can communicate to the user. Both methodologies allow design teams to optimize initial proposals through the application of UCD tools, direct user collaboration, and the analysis of user's perception of each product, through the different levels of response established by Norman [16]. The information provided by these analyses is of great value to industrial designers and companies and allows the design process to be optimized.

Current hairdresser washbasins are not only non-inclusive but can cause injuries that affect people health (disabled and non-disabled). These include neck pain, headaches, dizziness or strokes, and the "Beauty Parlor Stroke Syndrome", a rare disease that causes strokes due to the impingement of a vertebral artery during rotation or hyperextension of the neck over the sink basin [17].

This article shows the design process and the testing with users (clients and professionals) of a new concept of an inclusive hairdresser washbasing. The product has been designed based on various UCD tools and its evaluation and testing has been carried out based on the application of Emotional Design and through the behavioral level, related to the usability of the products.

2 Materials and Methods

The element studied is the area where hair washing takes place and therefore the most important. It has been designed under strict anthropometric and ergonomic standards to ensure that the client user's upper body rests comfortably. The perception of it has been studied on the basis of professional users and customers.

The design process used consists of a first phase of research (1), a second phase of conceptual design of solutions (2), and a third phase of detailed design (3). The first one includes the analysis of the evolution of head washing and the necessary elements for this process; the current washing process and the different types of washbasins available; and an intensive ergonomic study. This last aspect, of particular relevance, is developed on the basis of different areas of the human body: cervical, thoracic and lumbar; upper limbs and lower limbs. During the second phase, different alternatives were developed. These evolve from UCD techniques (interviews and trials with real users) from testing full-scale mock-ups in the early phase. Results of these will lead to their evolution and optimization during the third phase, which will focus on preparing the product for manufacturing. For the evaluation of this product, a 1:1 scale prototype has been used. This prototype was made using rapid prototyping techniques (3D print) from a CAD model designed with SolidWorks. The final results were achieved using up to 80 pieces of PLA bonded with epoxy adhesive. To make it as realistic as possible the structure is sanded and coated with a layer of plastic primer. It is then spray painted white. Also, two wedges made of wood with several incisions were needed, allowing the inclination of the element to be tested in different positions (Fig. 1). The inclination achieved guarantee the comfort from the 5th percentile for women to the 95th percentile for men (178º–262.5º).

Fig. 1. Prototype on adjustable wedge

2.1 Evaluation

The assessment of this prototype was based on the second level established by Norman (BL) and through observation. Here, product usability is analyzed for each user. To this end, 5 parameters of usability were established: ease of use, efficiency (the ability to accomplish something with the least amount of wasted time), effectiveness (the degree to which something is successful in producing a desired result), memorization and satisfaction, which were evaluated by customer and professional users on a 1–5 scale, based on basic semantic differentials. On these parameters, different product-specific issues will also be assessed. In order to detect these issues, a first controlled test is carried out with users with experience in industrial design and knowledge of the product. During this test, the level of comfort of the neck and the posture it generates in the client user are established as an analysis interest (1); the effectiveness of the mobile zone as well as its comfort for both the professional user (2) and the customer user (3); and the ease of cleaning the various components of the washbasin (4). Ergonomic comfort, posture and position of the professional user in general are also evaluated (5) as well as the adaptation of 4 collar designs, corresponding to the original design and three different designed alternatives (Table 1).

Table 1. Different collar options

	Original	*1st Alternative*	*2nd Alternative*	*3rd Alternative*
Rear edge	4 mm	10 mm	10 mm	7 mm
Front edge	4 mm	20 mm	10 mm	20 mm
Inner radius	4 mm	5 mm	10 mm	4 mm
Outer radius	4 mm	5 mm	10 mm	4 mm
Side view	Straight	Straight	Curved	Straight
Top view	Circular	Circular	Circular	Elliptical

3 Results

Despite the variety of services, there is currently a great lack of adaptation for people with functional diversity. Generally, the adaptation is based on the removal of the seat to avoid the client user having to change chairs. In these cases, the comfort depends on the user's chair and the adaptation of the work surface falls entirely on the washbasin, which is a key element in guaranteeing the safety and comfort of users with and without disabilities. It is important to point out that the position adopted by the user on this piece carries a risk regardless of the physical condition of each user, especially in people of a certain age.

For example, the position often used in beauty salons to wash clients' hair, reclining in an armchair with the neck resting on the basin, can cause severe pressure on the veins that carry blood to the brain, decreasing the flow. This can lead to strokes in some people. The well-known salon death syndrome directly affects customers of hairdressing salons. The definition, symptoms and consequences of this syndrome were published in 1993 by the Journal of the American Medical Association. According to physician Michael Weintraub [16] it is neither uncommon nor harmless. Table 2 summarizes the injuries and postures to be avoided for the user.

Table 2. Positions and injuries to avoid according to area (top) and limbs (bottom)

Cervical area	*Thoracic area*	*Lumbar area*
The positions that produce the greatest load are bending, flexion-extension and side-to-side movements.	*Excessive stretching or slight tears in the soft tissues can cause weakness, stiffness and/or deformity, and the pain may radiate to other areas of the body.*	*It is the usual focus of pain (lumbar injuries) due to overloading of the intervertebral discs.*

Upper limbs			***Lower limbs***
Shoulders and arms	***Elbow and forearms***	***Hands and wrists***	
Minimize abduction and flexion movements greater than 60° and elevations greater than 30°.	*Eliminate repetitive movements and sustained postures involving bending the elbows and lifting above the shoulders.*	*Avoid wrist work with prolonged postures*	*Muscle weakness, neurological deficits, joint instability or stiffness and bone changes, knee bursitis, etc. are common.*

Referring to the working user (hairdresser), hairdressing activities are characterized by the use of upper limbs and a standing working posture, which leads to static loads. They should be avoided because of slower activity, fatigue, awkward and unsafe movements, increased heart rate, pain and injuries. Taking these aspects into account, different alternatives were developed by exploring the adoption of new positions by the customer user. Figure 2a shows some sketches of the development of different alternatives. The first option is based on a seated washing posture (customer), with a fully vertical neck. The washbasin fits customer head, leaving his/her face on the outside and the hair on the inside. In the second proposal, the user puts his/her head in the pile in a similar way to the current concept. However, the towel (dropped on his neck) is held in place by two upper guides that slide along the pile. This can be tilted 90° and adjusted in height, which makes it easier to use for wheelchair users. In the last option, the user lies down and rests his/her face over a hole in the washbasin. The hair would remain inside it and the working user could start washing it. After analyzing these positions, it was concluded that proposal number one was the most suitable for product redesign. A fully upright posture means no risk of injury to the client, making it the safest option. Moreover, it is a more natural posture for all kinds of groups (elderly people, users with some kind of functional diversity, etc.). On the other hand, the hair remains entirely in the inner area, improving the experience for both users (customer and worker). For this reason, the final proposal design is characterized by the upright posture adopted by the client user, differing from the current posture (leaning backwards), eliminating any risk of injury and facilitating its use by wheelchair users. The product consists of two parts: a fixed zone and a movable zone (Fig 2b).

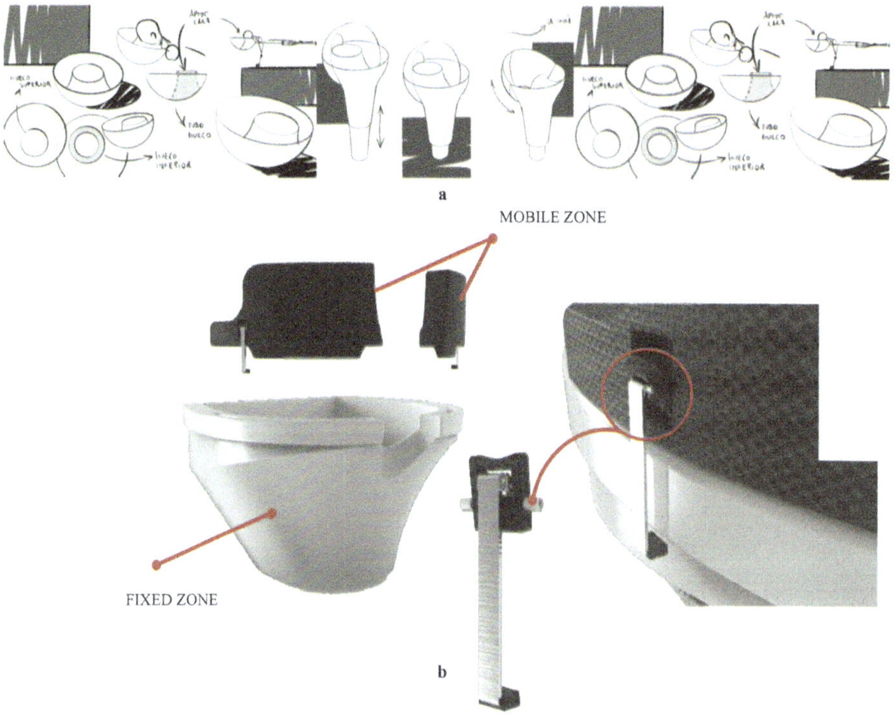

Fig. 2. Pile evolution and description.

The first is the most solid (and main) area, where the faucet is located. It contains two simple "U-shaped" guides on both sides, where the mobile zone slides improving user experience by preventing water filtration and splashing thanks to a foam material that fits the shape of user head. To use the product, the user inserts his/her neck into the washbasin "mouth", and the working user places the hair inside the basin. Once in place, the movable zone is adjusted to the user head. The side mechanism is then adjusted to hold the movable zone in place. Once the client's hair is washed, the each movable zone can be unlocked, allowing the user to leave.

3.1 Evaluation

A total of 14 users (12 women and 2 men) took part in the evaluation of the prototype. All working users were women aged between 27 and 48 years old, belonging to the 50th and 95th percentiles of body measures. Client users were aged between 17 and 66 years old and belonged to the 50th and 95th percentiles.

Table 3 shows the issues related to usability parameters analyzed. Ease of use was measured by making questions about product simplicity or intuitiveness. Efficiency and effectiveness were measured through observation of both type of users. Satisfaction was quantified making aesthetics, usability and cleanliness questions. For memorization, again based on observation, we checked whether after a certain period they remembered how to use the product.

Table 3. Product usability. Professional and customer.

	Original	Before or after explanation	Professionals score obtained (1–5)	Users
Ease of learning	Complex/Intuitive	Before	4	3.8
	Hard/Easy		2.8	4
Efficiency	More time spent in use /Less	After	4	1
	More time spent in use /Less	After	4.2	5
Effectiveness	More errors made/ Fewer	After	4	4.7
Satisfaction	Unattractive/ Very attractive	Before	2	3.89
	Difficult cleaning/Easy cleaning	After	3.6	–
	Useless/Practical	After	2.8	–
	Uncomfortable/Comfortable	After	3.8	3.33
Memorization	Difficult to remember to use/ Easy to remember	After	4.8	4.5

Generally, users found the product intuitive and easy to use. However, some users (especially professionals), could not identify how to use the movable area, placing the components in the opposite direction and causing some workers to find it difficult to use. Despite this, after a brief explanation, the efficacy in both groups was higher. In addition, we emphasize that customers who did not find the product intuitive tried to turn their heads backwards, which also affects the perceived comfort of the different customer users. For this reason, it took longer for customers to position correctly the washbasin when they first met it. After explaining how the product worked, a faster positioning was achieved. This explains the difference between the two parameters measured in the effectiveness of the product, at the user level. On the other hand, the time spent by professionals using the product did not vary significantly after the explanation. In terms of overall satisfaction, there was some discrepancy among professional about the depth of the pile, probably due to differences of working methods or techniques employed. Aesthetics were much higher for customers than for workers. Professionals reported easy cleaning, except for the cavity over which the mobile area slides, which should be improved (Fig. 3).

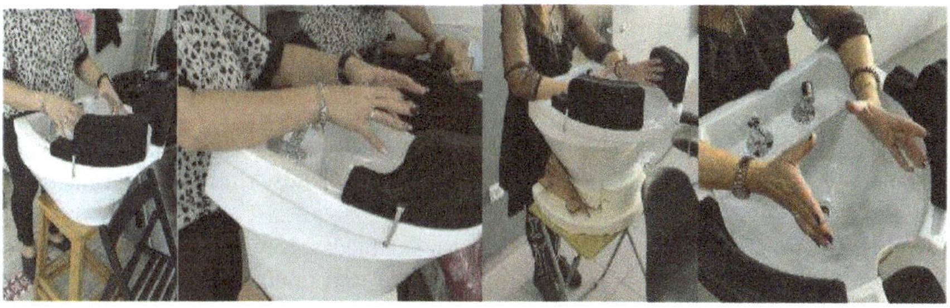

Fig. 3. Professional users subjected to the observation process

4 Conclusions

This study shows the design process and evaluation of a product for the beauty sector, specifically a hairdressing washbasin. This product has been designed under different UCD tools, considering users of different nature and condition, and looking for an inclusive element that can be used by different minorities. The main feature of the product is allowing its use by any type of user (including those who requires the use of a wheelchair for everyday life), respecting worker's posture and reducing injuries for clients and professional users.

Design simplicity makes the product easy to use and clean, allowing a quick assembly and disassembly. As a standard design, it can be placed on top of any hairdresser washbasin's structure maintaining a certain resemblance to a traditional washbasin to ensure a correct perception for users. Once the washbasin is in place, the moving part is easily inserted into the guides. Thanks to the rounded design and ease of disassembly, dirt cannot accumulate and it can be quickly cleaned. The outside of the movable area is accessible to the fingers for ease of use.

Also, it could be noted that the idea generated by the washbasin after the explanation was positive, as the users preferred the new posture adopted. However, the habits adopted by the customer about the conceptual washbasin caused a bad impression at first sight, and an evident lack of intuitiveness, which needs to be improved. However, the assessment of the final product through prototyping and with the participation of real users (customers and professionals) highlighted the need for final product testing to ensure its success. This showed that the analysis method used is a very useful tool for industrial designers, as it can predict usability errors through the involvement of real users under UCD tools.

References

1. Roy, R., Goatman, M., Khangura, K.: User centric design and Kansei engineering. CRIP J. Manuf. Sci. Technol. **1**(3), 172–178 (2009)
2. Aftab, M., Rusli, H.A.: Designing visceral behavioural and reflective products. Chin. J. Mech. Eng. **30**(5), 1058–1068
3. Rusli, H.A., Aftab, M.: Unbroken: rediscovering long-term value of products through change in perception. In: Proceedings of the International Workshop on Digital Design and Manufacturing Technologies—Embracing Industry 4.0 and Beyond. Conference 2016, New-castle Upon Tyne, North East England, UK (2016)

4. Forlizzi, J., Ford, S.: The building blocks of experience: an early framework for interaction designers. In Proceedings of the 3rd Conference on Designing Interactive Systems: Processes, Practices, Methods, and Techniques. Conference 2000, New York, NY, USA, pp. 419–423 (2000)
5. Draper, S.W., Norman, D.A.: User Centered System Design: New Perspectives on Human-Computer Interaction. L. Erlbaum Associates Inc., 365 Broadway Hillsdale, NJ, United States (1986)
6. Norman, D.A.: The Design of Everyday Things. Doubleday, New York (1988)
7. Veryzer, R.W., Borja de Mozota, B.: The impact of user-oriented design on new product development: an examination of fundamental relationships. J. Prod. Innov. Manag. **22**(2), 128–143 (2005)
8. Abras, C., Maloney-Krichmar, D., Preece, J.: User-Centered Design. W. Bainbridge, Ed. Sage Publications, Thousand Oaks (2004)
9. Schütte, S.T., Eklund, J., Axelsson, J.R., Nagamachi, M.: Concepts, methods and tools in Kansei engineering. Theor. Issues Ergon. Sci. **5**(3), 214–231 (2004)
10. Alonso-García, M., Pardo-Vicente, M.-Á., Rodríguez-Parada, L., Moreno Nieto, D.: Do products respond to user desires? A case study. Errors and successes in the design process, under the umbrella of emotional design. Symmetry **12**(8), 1350 (2020)
11. Veryzer, R.W., De Mozota, B.B.: The impact of user-oriented design on new product development: an examination of fundamental relationships, pp. 128–143 (2005)
12. van Boeijen, A., Daalhuizen, J., Zijlstra, J., van der Schoor, R.: Delft design guide: design methods (2014)
13. Morrison, R.L., Smollan, R.K.: Open plan office space? If you're going to do it, do it right: a fourteen-month longitudinal case study. Appl. Ergon. **82**, 102933. https://doi.org/10.1016/j.apergo.2019.102933
14. Goulding, C.: Grounded theory, ethnography and phenomenology strategies for marketing research (2003)
15. Bruseberg, A., Mcdonagh-philp, D.: Focus groups to support the industrial/product designer: a review based on current literature and designers' feedback. Appl. Ergon. **33**, 27–38 (2002)
16. Norman, D.A.: Emotional Design: Why We Love (or Hate) Everyday Things. Basic Books (2005)
17. Weintraub, M.I.: Beauty parlor stroke syndrome: report of five cases. Jama **269**(16), 2085–2086 (1993)

Manufacturing and Industrial Process Design

Fixing Elements Localization in Aircraft Large Structures Using Machine Learning Techniques

Leandro Ruiz[1,2](✉), Alejandro Gómez[2], Sebastián Díaz[2], José M. González[2], and Francisco Cavas[3]

[1] Doctorate Program in Industrial Technologies, International School of Doctorate, Technical University of Cartagena, 30202 Cartagena, Spain
leandro.ruiz@mtorres.com

[2] Innovation Division, MTorres Diseños Industriales SAU, Ctra. El Estrecho-Lobosillo, Km 2, Fuente Álamo, 30320 Murcia, Spain

[3] Department of Structures, Construction and Graphical Expression, Technical University of Cartagena, 30202 Cartagena, Spain

Abstract. The use of artificial vision systems is widespread in aircraft manufacturing. These systems usually rely on the detection and measure of drills, fasteners, and other elements to correctly perform the machining and assembly of aircraft parts, and also provide the equipment with the ability to report quality measures of its own performance (i.e., drill diameter, countersink depth). State of the art vision systems are based on classical processing algorithms as intensity segmentation, blob detection and circularity measurements. These tools work quite fine in optimal conditions (clean surfaces, controlled illumination) although they tend to fail in more severe environments, where metal chips, oil, lubricant, variable illumination will decrease the reliability of the measure, forcing human intervention and thus reducing the global manufacturing rate. A novel approach based on Convolutional Neural Networks (CNN) is introduced to classify the different objects detected in the scene. The image dataset has been created from real images taken from production machines. This approach achieved up to 94% classification accuracy needing only 0.8ms to classify the image.

Keywords: Artificial vision · Deep learning · Neuronal networks · Aircraft manufacturing · Machining

1 Introduction

In recent years, the aerospace sector has undergone a major transformation towards a fully automated assembly environment. Processes formerly carried out manually such as drilling, riveting or assembly of parts are currently carried out automatically by machine tools equipped with numerical controls (also called CNC), thus increasing both productivity and precision in manufacturing processes.

One of the most important stages of these processes is the correct referencing and positioning of the part to be machined within the working volume of the machine.

F. Cavas Martínez et al. (Eds.): INGEGRAF 2021, LNME, pp. 139–146, 2022.
https://doi.org/10.1007/978-3-030-92426-3_17

Normally the parts are installed as close as possible to its theoretical position defined in the program that will be executed by the numerical control. This is usually not sufficient to achieve the tight manufacturing tolerances required by aeronautical processes (generally around 0.1 mm).

It is therefore necessary for the machine to be able to measure the actual position of the part and adjust the CNC program accordingly. One of the most widely used methods in the industry for referencing in manufacturing processes is based on the use of artificial vision systems to inspect holes or temporary fixing elements on the part itself which are added in a previous phase. On other occasions, it is required to drill holes in a part whose positions have been previously marked by a previous drilling of a much smaller diameter (pre-drilling).

Currently the artificial vision techniques used in the industry lose reliability as the complexity of the captured scenes increases. On one hand, the casuistry of objects to be detected is quite wide, distinguishing between drills, blind drills, countersinks, rivets of different types, temporary fasteners… On the other hand, the presence of machining chips or drops of oil can "confuse" algorithms, requiring the introduction of both previous stages of image pre-processing (to work with an image as clean as possible) and the addition of numerous subsequent rules to detect and discard false positives. This processing can be computationally expensive, especially considering the high resolution of the images required to achieve accurate measurements.

This work introduces a new technique based on machine learning to classify the different objects present in the scene, eliminating the introduction of artificial rules and at the same time increasing the speed of image processing.

Currently, most of the aeronautical component manufacturing processes are carried out by numerically controlled machine tools (Fig. 1).

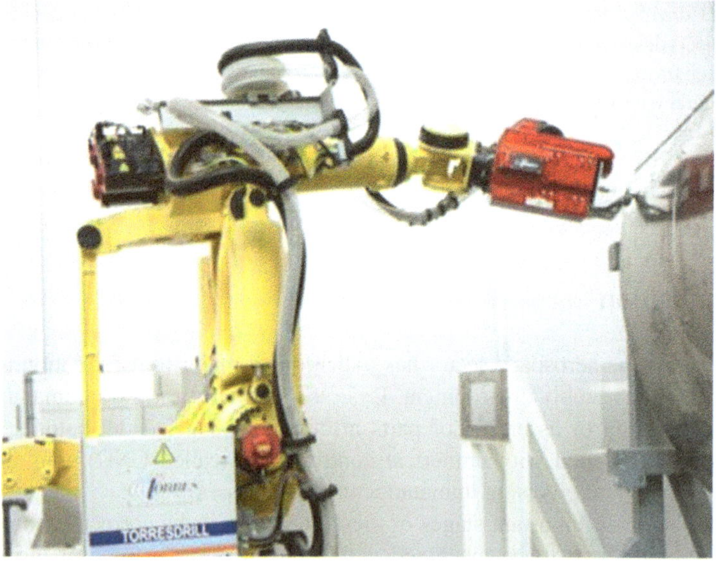

Fig. 1. Drilling machine

Vision systems are used in drilling and riveting machines during the different stages of the manufacturing process. There are three situations in which vision inspection plays a critical role for an accurate operation of the machine: inspection for part referencing and positioning, re-drilling and drill verification prior to riveting [1–3].

There are different computer vision processing algorithms that can be applied to the detection of circles, being the Hough transform [4] and the detection by labeling of connected regions (Blobs) [5, 6], two of the most widely spread.

The described algorithms, as well as other similar methods, work with high efficiency and processing speed in controlled environments, under "laboratory" conditions. When these algorithms are transferred to industrial environments, their robustness decreases, although they can be controlled with the inclusion of additional techniques to strengthen the algorithms, increasing the complexity of the system and therefore making it difficult to maintain.

2 Material and Methods

The image processing in this study has been carried out using different programming languages and libraries. The applications for the pre-processing of images, as well as the patch generation have been developed in C++, using the OpenCV computer vision library. This programming language has been chosen for its, given the high number of images to process. The development and training of neural networks has been based on the Keras and Tensorflow frameworks [7], using a GPU model GTX1070Ti.

2.1 Image Dataset Preparation

Raw Image Pre-processing

A set of 33400 images, captured in a total of 12 different machines has been used. Due to their diverse origin, the variability of the scenes captured in the images is very large. One of the first challenges of this work has been to transform this set of images from different sources, machines typologies and resolutions into a structured and homogeneous data set.

The processing algorithm carried out is based on applying a blob extraction algorithm to the images, which were previously thresholded. Subsequently, those regions with an aspect ratio in contained the interval [0.83, 1.20] have been selected. Finally, the extracted regions have been resized to a size of 64×64 pixels, obtaining a total of 134000 of these images, which will be called "patch" (Fig. 2).

Classification Category Definition

In aerospace manufacturing, different types of fixation elements are used. All these elements, defined by these manufacturing processes, have been categorized into eight different classes:

- Drill (D), which includes all straight holes (that is, without countersinking), whether blind or through.

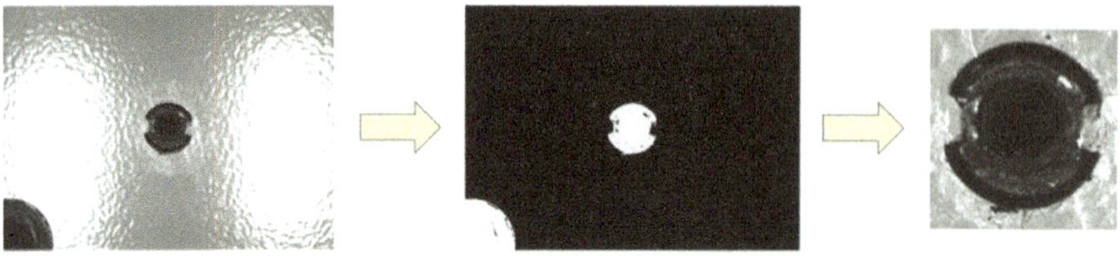

Fig. 2. Patch extraction process

- Countersink (Cs), which includes holes with countersink.
- Rivet (R), which includes rivets, either they are flush with the surface or not.
- Temporary fasteners (F), which will be drilled later to insert a rivet.
- Hexagonal (Hx), which includes all those objects with a hexagonal shape.
- Bolt (Bo), which includes images of bolts.

In addition to the categories defined by the process, an additional category is defined to denote the absence of referencing elements, which will be called Background (Fig. 3).

- Background (B), including all those images corresponding to surface areas, whether clean or stained with oil, machining chips…

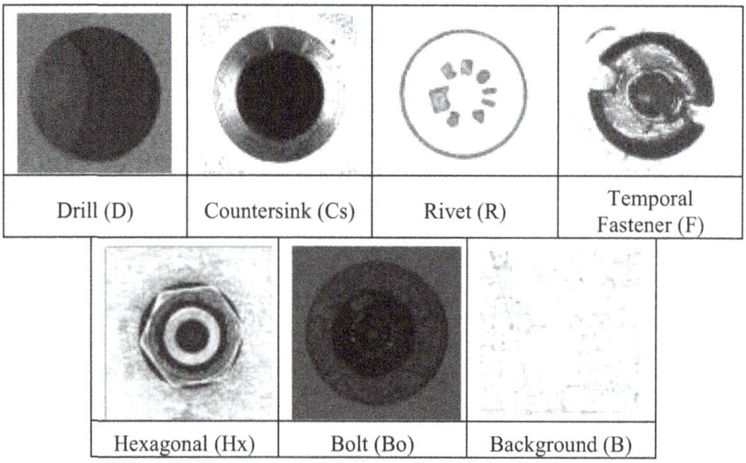

Fig. 3. Image category examples

Image Library Construction

For the construction of the image dataset, it was necessary to classify all the patches generated in the pre-processing stage. Manual review process has been accelerated by using an automatic labeling process based on Autoencoder-type neural networks together with clustering algorithms [8].

The images have been processed by the autoencoder that extracts a feature vector of reduced dimensionality. A clustering algorithm (k-means) has been then executed on these vectors. This algorithm arranges the given feature vectors into a predetermined number of categories so that the variance of the points in the same category is minimized [9]. In this case, the dataset has been analyzed with $k = 64$ classes, which have subsequently been manually assigned to each of the 9 defined categories. Once all the images have been assigned to a category, they have been reviewed manually, using the classification interface.

Subsequently, once 10k images were manually reviewed, a convolutional neural network (with an architecture similar to that described in Sect. 2.2) was trained to generate a new, more precise automatic labeling on the remaining images. Finally, a total of 21k images have been reviewed and classified.

Classification Interface
For the manual review and classification of the images, a graphical interface based on web technologies has been developed [10]. The web interface has been developed in HTML, PHP and JavaScript, using responsive design techniques so that the interface is accessible on all types of devices, both computers and smartphones.

The interface presents the patch image to be classified, with the category proposed by the automatic labeling. The user is then presented with 2 options: accept the proposed label, or reject it, in which case he is asked to select the correct category. Once one of these actions has been carried out, the interface refreshes showing a new image. Example images of each category are also shown to help the user select the category that best fits the image to be classified.

2.2 Convolutional Network

The neural network has been built and trained from scratch following a convolutional network scheme, with an architecture inspired by the LeNet family of neural models [11]. Network size has been kept small to achieve fast execution times, which have little impact on the overall machine processing time.

The network consists of an input layer of 64×64 pixels, connected to several stages of convolutional layers alternated with pooling layers. In this way, the filters contained in the weights of the convolutional layers are applied to the image, extracting information of an increasingly higher level of abstraction. A fully connected layer is added to the output of the last convolutional processing stage, which connects with the 10 possible outputs of the system, regularized according to a softmax distribution (Fig. 4).

The network includes an L2-Regularization factor equal to 0.001 on the convolutional layers, as well as a Dropout of 10% in both the first and last layers.

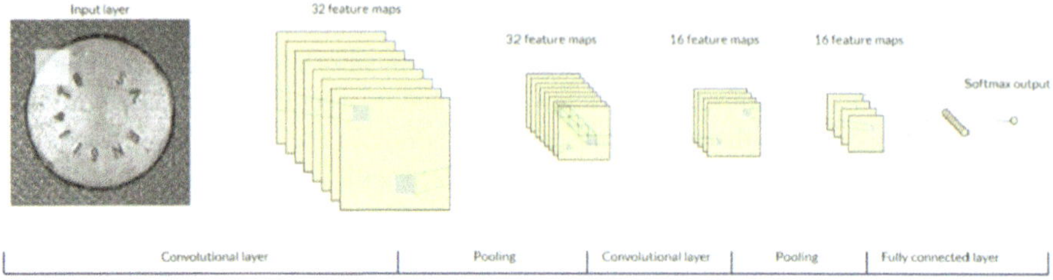

Fig. 4. Neuronal network structure

Given the spatial symmetry of the objects to be classified (they are always circular objects, or at most hexagonal), a data augmentation scheme has been applied on the dataset based on rotations and symmetries, thus obtaining a final dataset of 126k images, which has been randomized and divided into training and validation subsets following an 80%–20% ratio. A 5-iteration cross-validation has been carried out, dividing the set into 5 parts so that in each of the iterations the ith portion is used to validate and the remaining 4 to train. In this way, the influence of the distribution of categories in the training data on the results obtained can be analyzed (Fig. 5).

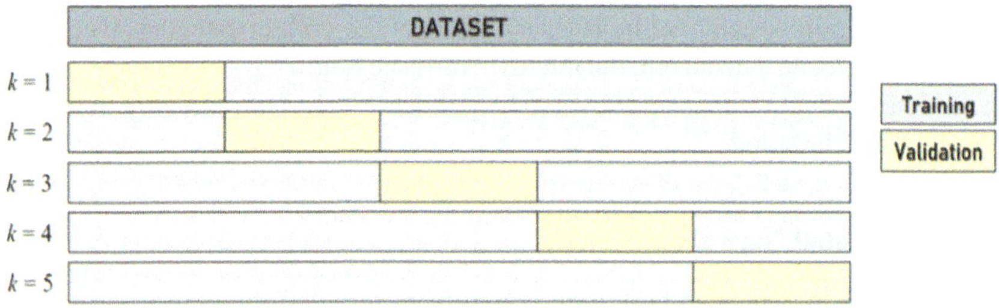

Fig. 5. Cross-validation training scheme

The network has been trained for 150 epochs using Adadelta optimizer and Categorical Cross-Entropy loss function. Class balancing has also been applied during training, in order to mitigate the differences between the number of images belonging to each category.

3 Results

After training, the categorical accuracies achieved by the network are between 94.5% and 96.5%, with an average accuracy of 95.7%. The confusion matrix of the system has been calculated by averaging among every trial and normalizing the results by class (Table 1).

Table 1. Averaged and normalized confusion matrix.

True Class \ Predicted Class	B	D	Cs	R	F	Hx	Bo
B	99.6	0.3	0.0	0.0	0.0	0.0	0.0
D	1.2	98.1	0.5	0.0	0.2	0.0	0.0
Cs	2.2	26.0	68.8	0.9	2.0	0.1	0.0
R	6.7	9.5	0.8	82.5	0.5	0.0	0.0
F	1.1	2.5	1.2	0.8	94.1	0.1	0.1
Hx	5.9	0.4	0.4	0.0	0.4	92.8	0.0
Bo	1.5	4.4	3.7	0.0	2.2	0.0	88.2

The main purpose of this network is to detect the presence of referencing elements, being the classification of these elements an additional functionality that serves to apply specific algorithms to measure the position of said elements. Therefore, it is interesting to analyze the effectiveness of the network in absolute terms of detecting objects, regardless of their category.

Table 2. Global confusion matrix.

True \ Predicted	Background	Ref. element
Background	99.6	0.4
Ref. element	1.7	98.3

In global terms of the detection of referencing elements, the results are much better: A false positive rate of 0.4% and a false negative rate of 1.7%.

4 Conclusions

In this work a neural network for the classification of fixation elements of aeronautical structures has been developed and trained, with an average classification time of 0.8ms on a GTX1070Ti GPU. The integration of this neural network in the vision systems of drilling/riveting machines has made it possible to strengthen the detection of holes in a real production environment.

The main challenge of this work lies in classifying elements that are very similar to each other with a high level of confidence, for example, holes and countersinks (Table 1). In global terms (Table 2), the designed network achieves high success rates, maintaining a low percentage of false positives and negatives (2.3% and 2.2% respectively). These results are similar to those obtained in [12, 13], where convolutional neural networks have been used for the classification of defects on thermal images of carbon fiber.

References

1. Baigorri Hermoso, J.: Improved automatic riveting system, 24 November 2005
2. Douglas, A., Jarvis, P., Beggs, K.: Micropositioning system, 19 April 2001
3. Baigorri, H.J.: System of machining by areas for large pieces. ES2303803A1, 16 August 2008
4. Djekoune, A.O., Messaoudi, K., Amara, K.: Incremental circle hough transform: an improved method for circle detection. Optik **133**, 17–31 (2017)
5. Ballard, D.H., Brown, C.M.: Computer vision. Prentice Hall Professional Technical Reference (1982)
6. Samet, H., Tamminen, M.: Efficient component labeling of images of arbitrary dimension represented by linear bintrees. IEEE Trans. Pattern Anal. Mach. Intell. **10**, 579–586 (1988)
7. Jiao, L., et al.: A survey of deep learning-based object detection. IEEE Access **7**, 128837–128868 (2019)
8. Song, C., Huang, Y., Liu, F., Wang, Z., Wang, L.: Deep auto-encoder based clustering. Intell. Data Anal. **18**, S65–S76 (2014)
9. MacQueen, J.: In some methods for classification and analysis of multivariate observations. In: Proceedings of the Fifth Berkeley Symposium on Mathematical Statistics and Probability, Volume 1: Statistics, pp 281–297. University of California Press, Berkeley (1967)
10. Ruiz, L., Torres, M., Gómez, A., Díaz, S., González, J.M., Cavas, F.: Detection and classification of aircraft fixation elements during manufacturing processes using a convolutional neural network. Appl. Sci. 10 (2020)
11. Lecun, Y., Bottou, L., Bengio, Y., Haffner, P.: Gradient-based learning applied to document recognition. Proc. IEEE 86, 2278–2324 (1998)
12. Schmidt, C., Hocke, T., Denkena, B.: Artificial intelligence for non-destructive testing of CFRP prepreg materials. Prod. Eng. Res. Dev. **13**, 617–626 (2019)
13. Schmidt, C., Hocke, T., Denkena, B.: Deep learning-based classification of production defects in automated-fiber-placement processes. Prod. Eng. Res. Dev. **13**(3–4), 501–509 (2019). https://doi.org/10.1007/s11740-019-00893-4

Design of an Autonomous Monitoring System for Oceans

Javier Sánchez(✉), Luis Mízquez, Roberto Casas, and Teresa Blanco

Aragon Institute of Engineering Research (I3A), University of Zaragoza, 50018 Zaragoza, Spain
javiersanchez@unizar.es

Abstract. The study describes the design, prototyping and prospective planning of an autonomous ocean monitoring system. This concept aims to harness the potential of some of the enablers from Industry 4.0 to address a complete process, ranging from a fleet of low-cost sensorized physical instruments to the communication and processing of the data obtained. The product design part assumes great relevance within the project, since it provides the system with many of the differentiating characteristics: the low cost owing to additive manufacturing techniques, the autonomous operation supported by wave energy, as well as scalability. By means of a modular structure, a set of sensors installed inside a capsule moving vertically from the surface to a specific depth are responsible for capturing data from various physicochemical variables that help to establish profiles of the surface layers of the sea, providing key information in ocean observation. This information is shared among a matrix of profilers offshore and communicating with a center onshore, where it is processed and spread to interested agents.

Keywords: Industrial design · Prototyping · Industry 4.0 · Sensing · Oceanography

1 Introduction

The ocean is home to 96% of life on Earth [1] and the importance of its continuous study and monitoring is evident. The impact of changes in the geochemical and biological processes of the ocean at all levels on the rest of the planet. One of the actions carried out to inform, alert and know the progress of these changes is the monitoring of certain water parameters. However, this work is hampered, on numerous occasions, by the scarcity of resources of the agents involved. The need for a reliable open information network -similar to that of international programs such as Argo [2]- is increasingly pressing and in demand. Decision-making at the international level, accompanied by reliable data and treated with the greatest possible ubiquity, is therefore essential.

The most frequent problem for carrying out research at sea is the lack of accessible and affordable instrumentation. It is not unusual to find testimonies from research groups dedicated to oceanography with data acquisition problems due to a lack of budget to equip themselves with the most common commercial elements. In many cases, there is a problem of scale: users do not need all the features offered by the complete monitoring

F. Cavas Martínez et al. (Eds.): INGEGRAF 2021, LNME, pp. 147–154, 2022.
https://doi.org/10.1007/978-3-030-92426-3_18

and profiling systems available. The cost should not be a barrier in scientific research and that is why projects like this aim to bring researchers with easy-to-get instrumentation alternatives, at an affordable cost, with open source and open data.

1.1 State of the Problem

There are numerous solutions to the monitoring problem, however, not all of them meet the initial requirements (low price, accessibility, and universal availability). Some of the most relevant are satellite imaging, data collection by vessels, specific measurements with ROVs (Remotely Operated Vehicle) and AVP's (Autonomous Vertical Profilers). The latter are the best option to develop a mesh of probes connected to each other. The vertical profiles generated in a cyclical manner provide very valuable information for the analysis of the state of the most superficial layers of seawater. In this group there are different types of products that serve as a starting point for the design and development of a product that meets the characteristics specified above. Gliders [3] and other autonomous systems, such as those propelled by bladders and pumps [4] are solutions obtained to satisfy different needs, profiles with long horizontal distances and profiles with great depth, respectively. However, the product that comes closest to specs is the Wirewalker [3]. This robust system can operate autonomously with wave energy, so that it does not require an energy input.

2 Methods and Tools

The result of an iterative design and research process is Argira: a global system that comprises a solution that combines the development of a product, an Autonomous Vertical Profiler, with its arrangement in the form of an array across the sea surface and communication, processing and analysis of data collected through an IoT platform.

Both product development and communications development are geared towards affordable data acquisition.

2.1 The System

The adoption of digital enablers and technologies related to Industry 4.0 is a challenge, because there are more factors that must be controlled in the face, especially, of cybersecurity, but also of maintenance and supervision tasks. However, the ideal Argira application is for users of one or more research organizations to obtain, use and share the information provided by the system. In the proposal, the system is defined using a Reference Framework used by the Industrial Internet Architecture. This reference architecture was created by the Industrial Internet Consortium to provide IoT deployments with interoperability between technologies.

The system holds the communications between AVP buoys (Fig. 1).

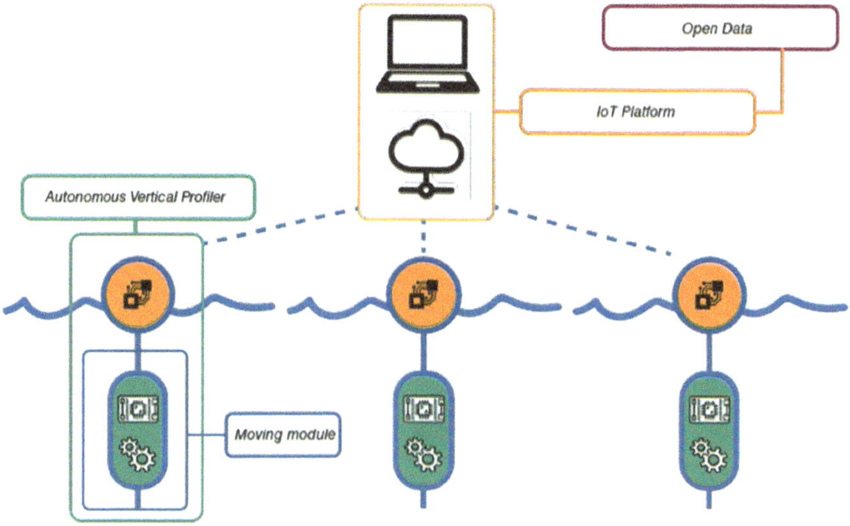

Fig. 1. Argira: array of buoys. Communications.

The Argira project aspires to generate an open data ecosystem. The data obtained by each profiler in the system will be processed and mined; however, the raw data -although structured and related- will be available for the required public or private use.

To have a data catalog that allows information to be accessible and reusable, tools such as Ckan were developed [5].

2.2 Autonomous Vertical Profiler

For the development of the product, the autonomous systems based on Pinkel's Wire-walker (WW) have been taken as a reference. Stand-alone vertical profilers work thanks to the impulse of various energy sources. In the case of Argira, a wave energy impulse system is proposed: the waves of the body of water cause the impulse that constantly causes the mechanism to slide through the cable, away from the surface. By the time the mechanism collides -at the last impulse- against the ballast or the bottom (there are two possible options, anchoring or hanging a ballast), the mechanism is released and begins to ascend by its own buoyancy. Upon reaching the surface buoy, the ratchet mechanism (Fig. 2) re-engages, initiating a new descent-ascent cycle. In this last step, communications are carried out between the microcontroller of the mobile module and that of the surface buoy, storing the data (Fig. 3).

This whole product consists, therefore, in three modules joined by a cable, whose length is around 20 m.

Fig. 2. AVP mechanism: the ratchet

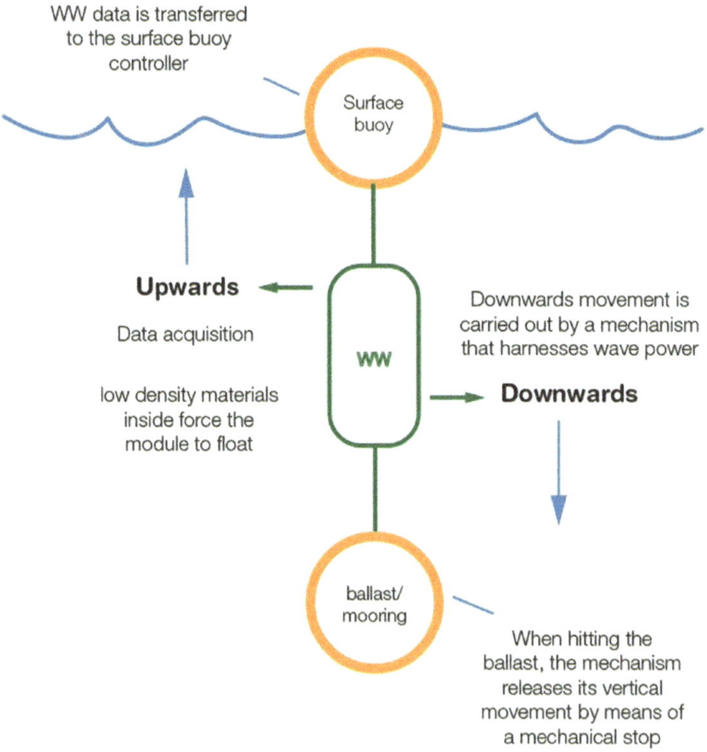

Fig. 3. Argira buoy modules and dynamics

As the moving module ascends, the sensors collect data over a certain period. The measurements are stored in an external memory (since there is not enough internal memory for which microcontrollers). As soon as the module reaches the upper limit (after the ascent), the collected information is transmitted via Bluetooth to the surface buoy microcontroller.

The surface buoy continuously acquires data on wind speed, direction, speed of water movement, collects solar energy, etc. (Fig. 4). The energy harvesting technique is counted as future development work, due to its potential in the durability of the product.

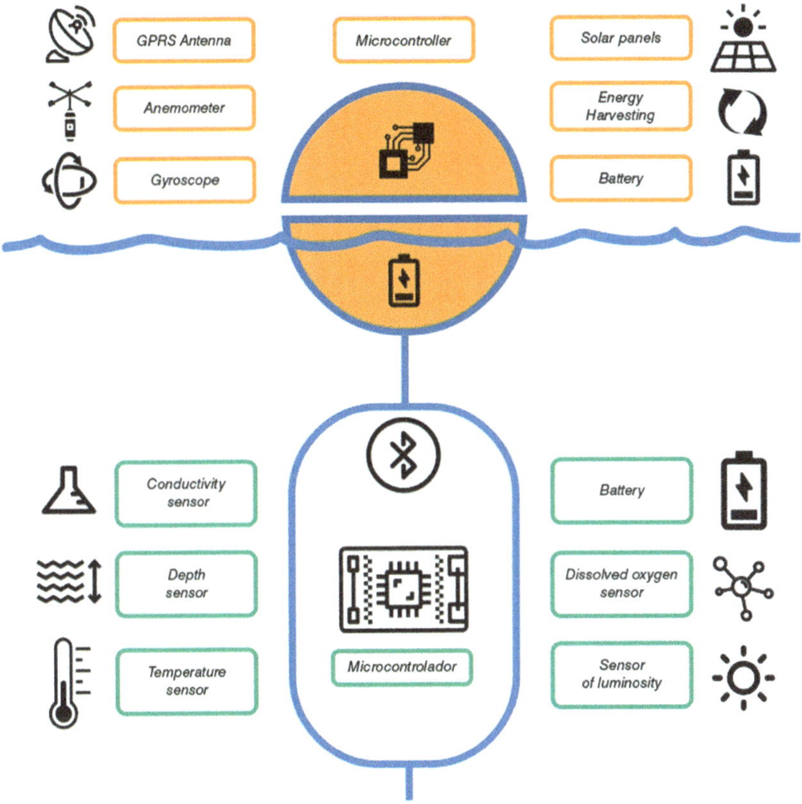

Fig. 4. Argira buoy architecture

For communication between profilers and users two possible applications have been considered.

The first one defines that all data is communicated by long-area wireless communications (LoRaWAN) from the AVPs to a floating module that will act as a HUB (functionally, as a gateway). This intermediary module will communicate through a GPRS antenna to a server, which will act as a bridge between the system and the Cloud platform. The second option is that each AVP has a built-in GPRS antenna for direct communication with the server. The first has the advantage that functionalities can be extended, adding interactions between the AVPs, however, each one of them may end up too far apart to communicate and physically joining them would be expensive. The second saves the mentioned problems and is simpler and easier to apply.

The communication by which the data is transmitted to the surface buoy from the moving module must meet, in order of importance, the following conditions:

a. Effective communication over short distances.
b. Low energy consumption.
c. Possibility of scalability.

It is necessary to bear in mind the conditions under which said communication is established. The distance between the antenna of the moving module and the receiver on the surface buoy will be contained in the range of 40 to 50 cm. This data limits the search margin for short-range communications such as NFC, whose range of action and its maximum transfer speed is 424 kbps.

The time that the mobile module will remain in the area closest to the surface buoy is indeterminate; not because of the difficulty of the calculation, but it depends, directly, on the rate of descent -related to the frequency and amplitude of the waves at the moment. In any case, it is estimated that it will not move more than 50 cm in less than two seconds. This situation allows opening the range of short and medium distance wireless communication technologies, since the minimum data transfer speed (Z-Wave, with 40 Kbps) is sufficient to send the estimated amount of data per cycle. The best valued options, therefore, are RFID, Z-Wave and Bluetooth LE.

3 Results

After some trials in the proof-of-concept phase, a functional prototype was built up. Thanks to additive manufacturing -in particular, FDM (Fused Deposition Modelling)- more than 75% of the parts were custom built. Some parts such as the outer casing, the inner connecting rods and the adaptations between the commercial elements have been manufactured using this technique.

For this proof of concept, non-definitive elements have been prepared to improve the adaptability of the prototype to the test environment. The mobile module was the centre of attention, since it is the one with the greatest technical difficulty, while the upper (surface buoy) and lower (ballast) modules have been simulated with dummy elements. The kinematics of the mechanism have been found to be valid since the ascent (by floating) and descent (by the wave-driven mechanism) occurs satisfactorily. Some sensors have also been included to start the testing of the electronics of the mobile module.

Overall, the results have been rewarding, as the tests have shed light on the way ahead (Fig. 5).

Fig. 5. Prototype being tested.

4 Conclusions and Future Work

The design of the AVP, both in its mechanical and electronic dimensions, must contemplate improvements in manufacturing and programming (respectively).

A big deal of calculations must be made to determine the autonomies of the batteries to be incorporated based on the energy efficiency of the communications applied to the final design of AVP.

Fault control algorithms must be used to detect bugs and looseness and reduce the uncertainty in the measurements. In addition, a deep effort should be done towards failure prevention by determining critical parts (mechanisms, sensors, communication channels, etc.) and its interaction with the environment.

Communication between modules is therefore the most demanding task due to the need for deployments at sea to experiment with underwater communications.

It could be concluded that the combination of the different digital enablers from Industry 4.0, applied together, help to achieve the objectives set: efficiency, autonomy, affordability, and open data.

References

1. Turley et al.: Ocean under stress (2011)
2. Riser, S.C., et al.: Fifteen years of ocean observations with the global Argo array. Nat. Clim. Change **6**(2), Article no. 2 (2016). https://doi.org/10.1038/nclimate2872
3. Rudnick, D.L., Davis, R.E., Eriksen, C.C., Fratantoni, D.M., Perry, M.J.: Underwater gliders for ocean research. Mar. Technol. Soc. J. **38**(2), 73–84 (2004). https://doi.org/10.4031/002533204787522703
4. Muthuvel, P., et al.: Development of deep sea profiling float: challenges and technology demand. In: 2018 OCEANS - MTS/IEEE Kobe Techno-Oceans (OTO), Kobe, pp. 1–4, May 2018. https://doi.org/10.1109/OCEANSKOBE.2018.8558832

5. Winn, J.: Open data and the academy: an evaluation of CKAN for research data management. Presented in IASSIST 2013, Cologne, May 2013. Accedido, 24 May 2021. [En línea]. http://eprints.lincoln.ac.uk/id/eprint/9778/

Graphical Bioengineering

José Lapayese's Silk-Twisting Lathe: Approach to Its Geometric Modelling and 3D Digital Restitution

José Ignacio Rojas-Sola(✉) and Manuel Hurtado-Expósito

Universidad de Jaén, 23071 Jaén, Spain
jirojas@ujaen.es

Abstract. This paper shows the geometric modelling and 3D digital restitution of the lathe for twisting silk by José Lapayese included in his work "Treatise on the art of spinning, winding, folding and twisting silk, according to the method of Mr. Vaucanson with some additions, and corrections to it. Beginning and progress of the Vinalesa Factory, in the Kingdom of Valencia, established under the protection of H.M." from 1779. For this, Autodesk Inventor Professional software has been used to obtain its geometric modelling, 3D digital restitution and dynamic simulation, as well as AutoCAD to obtain its detailed geometric documentation (exploded view, assembly plan and detail drawings of each element). The starting material for this research (specification and plan) has been downloaded from the Digital Library of the Royal Botanical Garden. Moreover, it should be noted that during the 3D digital restitution process, and in the absence of specific information in the treatise, it was necessary to make a series of hypotheses regarding the mechanisms that make up the lathe, as well as to apply a series of dimensional, geometric and motion (degrees of freedom) constraints. However, and despite this drawback, it has been possible to deduce the transmission systems of the movement of the lathe and thus its complete operation.

Keywords: Silk · Lapayese · Vaucanson · Geometric modelling · Digital restitution · Autodesk Inventor Professional · AutoCAD · Constraints

1 Introduction

Over the centuries, silk production has undergone a commendable technological evolution that makes the silk industry a motor for technical development.

The art of weaving natural silk began in China in the seventeenth century BC. After several centuries, it was transmitted to Western Europe at the end of the Middle Ages and its industrialization began in the nineteenth century. At the beginning of the thirteenth century, there was already a procedure to twist the silk thread and in the fourteenth century, the machine for winding silk appeared. In addition, at the end of the fourteenth century and the beginning of the fifteenth century the use of hydraulic yarn-spinners and Jean Le Calabrais-type looms expanded. Later in the seventeenth and eighteenth centuries, the first attempts to simplify and rationalize manufacturing took place with

F. Cavas Martínez et al. (Eds.): INGEGRAF 2021, LNME, pp. 157–165, 2022.
https://doi.org/10.1007/978-3-030-92426-3_19

the appearance of the perforated card machine for the semi-automatic loom of Basile Bouchon and Jean-Baptiste Falcon, which was improved by Jacques de Vaucanson in 1775. Joseph Marie Jacquard later developed the first programmable punched card loom (Jacquard loom) [1].

This paper focuses on José Lapayese's silk-twisting lathe, which he presents in his work "Treatise on the art of spinning, winding, folding and twisting silk, according to the method of Mr. Vaucanson with some additions and corrections to it. Beginning and progress of the Vinalesa Factory, in the Kingdom of Valencia, established under the protection of H.M." from 1779 [2].

This treatise reflects on the advantages, disadvantages and corrections of the new procedure that the Frenchman Guillermo Reboull had introduced into Spain in 1769 on the spinning and twisting method perfected in France by Jacques de Vaucanson. His method was put into practice in his factory in Vinalesa (Valencia), with 3419 looms in 1748 [3], where he carried out a mechanization process to wind, fold and twist the silk, which accelerated manufacturing by having various hydraulic mills, each with 70 thread winders that collected the silk from as many thread spools [4].

The importance of this research lies in the fact that there is no study of this invention, a milestone in the technological development of the time in Spain, from the perspective of engineering graphics.

The final objective pursued by this research is to obtain a good approximation to the 3D digital restitution of the model, as well as its detailed geometric documentation, which will allow a much better understanding of its complex operation and disseminate its knowledge in society. This entire complex process is based on three principles: authenticity, historical rigor and scientific transparency, as outlined in the Seville Principles [5] document on virtual archaeology, as well as in the London Charter [6] on the computer-aided visualization of cultural heritage.

2 Material and Methods

The starting material for this research has been downloaded from the Digital Library of the Royal Botanical Garden [2]. It is a book in .pdf format in which, unfortunately, the operation of the invention is not explained in detail, although it does indicate a series of corrections with respect to Jacques de Vaucanson's fabric folding machine.

Despite the fact that this book only presents a single plan related to the invention (Fig. 1), with a clear absence of detailed information on the systems that make up the same, it has been possible to deduce its operation for which it has been necessary to establish a series of hypotheses, modifying some elements of the movement transmission mechanisms. For this, similar mechanical systems have been designed that reproduce the same operation of the machine.

Therefore, it is necessary to indicate that the aim pursued in this research has been to obtain an approximate model of the invention that shows the conceptual design proposed by the author in his work. These difficulties have meant that both the geometric modelling of the invention and its subsequent assembly have turned out to be complex, particularly for parts with more complex geometry which could not be obtained through basic operations such as revolution, emptying or extrusion, among others. Nevertheless, the use of free design tools such as NURBS has not been necessary.

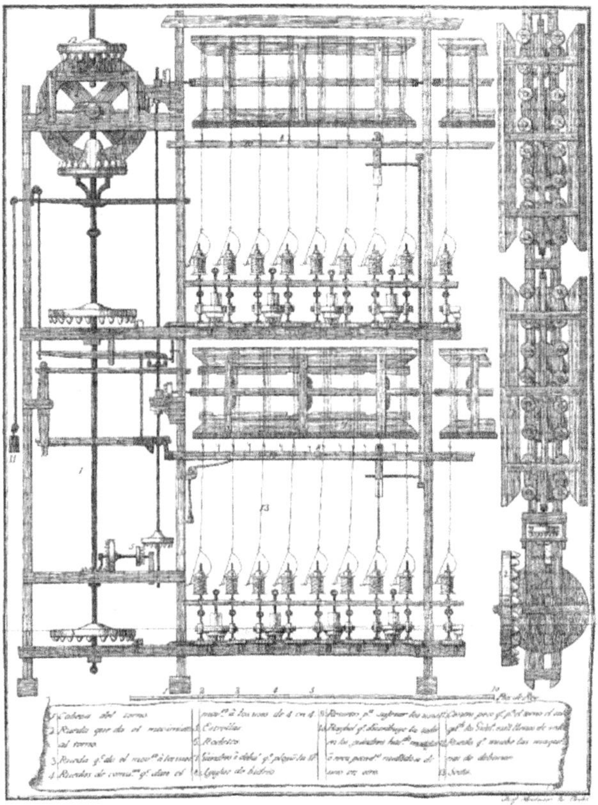

Fig. 1. José Lapayese's silk-twisting lathe (1779) [7].

As can be seen in the plane of Fig. 1, the unit of measurement has been the "king's foot", so it is necessary to convert this unit to units of the international system, taking into account the equivalence of one foot to 0.3048 m.

Subsequently, making use of the knowledge of descriptive geometry and after having applied the empirical technique of direct measurement on the plan by means of a millimetre ruler and the adoption of the corresponding graphic scale, we proceeded to model each and every one of the components of the invention with the aid of two software.

Thus, Autodesk Inventor Professional has been used to obtain geometric modelling, 3D digital restitution and dynamic simulation of its operation [8], and AutoCAD to obtain detailed geometric documentation of the invention (exploded view, assembly plan and detail drawings of each element). With Autodesk Inventor Professional the drawings were obtained in .dwg format (native AutoCAD format), and with AutoCAD they have been improved to obtain a higher normative quality. The justification for the use of these software lies in their functionality for exchanging information with other researchers. This research methodology has been previously described in several papers by the author on various historical inventions relevant to technological development in Spain and Europe [9–14], where each and every one of the steps to follow to obtain a functional and consistent geometric model is detailed.

3 Results and Discussion

To understand the operation of the invention Figs. 2 and 3 are shown, which show an axonometric view of the 3D CAD (computer-aided design) model of the lathe as well as an exploded or assembly views thereof, respectively.

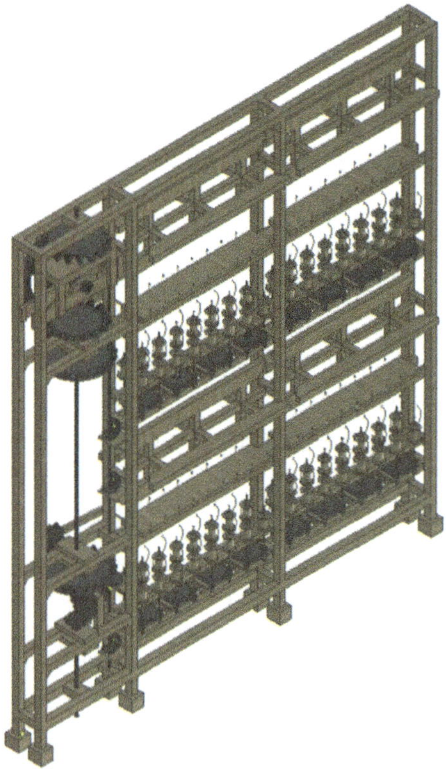

Fig. 2. Axonometric view of 3D CAD model.

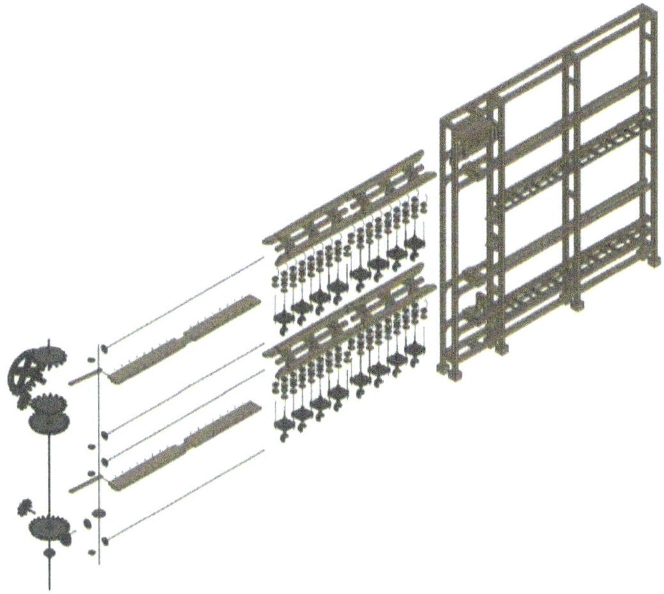

Fig. 3. Exploded view of the 3D CAD model.

Furthermore, to be able to describe the operation of the lathe the assembly plan of the invention (Fig. 4), and two sub-assembly plans (Figs. 5 and 6) are shown, where all the elements that form it appear with an indication of their number, name and material.

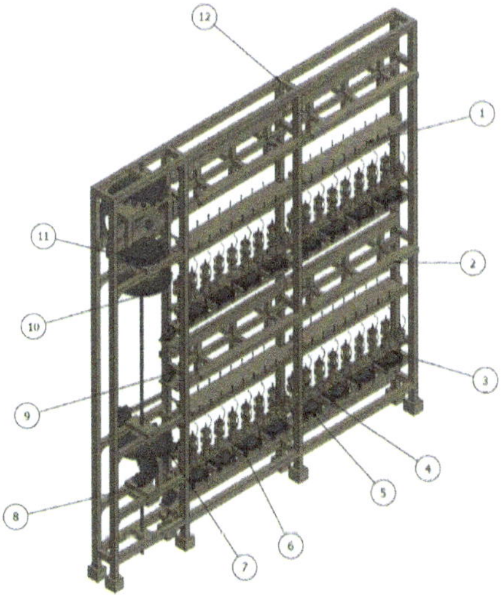

12	1	Main structure	Wood
11	1	Bevel gear system with main drive shaft	Cast Iron
10	2	Reciprocating bevel gear system	Cast Iron
9	4	12:10 Bevel gear system	Cast Iron
8	1	Translacional bevel gear system on main drive shaft	Cast Iron
7	1	Translational bevel gear system on secondary shaft	Cast Iron
6	16	10:8 Bevel gear system	Cast Iron
5	16	50-tooth straight gear	Cast Iron
4	64	8-tooth straight gear	Cast Iron
3	64	Thread spool with guide	Wood
2	4	Thread winder	Wood
1	2	Reciprocating system with pin	Wood
Number	Quantity	Name	Material

Fig. 4. Assembly plan with an indicative list of elements.

The invention is mainly composed of a fixed main structure (12), providing rigidity and stability to the assembly, and a bevel gear system with main drive shaft (11) that transmit the movement to the rest of the components.

The action of this bevel gear system (11) was carried out by means of the hydraulic energy provided by the water from a canal attached to the Vinalesa factory. This movement is transmitted to a reciprocating bevel gear system (10) and to the translational bevel gear system on the main drive shaft (8). In addition, the bevel gear system intended for reciprocating (10) is coupled with a reciprocating system with pin (1) that causes a continuous reciprocating movement of 5 cm, which seeks to distribute the thread along four thread winders (2), preventing it from overlapping. The function of the bevel gear system on the main drive shaft (8) is to transmit the movement to a second translational bevel gear system on a secondary shaft (7), which transmits the movement to four systems of 12:10 bevel gears (9). Two of the latter systems are intended to transmit the movement to a main straight gear with 50 teeth (5) and to another secondary straight gear with 8 teeth (4), by means of bevel gear systems with a 10:8 gear ratio (6) (Fig. 5), while the remaining ones transmit the turn to the thread winders (2). The main straight gear with 50 teeth (5) transmits the rotation to 4 secondary bevel gears with 8 teeth (4) that rotate a thread spool with a guide (3), which unwinds once the thread has been wound onto the thread winder (2), causing the twisting of the silk.

Figure 5 shows the sub-assembly that constitutes the bevel gear system with a 10:8 gear ratio (6).

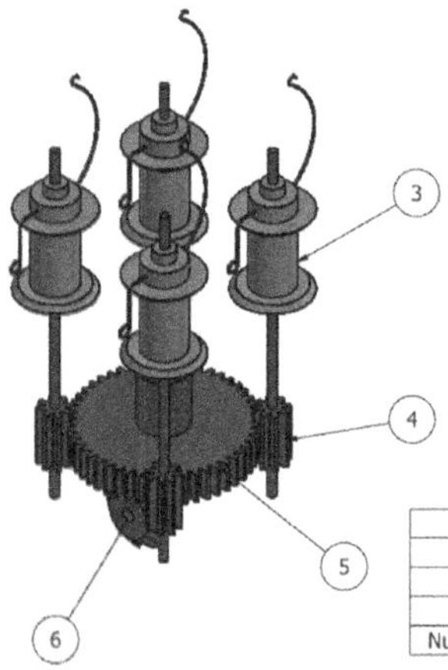

6	16	10:8 Bevel gear system	Cast Iron
5	16	50-tooth straight gear	Cast Iron
4	64	8-tooth straight gear	Cast Iron
3	64	Thread spool with guide	Wood
Number	Quantity	Name	Material

Fig. 5. Drawing of the '10: 8 bevel gear system' sub-assembly.

As previously stated there have been various drawbacks that have made it impossible to carry out a reliable modelling of the invention. These include the absence of detailed information both in the specification and in the single plane thereof, as well as the incoherent relationships between the different dihedral projections, as exact records of geometric and dimensional relationships were not kept.

Therefore, and due to these drawbacks, the reciprocating movement system (Fig. 6) has been slightly modified in this approach to obtaining the general operation of the lathe.

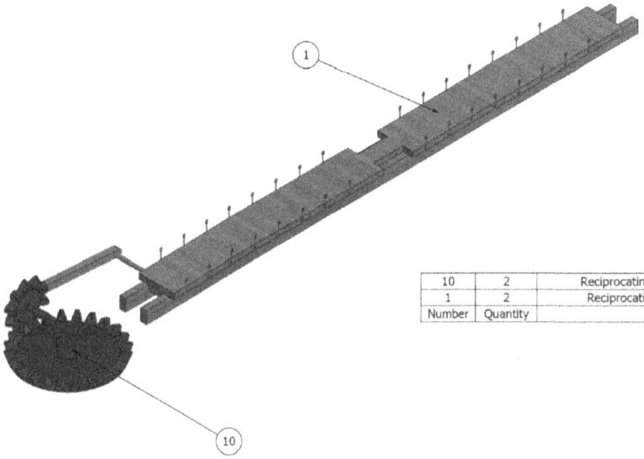

10	2	Reciprocating bevel gear system	Cast Iron
1	2	Reciprocating system with pin	Wood
Number	Quantity	Name	Material

Fig. 6. Drawing of the 'Reciprocating movement system' sub-assembly.

Nevertheless, most of the elements have been able to be correctly modelled following the usual operations such as extrusion or revolution, among others.

Finally, it should be noted that the greatest difficulty encountered has been the assembly process of the components of the invention, since this has implied the existence of numerous dimensional, geometric and movement (degrees of freedom) constraints in order for their operation to be coherent.

4 Conclusions

This paper presents the results of the process of geometric modelling and 3D digital restitution of José Lapayese's silk-twisting lathe from 1779, thanks to the parametric computer-aided design software Autodesk Inventor Professional.

In addition, and thanks to the AutoCAD software, the improved geometric documentation of the invention has been obtained (exploded view, assembly plan and detailed plans of each element). For this, knowledge of descriptive geometry has been essential, as well as the use of the empirical technique of direct measurement in the plan by means of a millimetre ruler and the adoption of the corresponding graphic scale, facilitating a better understanding of the set and its operation.

However, there have been numerous difficulties that have made this a complex process. The main one has been the absence of detailed information on the invention both in the specification and in the only existing plan, such as for example specific details of each of the components of the set or the incoherent relationships that exist between the different dihedral projections, since exact records of the geometric and dimensional relationships were not kept. This has obliged us to establish a series of hypotheses, modifying some elements of the movement transmission mechanisms as well as applying a series of dimensional, geometric and movement (degrees of freedom) constraints in order to obtain a coherent and functional 3D model.

As possible future work the 3D geometric model should be analysed using CAE (computer-aided engineering) techniques in order to perform a static analysis of the invention under operating conditions, assigning the original materials of the invention (cast iron and oak wood) as well as establishing the boundary conditions, the loads and obtaining a meshing of the 3D CAD model. This will enable us to determine the distribution of the von Mises stresses, the displacements, the equivalent deformations and the safety factor, which will in turn determine whether the invention was oversized, taking into account material resistance criteria.

Acknowledgements. The research carried out here has been possible thanks to the collaboration grant shared with the Department of Engineering Graphics, Design and Projects of the University of Jaén obtained in 2020 from the Ministry of Education and Vocational Training of the Government of Spain. Moreover, the authors sincerely appreciate the work of the reviewers of this paper.

References

1. Historia de la seda. https://es.wikipedia.org/wiki/Historia_de_la_seda. Accessed 20 July 2021
2. Lapayese, J.: Tratado del arte de hilar, devanar, doblar y torcer las sedas, según el método de Mr. Vaucanson con algunas adiciones, y correcciones a él. Principio, y progresos de la Fábrica de Vinalesa, en el Reyno de Valencia, establecida baxo la proteccion de S.M. Blas Román, Impresor de la Real Academia de Derecho Español y Público, Madrid (1779). Biblioteca Digital del Real Jardín Botánico. https://bibdigital.rjb.csic.es/idurl/1/16251. Accessed 20 July 2021
3. Calvo Calvo, A.: Transferencia internacional de tecnología y condicionamientos nacionales: la industria sedera catalana durante su transición al régimen liberal. Quaderns d'Història de l'Enginyeria **3**, 93–123 (1999)
4. Solaz Albert, R.: Instrucción para la hilanza del organsin o seda fina pelo según el nuevo método de Monsieur Vaucanson. In: Climent Martínez, J.D., Guerola Inza, A., Muñoz Feliu, M.C., Paricio i García, A. (eds.) Pasiones Bibliográficas II, pp. 193–195. Societat Bibliogràfica Valenciana Jerònima Galés, Valencia (2017)
5. Principles of Seville. http://smartheritage.com/wp-content/uploads/2015/03/FINAL-DRAFT.pdf. Accessed 20 July 2021
6. The London Charter. http://www.londoncharter.org. Accessed 20 July 2021
7. Plano del torno para torcer las sedas de José Lapayese. https://bibdigital.rjb.csic.es/idviewer/16251/225. Accessed 20 July 2021
8. Hansen, L.S.: Autodesk Inventor 2020: A tutorial introduction. SDC Publications, Mission (2019)
9. Rojas-Sola, J.I.: Ancient technology and computer-aided design: olive oil production in Southern Spain. Interdisc. Sci. Rev. **30**(1), 59–67 (2005). https://doi.org/10.1179/030801805X19690
10. Villar-Ribera, R., Hernández-Abad, F., Rojas-Sola, J.I., Hernández-Díaz, D.: Agustín de Betancourt's telegraph: study and virtual reconstruction. Mech. Mach. Theory **46**(6), 820–830 (2011). https://doi.org/10.1016/j.mechmachtheory.2011.01.008

11. Rojas-Sola, J.I., De la Morena-de la Fuente, E.: Agustin de Betancourt's wind machine for draining marshy ground: approach to its geometric modeling with Autodesk Inventor Professional. Technologies **5**(1), Article no. 2 (2017). https://doi.org/10.3390/technologies5010002
12. Rojas-Sola, J.I., De la Morena-de la Fuente, E.: Digital 3D reconstruction of Agustin de Betancourt's historical heritage: the dredging machine of the port of Krondstadt. Virtual Archaeol. Rev. **9**(18), 44–56 (2018). https://doi.org/10.4995/var.2018.7946
13. Rojas-Sola, J.I., Galán-Moral, B., De la Morena-de la Fuente, E.: Agustín de Betancourt's double-acting steam engine: geometric modeling and virtual reconstruction. Symmetry-Basel **10**(8), Article 351 (2018). https://doi.org/10.3390/sym10080351
14. Rojas-Sola, J.I., De la Morena-de la Fuente, E.: The Hay inclined plane in Coalbrookdale (Shropshire, England): geometric modeling and virtual reconstruction. Symmetry-Basel **11**(4), Article 589 (2019). https://doi.org/10.3390/sym11040589

José Lapayese's Silk Thread Winding Machine: Approach to Its Geometric Modelling and 3D Digital Restitution

José Ignacio Rojas-Sola(✉) and Manuel Hurtado-Expósito

Universidad de Jaén, 23071 Jaén, Spain
jirojas@ujaen.es

Abstract. This paper describes the geometric modelling and 3D digital restitution of the machine for folding silk by José Lapayese included in his work "Treatise on the art of spinning, winding, folding and twisting the silks, according to the method of Mr. Vaucanson with some additions, and corrections to it. Beginning and progress of the Vinalesa Factory, in the Kingdom of Valencia, established under the protection of H.M." from 1779. For this, the Autodesk Inventor Professional software has been used in order to obtain its geometric modelling, 3D digital restitution and dynamic simulation, as well as AutoCAD to obtain its geometric documentation (exploded view, assembly plan and detail drawings of each element). The starting material for this research (specification and plan) has been downloaded from the Digital Library of the Royal Botanical Garden. Furthermore, it should be noted that during the 3D digital restitution process, and in the absence of specific information in the treatise, it has been necessary to make a series of hypotheses regarding the mechanisms that make up the machine. However, and despite this drawback, it has been possible to deduce the transmission systems of the movement of the machine and thus, its complete operation.

Keywords: Silk · Lapayese · Vaucanson · Geometrical modelling · Digital restitution · Autodesk Inventor Professional · AutoCAD · Constraints

1 Introduction

José Lapayese wrote the "Treatise on the art of spinning, winding, folding and twisting silks, according to the method of Mr. Vaucanson with some additions, and corrections to it. Beginning and progress of the Vinalesa Factory, in the Kingdom of Valencia, established under the protection of H.M." in 1779 [1].

In this treatise, some considerations about the method of spinning silk of the French engineer Jacques de Vaucanson (1709–1782) of the Royal Academy of Sciences of Paris are reflected. The advantages, disadvantages and corrections of the new procedure that the Frenchman Guillermo Reboull had introduced into Spain in 1769 on the spinning and twisting method perfected in France by Vaucanson are also described.

Subsequently, a factory owned by José Lapayese was created in Vinalesa (Valencia) under the direction of Guillermo Reboull and his son Santiago, eventually installing 30

F. Cavas Martínez et al. (Eds.): INGEGRAF 2021, LNME, pp. 166–172, 2022.
https://doi.org/10.1007/978-3-030-92426-3_20

double spinning lathes on which two spinners could work at the same time [2]. This factory had 44 mills powered by hydraulic energy from the water from the Moncada ditch with which a mechanization process was carried out to wind, fold and twist the silk. This procedure accelerated the pace of work compared to the traditional system since each of the mills had 70 thread winders that collected the silk from as many skeins, while in the traditional home system each woman could only operate a single thread spool [2].

Finally, in the treatise there is information on the lathes for twisting silks and on the Vinalesa Factory along with six plans, one of which is the silk thread winding machine, the object of this paper, justified as it is an invention of great historical relevance in his time.

The importance of this paper lies in the fact that there has as yet been no study of this invention, which was a milestone in the technological development of the time in Spain, from the perspective of engineering graphics.

Therefore, the main objective of this research is to obtain the approximate geometric modelling of the machine, its 3D digital restitution and its geometric documentation. To do this, the objectives set out in the Seville Principles [3] document on virtual archaeology, as well as the London Charter [4] on the computer-aided visualization of cultural heritage have been followed. Therefore, the aim is to obtain a digital restitution based on research, documentation and dissemination that underlines the importance of authenticity, historical rigor and scientific transparency.

2 Material and Methods

The material used in this research has been downloaded from the Digital Library of the Royal Botanical Garden [1]. Although the operating process of the machine is not explained in detail in the specification, it does make a series of corrections with respect to the Jacques de Vaucanson fabric folding machine. Thanks to these small corrections and the only existing plan (Fig. 1), it has been possible to deduce the operation of the machinery thanks to a series of hypotheses on the mechanisms of transmission of movement together with a series of dimensional, geometric and motion (degrees of freedom) constraints. Thus, similar mechanical systems have been designed that reproduce the operation of the machine.

It must be taken into account that the purpose of this research has not been to obtain an exact modelling of the invention, but rather to obtain a conceptual design that would convey the idea of its operation, due to the absence of specific information on its construction details in the documentation available.

For all this, and due to these drawbacks, the geometric modelling process has been complex. The plane in Fig. 1 is dimensioned with the "king's foot" unit of measurement, so in order to obtain the dimensions in units of the international system its conversion has been carried out taking into account that one foot is equal to 0.3048 m. Thus, making use of the knowledge of descriptive geometry and the direct measurement of the dimensions in the plan, it was possible to determine its graphic scale and to proceed to its CAD (computer-aided design) modelling.

In addition, in the present research two computer-aided design software have been used: First, Autodesk Inventor Professional [6] which has been used to obtain geometric

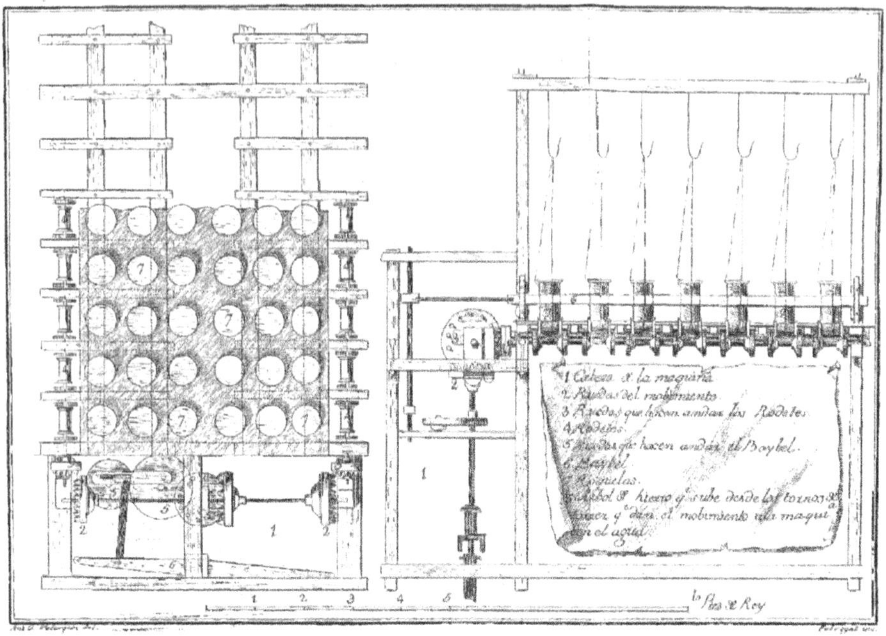

Fig. 1. José Lapayese's silk thread winding machine (1779) [5].

modelling, as well as the 3D digital restitution and dynamic simulation of its operation, and secondly, AutoCAD which has enabled us to correctly define the graphical results obtained in order to obtain the geometric documentation of the invention (exploded view, assembly plan and detail drawings of each element). The justification for their use is due to their functionality and the fact that they allow information to be shared with other researchers working in the Autodesk environment. The methodology followed in this research has been described in detail in previous papers by the author on various historical inventions relevant to technological development in Spain and Europe [7–12].

3 Results and Discussion

For a better understanding of the invention and its operation Figs. 2 and 3 are presented, which show an axonometric view of the 3D CAD model as well as an exploded and assembly views thereof, respectively.

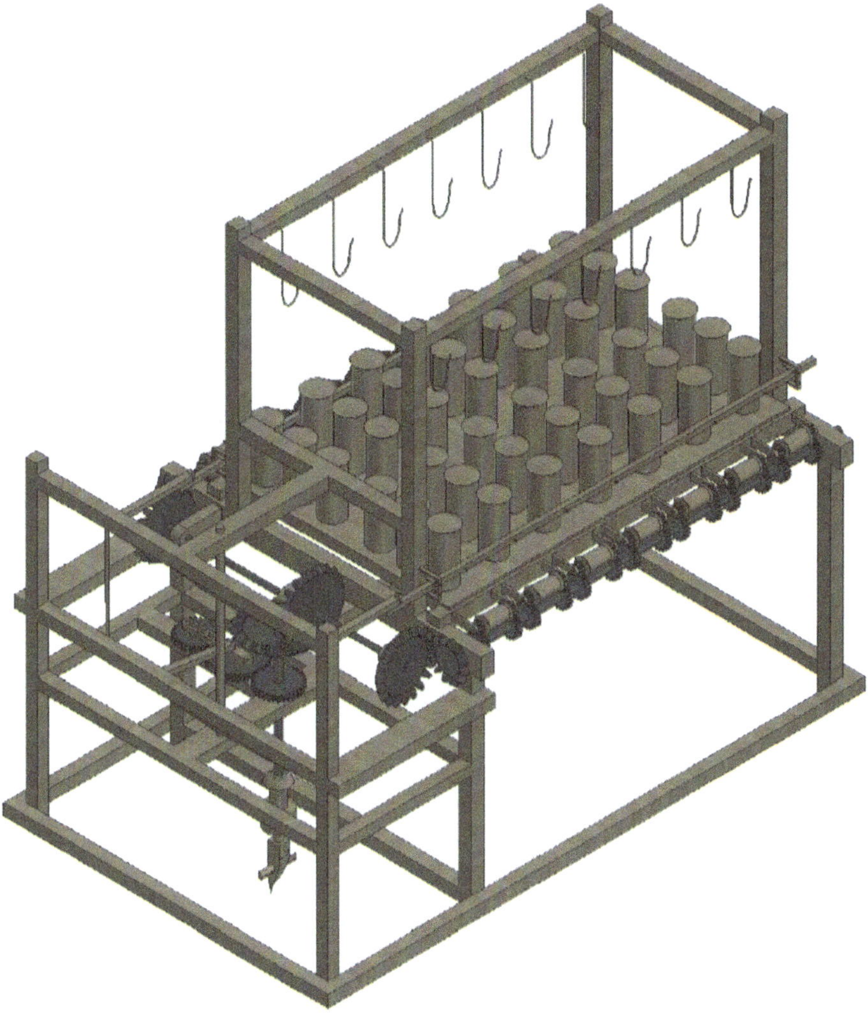

Fig. 2. Axonometric view of 3D CAD model.

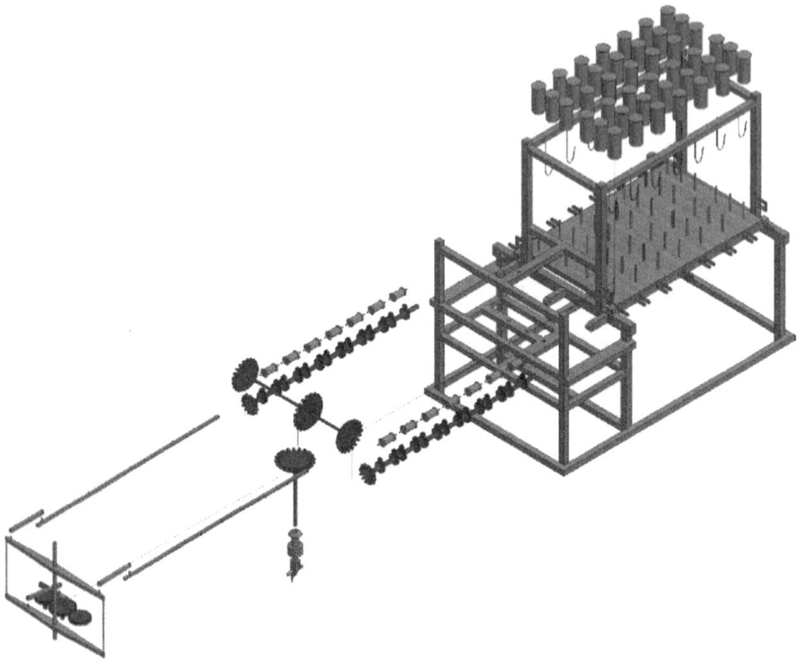

Fig. 3. Exploded view of 3D CAD model.

In addition, in order to describe the operation of the machine the assembly plan of the invention is shown (Fig. 4), where all the elements that comprise it appear with an indication of their number, name and material.

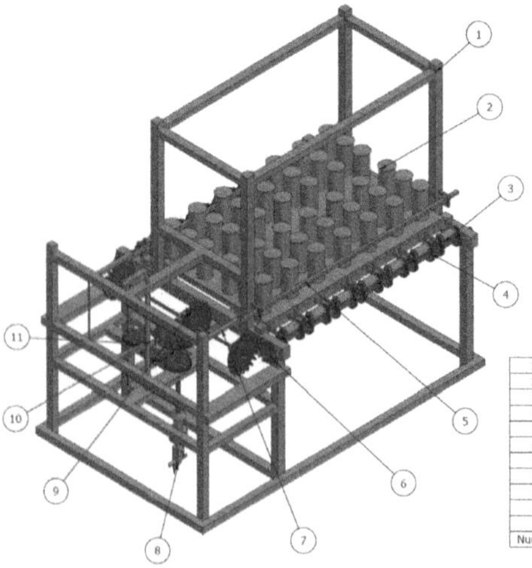

11	1	40:32 Straight gear system	Cast iron
10	1	Main reciprocating system	Wood
9	1	Bevel gear system with main drive shaft	Cast iron
8	1	Blade	Wood
7	2	Reciprocating movement transmission shaft	Wood
6	2	20:12 Bevel gear system	Cast iron
5	2	Reciprocating bar with pin	Wood
4	28	20:8 Straight gear system	Cast iron
3	14	Thread winder	Wood
2	42	Thread spool	Wood
1	1	Main structure	Wood
Number	Quantity	Name	Material

Fig. 4. Assembly plan with an indicative list of elements.

The invention is mainly composed of a fixed main structure (1), providing rigidity and stability to the assembly, and a bevel gear system with main drive shaft (9) and a blade (8) that transmit the movement to the rest of the components.

The drive of the bevel gear system (9) was produced through the hydraulic force of the water on the blade (8), which was submerged in a small channel. The bevel drive gear system (9) translated motion through the straight gear system (11) with a 40:32 gear ratio and the bevel gear system (6) with a 20:12 gear ratio. In addition, the straight gear system (11) produced the reciprocating movement through the main reciprocating system (10), the transmission shaft of the reciprocating movement (7) and the reciprocating bar with pin (5), causing the folding along the thread winders (3). On the other hand, the function of the bevel gear system (6) was to transmit the rotation to the thread winders through the straight gear system (4) with a 20: 8 gear ratio. These thread winders collected and folded the silk thread from the thread spools (2) which were placed on the central board of the main structure (1).

As previously stated, there have been various drawbacks that have complicated the geometric modelling process. The first is the absence of detailed information in the available documentation, as this documentation was only intended to describe the operation of the invention, not to allow a reliable restitution. Added to this are the incongruities between the different dihedral projections of the original plan, as exact records of geometric and dimensional relationships were not kept.

However, the geometric modelling of most of the elements has not been difficult, taking into account the usual operations of this process (extrusion, revolution and emptying, among others).

Nevertheless, the greatest difficulty of this geometric modelling has resided in the assembly process of the different elements of the set, since this has implied numerous movement constraints, which on occasions have caused operational conflicts.

4 Conclusions

This paper describes the results of the complex process carried out to obtain a reliable approach to the geometric modelling and 3D digital restitution of José Lapayese's silk thread winding machine included in his work "Treatise on the art of spinning, winding, folding and twisting silk, according to the method of Mr. Vaucanson with some additions, and corrections to it. Beginning and progress of the Vinalesa Factory, in the Kingdom of Valencia, established under the protection of H.M." from 1779, using Autodesk Inventor Professional software.

Our knowledge of descriptive geometry and the use of the empirical technique of direct measurement with a millimetre ruler of the dimensions in the plan with the determination of the corresponding graphic scale have been the pillars of the methodology used. As the only existing plan lacks detailed information and presents inconsistencies between the different dihedral projections, it has been necessary to establish dimensional, geometric and movement (degrees of freedom) constraints in order to ensure the correct operation of the historical invention and thus obtain a coherent 3D model.

The geometric documentation of the invention (exploded view, assembly plan and detail drawings of each element) has been obtained from the 3D CAD model, facilitating its correct interpretation and operation for the reader.

Finally, it should be noted that the 3D CAD model will be the starting point for a subsequent CAE (computer-aided engineering) analysis whereby the von Mises stresses, displacements and deformations will be analysed, as well as the safety factor of the model in operating conditions.

Acknowledgements. The research carried out here has been possible thanks to the collaboration grant shared with the Department of Engineering Graphics, Design and Projects of the University of Jaén obtained in 2020 from the Ministry of Education and Vocational Training of the Government of Spain. Moreover, the authors sincerely appreciate the work of the reviewers of this paper.

References

1. Lapayese, J.: Tratado del arte de hilar, devanar, doblar y torcer las sedas, según el método de Mr. Vaucanson con algunas adiciones, y correcciones a él. Principio, y progresos de la Fábrica de Vinalesa, en el Reyno de Valencia, establecida baxo la proteccion de S.M. Blas Román, Impresor de la Real Academia de Derecho Español y Público, Madrid (1779). Biblioteca Digital del Real Jardín Botánico. https://bibdigital.rjb.csic.es/idurl/1/16251. Accessed 20 July 2021
2. Solaz Albert, R.: Instrucción para la hilanza del organsin o seda fina pelo según el nuevo método de Monsieur Vaucanson. In: Climent Martínez, J.D., Guerola Inza, A., Muñoz Feliu, M.C., Paricio i García, A. (eds.) Pasiones Bibliográficas II, pp. 193–195. Societat Bibliogràfica Valenciana Jerònima Galés, Valencia (2017)
3. Principles of Seville. http://smartheritage.com/wp-content/uploads/2015/03/FINAL-DRAFT.pdf. Accessed 20 July 2021
4. The London Charter. http://www.londoncharter.org. Accessed 20 July 2021
5. Plano de la máquina para doblar las sedas de José Lapayese. https://bibdigital.rjb.csic.es/idviewer/16251/223. Accessed 20 July 2021
6. Hansen, L.S.: Autodesk Inventor 2020: A Tutorial Introduction. SDC Publications, Mission (2019)
7. Rojas-Sola, J.I.: Ancient technology and computer-aided design: olive oil production in Southern Spain. Interdisc. Sci. Rev. **30**(1), 59–67 (2005). https://doi.org/10.1179/030801805X19690
8. Villar-Ribera, R., Hernández-Abad, F., Rojas-Sola, J.I., Hernández-Díaz, D.: Agustín de Betancourt's telegraph: study and virtual reconstruction. Mech. Mach. Theory **46**(6), 820–830 (2011). https://doi.org/10.1016/j.mechmachtheory.2011.01.008
9. Rojas-Sola, J.I., De la Morena-de la Fuente, E.: Agustin de Betancourt's wind machine for draining marshy ground: approach to its geometric modeling with Autodesk Inventor Professional. Technologies **5**(1), Article no. 2 (2017). https://doi.org/10.3390/technologies5010002
10. Rojas-Sola, J.I., De la Morena-de la Fuente, E.: Digital 3D reconstruction of Agustin de Betancourt's historical heritage: the dredging machine of the port of Krondstadt. Virtual Archaeol. Rev. **9**(18), 44–56 (2018). https://doi.org/10.4995/var.2018.7946
11. Rojas-Sola, J.I., Galán-Moral, B., De la Morena-de la Fuente, E.: Agustín de Betancourt's double-acting steam engine: geometric modeling and virtual reconstruction. Symmetry-Basel **10**(8), Article no. 351 (2018). https://doi.org/10.3390/sym10080351
12. Rojas-Sola, J.I., De la Morena-de la Fuente, E.: The Hay inclined plane in Coalbrookdale (Shropshire, England): geometric modeling and virtual reconstruction. Symmetry-Basel **11**(4), Article no. 589 (2019). https://doi.org/10.3390/sym11040589

Analysis of Clinical Thermal Images for the Detection of Circulatory Pathologies and Venous Insufficiency

Francisco José Soto Lara[1](✉), Manuel Damián Marín Granados[2], Juan Franquelo Soler[3], and Francisco Javier Salgado Fernández[4]

[1] Ingeniero Técnico Industrial Termógrafo certificado nivel 2 según ISO 18436, Málaga, Spain
francisco-soto@intermal.es

[2] Ingeniero Industrial, Dtor Dpto Expresión Gráfica en la Universidad de Málaga, Málaga, Spain
mdmarin@uma.es

[3] Ingeniero Industrial, profesor del Dpto de Expresión Gráfica en la Ingeniería de la Universidad de Málaga, Málaga, Spain
jfranquelo@uma.es

[4] Ingeniero Técnico Industrial. Profesor del dpto. de Expresión Gráfica en la Ingeniería de la Universidad de Málaga, Málaga, Spain
javier.salgado@uma.es

Abstract. Clinical thermography is currently used in various aspects of medicine for the detection (in many cases prematurely) of mild pathologies.

Since thermography technology is based on the image captured by an infrared sensor, its analysis requires the use of the appropriate tools, which are usually provided by software that the manufacturers themselves create for these tasks.

Whilst capturing the thermal image is of vital importance, this is only part of the task, since there is a need to bring to light significant thermal patterns that allow for visualizing the areas to be measured or compared, using the appropriate software and analytical tools provided.

This article will demonstrate how to analyse a clinical thermal image in order to obtain information for detecting pathologies related to the circulatory system or to discern whether an element of the image could be pathological and related to the circulatory system.

Keywords: Clinical thermography · Pathology · Infrared · Venous insufficiency

1 Introduction

The discovery and development of infrared thermography generally began with the capture of infrared radiation emitted by any surface that has a temperature greater than absolute zero, which can subsequently be translated into an image.

Initially, and thanks to the Stefan-Boltzmann Law (1), the possibility of measuring temperatures at a distance is simple, given that this law is able to relate the emissive power of the radiant surface to the temperature by means of an equation that initially

F. Cavas Martínez et al. (Eds.): INGEGRAF 2021, LNME, pp. 173–181, 2022.
https://doi.org/10.1007/978-3-030-92426-3_21

characterized the emissive power of a black body. However, a minor correction, referred to as emissivity (the ability to emit radiation from a surface) provides us with the exact relationship between the radiant surface emissive power of a real body and its surface temperature:

$$W = \varepsilon \cdot \sigma \cdot Te^4 \tag{1}$$

Where ε is the emissivity, a value between 0 and 1; σ is the Stefan-Boltzmann constant, and Te is the surface temperature of the observed object.

The development of "portable" thermography, particularly in the 1970s, was based largely on the generation of images that can be studied to find "anomalies" that is, thermal patterns that would be unexpected either on the basis of the functioning and exposure to heat of the object studied or comparison with a similar object showing correct thermal behaviour.

Further, the current standardized training for thermographs, included in ISO 18436-7 [1], ensures that the skills acquired are not only transversal and useful for working with thermal cameras in any sector, but also include the concepts of "Qualitative thermography" and "Quantitative thermography".

Qualitative thermography aims to detect possible anomalies by comparison with a normal state, that is, the expected state according to the correct behaviour of a process or one with a symmetrical or other similarly "healthy" element.

Once the anomaly is detected, quantitative thermography is then used to measure the temperature and compare the difference between the anomalous and the correct state, which allows for evaluating the level of the detected problem.

It is very important to consider several software tools for the image analysis, as can be observed in the following image. The "raw" image does not enable the user to properly identify any patterns or remarkable elements with enough clarity, which makes it impossible to see any apparent alterations or to measure temperature (Fig. 1).

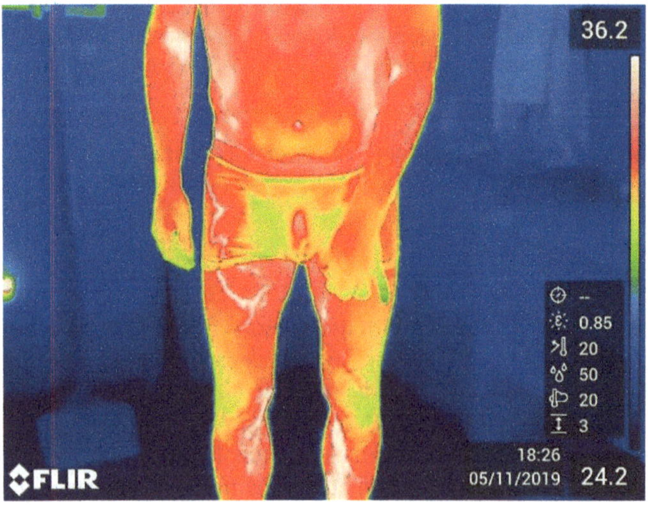

Fig. 1. Image recorded by a thermal camera before analysis

Whilst it is true that thermal cameras currently allow for analysing the images and the use of various analytical tools on their own screens, it is also the case that this task is inefficient, lengthening the time spent working with the patient and unnecessarily consuming battery resources.

All this leads us to consider the possibility of developing a common methodology that can be used in any clinical thermographic study along with a simple but specific methodology for the detection of possible circulatory problems.

Nowadays, the process of taking thermal images in the Health sector is carried out manually, and there are very few software tools that are able to analyse said images automatically. Even though there is currently some research in the field, "Thermohuman" [2] is the only commercially distributed software tool as of today.

On the other hand, the experience of these specialists dedicated to conduct clinical studies through thermography for doctors and physiotherapists, makes us realize that it is them who should carry out an specific diagnose, according to the examinations they may see adequate. Because of this, it seems to be that a correct analysis and data display for these specialists should include a clear and visible exhibition of the images, findings and data, which could potentially lead to a precise diagnosis.

Therefore, we consider the objective of the image analysis to be to show any possible pathologies which the doctor could later confirm and, if possible, determine their causes and treatments.

For this reason, this article focuses on the finding of an analysis tool able to visualize several particular pathologies such as angiopathy, as well as its obtaining and visualization methods for the end-user assessment.

2 Materials and Methods

At this point we will present all general and specific materials, methods, and conditions required for any clinical thermography study of a patient, as well as those needed for the purpose of this article.

The environmental conditions for thermal imaging are of vital importance since infrared radiation is emitted from the surface of the body and factors such as air currents, direct effects of the sun, or significant heat sources would be revealed in the image, which could reduce the reliability of the qualitative analysis.

Similarly, the need for repeatability of the test requires that all possible conditions be controlled.

When capturing images, the following requirements must generally be met [3, 4]:

- The patient must be in a room without drafts, whilst windows must be covered in such a way that the patient is not exposed to the direct effects of the sun or significant sources of heat.
- The suggested ambient temperature should be within a range of 21 to 23 ºC and in some studies it is proposed that this be sufficiently controlled so that it does not rise above these values whilst the images are being captured.
- The patient's skin must be acclimatized so that its surface temperature is regulated by removing any traces of clothing or accessories that could leave a thermal mark. It has

been shown that 10 min without clothing is sufficient for effective acclimatization of the skin [5].
- A further precaution that should be taken is to warn the patient that they should not take painkillers or anti-inflammatories at least 24 h before the study, and, if desired, should shower at least 2 h before the test. The last meal should be consumed at least 4 h before the test.
- According to the guidelines for this type of study, the patient is also asked to avoid treatments 24 h before the test and to also refrain from sunbathing or shaving the areas to be observed in that time interval.

With regard to the cameras used, it is clear that a higher resolution implies more pixels and therefore greater detail for detecting pathologies. Whilst studies have been conducted using cameras of varying degrees of resolution, it is also widely accepted that a resolution of 320 × 240 pixels is sufficient for obtaining good image quality and a high capacity for pathology detection (Fig. 2).

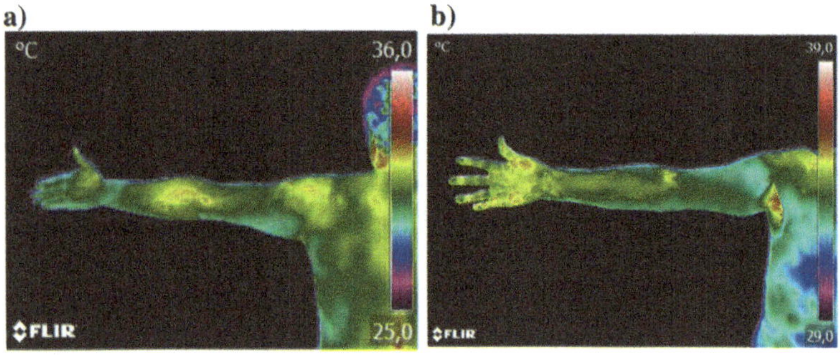

Fig. 2. (a) and (b) 320 × 240 pixel images (left) and 464 × 348 pixels (right). The images are taken from different patients.

It is obvious that whilst the resolutions shown here do not differ considerably, the ability to display details is greater as the resolution increases.

For the case study to be presented, the FLIR Systems T540 camera was used with a resolution of 464 × 348, and for analysis the Tool + software (version 6.4.18039.1003) was used.

This resolution offers a series of advantages for the measuring of temperature values. In this case, due to the qualitative nature of the detections, the recommended resolution is more than enough to visualize a pathology.

Ambient temperature and humidity were measured with the Extech MO297 thermo-hygrometer to ensure that the ambient temperature did not change significantly whilst capturing the images.

The previously mentioned software includes analytical tools such as various colour palettes, the ability to adjust colours in ranges of temperature fields according to needs, measurement points, and lines, along with other elements also designed for the same purpose. It is a widely used software, given that most thermal camera manufacturers provide it as a free and easy-to-use tool for taking images and carrying out analysis manually.

The case studied here is a 23-year-old female student who served as a model to test an ergonomic support system for sleeping. Thermal images were captured before lying down, after lying down without the supports, and after lying down with the supports.

3 Results and Discussion

The images of the patient were taken according to the instructions described in the previous section, obtaining the following thermograms of the lower body (Fig. 3):

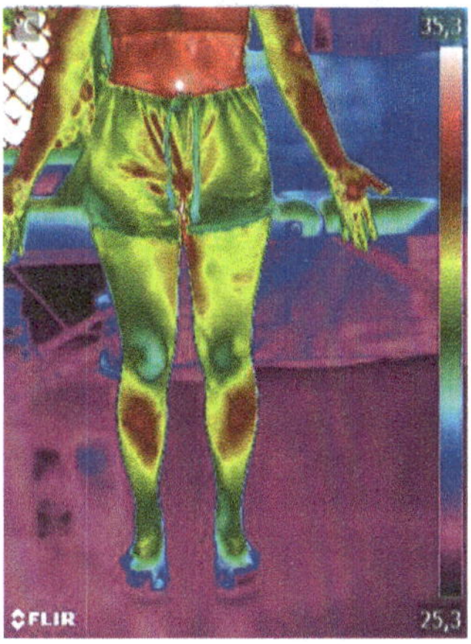

Fig. 3. Image of the patient before lying on the couch that was used for the trial

It is already possible to qualitatively observe thermal asymmetries between limbs, and different thermal patterns in symmetrical areas of the body, such as the knees.

However, there are no indications of circulatory problems, as we will see later.

After the patient had been lying for 20 min on a couch in a position similar to that used for sleeping, the following image was obtained:

In this image, thermal patterns reminiscent of veins or arteries can be observed in the upper part of the left quadriceps.

However, to a health specialist, an image like this poses more questions than answers, as the smallest difference in temperature could have many different causes, which could even lead up to reaching conclusions that could differ from an angyopathy.

Because of this, it is of great interest to be able to analyse and visually modify the image to study any possible anomalies with a higher precision and depth.

Changing the colour palette of the image to the "Inverted grey" of the software and adjusting the colour distribution in an interval that allows us to better visualize the area of interest produced the following (Fig. 5):

Using an appropriate analysis and employing the tools provided by the analytical software, the type of pathology can be estimated in many cases, although it is advisable

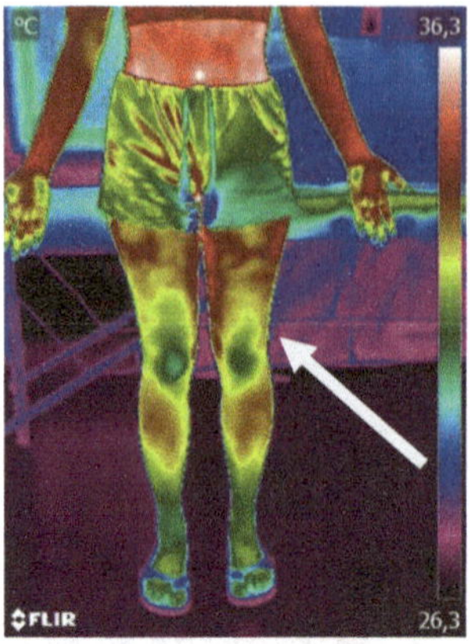

Fig. 4. Image of the patient having been in a sleeping position for 20 min on a couch

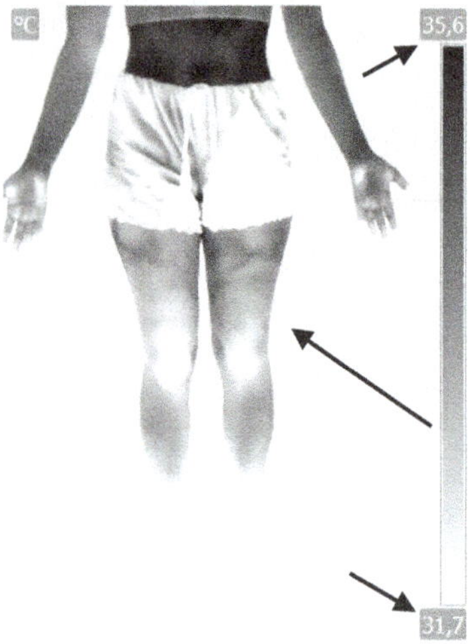

Fig. 5. Image from Fig. 4 with the "Inverted grey" colour palette and a thermal field adjustment that allows for a better visualization of the type of pathology.

to take into account that in many cases the specificity of this technique for making a diagnosis is low.

In image 6 we can see that veins and arteries have some type of alteration. Their position and even their shape enable us to have an idea of which veins or arteries we are looking at.

Furthermore, the general ability to visualize a patient as a whole allows thermography to be considered in many instances as an excellent tool for making an initial estimate or allowing for early detection of symptoms or signs of pathologies.

The latter can subsequently be determined using a more specific technique by studying an area that has been identified with an objective criterion such as a thermal asymmetry, a significant temperature, or a pattern that is recognizable as being potentially pathological.

The case study was designed to show thermal imaging after the patient had been lying down using the ergonomic leg supports.

In the area of interest, the following image was obtained (Fig. 6):

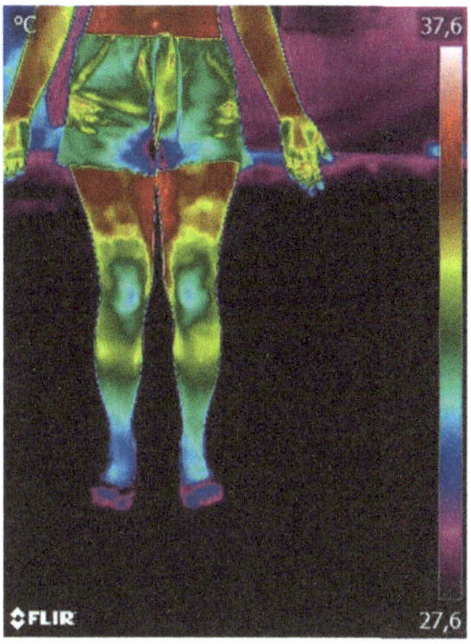

Fig. 6. Image of the patient after lying down for 25 min with ergonomic supports.

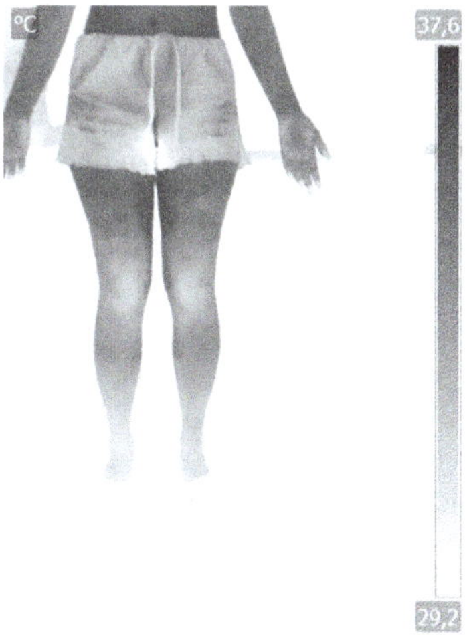

Fig. 7. Image of the patient after lying down for 25 min with the ergonomic supports and with the "Inverted grey" palette in order to visualize possible circulatory pathologies.

It can be observed that heat is proceeding to concentrate in the area of interest and is even suspected to be of a circulatory nature.

Again, after performing the correct analysis, the following was obtained (Fig. 7):

It can be seen that the heat concentration detected in the upper left quadriceps is not related to the circulatory system, and, even if this were so, this would be very mild.

This lets us verify that the followed methodology enables us to precisely identify significant changes that could occur to the circulatory system.

4 Conclusion

Infrared thermography continues to make progress in its applications and, in particular, in its use for imaging and analysis for the detection of pathologies in the health sector [6].

In this regard, experts have established the ambient conditions and equipment required for capturing images of the patients, including methodologies for the detection of pathologies (e.g., comparison of patterns and thermal asymmetries) [7].

Similarly, there is specialized training available on thermography, its uses, analytical methodology, and infrared behaviour, which is transversal since it is the professional, who uses thermography, who must be familiar with the meaning and usefulness of the information contained in each image. Because there are very few tools available for automatic analysis, the professional who analyses the images has a vital role to play in the analysis since they make the observations and then decide whether to use the different analytical tools to provide the maximum amount of data available regarding a possible pathology.

Based on the above, in the case presented in this article, the professional, being familiar with the analytical tools of the appropriate software, could inquire whether a thermal pattern suspected of being abnormal could be related to the patient's circulatory system.

The observed case shows how a thermal pattern with a colour palette that is considered anomalous is able to provide more information using analytical tools such as other colour palettes. Using grey scales, for instance, it is possible to determine whether the observed pattern has characteristics that indicate a pathology of circulatory origin.

Given that the exposed methodology is generic for any type of thermal image and that the result depends on the level of intensity of the pathology, it is not unreasonable to conclude that it is a method, which could, in a general way, enable the visualization of these types of anomalies. Even though in this study the increase in vein intensity was caused by the adoption of specific postures, real and more intense pathologies will be even more easily identifiable.

Therefore, as a result of this case, it is apparent that an analysis methodology with the correct colour palettes makes it possible for health specialists to visualize possible angyopathies, along with their location and affected veins or arteries. It is with this information that a professional can carry out a more precise diagnosis or ask for more specifics tests.

This opens new future paths of research that could determine what type of pathologies could be visible or not, as well as the necessary intensity for a correct visualization.

From there, new quantitative measures could be established, and even the possibility of automating the process.

Acknowledgements. The authors would like to thank the University of Malaga for their collaboration and support in providing the knowledge and assistance for developing the use of infrared thermography for health purposes.

References

1. Marins, B., Carlos, J., et al.: Time required to stabilize thermographic images at rest. Infrared Phys. Technol. **65**, 30–35 (2014)
2. Srini, G., Brioschi, M., et al.: Guidelines for neuromusculoskeletal infrared thermography sympathetic skin response (SSR) studies. The American Academy of Thermology (AAT) (2015)
3. American Academy of Thermology: Guidelines for neuro-musculoskeletal infrared medical thermography and sympathetic skin response (SSR) studies. American Academy of Thermology (2018). https://aathermology.org/organization-2/guidelines/guidelines-for-neuro-musculoskeletal-thermography/. Cited 5 March 2019, Citado el 5 de marzo de 2019
4. Thermohuman. Thermohuman (2014). https://thermohuman.com/es
5. Technical Committee: ISO/TC 108/SC 5 Condition monitoring and diagnostics of machine systems. ISO 18436-7:2014 Condition monitoring and diagnostics of machines — Requirements for qualification and assessment of personnel — Part 7: Thermography (2014)
6. Uematsu, S., et al.: Quantification of thermal asymmetry. J. Neurosurg. **69** (1988)
7. Hildebrandt, C., et al.: The application of medical infrared thermography in sports medicine. In: An International Perspective on Topics in Sports Medicine and Sports Injury, p. 534. InTech (2012)

Design of a Neurosurgery Training Simulator with Additive Manufacturing to Practice the Suture of the Dura Mater

Paz Morer-Camo[1(✉)], Jacobo Paredes-Puente[1], Marcos Llorente-Ortega[2], and Xabier Unamuno-Iñurritegui[2]

[1] School of Engineering, University of Navarra, Paseo Manuel de Lardizábal 13, 20018 San Sebastián, Spain
lucia.rodriguez@uca.es
[2] Lab Medical Image, School of Medicine, Campus University of Navarra, Pamplona, Spain

Abstract. The objective of this work has been the realization of a training simulator for neurosurgery operations using additive manufacturing. This work has been developed in collaboration with two neurosurgeons. The simulator developed contains two parts, a superficial skull inside which is the brain. Between both is the Dura Mater: a layer that cover the brain. The Dura Mater is a fibrous, solid, thick and not very flexible tissue, with a thickness of about 0.5 mm. In any brain operation, after having performed the craniotomy, the Dura Mater must be sutured. This membrane has a special consistency and the suture has to be made very close to the bone. Therefore, this complex skill must be trained. The Dura Mater must be replaced for each of the practices and satisfy with the thickness and consistency of human tissue so that it responds in a similar way to a suture performed in the operating room.

Keyword: Simulation based medical education · Neurosurgery · Additive manufacturing · 3D-printing · Skull surgery · Dura Mater

1 Introduction

The simulation, in general terms, can be understood as "A technique—not a technology—to replace or amplify real experiences with guided experiences that evoke or replicate substantial aspects of the real world in a fully interactive manner" [1]. The simulation applied to the medical environment, and therefore to health care, will have the ability to reproduce clinical situations. This practice helps to improve the quality of care given to patients and, in turn, helps to save lives [2]. Therefore, the Society for Simulation in Healthcare defines it, as "Simulation is the imitation or representation of one act or system by another. Healthcare simulations can be said to have four main purposes: education, assessment, research, and health system integration in facilitating patient safety" [3].

The use of simulators for the training of future neurosurgeons is of great help, since it requires prior experience before performing this type of operation on a patient. Even

F. Cavas Martínez et al. (Eds.): INGEGRAF 2021, LNME, pp. 182–190, 2022.
https://doi.org/10.1007/978-3-030-92426-3_22

neurosurgeons themselves need to be trained in these types of operations, since these are not necessarily frequent. They constantly must train dexterity and ability and be able to perform pseudo-operations in a safe, trusted environment. Therefore, the use of simulators in medicine has been a great revolution. It allows to better structure the time, facing a surgical intervention.

These of simulators have many advantages: it encourages students to become familiar with self-learning methods, adapting to their learning speed, allows continuous evaluation by professionals and error is used as a sample of learning and even allows treating correctly ethical problems and inconvenience caused to patients. In fact, its use has been increasing throughout these last years of teaching [4–7].

A medical simulator for training purposes is understood to be that system that imitates, in the most realistic way possible, the actions that a doctor must carry out when doing a concrete and specific process. We understand simulation as a technique that replaces and amplifies real experiences, evoking and replicating substantial aspects of the real world in an interactive way. Its origins can be found in the second half of the 20th century [8, 9]. Two key moments can be distinguished. First, The Resusci Anne simulator, by Asmund Laerdal, in which the techniques for cardio-respiratory resuscitation and heart problems were practiced. Later, the complex simulators created by Abrahamson and Denson in the 1960s known as "SimOne" of very high cost (so it was not successful in the market) which allowed real-time physiological responses to actions performed on it. From this moment on, a great technological advance is observed with increasingly sophisticated models [10]. Its use throughout the teaching of medicine has been established as a necessary path to acquire the skills associated with this profession in order to improve the abilities and skills of the student.

There is a great typology of simulators, depending on the use to which they are intended and the technology used. Some are used primarily as a learning strategy and others seek to develop skills and abilities when performing surgical operations. It can even find video games that invite you to perform simulations of surgical operations. Among the simulators for specific use, we find those used in laparoscopic surgery [11] or, as is the case presented in this article, those used in neurosurgery operations.

2 Work Objective

The simulator presented in this work corresponds to an anatomical simulator. Among the applications of this group of medical simulators are pre-surgical planning, pre- surgical simulation, selection of surgical equipment and education. In addition, the techniques used to manufacture anatomical simulators are also used for different applications within the medical field such as the creation of implants, prostheses, splints or external fixators. Anatomical models used in neurosurgery such as cranial and brain structures occupy a large part of this broad field. In fact, in 2016 about 60 articles of this class of prototypes were registered [10, 12, 13].

The objective of this project consisted in the realization of a prototype that practices the suture of the Dura Mater, a procedure that is carried out at the end of operations such as craniotomy. The dura mater is one of the layers between the skull and the brain, closely placed between the brain and the inner layer of bone. It is a fibrous, dense and inflexible tissue (Figs. 1 and 2). It is around 0.5 mm thick, characteristics that make it a piece with difficult-to-replicate properties. Different prototypes made with additive manufacturing have been used presenting great advantages in the processes of different training medical skills [14–18].

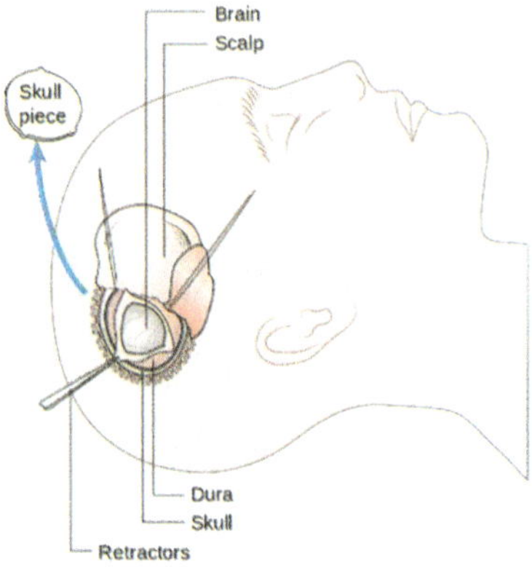

Fig. 1. Craniotomy © Cancer Research UK – CRUK (CC BY-SA 4.0)

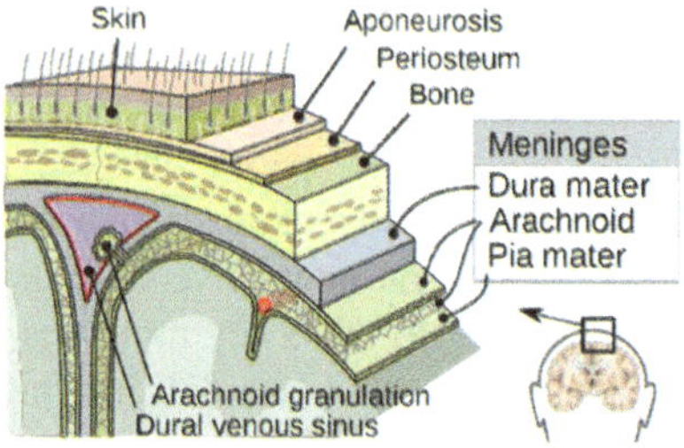

Fig. 2. Representation of the different layers of the brain © SEER Development Team (CC BY-SA 3.0)

Intracranial surgeries are complex due to the reduced space and the care that must be taken with the soft tissue (brain). For this reason, it is important to acquire the necessary skills in these procedures. The simulator that has been developed consists of two parts. In the first place, a skull has been reproduced with a removable opening, known as a craniotomy. The second piece is the brain. Inside the brain, a space must be available, destined to different games that can be positioned, to train the skill of the operation of a tumor lodged in that area. This area is accessed with a dynamic retraction microscope with instrumentation or with a spatula attached to a removable retractor. Therefore, it is necessary to coordinate the actions carried out inside without having a direct vision. In order to reach this inner area, it is also necessary to previously cut the Dura Mater and enter through Silvio's fissure (the corridor between the frontal and temporal lobes). The area of the skull through which the craniotomy is performed is known as the Pterional region area.

3 Procedure: Materials and Methods

From the medical images obtained by Computerized Axial Tomography (CT) and Magnetic Resonance (MR) (data from the equipment: TAC Somaton Drive/Force, RM Magnetom Aera 1.5 T) located in the CUN (Navarra), both the skull and the brain were reconstructed using free open source computer software available on multiple operating systems, with plug-in capabilities to add algorithms and applications. The CT study parameters were as follows: matrix - 512; exposure factors: 120 kV, 208 mA; slice thickness - 0.8 mm (Fig. 3).

The 3D models are obtained with the 3D Slicer software [19], and later, the models are worked on in Meshmixer [20] in order to align the brain inside the skull. 3D Slicer is a

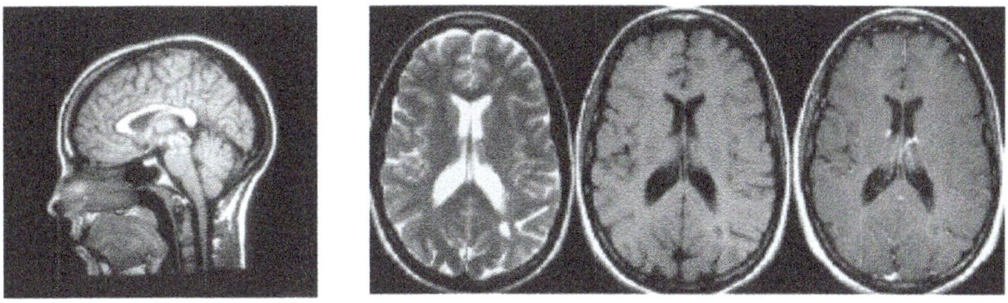

Fig. 3. MRI images processed in Slicer 4.10

software platform for the analysis and visualization of medical images; it is a free, open source software, available in multiple operating systems, with plug-in capabilities to add algorithms and applications. Although it was started and coordinated from Harvard University, 3D Slicer is the effort of a large group of developers and collaborators from multiple institutions throughout the world. This software uses slices captured by CT scanners and treats them as DICOM (Digital Imaging and Communications in Medicine) images to generate virtual 3D models corresponding to anatomical parts of the human body (Fig. 4).

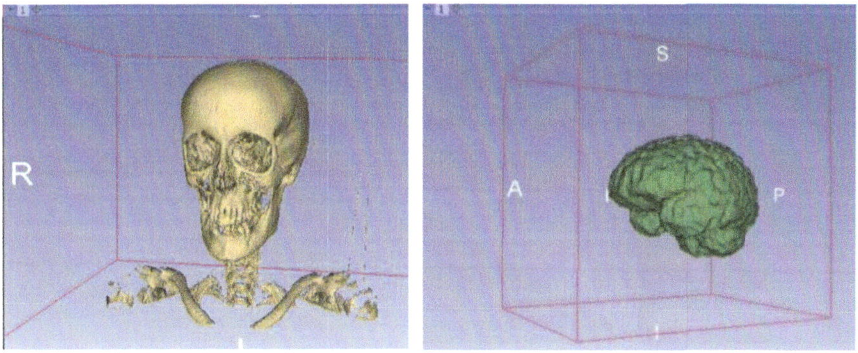

Fig. 4. 3D model with Slicer 4.0

Once the 3D model is obtained, it is exported to "STereoLithography", commonly known as *.stl format. The file contains information about the model made up of hundreds of triangles linked together. Meshmixer allows cleaning the previous model obtained with 3D Slicer, based on high-resolution dynamic triangle meshes, smooths the mesh and provides the final file to be printed by additive manufacturing (Fig. 5). This program allows cleaning discontinuities, detecting mesh errors that can cause problems when printing in 3D.

From these models, it was decided to replicate half of the skull, since it was sufficient for the simulation that it was desired to do. From there, the printing was carried out using additive manufacturing, obtaining the smock-ups in PLA. Layer resolution was up to 0.1 mm and extruder diameter was 1.75 mm (Fig. 6).

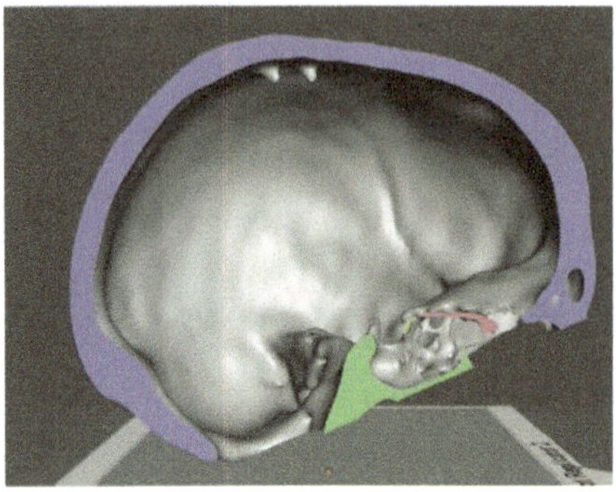

Fig. 5. Segmented image of the half of the skull obtained in Meshmixer

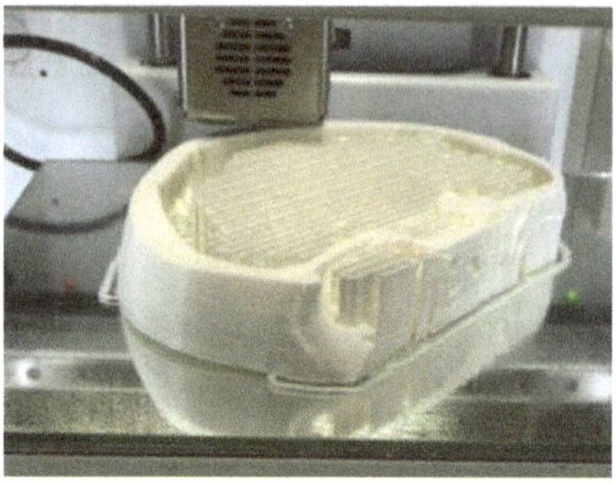

Fig. 6. 3D printing process of the skull in PLA

To create the brain model, a texture with biomechanical properties similar to those of the human brain is sought. Firstly, it is necessary to make a mold to contain the brain. The casting process is a manufacturing technique in which a liquid material is poured into a mold with the desired shape and subsequently cured and solidified. The mold is manufactured with additive manufacturing (Fig. 7).

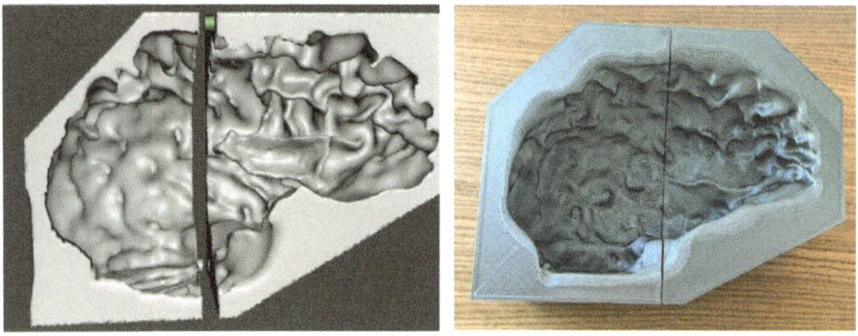

Fig. 7. Skull mold used to obtain the brain mock-up.

Secondly, it was decided that silicone would be used for brain development since it is the material whose texture most closely resembles that of the brain. This silicone was produced using two different components, Ecoflex 030 and Silicon Thinner (Table 1).

To separate the prototype from the mold, a spray was used that prevented the adhesion of the silicone to the mold (Fig. 8).

Lastly, for the creation of the Dura Mater, it is interesting to analyze the different studies that have been carried out to obtain substitutes for the human dura. The similarity of the properties of these substitutes with the outer meninges is the priority when choosing one material or another. The most relevant characteristics to achieve greater realism are the thickness of the layer, the Young's modulus and the tensile strength.

Table 1. Quantities and proportions for the manufacture of the brain.

	Quantity (gr)	Proportion
Ecoflex 030 PART A	201.63	36.66%
Ecoflex 030 PART B	201.63	36.66%
Silicon Thinner	146.63	26.66%

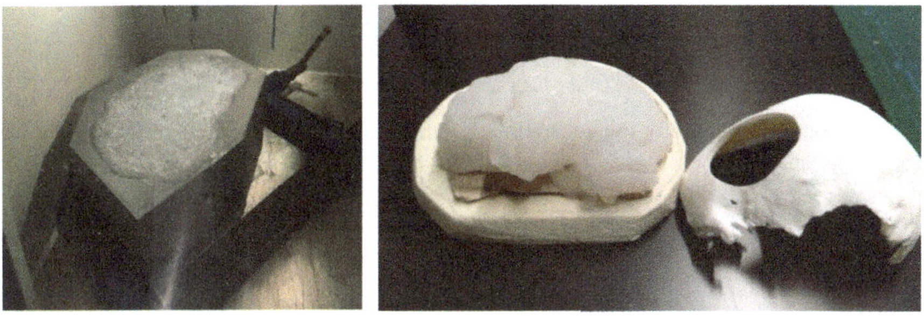

Fig. 8. Silicone brain and mold process in vacuum machine.

A fibrous paper was impregnated with silicone, manually. Subsequently, the excess was removed, leaving it on a wooden board and exerting light pressure with a piece of methacrylate. It was left in the oven for about an hour to cure. Although the initial idea was to fix the dura to the skull, it was decided to change strategy and fix it to the brain using a series of nails, ensuring that it remained tight (Table 2).

Table 2. Mechanical properties of the human Dura Matter and the silicones used

	Tensile strength (MPa)	Elongation at break
Duramadre	7.01 ± 0.77	116 ± 3%
Econ 80	6.58	127
Sorta – Clear 40	5.52	400

In this way, once the suture is in place, the tissue could be easily replaced by removing the pins and repositioning them (Fig. 9).

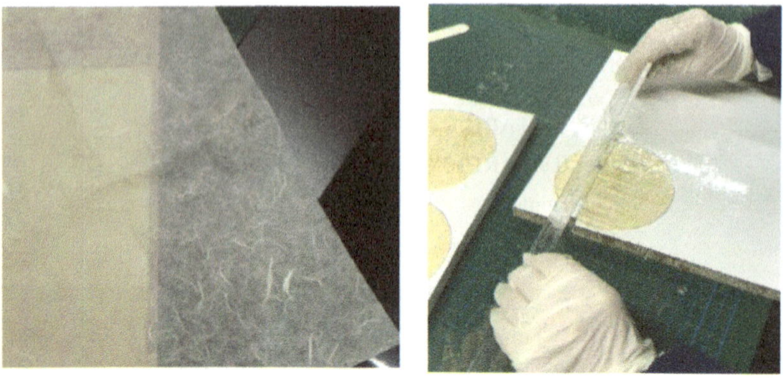

Fig. 9. Fibrous paper used for the generation of dura mater. And Simulation of dura mater from fibrous paper and silicone

4 Final Prototype and Validation

Finally, another important feature of the simulator was that it had a support to support the brain and skull. Due to the inclination that the head of a patient must have in relation to the posture of the neurosurgeon when he is going to perform the intervention. For this, a support was built, using the Inkscape program, leaning on the shape of the skull. This support was made of wood to be later cut with a laser. The skull was fixed with two Ls and screws to prevent movement. This base piece was raised over an identical one, gluing more material between the two to avoid the difference in height that the screws and the L's supposed. Instructions were drawn up on the back of the lower piece so that users could use the prototype (Figs. 10, 11 and 12).

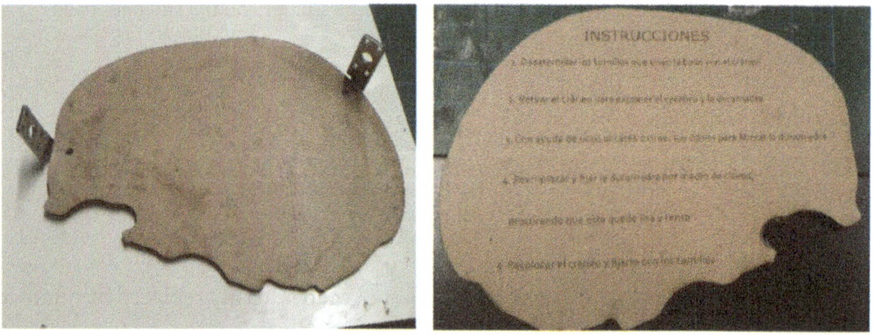

Fig. 10. Top and bottom of the bracket.

Validating is the last phase of any design process, it can be said that it is the moment of truth, the end of the journey. It is not enough to show the prototype. It is necessary to collect feedback from the user. The prototype must include the basic functionalities to solve the problem posed: train the ability to practice suturing the dura mater when neurosurgical operations that require trepanation of the skull, cut the dura mater, advance

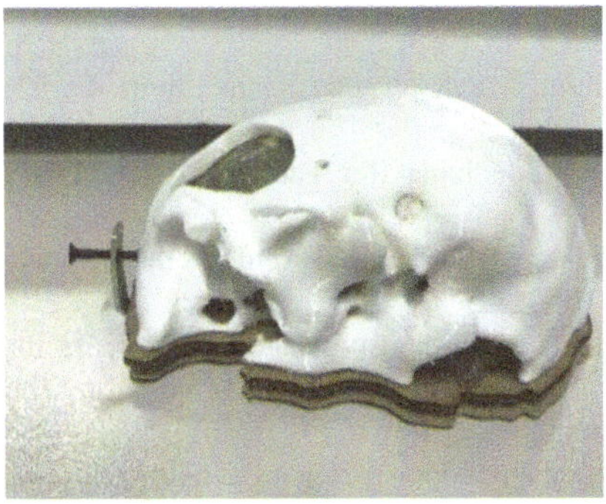

Fig. 11. Final prototype.

Fig. 12. The neurosurgeon proceeds to cut and subsequently suture the dura mater.

through the Silvio fissure, and re-suture the dura before reclosing the skull. The initial problem was raised by the doctors of the Department of Neurosurgery of the Clinic of the University of Navarra. It could be approached thanks to medical support. Therefore, they were asked to see for themselves the usefulness of the prototype and if it fit their needs. The prototype was found to be useful, desirable, feasible, and feasible. The success of the simulation depends on the existence of high physical fidelity between the prototype and what is to be reproduced in order to develop manual and dexterity skills. On the one hand, a high conceptual fidelity is sought in which clinical reasoning and the ability to solve problems are developed and, on the other hand, a high emotional fidelity in which the retention of information is favored through the management of complex processes that involve knowledge or emotions. A successful simulation is not based on the realism of the simulation per se, but on the commitment of the participants in their roles, an adequate connection between those involved in order for the student to be able to actively link social, psychological and social experiences. Previously lived clinics.

Acknowledgments. The research work reported here was possible thanks to the collaboration and support offered by the CUN simulation laboratory and the neurosurgeons who have followed this initiative.

References

1. Gaba, D.M.: The future vision of simulation in health care. Qual. Saf. Health Care **13**(Suppl. 1), i2–i10 (2004). https://doi.org/10.1136/qshc.2004.009878
2. Montgomery, J.: Medical simulation technology. J. Pediatr. Surg. Nurs. **5**(4), 107–111 (2016). https://doi.org/10.1097/JPS.0000000000000119
3. Society for Simulation in Healthcare. https://www.ssih.org
4. Carriel Mancilla, J., et al.: Prácticas de simulación en medicina: ventajas, limitantes, recuento histórico y perspectiva ecuatoriana. Rev. Med. FCM-UCSG **17**(4), 285–291 (2011)

5. Hammond, J.: Simulation in critical care and trauma education and training. Curr. Opin. Crit. Care **10**(5), 325–329 (2004)
6. Abdulmonhsen, H.: Simulation-based medical teaching and learning. J. Fam. Commu. Med. **17**(1), 35–40 (2010)
7. So, H.Y., et al.: Simulation in medical education. J. R. Coll. Physicians Edinb. **49**, 52–57 (2019)
8. Neri-Vela, R.: El origen del uso de simuladores en Medicina. Rev. Fac. Med. UNAM **60**(S1), 21–27 (2017)
9. Singh, H., et al.: History of simulation in medicine: from Resusci Annie to the Ann Myers Medical Center. Neurosurgery **73**(4), 9–14 (2013)
10. Hoang, D., Perrault, D., Stevanovic, M., Ghiassi, A.: Today surgical applications of three-dimensional printing: a review of the current literature & how to get started. Ann. Transl. Med. **4**(23) (2016). https://doi.org/10.21037/atm.2016.12.18
11. Garcia Murillo, J., et al.: Diseño de prototipo de simulador para entrenamiento en cirugía laparoscópica. Revista Ingeniería Biomédica **5**(9), 13–19 (2011)
12. Namba, K., et al.: Microcatheter shaping for intracranial aneurysm coiling using the 3-dimensional printing rapid prototyping technology: preliminary result in the first consecutive cases. World Neurosurg. **84**(1), 178–186 (2015). https://doi.org/10.1016/j.wneu.2015.03.006
13. Pacione, D., et al.: The utility of a multimaterial 3D printed model for surgical planning of complex deformity of the skull base and craniovertebral junction. JNS J. Neurosurg. **125**(5), 1194–1197 (2016). https://doi.org/10.3171/2015.12.JNS151936
14. Javan, R., et al.: A prototype educational model for hepatobiliary interventions: unveiling the role of graphic designers in medical 3D printing. J. Dig. Imaging **31**, 133–143 (2017)
15. Kondo, K., et al.: A neurosurgical simulation of skull base tumors using a 3D printed rapid prototyping model containing mesh structures. Acta Neurochir. **158**, 1213–1219 (2016)
16. Skrzat, J., et al.: Three-dimensional model of the skull and the cranial bones reconstructed from CT scans designed for rapid prototyping process. Folia Medica Crac. **LVI**(2), 45–52 (2016)
17. Anchietea, M.V.M., et al.: Skull reconstruction after resection of bone tumors in a single surgical time by the association of the techniques of rapid prototyping and surgical navigation. Int. J. Comput. Assist. Radiol. Surg. **11**, 1919–1925 (2016)
18. Favier, V., et al.: Geometric and mechanical evaluation of 3D-printing materials for skull base anatomical education and endoscopic surgery simulation – a first step to create reliable customized simulators. PLoS ONE **12**(12) (2017)
19. Página oficial del Softwre 3D Slicer. https://www.slicer.org/
20. Página oficial del Software Meshmixer. http://www.meshmixer.com/
21. Pacione, D., et al.: The utility of a multimaterial 3D printed model for surgical planning of complex deformity of the skull base and craniovertebral junction. JNS J. Neurosurg. **125**(5), 1194–1197 (2016)
22. Javaid, M., et al.: Additive manufacturing applications in medical cases: a literature based review. Alex. J. Med. **54**(4), 411–422 (2018)

Semiautomatic Modeling of Bone Tissue from Medical Image for Finite Element Method Based Biomechanical Studies

Álvaro Pérez-Sala[1](✉), Rafael Peláez[1], Fátima Somovilla Gomez[2], María Ángeles Martínez Calvo[2], Marina Corral Bobadilla[2], Saul Íñiguez Macedo[3], Asier Rodríguez San Miguel[3], Enrique Fernández Martínez[3], Manuel Rubio Sampedro[3], Rubén Lostado Lorza[2], and Ignacio M. Larráyoz[1](✉)

[1] Grupo de Biomarcadores y Señalización Molecular, Centro de Investigación Biomédica de La Rioja (CIBIR), Logroño, Spain
{aperez,ilarrayoz}@riojasalud.es

[2] Departamento de Ingeniería mecánica, Universidad de la Rioja, C/ Luis de Ulloa, 4, 26004 Logroño, La Rioja, Spain

[3] Centro Tecnológico del Calzado de La Rioja (CTCR), Calle Raposal, 65, 26580 Arnedo, La Rioja, Spain

Abstract. Finite element method (FEM) biomechanical analyses have proved their effectiveness on prosthesis set up and bone tissue biomechanical behavior research through different studies. Over the past few years, new transactional appliances of performing biomechanical *in silico* studies have been found highly interesting for orthopedic surgery and traumatology. One of the main drawbacks for swift FEM model building on joint, and complex structure scenarios, is the high workload derived from 3D model building and subjacent medical image processing. As a consequence, new breakthrough segmentation methodologies development would significantly help to improve the contributions of FEM models to medical daily practice reaching more effective workflows to build the models. For this reason, a new approach for an automatic, semi-assisted segmentation methodology is proposed centered on FEM ready mesh modeling from bone tissue region isolation defined on medical images. Generated meshes from proposed methodology have proved to be coherent for FEM remeshing and they were able to generate a robust node and elements model, avoiding mesh holes and errors. Moreover, those meshes showed mechanical behavior similarities on FEM essays with manually created models, proving their reliability. To conclude, this methodology is a valid alternative for bio-mechanical model development from medical image tissue region isolation, allowing researchers and engineers to reduce time needed for 3D model generation processes.

Keywords: FEM · Segmentation · Medical image · Biomechanics

1 Introduction

Finite element method (FEM) is a numerical resolution algorithm which can be applied to a wide range of problems in different disciplines. Finite element analysis (FEA)

F. Cavas Martínez et al. (Eds.): INGEGRAF 2021, LNME, pp. 191–200, 2022.
https://doi.org/10.1007/978-3-030-92426-3_23

are suited to solve complex problems that given their nature cannot be handled by analytical procedures, for instance, biomechanical modeling [1]. FEA models *in silico* allow researchers to study biomechanical behaviors of bones, joints, ligaments, and others. Those analysis consider a high number of variables, turning them into a highly valuable tool for analysis and research on orthopedical surgery and traumatology fields [2–5].

Usually, the required process to build an *in silico* biomechanical FEA is divided in eight different phases: geometry generation of each part, remeshing for final elements mesh, mechanical properties set up for different materials involved, boundary and contact interactions definition between element meshes, displacements and force set up, simulation, experiment data retrieving and result validation [6].

Regarding biomechanical applications, geometry generation process consists in performing a segmentation step on the bidimensional registered image set which constitute either the full magnetic resonance image (MRI) or the computerized axial tomography (CT) image. This segmentation process target is building a binary mask over the pixels of each image slice that belongs to the tissue of interest. That binary mask encapsulates all the tissue of interest whereby surfaces and polygonal meshes will be generated. Different algorithmic approaches have been described in different categories of reliability, complexity, computational cost and manual processing cost to obtain the full required geometry. One of the most common approaches is thresholding algorithms over Hounsfield units (standard scale for CT image density values), which allows to isolate those voxels whose intensity is between the given value range. Models produced by thresholding have shown its robusticity and accuracy in many studies, however, artifacts and external noise are common in this process and a manual revision and cleaning process is needed involving a high cost [7]. Therefore, simplifying and automatizing the geometry generation processes from CT images would considerably facilitate the development of biomechanical FEAs, opening up new research applications of those analysis, and increasing significantly clinical cases assessed by this technology. On this basis, we propose a semi-assisted segmentation methodology based on improved thresholding algorithms, specially tuned up to produce finite element remeshing-ready geometries for their addition to FEA *in silico* models of biomechanical systems.

2 Objectives

The aim of this study is to analyze the geometry accuracy obtained by the proposed segmentation methodology against a reference model geometry obtained by manual segmentation process considered as control model. The described manual segmentation process of the reference geometry was performed by an expert and exported to standard STL format.

3 Materials and Methodology

The proposed methodology is based on an improved thresholding algorithm developed with the open-source medical and scientific image processing libraries "Insight segmentation and registration toolkit" (ITK) and "Visualization Toolkit" (VTK) [8]. Those libraries provide powerful image processing tools and its use is widely extended on the field [9].

Indeed, many developments are based on those libraries and they are part of the most important open-source medical image visors kernel such us 3D Slicer [10].

The proposed geometry generation technic is based on thresholding isolation combined with a region growing seed to identify the desired region. Final mesh generation criteria are based on contour threshold of the region grown from the seed matching input threshold values. While the first version was designed for general body tissue 3D mesh generation [11], the algorithm could include many different parts semi connected or with similar CT density in the same model. Thus, this version has been improved by configurable region of interest (ROI) in order to discriminate different bones or regions.

In addition to the DICOM (Digital imaging and communications in medicine) image files, final algorithm requires the following values as user input:

- Region of interest
- Seed point belonging to the desired tissue to extract
- Hounsfield units range for the matched tissue

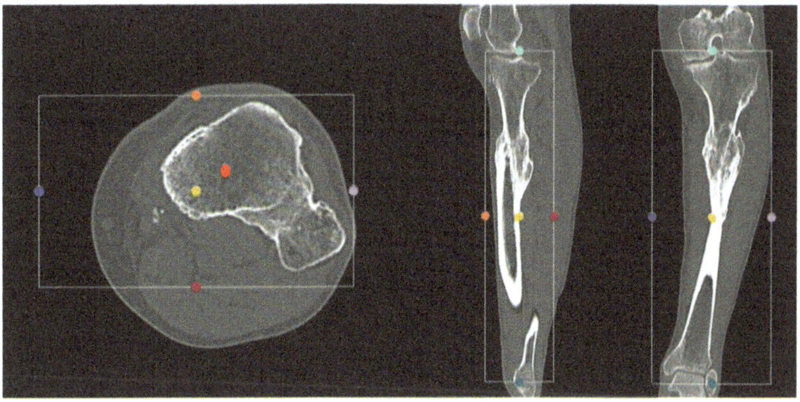

Fig. 1. Axial, sagittal and coronal slices (left to right) of the lower limb containing the defined ROI and selected seed for tibia isolation.

Since our goal was to extract solid models including cortical and trabecular bone for biomechanical studies, thresholding values for bone tissue were defined between 308 and 2400 Hounsfield units. Therefore, we fixed this input parameter and reduced the input parameters to two: seed and ROI, simplifying the required user input to assist the system.

To test the performance of our proposal, we selected a clinical case of lower limb with geometry different from the standard anatomy. Thus, we were able to distinguish this type of algorithm from atlas or statistical inference-based ones, because those could achieve high efficacy with standard anatomy but are useless on clinical practice when bone geometry is altered [7, 12]. The selected study case is a knee joint, including the whole tibia, which presents a bad healed fracture turning out on a dealignment of the support axis, along with a callous linking proximal tibia with the fibula. Since that model is considerably different of a normal tibia, probabilistic and atlas-based algorithms would find strong difficulties to generate the matching geometry. A good outcome of the proposed methodology would postulate it as a better option for clinical biomechanical FEAs than previous cited ones.

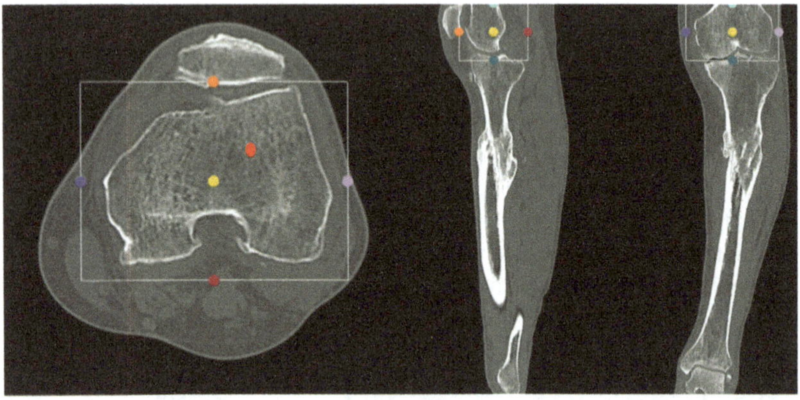

Fig. 2. Axial, sagittal and coronal slices (left to right) of the lower limb containing the defined ROI and selected seed for femur isolation.

The designed *in silico* model was centred on test load point on knee joint derived from the misalignment produced by tibia fracture. In order to build the model, full tibia and distal femoral epiphysis geometries were required to be 3D-modeled extracting the bone tissue geometry from the CT-scan. Giving that hip joint is not present, we decided to design FEA model applying forces in the theoretical support axe distributed over a perpendicular cut in the femur.

Firstly, ROIs were defined for each geometry as described in Fig. 1 for the tibia and Fig. 2 for distal femur. Then, growing seeds were defined inside the desired tissue to assist algorithm during segmentation choosing a random point inside the tissue area. Using that setup, poligonal meshes were generated and saved in STL standard format, commonly used for meshed 3D surface models storage [7].

The accuracy of the model generated with the proposed methodology was tested by surface difference analysis versus manually-generated control geometry, point by point. Distance map was calculated by Artec Studio 15 Professional software measure tools (Artec Inc., Luxemburgo), and represented by the colored distance map over the model and model point-to-point distance table. From point-to-point distance data, descriptive statistic measures were calculated to establish mean difference of proposed model versus the reference, distance variability and highest deviation. Next step was the inclusion of all generated meshes into the FEM simulation software MSC Marc [13] testing their compatibility with the remeshing processes from polygonal surface to elements and nodes.

Reference meshes were tuned up and checked for incoherence to be included in Marc scenario. STL mesh was imported as faceted surface and a remeshing process was performed to generate matching element meshes. Surface remeshing was set up using an element size of 3 mm, Results were checked by surface tester to ensure mesh was coherent and free of breaks or holes. Once we obtained element surfaces, volume element remeshing was performed to generate a tetrahedral element mesh using a grossing factor of 10 for internal elements. From that process, we obtained the final element meshes we would use to run the simulation. For semiautomatic-built models, same process is followed testing that auto-generated meshes are fully compatible with Marc remeshing workflow needed to develop the *in silico* model.

4 Results

Firstly, two STL files containing reference geometries were provided by an expert, one of them for tibia and the other for distal femur. Meanwhile, following our proposed methodology we generated two matching STL files one for each part geometry. As a first analysis, each file was visually inspected on the mesh treatment software Autodesk Meshmixer, where many artifacts could be observed on semiautomatically generated models. Those artifacts appeared around the joint area and, as we expected, especially around fibula bone callous.

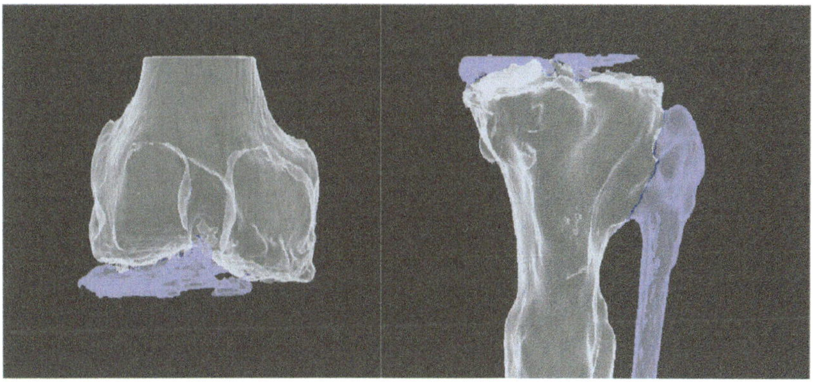

Fig. 3. A. Artifacts found on semiautomatic result of femur segmentation. B. Artifacts found on semiautomatic result of tibia segmentation.

Therefore, a quick cleaning step was performed using the same visualization tool to remove artifacts and saving final meshes. In Fig. 3, the removed artifacts can be identified by a blue shadow mark, and the final models with a grey shadow mark. Models kept the same mesh density and characteristics and no conversion or remeshing was applied while saving the files. Most jeopardizing point was removing fibula callous because in the reference model this area was carefully divided manually by the expert based on little features on callous intensity. Meanwhile, semi-automatic model included whole fibula and it was removed manually with the other artifacts following the surface trend. Due to this correction, callous surrounding area is expected to be significantly less accurate than the rest of the model.

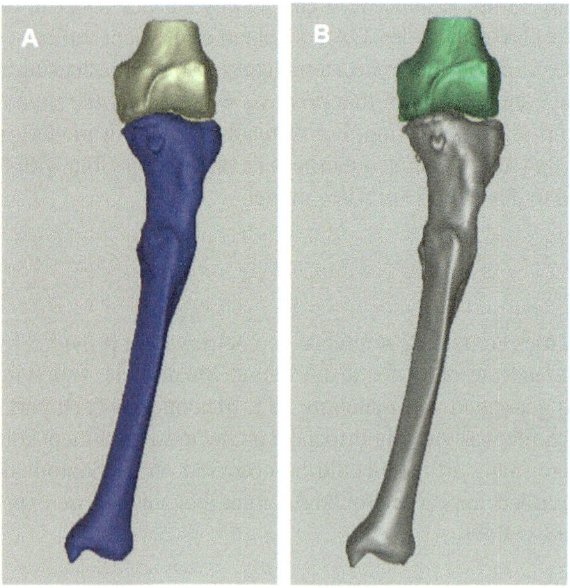

Fig. 4. A. Outcome geometries from manually segmented models. B. Outcome geometries from semiautomatic segmented models.

Once models were cleaned of artifacts, they were ready to check the performance of the algorithm comparing both model sets. Manually generated models did not require further action after the visual inspection, as it can be appreciated on Fig. 4 A (Table 1).

Table 1. Model data of the different geometries obtained. V = Volume, A = Area, Polig. = Polygon number of generated mesh. Cost Seg/Post = Human time effort for segmentation (Seg) y postprocessing (Post) task of the model.

Geometry	Component	V (cm^3)	A (cm^2)	Polig.	Seg. Cost (min)	Post. Cost (min)
Reference	Femur	212.49	217.8	75864	35	0
Reference	Tibia	437.73	509.88	175816	60	0
Study	Femur	210.18	221.99	97402	5	10
Study	Tibia	444.38	527.76	61672	5	10

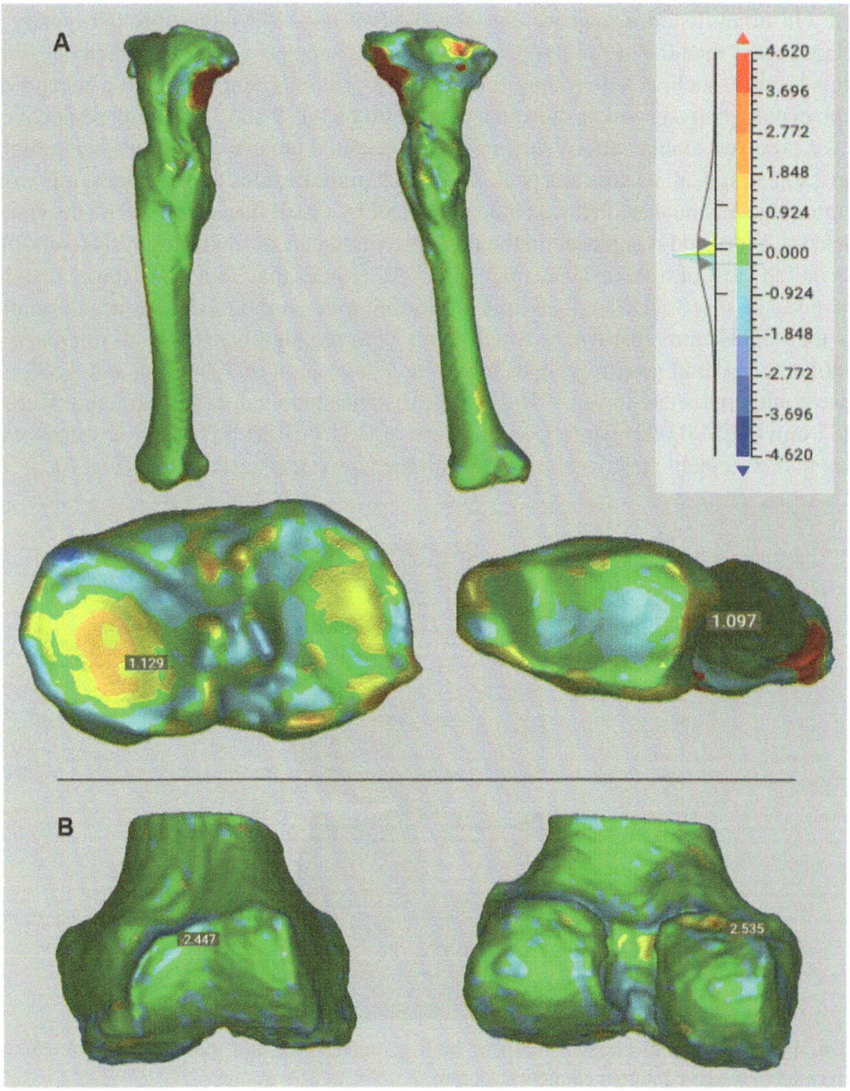

Fig. 5. Distance maps of semi-automatic generated models against reference models. A. Tibia map. B. Femur map

Model data was analysed reporting minimal desviations of 1.14% on volume and 3.5% on model area over control. As a result of using different tools to create the meshes, polygon counts are quite different between both methodologies but it does not necessarily compromise the accuracy of the method. Regarding file properties, all meshes have a normal polygon count and a moderated weight of 8 MB, being semiautomatic-generated meshes lighter than manually-generated matches. A separate mesh deviation analysis was performed on the tibia and femur and different distance maps were generated for each. On tibial distance map we can observe that mesh deviation through mostly all surface is less than 0.924 mm (yellow green and cyan zones). However, there are some points to note, such as a desviation on the tibial proximal epiphysis and a second one in the distal epiphysis which showed a desviation of 1.129 and 1.097 mm, respectively. As we expected, fibula callous surroundings presented greater deviations, being higher than 2.772 mm in all its area and peaking at 4.620 mm. Besides those regions, it is worth mentioning a residual artifact we could not detect by visual inspection whose desviation reaches 4.1 mm and it is placed in the most proximal part of tibial epiphysis (Dark blue spot on top view of tibia distance map, Fig. 5 A. Besides that, femoral distance map has showed to be more accurate than tibial, reporting most of the model surface desviation less than 0.924 mm. Just two isolated points deferred significantly from reference on condilar frontal and posterior edges reporting a desviation of 2.447 mm and 2.535 mm as we could appreciate in Fig. 5 B. Average distance map values for the femur were of 0.257 mm (SEM 0.002) while for tibia were of 0.315 (SEM 0.003). Both values were more than acceptable results compared with the equivalent surface area (Fig. 6).

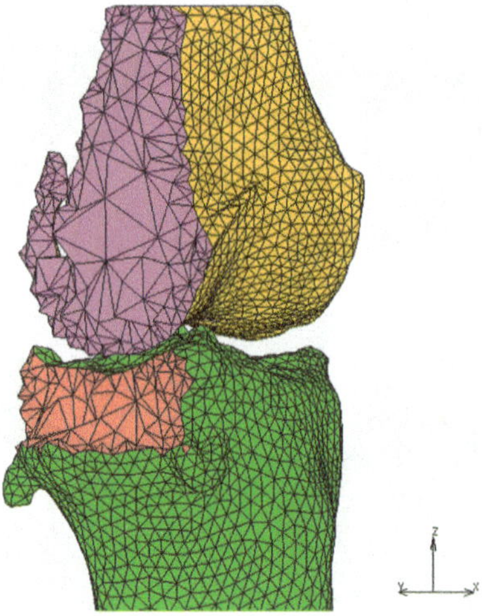

Fig. 6. Final tetrahedral element mesh of both geometries around the joint area. A cut was performed to observe internal element structure.

To conclude, remeshing workflow described above was performed over all geometries, reporting fully connected element meshes in addition to fulfilling all test-checking criteria. Moreover, no overlapping or flipping effects were found near the joint, which could interfere on the FEM simulation. Final element meshes showed 36 845 and 20 745 elements for the femur and the tibia, respectively. Both models got MSC Marc experiment designer's approval and satisfied all requirements to set up load cases and simulation.

5 Conclusions

Geometry extraction step from medical images is a key point to achieve a dynamic development of accurate models to study biomechanical behaviors on clinical practice. Therefore, new methodologies and workflows for that purpose needs to be developed in order to improve their efficacy, reducing the time and effort required. Those improvements could introduce *in silico* FEAs as a new tool for orthopedics allowing a new era of personalized medicine [14]. The studied approach looks forward to reach a balance between effort and precision. Specifically, it is focused on reducing the overall human dedication needed to perform the segmentation, model generation and preparation processes. Despite of reducing the required dedication for those processes, the presented solution still requires some knowledge in mesh edition software and it is still a semi-attended process. However, it has proved to be a good entry point to achieve a fully automatic extraction process providing consistent and accurate models.

To sum up, we conclude that the proposed methodology meets its objective of generating good quality models, fully compatible with standard FEM design software. Meanwhile it has many limitations, this system satisfies all process requirements better than other computational options, which makes it a promising approach to automated geometry extraction from medical image.

Acknowledgements. This research was funded in part by a grant (PI19/01805) from the Instituto de Salud Carlos III, co-funded by European Regional Development Fund (ERDF) "A way to build Europe" and by Fundación Rioja Salud. I.M.L. is supported by a Miguel Servet contract (CPII20/00029) from the Instituto de Salud Carlos III, co-funded by European Social fund (ESF) "Investing in your future".

References

1. Erdemir, A., Guess, T.M., Halloran, J., Tadepalli, S.C., Morrison, T.M.: Considerations for reporting finite element analysis studies in biomechanics. J. Biomech. **45**(4), 625–633 (2012)
2. Mattei, L., Campioni, E., Accardi, M.A., Dini, D.: Finite element analysis of the meniscectomised tibio-femoral joint: implementation of advanced articular cartilage models. Comput. Methods Biomech. Biomed. Eng. **17**(14), 1553–1571 (2014)
3. Gozar, H., Derzsi, Z., Chira, A., Nagy, Ö., Benedek, T.: Finite-element-based 3D computer modeling for personalized treatment planning in clubfoot deformity. Medicine **97**(24), e11021 (2018)
4. Chu, Y.-L., Chen, C.-H., Tsuang, F.-Y., Chiang, C.-J., Wu, Y., Kuo, Y.-J.: Incomplete insertion of pedicle screws in a standard construct reduces the fatigue life: a biomechanical analysis. PLoS ONE **14**(11), e0224699 (2019)
5. Baldawa, R., Bhad, W.: Stress distribution analysis during an intermaxillary dysjunction: a 3-D FEM study of an adult human skull. Ann. Maxillofac. Surg. **1**(1), 19 (2011)
6. Pfeiffer, F.M.: The use of finite element analysis to enhance research and clinical practice in orthopedics. J. Knee Surg. **29**(2), 149–158 (2016)
7. van Eijnatten, M., van Dijk, R., Dobbe, J., Streekstra, G., Koivisto, J., Wolff, J.: CT image segmentation methods for bone used in medical additive manufacturing. Med. Eng. Phys. **51**, 6–16 (2018)

8. Schroeder, W., Martin, K., Lorensen, B.: The Visualization Toolkit, 4 edn. Kitware (2006)
9. McCormick, M., Liu, X., Jomier, J., Marion, C., Ibanez, L.: ITK: enabling reproducible research and open science. Front. Neuroinform. **8** (2014)
10. Fedorov, A., et al.: 3D slicer as an image computing platform for the quantitative imaging network. Magn. Reson. Imaging **30**(9), 1323–1341 (2012)
11. Mirpuri, E., Pérez-Sala, A., Setién, A., López Benito, J., Artetxe, E.: Sistema de visualización de imágenes médicas. ES2649789A1 (2016)
12. Balestra, S., Schumann, S., Heverhagen, J., Nolte, L., Zheng, G.: Articulated statistical shape model-based 2D-3D reconstruction of a hip joint. In: Stoyanov, D., Collins, D.L., Sakuma, I., Abolmaesumi, P., Jannin, P. (eds.) IPCAI 2014. LNCS, vol. 8498, pp. 128–137. Springer, Cham (2014). https://doi.org/10.1007/978-3-319-07521-1_14
13. Öchsner, A., Öschner, M.: The Finite Element Analysis Program MSC Marc/Mentat: A First Introduction. Springer (2016). https://doi.org/10.1007/978-981-10-0821-4
14. Chase, J.G., et al.: Next-generation, personalised, model-based critical care medicine: a state-of-the art review of in silico virtual patient models, methods, and cohorts, and how to validation them. Biomed. Eng. Online **17**(1), 24 (2018)

Analysis of the Use of Genetic Algorithms in the Design of Models and Graphical Techniques for Early Detection, Diagnosis, and Characterization of Clinical Pathologies

Francisco L. Sáez-Gutiérrez[1](✉), José S. Velázquez[1], Jorge L. Alió del Barrio[2,3,4], Jorge L. Alió[2,3,4], and Francisco Cavas[1]

[1] Department of Structures, Construction and Graphical Expression, Technical University of Cartagena, Cartagena, Spain
francisco.saez@upct.es
[2] Division of Ophthalmology, Department of Pathology and Surgery, Faculty of Medicine, Miguel Hernández University, Alicante, Spain
[3] Department of Research and Development, VISSUM, Alicante, Spain
[4] Cornea, Cataract and Refractive Surgery Department, VISSUM, Alicante, Spain

Abstract. Recent developments of diagnosis procedures and graphical techniques in medicine have stimulated the increasing need for computational analysis tools. Nowadays, there is a growing interest in using computational algorithms in graphic diagnostic and model analysis techniques. A clear example would be computed tomography (CT) or magnetic resonance imaging (RMI), which are applied to various medical specialities. This document aims at providing a better understanding of the use of genetic algorithms (GA) as a tool in graphic diagnostic techniques and model analysis. GA have been used in different disciplines, for example, for template matching on CT images or 3D images reconstruction by fitness function optimization by an error-based method. This work focuses on studying the different methods used and presents a comparative evaluation of the results achieved in the literature.

Keywords: Medicine · Optimization · Graphical techniques · Imaging diagnostic

1 Introduction

Nowadays, the use of computational optimization methods is more important than ever in Medicine to improve diagnosis procedures and graphical techniques. The methods found in scientific literature can be of different types, habitually including neural networks [1] and deep learning [2]. These methods require for their implementation specific hardware such as cloud computing or vision processing units (VPU). Thus, many authors are turning their eyes to the use of genetic algorithms (GA) (Fig. 1), as we currently know better the evolutionary processes and their computational implementation, providing more efficiency and less computational cost than some years ago [3].

F. Cavas Martínez et al. (Eds.): INGEGRAF 2021, LNME, pp. 201–207, 2022.
https://doi.org/10.1007/978-3-030-92426-3_24

One of the most critical discussions in this field is which should be the optimal set up of the optimization method and its implementation. In the use of genetics algorithm in medicine, we can distinguish two categories of research: GA in graphic diagnostic techniques and GA in biological structures modelling. In the first case, specific techniques use genetic algorithms to analyse graphical tools such as MRI and ECG, and GA support in other analysis tools as neural networks and state machines. In biological structures modelling, techniques use GA as a method to fit surfaces, curves and plots.

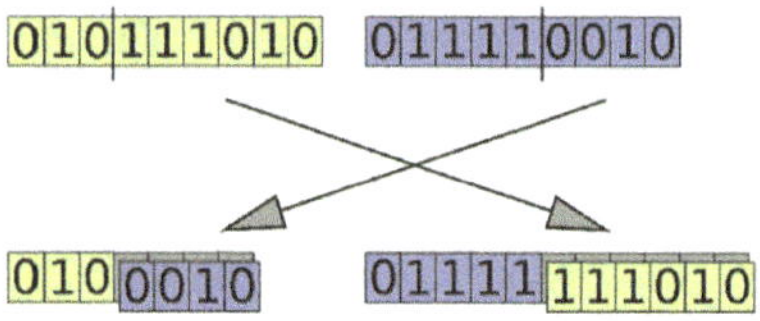

Fig. 1. Genetic algorithm. Crossover, cut and splice. CC BY-SA 3.0 by User: Yearofthedragon

In this work, it will be analyzed the result of applying genetic algorithms to modelling and graphical techniques in medicine. First, the application of GA in graphic medical procedures has been explored, searching for differences and features between diverse approaches. Then, a study of the GA application to model the biological structure was made, exploring different fitting methods used in medicine. Finally, an analysis was conducted, comparing and contrasting the studied genetic algorithm.

For our study, a literature review in the SCOPUS database was made. Some of the terms of reference were "genetic algorithm", "evolutionary computation", "graphic fitting", "medical image segmentation" or "computer-aided detection". Some search subject area filters such as "medicine" or "engineering" were also applied. The search did not exclude "conference paper" document types, to increase the number of studied cases. The target literature were the most cited and most relevant documents.

2 GA in Medical Graphic Techniques

In the field of Medicine, there are a significant number of graphic diagnostic techniques. A clear example of imaging and visual techniques of diagnostic are X-ray, computed tomography (CT or MRI scan), and histogram techniques as the electrocardiogram (ECG or EKG). Image processing has considerable relevance in Medicine because it aims to categorise regions of the image susceptible to a clinical diagnosis. Researchers have traditionally applied several techniques to make the data analysis, including statistical and structural and morphological methods. More recently, techniques such as fuzzy logic and neural networks have also been used. However, genetic algorithms have a more significant potential, due to a lower cost of implementation, as previously mentioned. Along this research, several investigations carried out in the field of Medicine using graphic diagnostic tools and genetic algorithms will be described.

In X-ray (RX) Radiodiagnosis, the application of genetic algorithms in the analysis of histograms and radiograms is interesting. Huang et al. [4] used the segmentation of the 2D histogram domain by GA in RX images obtaining 46% more efficacy than an auto-adapted threshold segmentation method. Ballerini et al. [5] used colour contour segmentation by genetic snake algorithm, thus improving RX diagnostic in bones. Even though it is a deferred method, GA in RX image histogram analysis is significantly advanced compared to other methods.

MRI scan analysis has experienced exciting applications of genetics algorithms. One of these applications consisted in implementing a categorization (clustering) of the MR images [6]. This study makes the segmentation of brain MR images through a GA instead of using a fuzzy c-means function. This method improves diagnostic in noisy images, thus obtaining 45% more segmentation accuracy. Another study takes the same approach, but with a hierarchical GA [7]. This study compares this method with neural networks, dynamic thresholding, k-means and fuzzy c-means with positives results. Another application in MR imaging is using an ant colonies genetic algorithm (ACGA) for object reference identification [8]. This method improved the fuzzy c-means over 20% more accuracy in MRI segmentation. Edge detection in diagnostic MR images is also interesting. A study applies this to detect glomerular borders of MR images of a kidney [9]. In this case, edge patching with GA was 33% more accurate than a Spline function fitting with the watershed method. Another example would be the detection of the edges of the pulmonary ventricles on MRI images using a GA with genetic subspecies [10].

The use of GA in Medicine is not limited to 2D or 3D graphics. It has also been used in char bar and histogram tools as electrocardiograms (EKG or ECG) or encephalograms (EEG). Arabasadi et al. [11] released a hybrid neural network-GA to detect coronary artery disease (CAD) in ECG data. This method increased the sensitivity by 11% when compared with the previous neural network method. Another research [12] studied a GA method that identified biometrical signals in EEGs. This study helped to reduce the number of required EEG electrodes for biometric identification.

The cases described above allowed us to verify the genetic algorithms methods as feasible methods, as well as they helped us to improve graphic diagnostic tools. In the conclusions section, a summary table of the characteristics of the cases studied herein will be presented.

However, there is a GA application of paramount interest in the field of Graphical Engineering applied to Medicine: the geometrical reconstruction of biological and medical structures.

3 GA in Medical Graphic Modelling Reconstruction

For different medical techniques, a geometric reconstruction of biological structures from images is required. This reconstruction allows obtaining a model that can be later used for diagnosing various pathologies. Usually, this reconstruction is based on geometric fitting with an optimization method. GA has proven to be an adequate method for this optimization.

Several examples of application of GA in research can be studied, and partial images reconstruction based in GA application is one of them. In this area, Zhang et al. [13] proposed the use of an adaptative GA in a novel image registration from further analysis (mutual information from CT, MRI, etc.). In this case, the evolutionary algorithm uses the real-time modification of mutation and crossover rates. With the same aim, Mañana et al. [14] used a parallel GA for dental structures reconstruction. The algorithm reduced the error by 20% in comparison with the downhill simplex method. Another study [15] used an elitist GA in cranial CT representation, improving the radial based representation. Other option is the use of physical logic gates for this purpose, as Dandekar et al. did in [16], but this is a less advisable option as this generates increasing computational time cost.

For further research in the group of this article's authors, the application of computational evolutionary algorithms in 3D reconstruction has a great interest. Some essays with GA prove to be effective in 3D biostructures reconstruction in some clinical specialities, like bony and head structures. The European Centre for Soft Computing [17] used a binary GA to transform the geometric parameters matrix for the 3D head bones reconstruction in forensic identification. This study was the target of a research for the detection via radiography reconstruction [18]. The method reached over 99% of accuracy. Another study [19] used a fly algorithm real-time-based strategy for MRI 3D projection. With a Parisian evolution method, this study got 33% more accuracy than other marginal fitting techniques. Also, another study used GA to reduce the correction error in cerebral cortex surface reconstruction in MRI with the same technique [20].

The research field in which the authors of this article are focused is Ophthalmology. In this discipline, several traditional geometric eye modelling methods exist. In this case, a fitting optimisation method is used for the surface reconstruction model. As there are few researches that aim to apply GA as an optimisation method, this application is an opportunity to improve the graphic modelling reconstruction methods in Ophthalmology.

Several attempts have been made in this regard. In 2016, Williams et al. [21] used a gradient function in corneal CTI to minimize an energy function. This energy function represented the focal distance between the surface points. The obtained surface was compared to the one received by the Zernike polynomial via a GA. This method obtained great accuracy but with a high computational cost because it used an initial 2D segmentation, which caused that some time was spent to segment the images. Another attempt in geometry reconstruction in Ophthalmology was made by Bevilacqua et al. in 2017 [22], obtaining a diagnostic method by a neural network in ocular CTI. In this study, the neural network was computer-aided designed by a GA. Interestingly, the article explains how the method is used for the diagnostic, and not only for the geometry reconstruction.

The above analysis has clearly demonstrated that the geometrical reconstruction of biomedical structures is feasible with GA optimization methods. This method allows improving the 3D reconstruction geometric fitting.

4 Results and Discussion

This paper studies the use of computational genetic algorithms in graphical techniques in Medicine. The use of this method improves the graphical diagnostic procedures and the biological structures reconstruction. Our research is focused on implementing a GA to reconstruct an ocular structure. As we have already seen, unlike other medical specialties, there are a few research works in the field of Ophthalmology. In fact, various applications in other disciplines could be easily found, something that suggests that there is margin for enhancement.

As it has been previously noted, GA implementation improves medical graphic techniques. As a replacement of clustering methods as c-fuzzy means, GA direct application provides the most significant improvement. On contrary, as part of a hybrid neural network-GA, the GA indirect application has the lowest ranking in terms of improvement. Moreover, in biostructures reconstruction, GA improves the methods by more than 20% in all cases. Table 1 illustrates the comparison between the fundamental studied investigations.

Table 1. Comparative table of the main studied cases

Reference	Application	GA method	Comparative method	Estimated improvement
Huang 2008 [4]	RX histogram	GA domain segmentation	Auto-adapted threshold segmentation	46%
Sasikala 2006 [6]	Brain MRI segmentation	Multi-objective GA (MOGA)	Fuzzy c-means	45%
Lai 2009 [7]	CT head image segmentation	Hierarchical evolutionary algorithm (HEA)	Neural networks (CHNN)	11%
”	Abdominal MRI segmentation	Hierarchical evolutionary algorithm (HEA)	Dynamic thresholding	15%
”	Knee MIR segmentation	Hierarchical evolutionary algorithm (HEA)	k-means	11%
”	Simulated image segmentation	Hierarchical evolutionary algorithm (HEA)	Fuzzy c-means	21%
Huang 2008 [8]	MRI object identification	Ant colonies GA	Fuzzy c-means	20%
Ma 2009 [9]	Glomerular borders id. in kidney MRI	Ant colonies GA	Watershed spline fitting method	33%
Arabasadi 2017 [11]	EKG and ECG analysis	Hybrid neural network-GA	Neural network	11%
Mañana 2005 [14]	Dental structures reconstruction	Parallel GA	Downhill simplex method	20%
Ibáñez 2009 [17]	3D head bone reconstruction	Fly RTS Parisian GA	Marginal fitting methods	33%

As it can be seen, in all cases the application of GA improves previous methods. In particular, GA performs better in direct application [4, 6, 9]. The neural network method

[7, 11] has a near return, but, as we said in the introduction, this method requires VPU hardware. In biostructure reconstruction, GA performance is much better than other methods. In this case, the evidence suggests that GA geometric fitting-based method [14, 17] is a suitable procedure. It has been suggested [21] that GA may set the polynomial expansion parameters, and this seems to be a reliable approach to biostructures reconstruction. The method given in this study is comprehensive and provides a plausible methodology. In [22], the authors analyzed GA biostructure reconstruction and its use for clinical diagnostic. This result means that the obtained geometry has a highreliability level. We can say that GA provides a powerful tool for geometry fitting in biostructures modelling.

5 Conclusions

In summary, this paper has given account of GA techniques and their applications, proving that genetic algorithms play an essential role as an optimization method in both modelling and graphical techniques in Medicine. GA have demonstrated their effectiveness, and it can be assured that they can be applied to this article's authors research field: optical structures modelling. This study provides baseline information that can be used in future research.

References

1. Issarti, I., Consejo, A., Jiménez-García, M., Hershko, S., Koppen, C., Rozema, J.J.: Computer aided diagnosis for suspect keratoconus detection. Comput. Biol. Med. **109**, 33–42 (2019). https://doi.org/10.1016/j.compbiomed.2019.04.024
2. Dos Santos, V.A., et al.: CorneaNet: fast segmentation of cornea OCT scans of healthy and keratoconic eyes using deep learning. Biomed. Opt. Express **10**, 622–641 (2019). https://doi.org/10.1364/BOE.10.000622
3. Pooja, Chaturvedi, P., Kumar, P.: Control parameters and mutation based variants of differential evolution algorithm. J. Comput. Methods Sci. Eng. **15**, 783–800 (2015). https://doi.org/10.3233/JCM-150593
4. Huang, T.L., Bai, X.: An improved algorithm for medical image segmentation. In: Proceedings of the 2nd International Conference on Genetic and Evolutionary Computing, WGEC 2008, Jingzhou, Hubei, pp. 289–292 (2008)
5. Ballerini, L., Bocchi, L.: Multiple genetic snakes for bone segmentation. In: Cagnoni, S., et al. (eds.) EvoWorkshops 2003. LNCS, vol. 2611, pp. 346–356. Springer, Heidelberg (2003). https://doi.org/10.1007/3-540-36605-9_32
6. Sasikala, M., Kumaravel, N., Ravikumar, S.: Segmentation of brain MR images using genetically guided clustering. In: Proceedings of the 28th Annual International Conference of the IEEE Engineering in Medicine and Biology Society, EMBS 2006, New York, NY, pp. 3620–3630 (2006)
7. Lai, C.C., Chang, C.Y.: A hierarchical evolutionary algorithm for automatic medical image segmentation. Expert Syst. Appl. **36**, 248–259 (2009). https://doi.org/10.1016/j.eswa.2007.09.003
8. Huang, P., Cao, H., Luo, S.: An artificial ant colonies approach to medical image segmentation. Comput. Methods Programs Biomed. **92**, 267–273 (2008). https://doi.org/10.1016/j.cmpb.2008.06.012

9. Ma, J., Zhang, J., Hu, J.: Glomerulus extraction by using genetic algorithm for edge patching. In: Proceedings of the 2009 IEEE Congress on Evolutionary Computation, CEC 2009, Trondheim, pp. 2474–2479 (2009)
10. Levman, J., Alirezaie, J., Khan, G.: Magnetic resonance based ventricle system classification by multi-species genetic algorithm. In: Proceedings of the 2nd International IEEE EMBS Conference on Neural Engineering, 2005, Arlington, VA, pp. 2–5 (2005)
11. Arabasadi, Z., Alizadehsani, R., Roshanzamir, M., Moosaei, H., Yarifard, A.A.: Computer aided decision making for heart disease detection using hybrid neural network-genetic algorithm. Comput. Methods Programs Biomed. **141**, 19–26 (2017). https://doi.org/10.1016/j.cmpb.2017.01.004
12. Albasri, A., Abdali-Mohammadi, F., Fathi, A.: EEG electrode selection for person identification thru a genetic-algorithm method. J. Med. Syst. **43**(9), 1–12 (2019). https://doi.org/10.1007/s10916-019-1364-8
13. Zhang, H., Zhou, X., Sun, J., Zhang, J.: A novel medical image registration method based on mutual information and genetic algorithm. In: Proceedings of the 2nd Conference on Computer Graphics, Imaging, and Vision: New Trends 2005, Beijing, pp. 221–226 (2005)
14. Mañana, G., González, F., Romero, E.: Distributed genetic algorithm for subtraction radiography. In: Proceedings of the Proceedings of the 7th Annual Workshop on Genetic and Evolutionary Computation, Washington, D.C., pp. 140–146 (2005)
15. Wen, P., Ruofeng, T., Guiping, Q., Jinxiang, D.: A local registration approach of medical images with niche genetic algorithm. In: Proceedings of the 2006 10th International Conference on Computer Supported Cooperative Work in Design, CSCWD 2006, Nanjing, pp. 1099–1104 (2006)
16. Dandekar, O., Plishker, W., Bhattacharyya, S., Shekhar, R.: Multiobjective optimization of FPGA-based medical image registration. In: Proceedings of the 16th IEEE Symposium on Field-Programmable Custom Computing Machines, FCCM 2008, Stanford, CA, pp. 183–192 (2008)
17. Ibáñez, O., Ballerini, L., Cordón, O., Damas, S., Santamaría, J.: An experimental study on the applicability of evolutionary algorithms to craniofacial superimposition in forensic identification. Inf. Sci. **179**, 3998–4028 (2009). https://doi.org/10.1016/j.ins.2008.12.02910.1007/s00500-008-0351-7
18. Gómez, O., Ibáñez, O., Valsecchi, A., Bermejo, E., Molina, D., Cordón, O.: Performance analysis of real-coded evolutionary algorithms under a computationally expensive optimization scenario: 3D–2D comparative radiography. Appl. Soft Comput. J. **97** (2020). https://doi.org/10.1016/j.asoc.2020.106793
19. Bousquet, A., Louchet, J., Rocchisani, J.-M.: Fully three-dimensional tomographic evolutionary reconstruction in nuclear medicine. In: Monmarché, N., Talbi, E.-G., Collet, P., Schoenauer, M., Lutton, E. (eds.) EA 2007. LNCS, vol. 4926, pp. 231–242. Springer, Heidelberg (2008). https://doi.org/10.1007/978-3-540-79305-2_20
20. Ségonne, F., Grimson, E., Fischl, B.: A genetic algorithm for the topology correction of cortical surfaces. In: Christensen, G.E., Sonka, M. (eds.) IPMI 2005. LNCS, vol. 3565, pp. 393–405. Springer, Heidelberg (2005). https://doi.org/10.1007/11505730_33
21. Williams, D., Zheng, Y., Davey, P.G., Bao, F., Shen, M., Elsheikh, A.: Reconstruction of 3D surface maps from anterior segment optical coherence tomography images using graph theory and genetic algorithms. Biomed. Signal Process. Control **25**, 91–98 (2016). https://doi.org/10.1016/j.bspc.2015.11.004
22. Bevilacqua, V., et al.: A computer aided ophthalmic diagnosis system based on tomographic features. In: Huang, D.-S., Hussain, A., Han, K., Gromiha, M.M. (eds.) ICIC 2017. LNCS (LNAI), vol. 10363, pp. 598–609. Springer, Cham (2017). https://doi.org/10.1007/978-3-319-63315-2_52

A New Method for Measuring Angular Variations Caused by High Heels in Sagittal Plane of Tibiotalar and Metatarsophalangeal Joints During Gait

Jose S. Velázquez[1](✉), Francisco L. Sáez-Gutiérrez[1], Amanda Robau-Porrúa[2], Arsenio M. Iznaga-Benítez[3], and Francisco Cavas[3]

[1] Departamento de Estructuras, Construcción y Expresión Gráfica, Universidad Politécnica de Cartagena, 30202 Cartagena, Spain
jose.velazquez@upct.es

[2] Tecnología en Construcción de Maquinarias, Facultad de Ingeniería Mecánica, Universidad Tecnológica de La Habana, José Antonio Echeverría, Calle 114 #11901, entre 119 y 127, CUJAE, Marianao, La Habana, Cuba

[3] Centro de Estudios de Ingeniería en Mantenimiento, Facultad de Ingeniería Mecánica, Universidad Tecnológica de La Habana, José Antonio Echeverría, Calle 114 #11901, entre 119 y 127, CUJAE, Marianao, La Habana, Cuba

Abstract. Human gait varies slightly from one individual to another because of many factors, such as age or gender, but the use of high heeled shoes (HH) is among the most prominent, ones. This paper presents a novel low-cost procedure that effectively determines angular variations in the joints of the foot (tibiotalar and metatarsophalangeal) when walking barefoot and with HH (2, 6 and 10 cm high). Video analysis and geometry analysis techniques were used to stablish the position of the foot at every moment. Results confirmed that as heel height increased both the length of step and the length of gait cycle shortened, although cadence remained unaffected. The Range of Motion (RoM) in tibiotalar joint reduced progressively from 28° when walking barefoot to 9° when a 10 cm heel was used, while for metatarsophalangeal joint, these reductions ranged from 30° to 5°, respectively. Thus, the proposed methodology can be used to characterize kinematic changes in ankle-foot set along the gait cycle in a more accessible and cost-effective way.

Keywords: Human movement analysis · Foot geometry · Optical motion tracking

1 Introduction

Human gait varies slightly from one individual to another, mainly because of the differences in age and gender [1, 2]. However, when healthy individuals walking barefoot are compared from an anatomical point of view, these variations fall within a narrow margin. Several models and expressions have been proposed in scientific literature to characterize these slight differences [3–5].

F. Cavas Martínez et al. (Eds.): INGEGRAF 2021, LNME, pp. 208–216, 2022.
https://doi.org/10.1007/978-3-030-92426-3_25

The fact that using high heeled shoes (HH) have a great impact in gait and posture is undeniable [6], but its influence can be studied from multiple points of view, such as biomechanical [7, 8], energetical [9], clinical [10], kinetical [1, 11, 12] or kinematical [1, 13, 14].

Focusing in the kinematical approach, many technologies have been used to analyze the joints of the human body [15, 16], but optical motion tracking [17] can be considered as one of the most commonly used in the fields of biomechanics, sports and clinical diagnosis. Optical motion tracking is based in the use of several stickers attached to the body, known as "markers", that allow to identify the geometry of anatomical parts by their contrast over the skin color, so their movement can be captured and later reproduced on a computer screen [18]. However, these systems are usually composed by an exclusive software package and several high-quality cameras (ranging from 2 to 8), something that makes them inaccessible for the labs of underdeveloped countries. Thus, the search of alternative low-cost optical motion tracking systems has a rising interest in the achievement of the Sustainable Development Goals (SDG) [19].

Kinematics has been widely used to study not only healthy joints, but also pathological and even operated ones, when patients walk barefoot [20]. Scientific literature around hip, knee and ankle joints when high heeled shoes are used is also extensive, but the studies centered around metatarsophalangeal joint [12] are scarce, and to the authors knowledge, none explains the variations experienced in its sagittal plane from a kinematic point of view.

This paper exposes a low-cost methodology to evaluate the angular variations in the sagittal plane that take place in the foot joints when HH are used, using geometrical models and a free tracking and video analysis software. The procedure herein exposed may help to reduce the cost of the system and therefore make it more accessible.

2 Materials and Methods

2.1 Subjects

Three healthy women (26 + 2.5 years, 1.60 + 0.2 m, 75 + 1.3 kg) with the same foot size (6 US/38 EUR) were recruited from Universidad Tecnológica de La Habana - José Antonio Echeverría (CUJAE). Presence of anomalies, injuries or deformities in their feet that could affect their gait was checked before the experiment was initiated.

All participants had no previous experience with high heeled shoes. Participants also freely signed an informed consent form stating their willingness to be included in the experiment. This research followed the tenets of the Declaration of Helsinki (7th revision, October 2013, Fortaleza, Brazil).

2.2 Geometrical Model

Feet movement takes place around three axes and in three planes (frontal, transversal and sagittal), so the first approximation usually adopted to reduce the complexity of the kinematic study of these joints, is centering on the sagittal plane, as the highest variations of the variables take place on it.

To do so, a geometrical study centered around the stance phase was made, in order to determine the influence of high heels in angular variations in the joints. Previous studies have shown that as heel height increases, so does insecurity during gait, thus monopodal support reduces progressively [2]. In Fig. 1, points A and B represent tibiotalar and metatarsophalangeal joints respectively, located one at the ankle level, and the other at the level of the heads of the metatarsals. θ_A and $\theta_B$represent the angle of rotation of these joints.

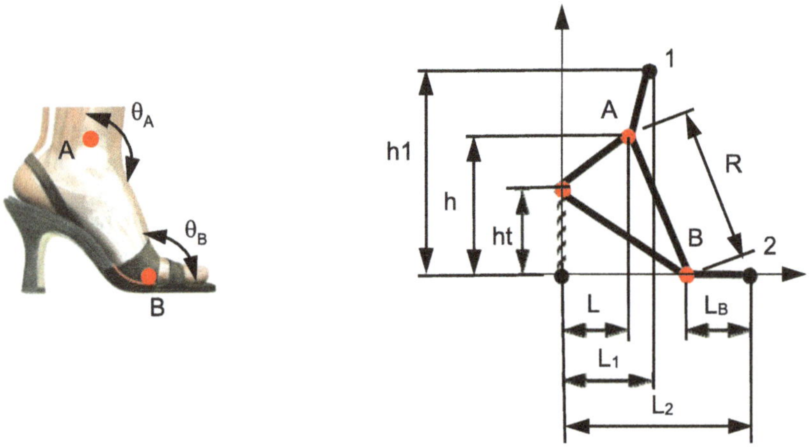

Fig. 1. Simplification of the set foot-ankle in the sagittal plane when the heel height increases.

2.2.1 Joint Rotation Determination

The geometrical expression of the scalar product allows to calculate the cosine of the smallest angle between two vectors. When applied to the feet joints, the smallest angle can be obtained by Eqs. (1) and (2):

$$\varphi_A = \cos^{-1}\left[\frac{(L_1-L)*(L_2-(L+L_B))-h*(h_1-h)}{((L_1-L)^2+(h_1-h)^2)^{1/2}*((L_2-(L+L_B))^2+h^2)^{1/2}}\right] \quad (1)$$

$$\varphi_B = \cos^{-1}\left[\frac{L_B*(L+L_B-L_2}{L_B((L+L_B-L_2)^2+h^2)^{1/2}}\right] \quad (2)$$

The result of cross product is a vector which is perpendicular to other two, and whose sense varies according to the angle formed by them. The application of this knowledge allows the calculation of two magnitudes that indicate if the angle obtained with the previously mentioned expressions correspond to the desired ones. The equations that allow calculating for each point are (3) and (4):

$$N_A = (-h*(L_1-L)) - ((h_1-h)*(L_2-(L+L_B))) \quad (3)$$

$$N_B = L_B * h \quad (4)$$

The final values for the angles are obtained when some of the following logical conditions are met:

$$\text{If } N_X > 0 \text{ then } \theta_X = \varphi_x;$$
$$\text{If } N_X = 0 \text{ then } \theta_X = \pi;$$
$$\text{If } N_X < 0 \text{ then } \theta_X = 2\pi - \varphi_x;$$

2.3 Equipment and Procedure

The experiment was designed for measuring kinematic variables along three full gait cycles, for each participant and for each heel height (barefoot, 2, 6 and 10 cm). The procedure was composed of several stages:

- Stage 1. Selection of the experiment location and track length: a sports pavilion with regular daylight was selected for the tests. Then, a track length (TL) of 6 m was selected, so spatiotemporal variables "length of the gait cycle" (LC), "time spent for a full cycle" (TC) and "speed" (SC) could be measured (Fig. 2).

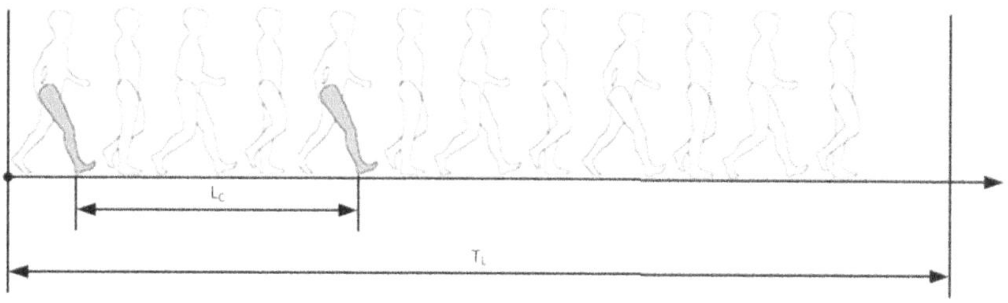

Fig. 2. Spatiotemporal variables.

- Stage 2. Camera's placement: two GE A1050 digital cameras of 10,1 megapixels were mounted on the horizontal plane, both at 12 cm height and 32 cm far from the track's start line, oriented to the sagittal plane and mutually opposed, with 4 m of separation. The participants walked along the line between them, so both medial and lateral sides of the feet were recorded for each walk.
- Stage 3. Marker's location: pieces of adhesive cloth were used as markers. Biggest ones (2 cm of diameter) were fixed over tarsals and metatarsals, while the smallest ones (1 cm of diameter). marked the phalanges. A white sock firmly fitted to the foot was used to fix them all (Fig. 3).

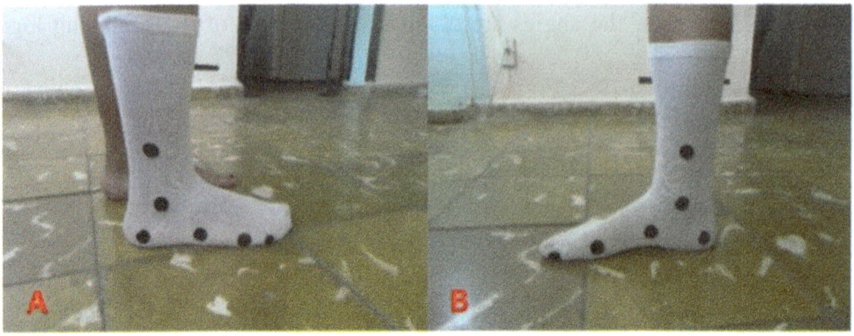

Fig. 3. Marker's fixation: a) in the lateral side of the foot: on lateral malleolus, 5 cm over the lateral malleolus, calcaneus lower part, head and base of the fifth metatarsal and distal phalange of the fifth toe; b) in the medial side of the foot: medial malleolus, 5 cm over the medial malleolus, calcaneus lower part, head and base of the first metatarsal and distal phalange of the hallux.

- Stage 4. Video recording and uploading to PC: recording was made at a resolution of 640 × 480 p and a framerate of 30 fps. The event of reference selected to synchronize both cameras was the moment in which the heel contacted with the ground. The USB port of the camera was used to upload the videos to the PC.
- Stage 5. Image processing, data extraction and angles calculation: Tracker software was used to manually extract data from each walk. This tool is a free software built on the Java Open-Source Physics (OSP) platform [21], that allows measuring angles and other kinematic variables.

3 Results

Averaged results for each heel height configuration (barefoot and 2, 6, and 10 cm heels) are presented in Table 1.

Table 1. Average results for the spatiotemporal variables. NC: number of cycles; TNC: time to complete NC cycles; LNC: distance covered in NC cycles; LC: length of the gait cycle; TC: time spent for a full cycle; SC: speed

Variable	Unit	Heel height [cm]			
		0	2	6	10
NC	cycles	5	5	5	5
TNC	seconds	5.80	5.80	5.50	5.50
LNC	metres	5.85	5.80	5.45	5.15
LC	metres	1.17	1.16	1.09	1.03
TC	seconds	1.16	1.16	1.10	1.10
SC	m/s	1.01	1.00	0.99	0.94

Figure 4 shows how the angle in tibiotalar joint (θA) changes along the gait cycle for each of the heel heights considered in the study. As it can be observed, total range of motion (RoM) progressively decreased from 28° in barefoot walking to just 9° when a heel of 10 cm was used.

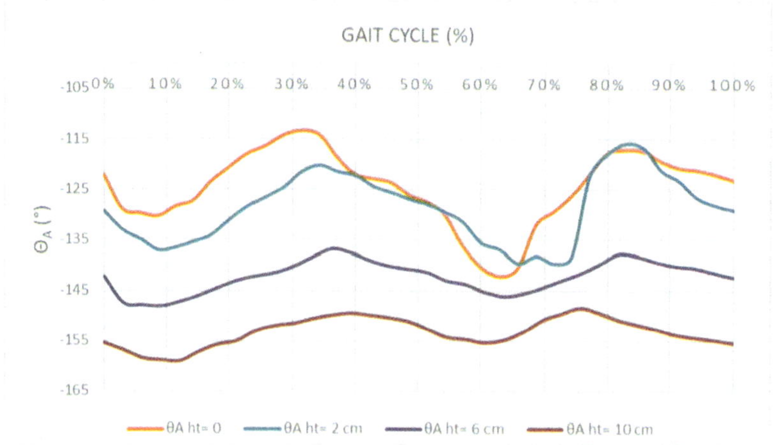

Fig. 4. Sagittal plane's angular variation in tibiotalar joint (°) in a full gait cycle for heel heights of 0, 2, 6 and 10 cm.

Angular variations in the metatarsophalangeal joint (θB) were also monitored and have been depicted in Fig. 5. In this case, RoM progressively decreased from 30° in barefoot walking to slightly above 5° with heels of 10 cm.

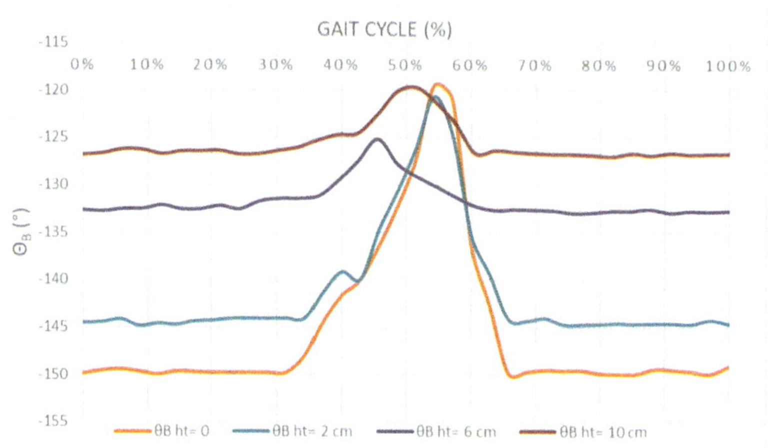

Fig. 5. Sagittal plane's angular variation in metatarsophalangeal joint (°) in a full gait cycle for heel heights of 0, 2, 6 and 10 cm.

4 Discussion

The results presented in Table 1 suggest that as heel height increases both the length of step and the length of gait cycle shorten, although cadence remains unaffected. These results coincide with the ones presented by other authors [22–24].

In tibiotalar joint (Fig. 4), and in barefoot conditions, results confirm Kommu et al. observations, which proved that dorsiflexion maximum angle reaches 8° in the stance phase, and plantar flexion reaches a minimum angle −20° right after take-off, with a total RoM of 28° [25].

Moreover, similarities found in the shape of the graph's and the average values shown in them between articles [2, 13, 25] and present one, support the validity of the measuring procedure, and suggest that it not only can be used for the rest of heel heights, but also for metatarsophalangeal joint.

This research also proves that increments in the heel height cause evident reductions in RoM in the ankle, something that has been already observed by other authors [2]. Stefanyshyn et al. [26] found that plantarflexion increases when HH are worn, something that has also been noted in present study. Mika et al. also noticed increments in ankle's angles at heel strike when HH were used [2]. It was also observed that HH force the foot into a position with a higher plantarflexion, so the ankle's angle differ in stance phase and after toe off [5].

When metatarsophalangeal joint kinematics are observed (Fig. 5), as other authors have appointed [12], great reductions in RoM were noticed when heel height increase. Scientific literature covering dorsiflexion angle values reached in metatarsophalangeal joint is, to the authors knowledge, inexistent for 2 cm and 10 cm high shoes. However, values listed for barefoot and 6 cm high shoes align with other authors results [12].

As a limitation of present study, it could be noted that as participants did not use their own shoes. This may have had an influence in their gait characteristics, and to minimize possible biases, they were given enough time to familiarize with the shoes before any test was made.

Another possible limitation would be that all participants were inexperienced HH users, something that could in theory affect their gait. However, precursory studies suggest that its influence in kinematics in the sagittal plane below the knee joint are marginal [2].

5 Conclusions

This paper presents a low-cost system capable of measuring angular variations in the joints of the feet when HH are worn, using a free video analysis software and a geometrical model. The procedure proposed can be of use in the study of several gaitrelated pathologies, and data reported here can be used as baseline for future studies.

References

1. MacWilliams, B.A., Cowley, M., Nicholson, D.E.: Foot kinematics and kinetics during adolescent gait. Gait Posture **17**, 214–224 (2003)

2. Mika, A., Oleksy, Ł, Mika, P., Marchewka, A., Clark, B.C.: The influence of heel height on lower extremity kinematics and leg muscle activity during gait in young and middle-aged women. Gait Posture **35**, 677–680 (2012)
3. Leardini, A., Benedetti, M.G., Catani, F., Simoncini, L., Giannini, S.: An anatomically based protocol for the description of foot segment kinematics during gait. Clin. Biomech. **14**, 528–536 (1999)
4. Cronin, N.J.: The effects of high heeled shoes on female gait: a review. J. Electromyogr. Kinesiol. **24**, 258–263 (2014)
5. Wiedemeijer, M.M., Otten, E.: Effects of high heeled shoes on gait. A review. Gait Posture **61**, 423–430 (2018)
6. Cowley, E.E., Chevalier, T.L., Chockalingam, N.: The effect of heel height on gait and posture: a review of the literature. J. Am. Podiatr. Med. Assoc. **99**, 512–518 (2009)
7. Hong, W.H., Lee, Y.H., Chen, H.C., Pei, Y.C., Wu, C.Y.: Influence of heel height and shoe insert on comfort perception and biomechanical performance of young female adults during walking. Foot Ankle Int. **26**, 1042–1048 (2005). https://doi.org/10.1177/107110070502601208
8. Snow, R.E., Williams, K.R.: High heeled shoes: Their effect on center of mass position, posture, three-dimensional kinematics, rearfoot motion, and ground reaction forces. Arch. Phys. Med. Rehabil. **75**, 568–576 (1994)
9. Ebbeling, C.J., Hamill, J., Crussemeyer, J.A.: Lower extremity mechanics and energy cost of walking in high-heeled shoes. J. Orthop. Sports Phys. Ther. **19**, 190–196 (1994)
10. Foster, A., Blanchette, M.G., Chou, Y.-C., Powers, C.M.: The influence of heel height on frontal plane ankle biomechanics: implications for lateral ankle sprains. Foot Ankle Int. **33**, 64–69 (2012)
11. Barkema, D.D., Derrick, T.R., Martin, P.E.: Heel height affects lower extremity frontal plane joint moments during walking. Gait Posture **35**, 483–488 (2012)
12. McBride, I.D., Wyss, U.P., Cooke, T.D.V., Murphy, L., Phillips, J., Olney, S.J.: First metatarsophalangeal joint reaction forces during high-heel gait. Foot Ankle **11**, 282–288 (1991)
13. Wang, M., Gu, Y., Baker, J.S.: Analysis of foot kinematics wearing high heels using the Oxford foot model. Technol. Health Care **26**, 815–823 (2018)
14. Lee, C.: The effects of lower extremity angle according to heel-height changes in young ladies in their 20s during gait. J. Phys. Ther. Sci. **26**, 1055–1058 (2014)
15. Al-Zahrani, K., Bakheit, M.: A historical review of gait analysis. Neurosciences (Riyadh, Saudi Arabia) **13**, 105–108 (2008)
16. Vázquez, S.C.L.: marcha: Historia de los procedimientos de análisis. Biociencias **2**, 13 (2004)
17. Kitagawa, M., Windsor, B.: MoCap for Artists: Workflow and Techniques for Motion Capture. CRC Press, Boca Raton (2020)
18. Davis Iii, R.B., Ounpuu, S., Tyburski, D., Gage, J.R.A.: gait analysis data collection and reduction technique. Hum. Mov. Sci. **10**, 575–587 (1991)
19. UNDP. Sustainable Development Goals | UNDP. http://www.undp.org/content/undp/en/home/sustainable-development-goals.html. Accessed 8 Mar 2021
20. Saari, T., Tranberg, R., Zügner, R., Uvehammer, J., Kärrholm, J.: Total knee replacement influences both knee and hip joint kinematics during stair climbing. Int. Orthop. **28**, 82–86 (2004)
21. Physlets.org. Tracker Video Analysis and Modeling Tool for Physics Education. https://physlets.org/tracker/. Accessed 08 Mar 2021
22. Di Sipio, E., et al.: Walking variations in healthy women wearing high-heeled shoes: shoe size and heel height effects. Gait Posture **63**, 195–201 (2018)
23. Cronin, N.J., Barrett, R.S., Carty, C.P.: Long-term use of high-heeled shoes alters the neuromechanics of human walking. J. Appl. Physiol. **112**, 1054–1058 (2012)

24. Esenyel, M., Walsh, K., Walden, J.G., Gitter, A.: Kinetics of high-heeled gait. J. Am. Podiatr. Med. Assoc. **93**, 27–32 (2003)
25. Kommu, S.S.: Rehabilitation Robotics; BoD–Books on Demand (2007)
26. Stefanyshyn, D.J., Nigg, B.M., Fisher, V., O'Flynn, B., Liu, W.: The influence of high heeled shoes on kinematics, kinetics, and muscle EMG of normal female gait. J. Appl. Biomech. **16**, 309–319 (2000)

Studying the Fluid-Structure Interaction in a Computational Model of the Human Eye During Non Contact Tonometry Tests

Osiris de la Caridad Núñez-Chongo[1(✉)], Claudia Muñoz-Villaescusa[1], Alfo José Batista-Leyva[1], and Francisco Cavas-Martínez[2(✉)]

[1] Higher Institute of Technologies and Applied Sciences, University of Havana, Ave. Salvador Allende 1110, Plaza de La Revolución, 10400 Havana, Cuba
osirisnunez93@gmail.com, abatista@instec.cu
[2] Technical University of Cartagena, Cartagena, Spain
francisco.cavas@upct.es

Abstract. Non contact tonometry (NCT) is a non invasive technique that measures intraocular pressure (IOP), which is one possible way to evaluate the biomechanical behavior of the cornea in vivo and an alternative to determine corneal material parameters. An accurate and reliable numerical model of the non contact tonometry test is required to further understand the eyes response mechanism in accordance with IOP action and the tonometer's air puff. This work developed a model of the human eye based on the finite element (FE) method, a Computational Fluid Dynamics (CFD) simulation of air jet and a fluid-structure interaction (FSI) model by Two-Way dynamic coupling between both systems. A mesh with optimal element densities and good meshing quality was generated. Then, the stress-free geometry of the eyeball was obtained according to IOP loading. As the airflow was selected as incompressible and turbulent during testing, the Spalart-Allmaras turbulence model was included. During the CFD simulation of the air domain, a mesh was generated, with good meshing quality and an estimated time step of 0.0001 s, which allowed reliable results to be obtained at a lower computational cost during this transient analysis. A FSI model was implemented that reproduces the air jet outflow up to its incidence with the corneal surface. A qualitative correspondence was found of the clinical CorVis ST images and the different instances of cornea deformation during simulation. This FSI model between the air puff and the human eye can be considered a starting point to analyze some biomechanical corneal properties during NCT testing for future research purposes.

Keywords: Ocular biomechanics · CorVis ST · Intraocular pressure · Finite element modeling · Computational Fluid Dynamics · Fluid-structure interaction

1 Introduction

Intraocular pressure is the force exerted by humors against inner eye walls. This pressure has been directly and indirectly related to some ocular diseases, such as retinal detachment, ocular hypertension and glaucomatous optic neuropathy. Glaucoma is manifested

F. Cavas Martínez et al. (Eds.): INGEGRAF 2021, LNME, pp. 217–228, 2022.
https://doi.org/10.1007/978-3-030-92426-3_26

by loss of vision with no other apparent symptoms. It may develop due to optic nerve deterioration as a result of either high IOP or other factors in the presence of normal pressure. IOP assessments are crucial factor to diagnose and treatment of this disease [1].

Tonometry is a technique employed to measure IOP that can be performed by contact or non contact. NCT is a non-invasive technique that uses a rapid air pulse to flatten the cornea. During this test, air puff force determines IOP at the applanation time. Advanced optical instruments, such as CorVis ST and the Ocular Response Analyzer (ORA), capture dynamic cornea changes. Although some biomechanical markers of the cornea have been successfully monitored *in vivo*, the main challenge lies in the difficulty of separating the corneal deformation effects that result from IOP action and the deformation effects caused by mechanical stimuli during examinations. For this reason, the design and implementation of mathematical models to predict and characterize the dynamic response of the eyeball have been considered [2].

Some papers have studied NCT theoretically, numerically and clinically. An analytical description consists of representing a dynamic model of the cornea subjected to air puff and IOP loading as a differential equation of a simple harmonic oscillator [3–5]. During computational simulation, the finite element (FE) method is a widely used tool to analyze eye's biomechanical response and the CFD simulation for the tonometer air puff. Two-dimensional (2D) corneal simulations [6, 7] and three-dimensional (3D) simulations to investigate cornea [8] or the entire eyeball's [9, 10] mechanical response are found in the literature. However, these works have some limitations; e. g., they do not include some internal eye structures, or they do not consider internal humors or they ignore the FSI between the air puff and the ocular structure. Other works indicate the importance of considering the fluid-structure interaction effect, but their geometry is based on porcine eyes, 2D and axial symmetry [11]. A more recent study has focused on quantifying the influence of FSI on corneal behavior predictions for patient-specific 3D eye models. However, its limitation lies in it representing only the external eyeball layer [12].

Despite the advances made in numerical models, the studies we found present major simplifications. However, an advanced model that integrates all the previously seen factors is needed to interpret and take advantage of the results. This work proposes performing a dynamic 3D simulation of an NCT test by means of coupling between a CFD model of the air puff and a biomechanical analysis of the human eye by the FE method. The influence of the fluid-structure interaction between air and mainly the anterior surface of an aspheric cornea was considered to achieve a better approximation to the real visual system.

2 Materials and Methods

This study is based on numerical models of the whole eyeball subjected to both IOP and a tonometer air-puff velocity field obtained by CorVis ST. Three important pillars are developed to evaluate the different numerical simulation strategies of a NCT test:

– Human eye numerical model using the FE method.

- A transient and turbulent CFD model that reproduces the air jet during the NCT test.
- Coupling of the fluid-structure interaction between both models.

In order to achieve a correct problem formulation, a series of simulation steps is followed and implemented by the ANSYS Multi-field packages. This program contains the numerical methods and algorithms needed for calculating the separate domains and, finally, for the fluid-structure interaction.

2.1 Human Eye Model Description

A simplified model of the complete human eye was constructed (Fig. 1). The model comprised tissues, such as cornea, limbus, sclera and adipose tissue. This allowed a comparison to be made with similar numerical simulations found in the literature. The geometric parameters of structures were the mean values obtained from the literature. The cornea was considered to be two aspheric surfaces with revolution symmetry around the central axis [13]. The other tissues were described as spherical surfaces of revolution [14].

In addition, boundary conditions were established, and their location, nature and value were fixed. The normal pressure that simulated the IOP caused by aqueous and vitreous humor on inner surfaces was applied to the inner wall. The IOP value was assumed to be 2 kPa (15 mmHg) [13]. A fixed boundary condition (no displacement) was used to limit eye movement on adipose tissue walls. This was where the bony contact and the contact with the six extraocular muscles had to be located [14]. An interface region between the air fluid and anterior eye surface structures was defined.

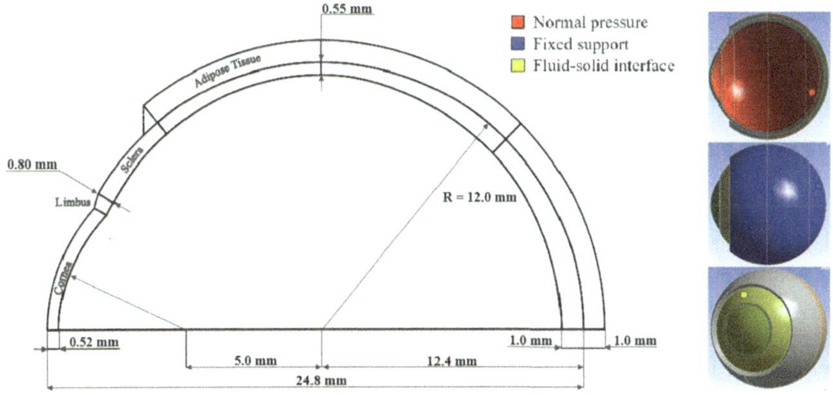

Fig. 1. Lateral view of the simplified eyeball model and the boundary conditions.

Biomechanics behavior was modeled by linear and non linear elastic materials. The material models and parameters for each eye model structure are found in Table 1. Cornea, limbus and sclera were represented by a hyperelastic and isotropic Mooney Rivlin five parameters model [14].

Table 1. Material parameters for the eye model.

Structure	Material Model	Young's modulus	Poisson's ratio	Density	Ref.
Cornea	Hyperelastic	–	–	1400 kg/m^3	[13]
Limbus	Hyperelastic	–	–	1400 kg/m^3	
Sclerotic	Hyperelastic	–	–	1400 kg/m^3	[15]
Adipose Tissue	Isotropic elastic	0.047 MPa	0.47	999 kg/m^3	[16]

To this end, stress-strain curves of cornea [17] and sclera [15] were extracted from the experimental data in the literature to determine the five parameters by the Mooney Rivlin equation [18]. These five parameter were obtained fitting stress-strain curves using ANSYS software, and are shown in Table 2.

Table 2. Parameters of the Mooney Rivlin hyperelastic model obtained in the ANSYS software.

Structure	$C_{10}(Pa)$	$C_{01}(Pa)$	$C_{20}(Pa)$	$C_{11}(Pa)$	$C_{02}(Pa)$
Cornea	3.15117E + 05	3.44546E + 05	1.50620E + 08	2.92175E + 08	1.42642E + 08
Anterior Sclerotic	5,10142E + 07	4,77710E + 07	4,62119E + 10	9,39490E + 10	4,79293E + 10
Equator Sclerotic	2,63660E + 07	2,45856E + 07	9,09079E + 10	1,86423E + 11	9,57846E + 10
Posterior Sclerotic	2,86205E + 07	2,67382E + 07	8,47567E + 10	1,73692E + 11	8,91934E + 10

The technique followed to determine a structure's dynamic response according to the action of any type of time-varying loads was a transient dynamic analysis. This analysis type has been used in ANSYS to establish time-dependent displacements, strains, stresses and forces in a structure [18].

2.2 Air Puff Model Description

A CFD model of the tonometer air puff was represented in a 3D symmetry domain. The domain boundaries were characterized by (Fig. 2): inlet orifice, wall condition, outlet condition (gauge pressure value was zero) and the projection of the anterior surface of the eye. The inlet nozzle was assumed to be aligned with the principal axis of the eye. Symmetry of revolution was considered to simplify the model. The air puff from the inlet orifice lasted 30 ms [12]. The value and direction of the velocity at the entrance to the domain were established by this boundary condition. In Fig. 2, the chart shows the function characterizing the amplitudes of velocities per time by varying the air puff on the nozzle plane, which was obtained from the experimental data fit of the CorVis ST device [12].

The interior domain was occupied by air. The material properties of air are defined in terms of its density (1.204 kg/m^3) and dynamic viscosity (1.83 10^{-5} kg/m·s^{-1}) [12]. Non

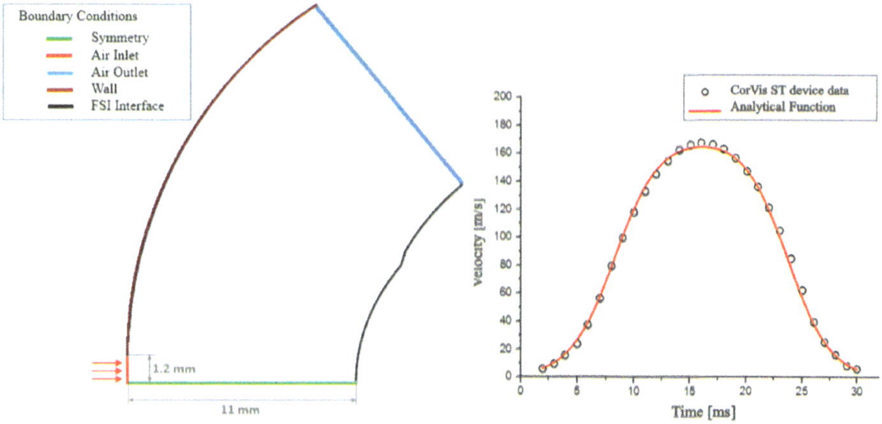

Fig. 2. Boundary conditions for the air jet domain and time velocity profile at the inlet nozzle.

dimensional numbers such as Mach and Reynolds numbers were calculated to define fluid compressibility and motion, respectively. As the Mach number was 0.48, air puff was taken as an incompressible flow. The obtained value of Reynolds number was 23647.05, which meant that a highly disturbed flow with eddies and turbulent vorticities had to be simulated. The Spalart-Allmaras turbulence model has been used in the literature to describe the turbulence of the air flowing out of a tonometer [12].

2.3 Spatial Discretization and Temporal Analysis of the Models

The spatial discretization of both domains was classified as a structured mesh composed of first order prismatic elements. In the air domain, the mesh was generated by a pyramid element with a triangular base and six nodes in a Eulerian configuration (Fig. 3a). A refined mesh was created in the area adjacent to the interface between the fluid and the eye because this is a region of much interest in accuracy terms (Fig. 3b).

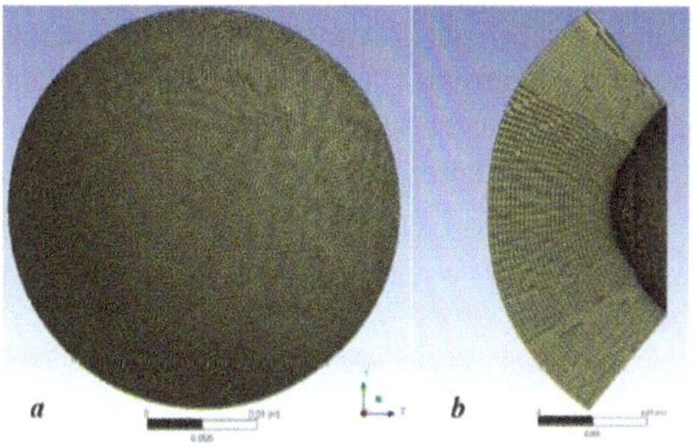

Fig. 3. Spatial discretization in numerical models: air domain mesh.

Eye domain elements were formulated as hexahedral-shaped solid volumes based on a Lagrangian configuration (Fig. 4a, b). Meshing quality was evaluated with the Aspect Ratio, and the Skewness and Orthogonal Quality Metrics. This value was compared to the optimal criteria, as indicated in ANSYS Help [18].

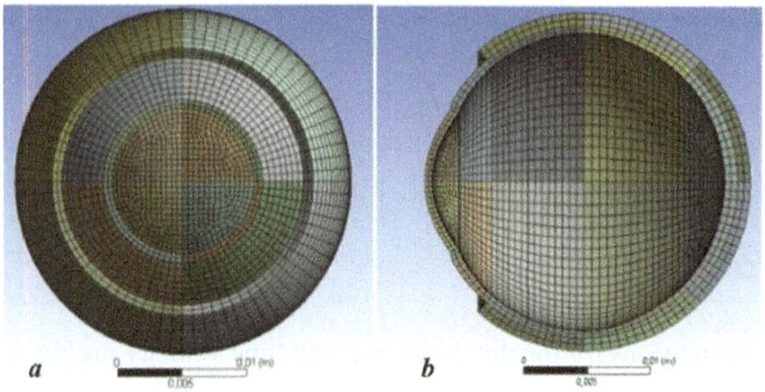

Fig. 4. Spatial discretization in numerical models in the eye domain. a) front view, b) lateral view.

The algorithm employed to obtain the tension-free configuration was applied to the eyeball surface [19]. It was necessary to define the optimal time step size (Δt) for the selected spatial discretization in the air domain. Several transient studies were calculated gradually by decreasing the time step size to reach a balance between the accuracy of the results and the computational cost. The residuals, the number of internal iterations needed for convergence and the Courant Friedrichs-Lewy (CFL) number were monitored during this process [20].

2.4 Fluid-Structure Interaction Models

A dynamic model of the fluid-structure interaction between eye structures and the air puff was constructed and divided into three simulation moments (Fig. 5). First, a stationary analysis (up to 20 ms) was conducted to increase IOP to 2 kPa (black line). In the process a transfer of displacements and forces occurs between the eye domain and the filling air. Secund, a transient analysis (from 20 ms to 30 ms of the simulation) was modeled to stabilize the motion in both domains and IOP is kept constant. Finally, a transient analysis (from 30 ms to 60 ms of the simulation) was done to represent of the NCT air jet. The red line in the graph represents the velocity profile established at the inlet nozzle.

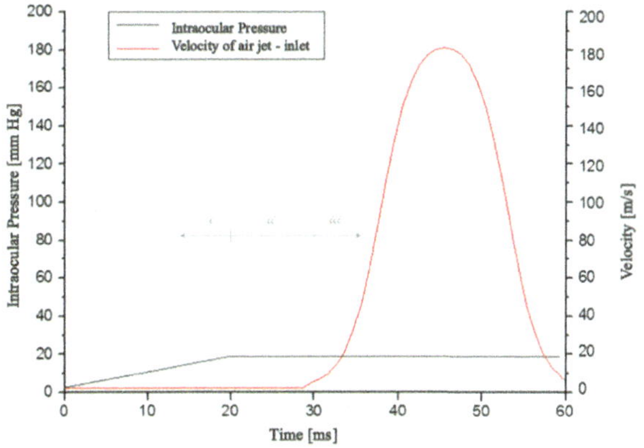

Fig. 5. The numerical simulation FSI moments of the NCT test.

3 Results and Discussion

Modern clinical methods followed to determine IOP and to assess the biomechanical properties of the cornea *in vivo* are based on corneal deformation studies by the air puff effect. A numerical simulation was herein performed by a dynamic analysis to reproduce an NCT test.

3.1 Sensibility Mesh and Free–Stress Geometry Results

The FE method based on mass, force and rigidity matrices was solved to calculate material structure deformations. The mesh density effect on the eye domain's convergence and accuracy was studied across 35 configurations. During the analysis, the size of the generated mesh elements is gradually and uniformly changed. The size of these elements was changed by the surface area (base faces of the prismatic elements in Fig. 4. *a*) in 29 different configurations, and also by the direction of the longitudinal eye slice (lateral edges of the elements in Fig. 4. *b*) in six different configurations, which allowed performing a sensitivity analysis of the mesh per structure. The change in cornea was controlled by varying the number of divisions on edges and meridians to ensure the same length. This was true for the limbus and sclera, but not for the adipose tissue elements, which acted as subordinate elements.

Variation in apical displacement, Von-Mises stress and maximum principal stress due to the IOP action were studied in the different configurations. The radial expansion at the scleral equator was also measured [9].

Table 3 shows the meshing arrangement obtained for each eye structure. The results showed that for a mesh size of approximately 43607 nodes, the changes in apical displacement and radial expansion at the sclera equator were less than 0.05%, and were below 2.0% for the maximum principal stress and Von-Mises stress. This allows us to select the mesh with the values shown in Table 3.

Table 3. Mesh composition for each eye structure.

Eye structure	Cornea	Sclera	Limbus	Adipose Tissue	Total
Number of elements	3888	9288	1080	14328	36144
Number of nodes	5760	24148	1920	20308	43607

The effect of mesh density on the accuracy of the strain values and computational cost can also be observed by analyzing the displacement field. Figure 6 compares the displacements obtained by an increase in IOP up to 2 kPa (15 mmHg) for different mesh densities. The results showed a significant difference in the displacements between the mesh of 2038 elements and that of 36144 elements, which was not the case when the number of elements was increased up to 143625.

In the coarse mesh, cornea apical displacement was 0.2574 mm, which did not coincide with the maximum anterior cornea surface displacement in relation to the X-axis direction (0.2617 mm). However, the results were similar when comparing the optimal mesh (configuration of Table 3) and a much more refined mesh. The maximum displacements coincided with the apical displacements on the cornea with values of 0.2820 mm and 0.2837 mm for each mesh, respectively. The displacement values in the normal direction to the surface were 0.002848 mm, 0.002730 mm and 0.002712 mm for each mesh, respectively. Also the displacement values in the normal direction to the surface were 0.002848 mm, 0.002730 mm and 0.002712 mm for each mesh when refining respectively. The results showed minor differences between magnitudes for the last two mesh configurations (~20 nm). The results obtained with the most refined mesh were 23-fold slower than the mesh selected as the optimal one. After this analysis, the mesh of 36144 elements achieved an adequate balance among the minimum configuration of the number of elements, accuracy and computational time.

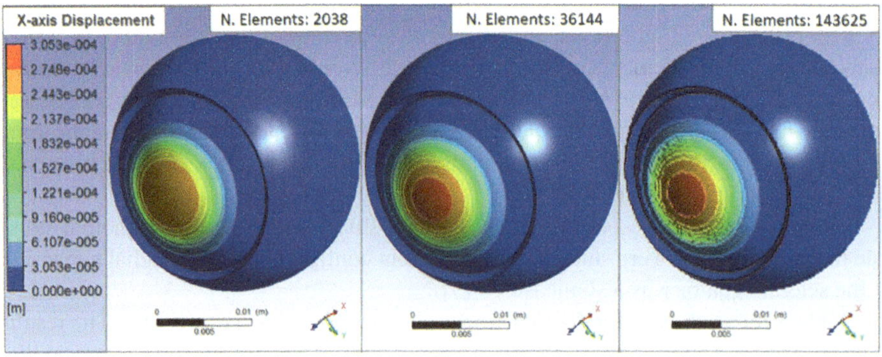

Fig. 6. Displacement field contours of the eye model for different mesh densities, the number of elements for each configuration are shown.

The mesh quality results were adequate. The maximum aspect ratio metric value was 3.4, but its values were distributed with a mean value of 2.04, agreeing with the optimal metric criterion of the elements. The skewness results showed that more than 80% of the elements had excellent metrics according to the criterion. The orthogonal quality of elements came close to 1 and the minimum reported value was 0.49, a criterion that indicates a good mesh [18].

The displacement-based algorithm employed to obtain a stress-free configuration was applied to the anterior and posterior corneal surfaces, the limbus surfaces and the sclera surfaces. In these geometries, an estimated maximum error of 0.005 mm was found in the third iteration of the algorithm, which was lower than the tolerance chosen to finalize the process. The new geometric parameters of the aspheric cornea surfaces were found for the model. However, the changes in the geometric parameters of the sclera surfaces were not significant compared to the values obtained for the cornea. This result can be explained by the greater sclera stiffness.

A CFD model of the air puff governed by momentum, continuity and turbulence equations was developed to calculate the pressure and velocity fields of flow. Several transient calculations for Reynolds-averaged Navier-Stokes equations (RANS) were conducted. The difference from one complete calculation to the other was based on the gradual decrease of the time step size (Δt). The analysis evidenced that the maximum number of iterations required for convergence of equations decreased approximately with a lowering Δt. In addition, the maximum CFL at the flow-sensitive locations decreased and computational time increased as Δt lowered. The results showed that for a time step size of 0.0001 s, a balance between the accuracy of the results and the computational cost was achieved. This result can guarantee the stability and efficiency of the convergence process of the solutions over the transient region of the air-fluid domain.

3.2 Qualitative Analysis of the Fluid-Structure Interaction Model

To be able to account for the interaction effect of air on the eye structure's response to loading, the two domains had to be solved simultaneously in a *Two-Way* coupling, which allowed data to be interchanged in each time step of the solution. Finally, a dynamic model of the fluid-structure interaction between eye structures and air puff was constructed. During the FSI numerical simulation, the main corneal states while testing were identified. Figure 7 shows two of these states: the first applanation (*a*) at about 39 ms and the maximum concavity (*b*) at roughly 46.1 ms.

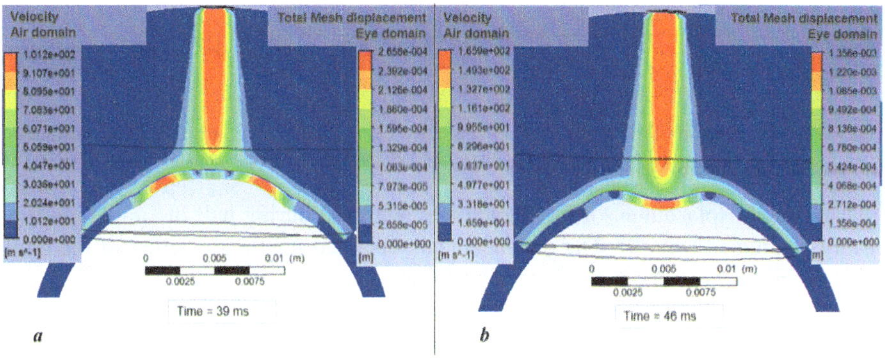

Fig. 7. The FSI model at the first applanation moment and upon maximum concavity.

The NCT test was successfully represented by making a qualitative comparison to the corneal deformations acquired during the clinical Corvis ST tests for human eyes *in vivo*. In Fig. 8, clinical images are observed at six time points of the FSI numerical simulation. The results showed a good agreement of corneal deformations, but they were not perfectly equal. This could be due to differences in geometry, IOP and material properties. Consequently, tests done with real patients differ from the average values assumed in the model. Some studies have also shown that the same displacements can be obtained for different combinations of these parameters [9].

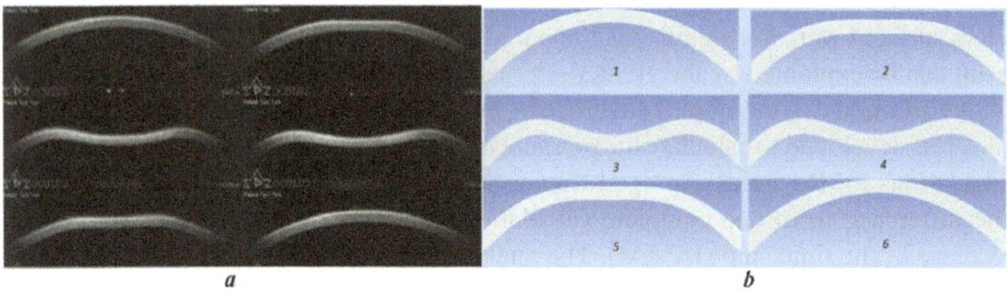

Fig. 8. The CorVis ST images [2] recorded during NCT testing (*a*) and the numerical corneal configuration (*b*). Time of comparison: 1) Physiological configuration, 2) First applanation, 3) Configuration at maximum air jet pressure, 4) Configuration at maximum apex displacement, 5) Second applanation, 6) New physiological configuration after testing.

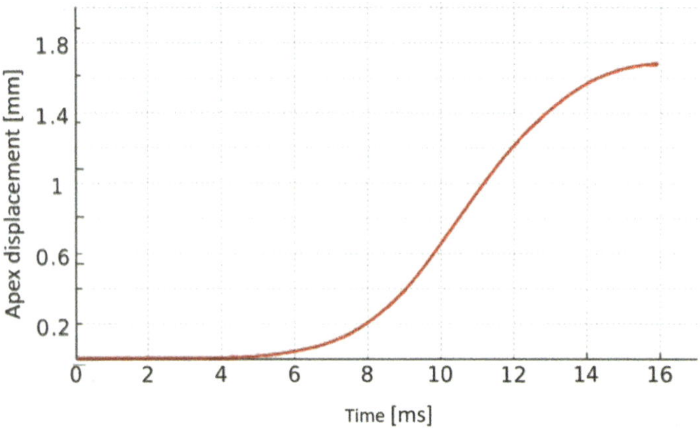

Fig. 9. Time evolution of the maximum corneal apex displacement for the simulations conduced.

The apex displacement of the cornea was analyzed during half of simulation time history conducted (Fig. 9). The displacements obtained for the modeling of the fluid-structure interaction model were within the ranges reported in the numerical models in the literature [9]. In some numerical models presented in the literature, the corneal apex deformation may have diverse responses during a numerical simulation of the air puff experiment. *Ariza et al.* [9] demonstrated that the same apical displacement can occur for different combinations of corneal material properties and IOP values. Then, an exact numerical value may vary depending on the properties and parameters assumed for each numerical model.

4 Conclusions

A numerical simulation was developed to reproduce an NCT test and to represent some of the dynamic response parameters of the human eye in accordance with IOP action and the tonometer air puff. The FSI between the human eye and air puff fluid was considered in the numerical model. For this purpose, an FE model of the human eye with internal structures and a CFD model of air fluid were developed.

The main results were: a model of the eyeball using FE analysis with optimal spatial discretization and a good mesh quality for the chosen eye geometries and material properties. The stress-free geometry of the eyeball affected by IOP action was obtained.

The CFD model was implemented by a transient analysis that reproduced the air puff outflow until its incidence on the corneal surface. Air fluid was considered to be incompressible and turbulent during testing and was numerically characterized by the Spalart-Allmaras turbulence model. In the air domain, a mesh with both good meshing quality and an estimated time step of 0.0001 s was generated, which gave reliable results at a reasonable computational cost during this transient analysis.

The NCT test was reproduced by a numerical simulation of the fluid-structure interaction by coupling among the above models: the information exchange between the two domains was framed within a dynamic interaction model with strong *Two-Way* coupling.

The six main experimental test states were reproduced by the numerical simulation. A correspondence was noted in the qualitative comparison of the clinical images acquired from Corvis ST and cornea deformation during the numerical simulation. A non contact tonometry test was reproduced by numerical simulation, including a dynamic analysis of the fluid-structure interaction between the test air pulse and the ocular structure model. This FSI model between air puff and the human eye will be taken as a starting point to analyze some biomechanical cornea properties during NCT testing performed for future research purposes.

Acknowledgements. We wish to thank the assistance and cooperation between the Polytechnic University of Cartagena, Spain, and the Higher Institute of Technologies and Applied Sciences, University of Havana, Cuba. This collaboration was funded by the Fundación Carolina with the 2020SEGIB mobility program.

References

1. Conti, F.: Fisiología de la visión, en Fisiología médica, McGraw-Hill, Ed., ed. España, pp. 379–408 (2010)
2. Montanino, A., Angelillo, M., Pandolfi, A.: Modelling with a meshfree approach the cornea-aqueous humor interaction during the air puff test. J. Mech. Behav. Biomed. Mater. **77**, 205–216 (2018)

3. Han, Z., et al.: Air puff induced corneal vibrations: theoretical simulations and clinical observations. Refractive Surger **30**, 208–213 (2014)
4. Simonini, I., Angelillo, M., Pandolfi, A.: Theoretical and numerical analysis of the corneal air puff test. J. Mech. Phys. Solids **93**, 118–134 (2016). https://doi.org/10.1016/j.jmps.2016.04.012
5. Kaneko, M., Tokuda, K., Kawahara, T.: Dynamic sensing of human eye. In: Presented at the Conference of Robotics and Automation, Barcelona, Spain (2005)
6. Kling, S., Bekesi, N., Dorronsoro, C., Pascual, D., Marcos, S.: Corneal viscoelastic properties from finite-element analysis of in vivo air-puff deformation. PLoS ONE **9**(8), e104904 (2014)
7. Roy, A.S., Kurian, M., Matalia, H., Shetty, R.: Air-puff associated quantification of non-linear biomechanical properties of the human cornea in vivo. J. Mech. Behav. Biomed. Mater. **48**, 173–182 (2015)
8. Montanino, A., Angelillo, M., Pandolfi, A.: A 3D fluid-solid interaction model of the air puff test in the human cornea. J. Mech. Behav. Biomed. Mater. **94**, 22–31 (2019). https://doi.org/10.1016/j.jmbbm.2019.02.030
9. Ariza-Gracia, M., Zurita, J.F., Piñero, D.P., Rodriguez-Matas, J.F., Calvo, B.: Coupled biomechanical response of the cornea assessed by non-contact tonometry. A simulation study. PLoS One **10**(3), e0121486 (2015)
10. Muench, S., Roellig, M., Spoerl, E., Balzani, D.: Numerical and Experimental Study of the Spatial Stress Distribution on the Cornea Surface During a Non-Contact Tonometry Examination. Exp. Mech., 59, 1285–1297 (2018)
11. Ariza-Gracia, M., Wu, W., Calvoa, B., Malv, M., Büchlere, P., Rodriguez Matas, J.F.: Fluid-structure simulation of a general non-contact tonometry. A required complexity? Comput. Methods Appl. Mech. Eng. **340**, 202–215 (2018)
12. Maklad, O., Eliasy, A., Chen, K.J., Theofilis, V., Elsheikh, A.: Simulation of air puff tonometry test using arbitrary lagrangian-eulerian (ALE) deforming mesh for corneal material characterisation. Int. J. Environ. Res. Public Health **17**(1), 54 (2019)
13. Muñoz-Villaescusa, C.: Modelo biofísico del ojo humano: Su aplicación en cirugía LA- SEK. Tesis de diploma, Instituto Superior de Tecnologías y Ciencias Aplicadas, La Habana (2014)
14. Núñez-Chongo, O.: Modelación en elementos finitos del ojo humano. Tesis de Diploma, Instituto Superior de Tecnologías y Ciencias Aplicadas, La Habana (2016)
15. Elsheikh, A., Geraghty, B., Alhasso, D., Knappett, J., Campanelli, M., Rama, P.: Regional variation in the biomechanical properties of the human sclera. Exp. Eye Res. **90**(5), 624–633 (2010)
16. Uchio, E., Ohno, S., Kudoh, J., Aoki, K., Kisielewicz, L.T.: Simulation model of an eyeball based on finite element analysis on a supercomputer. Brit. J. Ophthalmol. **83**(10), 1106–1111 (1999)
17. Elsheikh, A., Alhasso, D., Rama, P.: Biomechanical properties of human and porcine corneas. Exp. Eye Res. **86**(5), 783–790 (2008)
18. ANSYS: Mechanical APDL Theory Reference. Release 18.0. ANSYS, Inc. South- pointe, 275 Technology Drive Canonsburg, PA (2018)
19. Elsheikh, A., Whitford, C., Hamarashid, R., Kassem, W., Joda, A., Büchler, P.: Stress free configuration of the human eye. Med. Eng. Phys. **35**, 211–216 (2013)
20. ANSYS: Fluent Theory Guide. Release 15.0: ANSYS, Inc. Southpointe, 275 Technology Drive Canonsburg, PA (2013)

Euclid's Elements: Geometric Figures in the Copy of the Yuso Monastery, in San Millán de la Cogolla (Spain)

Sergio Rojo-Vea(✉), Jacinto Santamaría-Peña, and Félix Sanz-Adán

Department of Mechanical Engineering, University of La Rioja, 26004 Logroño, La Rioja, Spain
sergio.rojo@unirioja.es

Abstract. Euclid's treatise The Elements, first printed in the 15th century, is unanimously recognized as one of the most widely disseminated books in history as a whole, along with the Bible. The publisher of the first printing, Erhard Ratdolt (1442–1528), acknowledged in the preface that the real reason for the absence of publication of mathematical texts at the time lay in the difficulty of printing their geometric figures. And it was this master printer who contributed to transcribing geometric figures with the same ease as letters.

The mathematician Luis Javier Hernández Paricio, speaks, in his "On the Fundamental Principles of Geometry: inaugural opening lesson of the academic year 20002001", and in relation to the incunabula of the Yuso Monastery, of "the admirable geometric figures that appear in its margins". And it also talks about how well preserved it is compared to the other six copies that are kept in Spain, some of which would have their margins cut the observation of these figures and their analysis, by comparison to the text written in the volumes, is the objective of the publication presented here. And also stand out in having a copy in the small town of San Millán de la Cogolla.

Keywords: Treatise · Elements · Euclid · Geometry · Yuso

1 Introduction

The opinion held on the Elements of Euclid is universally accepted as the first fundamental mathematical work, founder of Geometry as a scientific discipline and, above all, the one that has had the most influence to this day. Whichever reference is sought, in all of them the accent is placed on it being the most published book in history as a whole, along with the Bible. Its compiling nature is undoubtedly one of its main values (not the only one), along with the fact that, the putting into crisis of some of its postulates generated various scientific controversies around geometry, and enriched it even the point of this treatise being the real origin of the appearance, in the 19th century, and through various channels, of the so-called Non-Euclidean Geometries. Whichever reference is sought, in all of them the accent is placed on it being the most published book in history as a whole, along with the Bible. Its compiling nature is undoubtedly one of its main values (not the only one), along with the fact that, the putting into crisis of some of its

F. Cavas Martínez et al. (Eds.): INGEGRAF 2021, LNME, pp. 229–236, 2022.
https://doi.org/10.1007/978-3-030-92426-3_27

postulates generated various scientific controversies around geometry, and enriched it even the point of this treatise being the real origin of the appearance, in the 19th century, and through various channels, of the so-called Non-Euclidean Geometries.

Let us think that Non-Euclidean Geometries are, in the words of Professor Hernández Paricio,

"one of the most spectacular advances that has been made during the second millennium in the world of mathematical thought."

Indeed, the controversy that generated, for centuries, the attempts to demonstrate the fifth postulate, better known as the "axiom of parallels", with its complex and intricate wording, opened a history of more or less successful attempts that led, twentyone centuries later, in the appearance of non-Euclidean Geometries, that is, those that do not satisfy the postulates of Euclid of Alexandria.

Euclid's treatise, insofar as it is a text dated around 300 BC. and is, therefore, twenty-three centuries old, it has a history rich in historical vicissitudes, which have to do with its translation into different languages and with the legacy of one generation to another.

2 Relevance of the Venetian Edition of Euclid's Treatise

The copy of the monastery of the Augustinian Recollects (of which there are only five more copies in Spain), corresponds to the edition of the master printer and typographer Erhard Ratdolt, of German origin and installed in Venice since 1476. It is one of the scientific books most beautiful prints, and their importance is advanced by himself in the preface to Euclid's Treatise. Thus, the editor describes how the impossibility of typographically representing technical drawings was an obstacle when it came to publish and disseminating mathematical texts. His contribution consisted in giving the sheets an eight-centimeter margin on the sides so that the graphics that accompanied the mathematical texts could be clearly represented with the same printing press. The resolution of this typographic problem viralized the printed edition of the Elements, to the point that it thus became the main mathematical book of the Western European schools until the 16th century.

This edition corresponds to an Arabic-Latin version commented in the middle of the 13th century by the Italian mathematician and chaplain of Pope Urban IV Campano de Novara, who in turn reworked the translation from Arabic to Latin by the English monk Adelardo de Bath, in the 12th century. In other words, we are talking about a text whose transmission throughout history corresponds to that of many other academic texts of Greek origin. These were first translated from Greek to Arabic in the 9th century; later translated into Latin in the 12th century; later, after the appearance of the printing press, these Latin versions were printed in the 15th century and later, editions of Latin versions were promoted from the Greek, thus facilitating the subsequent appearance of the "editio princeps" of the text itself Greek. The vernacular versions would come later.

The propagation of the Ratdolt edition of 1482 therefore predates even the prince edition, a fact that further highlights the significance of this edition. Despite the fact that Professor López Pellicer, in his speech at the commemorative session of the Instituto España Foundation (2003), indicated the edition corresponding to the Yuso incunabula as the princeps edition, other sources consider that the one published in Basel really is.

half a century later, in 1533 by the German theologian Simon Grynaeus the Elder and which is now written in Greek. The existence in Spain of this princeps version has been confirmed in, at least, the library of the University of Seville and in the library of the Monastery of El Escorial.

Finally, and going back to the Venetian edition, it is worth highlighting the beauty of its cover (Fig. 1), where there is a stormy three-edge engraving (upper, lower and left side) that would be used in other editions, becoming like this a kind of corporate brand. This engraving includes the title inside, in red ink, which reads: "Preclarissimus liber elementorum Euclidis perspicacissimi in artem Geometrie incipit qua [m] foelicissime".

Throughout the book (there are actually thirteen, but in a single binding), you can see how to achieve great precision and delicacy in printing the drawings, even with curved lines, especially if they are circumferences. In the elliptical arches, however, it does not seem to have reached such a perfect typeface.

Fig. 1. Handwritten and color annotations, both on the first page (left) and right on the graph of the axiom of parallels (right). Copy of Yuso.

3 Spanish Copies and the Uniqueness of Yuso

It is still an anecdotal fact that Yuso's copy rests in a place called "el infiernillo" along with other books that were banned or censored by the Holy Inquisition. Indeed, alongside these apocrypha, our treatise occupies a place. The objective of the analysis presented here is clear. Of the version that was printed for the first time in Venice in 1482, there are only six copies in Spain: three in Madrid, one in Barcelona, one in Valladolid and another in San Millán de la Cogolla, in La Rioja. According to the illustrious institutions that house a copy of such a valuable copy (see Table 1), it can be clearly said of the

San Millan copy that it is the most remote. And the presence of this incunabula in the monastery library deepens the idea that it is the owner of books of special historical singularity.

Table 1. Parties to the treaty elements

Parties to the treaty		Thematic
Basic plane geometry	Book I	Properties of triangles, theory of parallels, polygons, Pythagorean Theorem and its inverse
	Book II	Areas and geometric algebra of the Pythagorean academy. Cosine law
	Book III	Circumference, chords, tangents and angle measurement
	Book IV	Regular polygon constructions with ruler and compass
	Book V	Commensurable and incommensurable magnitudes. Irrational numbers
	Book VI	Similar triangles, proportional measures. Quadric equations. Proposition of the bisector
Prime and radical numbers, divisibility	Book VII	Prime numbers. Arithmetic Legacy of Pythagoras
	Book VIII	Series of numbers in continuous proportion and geometric progression
	Book IX	Even, odd, relationships Prime factors
	Book X	Irrationals
Solids, polyhedra, circumstantial spheres	Book XI	Solid geometry. Platonic solids
	Book XII	
	Book XIII	

It has been possible to verify on the site that Yuso's copy corresponds to that of Erhard Ratdolt, and it has also been possible to compare, thanks to the digitization of the institutions' funds, with other copies that exist not only in Spain but with some from other countries. Two characteristics of it are that it is restored, and that it presents handwritten comments, presumably in Latin, accompanying the drawings, during the first half of the book (Table 2).

Table 2. The 6 copies in Spain of the 1st printed edition of Erhard Ratdolt (1482).

Year	Edition	City	Institution	Link	Observations
1482	Ratdolt Venice	San Millán de la Cogolla	Yuso Monastery	Yuso	Latin annotations in margins in the first half of the book
		Madrid	National Library	B.Nacional	Digitization in b/w. Handwritten footnotes, different from Yuso's
			Spanish Royal Palace	Available upon request	Ex libris manuscript of Fouquier Brunet, S XVIII
			Casa de Alba	In-person consultation by appointment at Palacio de Liria	Bought in London in 1950 by the Duke of Alba
		Barcelona	Athenaeum	Barcelona	
		Valladolid	University	Valladolid	

Pérez Barriocanal and Sacristán Marín, authors of the catalog of the collection of the library of the monastery between the 15th and 18th centuries, clarify that Yuso's copy is printed with Gothic letters, with 45 lines per page and with the cover in two inks, red and black. (There are at least 7 copies in Germany with dedications in gold letters). But, unfortunately, they neither verify the possible origin of the specimen, nor do they identify the handwriting with any monk of the monastery (Tables 3 and 4).

Table 3. Some international copies of the 1st printed edition of Erhard Ratdolt (1482).

Año	Edición	Ciudad	Institución	Enlace	Observaciones
1482	Ratdolt Venecia	Munich	German National Library	Munich	Dedication in gold letters
		Berkeley, California	Bancroft Library	Berkeley	Gift of bibliophile Julius Wangenheim

Tabla 4. Copies of the princeps edition in Spain (1533).

Year	Edition	City	Institution	Link	Observations
1533	Simon Grynaeus Basilea	El Escorial	El Escorial Monastery	Escorial	In greek
		Seville	Sevilla University	Sevilla	

The monks have not dated or identified how this specimen could reach the Cárdenas valley. After being consulted about the incunabula, they also confirmed the difficulties of knowing its origin, alluding only to the interest of the friars, as forerunners of culture, in possessing books of the highest intellectual capacity; apparently it was common between monasteries the exchange of the same. It is important to note at this point that the Yuso monastery, as we know it today, was completely rebuilt in a style close to the Herrerian or Escurialense, in the 16th century. And it could be considered, therefore, that the Benedictine Order had tried to have books of notable reference in its new library. The preservation of this and other essential references in its library is also surprising, even knowing the multiple expulsions, plunder, confiscation and abandonment that the Royal Monastery has suffered throughout its history. Currently, it is a World Heritage Site in Europe, due to various cultural and historical facts of interest, the most relevant being the appearance of the codex with the Emilian Glosses, considered the first words written in Spanish (Figs. 2 and 3).

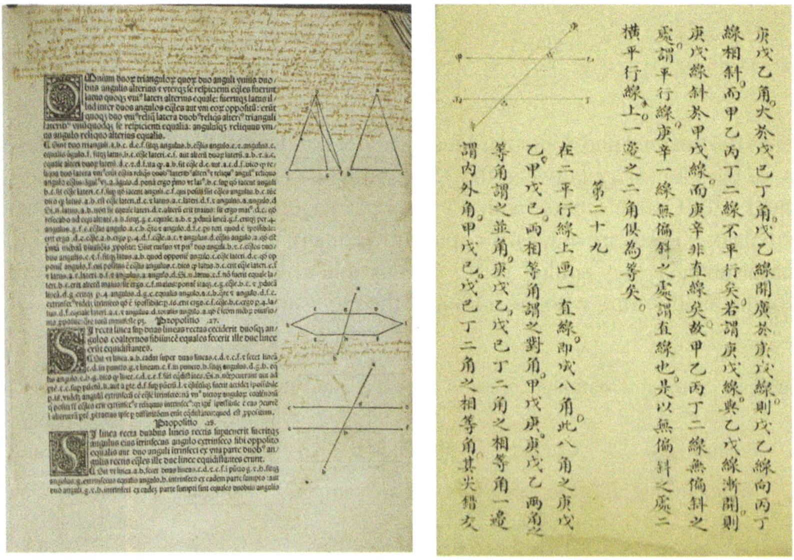

Fig. 2. Book I. Proposition 28. If a segment when incident on two lines makes the alternate angles equal to each other, the two lines will be parallel to each other. Yuso copy (left) and Chinese handwritten version of the s. XVII with the same graphic.

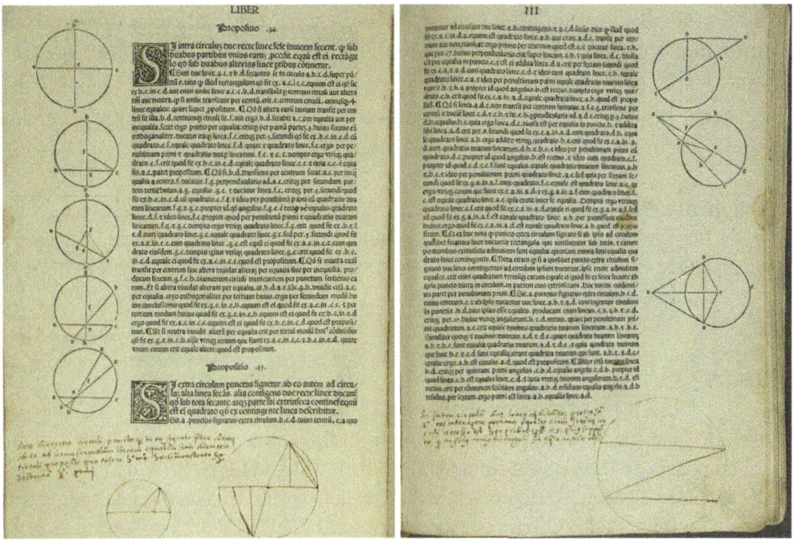

Fig. 3. Book III, proposition 35 (left and right): If two lines intersect each other in a circle, the rectangle comprised by the segments of one is equal to the rectangle comprised by the segments of the other. (Power of a point with respect to a circle). You can check the amount of handwritten annotations in red, in Yuso's copy.

Table 5. Other examples of interest

Year	Edition	City	Institution	Link	Observations
1576	Rodrigo Zamorano - Sevilla	Seville	Seville University Library	Sevilla	First edition in Spanish
1690	I.G., Pardies, J.F. Gerbillon, J. Bouvet	Beijing	Beijing National Library	Pekín	Chinese translation of two French Jesuit missionaries for the Kangxi Emperor
2ª half XVIII	Juan González Sysniega,	Logroño	Local library	Logroño	Handwritten in Spanish. All drawings at the end

Another curiosity worthy of mention is in the latest handwritten version, in Spanish, that is kept in the Public Library of La Rioja, corresponding according to the figure on its cover to Juan González de Sisniega, who died in 1805 as captain of the ship in battle from Trafalgar. On the first page of the book, you can read "This book is by Dn (sic) Juan Gonzalez de Sysniega, Royal Navy frigate ensign" (Table 5).

In this article, up to four tables are presented with the editions consulted. Due to the facilities that the consultation of funds that have been digitized presents, the link of the same is also provided, so that it is very immediate to go to the specific edition. The comparison between some editions and others is of enormous interest, taking into account their origin and their country of destination.

Acknowledgments. We want to especially thank the order of the Augustinian Recollects of the monastery for the facilities provided for the consultation of the incunabulum. Likewise, to the institutions that have digitized their funds and made them freely available to researchers.

References

1. Clagett, M.: The medieval latin translations from the arabic of the elements of euclid with special emphasis on the versions of adelard of bath. ISIS J. **43**(1/2), 16–42 (1953)
2. García Turza, J.: El Monasterio de San Millán de la Cogolla. Una historia de santos, copistas, canteros y monjes. 1 edn. Ediciones Everest. Madrid (2013)
3. Hernández Paricio, L.J.: Sobre los principios fundamentales de la geometría: lección inaugural del curso académico 2000–2001. Universidad de La Rioja, Logroño (2001)
4. López Pellicer, M.: Vida de los trece libros de Euclides, 1st edn. Fundación del Instituto de España, Madrid (2003)
5. González Urbaneja, P.M.: Los elementos de Euclides. Biblia de la geometría griega. 1era edición. Editorial Plaza Abierta, Barcelona. (2018)
6. Pérez Barriocanal, C., Sacristán Marín, E.: Catálogo de impresos s.XV-XVIII Monasterio de Yuso. Parlamento de La Rioja. Logroño (1999)
7. www.euclides.org
8. www.euler.us.es https://euler.us.es/~libros/griegos.html. recuperado por última vez en 2021/03/27
9. www.todolibroantiguo.es http://www.todolibroantiguo.es/libros-raros/elementoseuclides.html. recuperado por última vez el 27 Mar 2021
10. www.wdl.org https://www.wdl.org/es/item/7103/ recuperado por última vez en 27 Mar 2021
11. https://www.christies.com/en/lot/lot-6296916

Variable Complexity Corneal Surfaces Characterization by Modal Geometrical Reconstruction Methods: Comparative Study

Alejandro Ballesta[1(✉)], Jorge Alió[2], Jose Miguel Bolarín[3], and Francisco Cavas[4]

[1] International School of Doctorate, Technical University of Cartagena, 30202 Cartagena, Spain
alejandro.ballesta@edu.upct.es
[2] Cornea, Cataract and Refractive Surgery Department, VISSUM, Alicante, Spain
[3] Technological Centre of Information Technologies and Communications (CENTIC), Science Park of Murcia, 30100 Murcia, Spain
[4] Department of Structures, Construction and Graphical Expression, Technical University of Cartagena, Cartagena, Spain

Abstract. Validation of new methods for keratoconus diagnosis has generated a great interest for the last years in the field of ophthalmology. Corneal surface reconstruction by discrete 3D points creates the possibility to apply modal or zonal geometrical methods. The goal of this work is to determine the geometrical behavior of the cornea in each keratoconus stage by different modal geometrical models, mainly, those derived from the general equation of the biconical and its associated parameters. This study included a total amount of 75 eyes, 20 of these were healthy eyes and 55 classified with keratoconus pathology by using Amsler-Krumeich classification. After studying the proposed models, the biconical model showed the best fit model while spherical model showed the worst fit one. By assessing geometrical parameters from biconical model, curvature radius and asphericity in the main direction, the geometrical corneal behavior can be assessed in each keratoconus stage in both anterior and posterior cornea. Using a variable reduction in two parameters, mean radius and mean asphericity, a decreasing trend in both parameters with keratoconus progress is found. Therefore, this study proposes to validate the biconical model for variable complexity surfaces with keratoconus pathology as a tool for diagnosis in clinical practice.

Keywords: Geometrical modelling · Keratoconus · Asphericity · Curvature radius · Corneal topography

1 Introduction

The knowledge of geometrical characteristics in surfaces created by discrete 3D points is an essential factor in several engineering fields. [1–6]. In particular, in ophthalmological field is essential to know the geometrical characteristics in corneal surfaces created by elevation data given by corneal topographers [7, 8] for corneal pathologies characterization [9–12].

F. Cavas Martínez et al. (Eds.): INGEGRAF 2021, LNME, pp. 237–247, 2022.
https://doi.org/10.1007/978-3-030-92426-3_28

Corneal surfaces can be reconstructed by zonal [13, 14] or modal [15–20] mathematical models. With regard to zonal methods, the reliability of the fit estimated parameters results difficult to test, thus their implementation and computational cost could be higher than modal methods. [21]. Furthermore, modal reconstruction methods based on biconical geometry allow to identify global irregularities in each 3D dataset by the fit residuals. [15–20]. However, these methods have been only referred to the anterior surface thus, in this study, they could be useful to study diseases like keratoconus in both anterior and posterior corneal surface.

Keratoconus is a corneal degenerative pathology which generates a local thinning in the corneal thickness and in consequence, a curvature increase in corneal surfaces is detected. Keratoconus produces a damage in visual acuity to the patients, with the possibility of developing a soft stage, which could be corrected by lens, or a severe stage, which could only be solved by corneal transplantation [22].

In this investigation article, some modal algorithms are proposed to represent and characterize corneal surfaces with variable complexity in keratoconus pathology.

2 Materials and Methods

2.1 Patients

Investigation was designed by a retrospective, random and multicentric study started with official data in BioBanco IBERIA (RETICS-OFTARED). The study included a total of 75 eyes divided in two different groups: 20 healthy eyes and 55 eyes diagnosed with keratoconus pathology using Amsler-Krumeich classification method [23]. With regard to diagnosed eyes, they were divided in 19 Stage I eyes (35.00 ± 14.14 years), 17 Stage II eyes (34.45 ± 12.49 years), 10 Stage III eyes (33.63 ± 10.21 years) and 9 Stage IV eyes (31.5 ± 9.8 years). Control or healthy group presented an age 38.19 ± 13.95 years. In the total sample, 53.3% of the population were female eyes while 46.7% were male eyes.

As an exclusion approach in healthy group, eyes without any visual pathology or surgical procedure were considered [23].

The same exclusion approach was considered in keratoconus group. Moreover, as an inclusion criterion, the presence of a located bending, an asymmetric topography and also other biomicroscopical signals, such as stroma thinning or corneal apex protrusion, were considered.

In this study, one eye for each patient were selected by a random numerical section with the goal of avoiding potential correlation between eyes of the same patient [24]. The study was developed following the ethics standards defined in Helsinki declaration (Fortaleza, Brasil, 2013).

Although each patient presented a substantial number of variables in Biobanco, socio-demographic variables and corneal topographies were only considered in this study. An expert optometrist performs three topographical measures (Sirius, CSO, Italia) to each patient (random selection), choosing the best topography based on the quality in data acquisition (centered points and mapping $> 90\%$).

2.2 Point Cloud Generation

Surface reconstruction models are based in a discrete and finite data set which are representative of the surface to reconstruct [25]. In ophthalmologic field, these raw data can be obtained by corneal topographers based in Scheimpflug geometric method [23].

Corneal topographers are able to map all the corneal region shaped like concentric rings with radius $r = 0 \ldots 6$ mm. However, there are some areas in the corneal surface where raw points cannot be acquired due to extrinsic errors [26]. These points are treated with a value of -1000, being totally excluded from the data acquisition matrix in this study.

2.3 Geometrical Reconstruction Methods

Modal surface reconstruction methods are based in the approach to the surface by a combination of basic functions globally defined in all the discrete and finite data set [25], which could depend on a certain number of enough parameters to represent the relevant information from the surface to reconstruct.

In this study, modal reconstruction methods based in biconical geometry have been applied. Specifically, for the anterior and posterior corneal modelling, biconical, corneal ellipse, torical, spherical and non-revolution ellipsoid models have been implemented (Table 1). All of them derived from the general biconical equation.

$$\bar{z} = \frac{C_x \cdot x^2 + C_y \cdot y^2}{1 + \sqrt{1 - (1+Q)_x \cdot C_x^2 \cdot x^2 - (1+Q)_y \cdot C_y^2 \cdot y^2}}$$

Where the coefficients C_x and C_y are defined as the main direction curvatures and the coefficients Q_x and Q_y are defined as the main direction asphericities. The equation term $(1 + Q)$ is named as shape factor, which allows to analyze corneal curvature variation from corneal apex to periphery [16].

By implementing all these models for each cornea, a reproduction of the elevation corneal data given by a corneal topographer is pretended. In this way:

$$z = S_b(x, y) + r(x, y) \qquad (2)$$

Where $S_b(x, y)$ is the reconstruction model function and $r(x, y)$ is the residual function derived from the applied fit model.

2.4 Statistical Analysis

Every model's fit is implemented by a nonlinear minimum square fit method. Both models and results are programmed and obtained by using the software Matlab R2017 (Mathworks, EEUU). Fit parameters are obtained and the goodness of fit is analyzed by the RMSE (root-mean-square-deviation) parameter. For every parameter range, confidence 95% intervals for the mean are used.

Table 1. Reconstruction models for corneal geometry.

Spherical Model ($Q_x = Q_y = 0$; $C_x = C_y$)

$$\bar{z} = \frac{C_x \cdot x^2 + C_x \cdot y^2}{1 + \sqrt{1 - C_x^2 \cdot x^2 - C_x^2 \cdot y^2}} \quad (3)$$

Corneal Ellipse Model ($Q_x = Q_y$; $C_x = C_y$)

$$\bar{z} = \frac{C_x \cdot x^2 + C_x \cdot y^2}{1 + \sqrt{1 - (1 + Q_x) \cdot C_x^2 \cdot x^2 - (1 + Q_x) \cdot C_x^2 \cdot y^2}} \quad (4)$$

Torical Surface ($Q_x = Q_y = 0$)

$$\bar{z} = \frac{C_x \cdot x^2 + C_y \cdot y^2}{1 + \sqrt{1 - C_x^2 \cdot x^2 - C_y^2 \cdot y^2}} \quad (5)$$

Non revolution ellipsoid Model

$$\frac{x^2}{a^2} + \frac{y^2}{b^2} + \frac{z^2}{c^2} = 1 \quad (6)$$

$$R_x = \frac{a^2}{c};\ R_y = \frac{b^2}{c};\ Q_x = \frac{a^2}{c^2} - 1;\ Q_x = \frac{b^2}{c^2} - 1 \quad (7)$$

3 Results

3.1 Anterior Corneal Surface

After applying the five proposed reconstruction models, the results are shown in Table 2. With regard to healthy eye group, both biconical and ellipsoid models are the best fit models to the data set (0.026 ± 0.007) while the spherical model is the worst fit one. (0.036 ± 0.008).

These results are aligned with the obtained for group I, where biconical model (0.037 ± 0.013) and spherical model (0.053 ± 0.014) are the best and worst fit models respectively.

With regard to severe stage group, biconical model was also the best fit model, specifically for the stage II (0.104 ± 0.037), Stage III (0.138 ± 0.076) and Stage IV (0.130 ± 0.050). Spherical model was also the worst fit model (Table 2).

Table 2. RMSE comparison for every proposed model in anterior corneal surface.

Clinical stages	Normal	Stage I	Stage II	Stage III	Stage IV
Model	RMSE Mean ± SD	RMSE Mean ± SD	RMSE Mean ± SD	RMSE Mean ± SD	RMSE Mean ± SD
Biconical	0.026 ± 0.007	0.037 ± 0.013	0.104 ± 0.037	0.138 ± 0.076	0.130 ± 0.050
Spherical	0.036 ± 0.008	0.053 ± 0.014	0.129 ± 0.036	0.170 ± 0.073	0.178 ± 0.047
Torical	0.034 ± 0.008	0.046 ± 0.014	0.126 ± 0.036	0.162 ± 0.073	0.171 ± 0.048
Ellipse	0.028 ± 0.007	0.043 ± 0.014	0.106 ± 0.037	0.142 ± 0.077	0.134 ± 0.051
Ellipsoid	0.026 ± 0.007	0.038 ± 0.013	0.106 ± 0.037	0.140 ± 0.077	0.135 ± 0.050

Similarly, with the keratoconus progression, the goodness of fit is getting worse (Table 2). This fact is due to the local curvature and asymmetric irregularities in the corneal surface that increase with the pathology progression.

3.2 Posterior Corneal Surface

Results obtained for posterior surface can be observed in Table 3. According to the results, the goodness of fit is quite worse than those obtained in anterior surface. Similar to the anterior surface, biconical model is also the best fit model in posterior surface. In a clinical perspective, a healthy eye and Stage I could be easily distinguished from Stage II, Stage III and Stage IV. However, RMSE parameter is not enough to assess and discriminate all keratoconus stages. In consequence, more parameters are required.

Table 3. RMSE comparison for every proposed model in posterior corneal surface.

Clinical stages	Normal	Stage I	Stage II	Stage III	Stage IV
Model	RMSE Mean ± SD	RMSE Mean ± SD	RMSE Mean ± SD	RMSE Mean ± SD	RMSE Mean ± SD
Biconical	0.034 ± 0.013	0.062 ± 0.020	0.124 ± 0.037	0.173 ± 0.064	0.136 ± 0.039
Spherical	0.057 ± 0.019	0.093 ± 0.028	0.159 ± 0.040	0.236 ± 0.066	0.191 ± 0.037
Torical	0.054 ± 0.021	0.083 ± 0.030	0.155 ± 0.040	0.229 ± 0.067	0.181 ± 0.039
Ellipse	0.037 ± 0.013	0.069 ± 0.017	0.128 ± 0.037	0.177 ± 0.065	0.140 ± 0.038
Ellipsoid	0.036 ± 0.014	0.063 ± 0.020	0.128 ± 0.037	0.190 ± 0.068	0.142 ± 0.041

3.3 Asphericity and Curvature Radius

After resulting the biconical model the best fit one, all biconical parameters were analyzed. These are the asphericity and curvature radius in each main direction both in anterior and posterior corneal surface (Fig. 1).

In the anterior corneal surface, the mean curvature radius in X axis for healthy eyes was between 7.69 and 7.88, and for Stage IV eyes between 5.15 and 5.17. Regard to curvature radius in Y axis, for healthy eyes mean value between 7.37 and 7.62 were obtained in comparison with values between 5.25 and 5.58 obtained for Stage IV eyes.

By considering asphericities in X and Y main directions, a decreasing trend was observed with the increase of the disease severity both in healthy and pathological eyes.

In the posterior surface (Fig. 2), curvature radius in X and Y direction were also decreasing with the disease evolution, as it can be shown by Rx in Stage 0, (6.47 ± 0.39) compared to (4.48 ± 0.55) for Stage IV, and by Ry, in the Stage 0 (6.06 ± 0.38) compared to (3.84 ± 0.30) in Stage IV. Comparing to anterior surface, curvature radius is significantly lower, thus this surface is more bended.

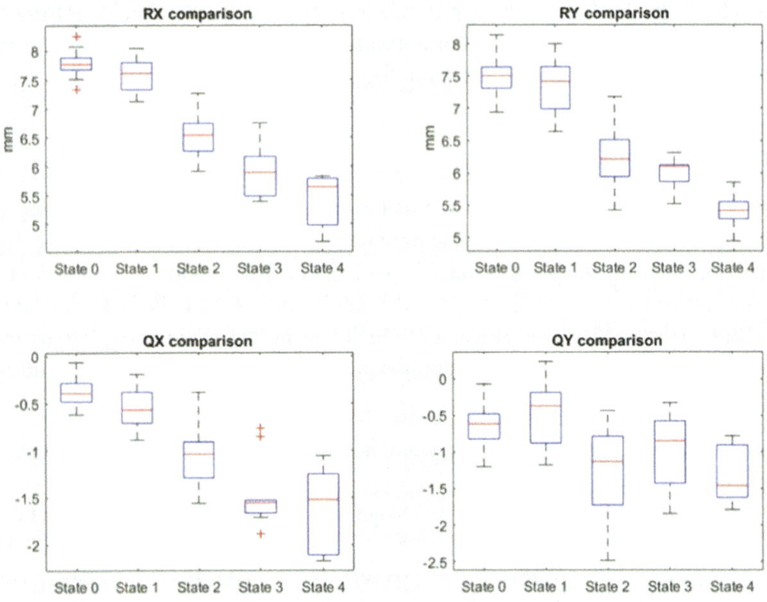

Fig. 1. Asphericity and curvature radius comparison in main directions for anterior corneal surface.

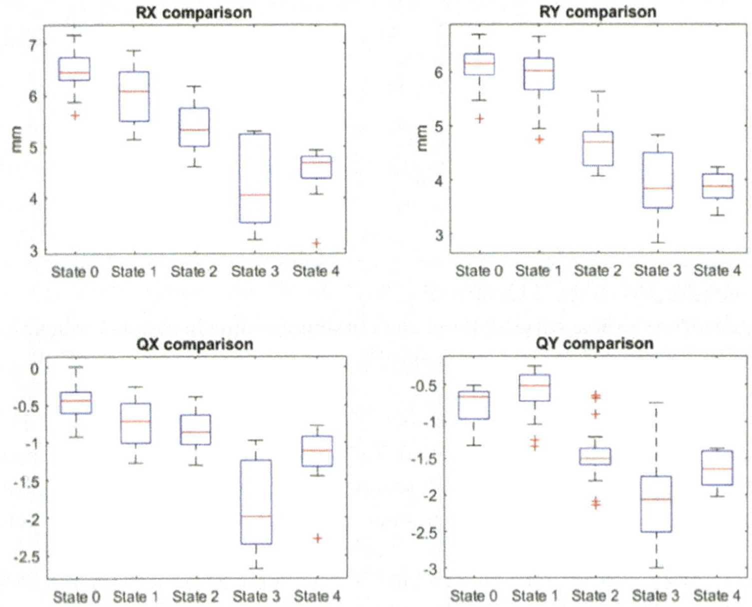

Fig. 2. Asphericity and curvature radius comparison in main directions for posterior corneal surface.

By considering main direction asphericities, Qx seem to present a slight decreasing tendency from Stage 0 (−0.566, −0.363) compared to Stage IV (−1.51, −0.89). Regard to Qy, a clear tendency cannot be concluded although it seems to present lower values with the pathology progression.

Conversely, considering these four parameters and reducing them into two parameters, mean curvature and mean asphericity were created. Using these two parameters, a fit equation which allows to parametrize differences between healthy, mild and severe stages is obtained (Fig. 3).

The connection between mean radius and mean asphericity for both anterior and posterior cornea is:

$$\text{Anterior Surface Fit: } Q_{\text{prom}} = -4.128 + 0.4774 \cdot R_{\text{prom}} \tag{8}$$

$$\text{Posterior Surface Fit: } Q_{\text{prom}} = -3.892 + 0.5334 \cdot R_{\text{prom}} \tag{9}$$

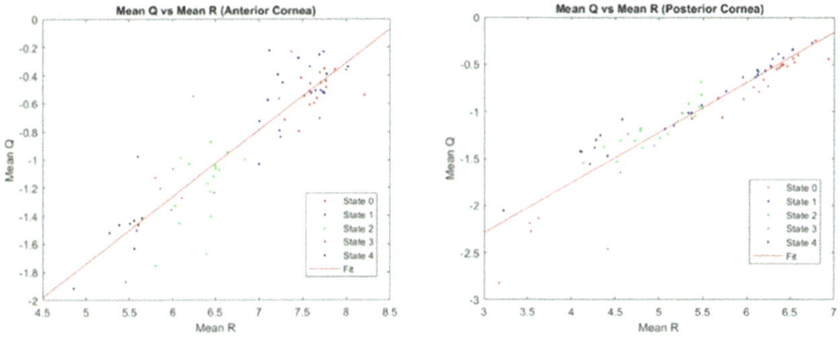

Fig. 3. Mean radius and mean asphericity connection for both anterior and posterior corneal surfaces.

4 Discussion

The progressive thinning suffered from the corneal structure during keratoconus pathology, creates morphological local changes in specific areas due to the presence of asymmetric irregularities. These changes report a more defined pattern with the increasing of keratoconus severity. Variable geometry characterization in this pathology using geometrical reconstruction models generate a high expectation in opthalmological field as the goal is to define the geometrical pattern in the cornea related to keratoconus severity.

Some studies have used computerized topographies to characterize topographical patterns typical from keratoconus [27–31]. However, these studies included treated data by non-controlled private algorithms. This investigation study has characterized the corneal geometry by geometrical reconstruction models using raw data without modification [23]. Data was obtained by artificial vision systems included in corneal topographers during the corneal surface scan process.

For the proposed models in this study, biconical model presented the best fit, both for anterior and posterior reconstruction surface and in all stages analyzed. This fact is because the biconical model is a generical model without restrictions unlike torical, spherical or elliptical models, which present restrictions related to the degrees of freedom. In detail, the spherical model was the worst fit model as it only considers two degrees of freedom.

These results are matching with others authors [32, 33]. Elsheikh et al. [32] applied the biconical model to define the base geometry in its biomechanical characterization model. Gatinel et al. [33] analyzed clinical and geometrical parameters with elevation data for the anterior surface. Both studies used the biconical model, however, only the anterior surface where reconstructed while in our study, biconical model is used for both anterior and posterior corneal surface reconstruction for the first time.

By using the proposed model in our study, and by considering the goodness of fit, biconical model reproduces the global irregularities of each 3D data set during keratoconus progression. These results are matching with the authors [24, 34, 35].

Furthermore, asphericity and curvature radius parameters in each main direction represent the curvature variability from the center to the corneal periphery [35]. In our study, regarding to the anterior surface, both curvatures radius and asphericity in X axis, developed a decreasing tendency with keratoconus progression while asphericity in Y axis seems to be more stable. This fact is generated because a lower curvature radius creates a higher curvature, thus the corneal surface is bending with keratoconus progression (oblate cornea) (Fig. 4). These results are matching with others authors with regard to anterior corneal surface [36–38]. In posterior surface, curvature radius presented a clear decreasing tendency, while asphericity values presented high variability. This fact is produced due to the intrinsic errors [23] which identify less amount of data in posterior surface compared to anterior surface. This conclusion is also explained in other clinical studies [39, 40].

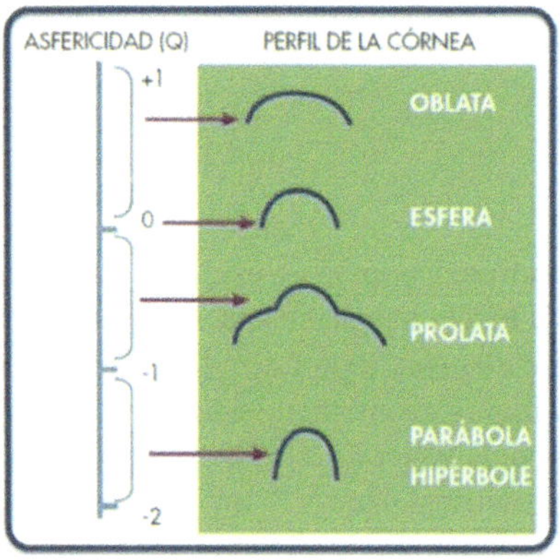

Fig. 4. Corneal profile based in asphericity. Healthy cornea with prolate profile compared to keratoconus cornea with oblate profile.

Mean curvature radius obtained in this study shows values for healthy eyes in anterior cornea between 7.53 and 7.75 being these values similar to other studies [37, 41, 42]. In the posterior surface values obtained for mean curvature radius are slightly lower than others existing in the literature [36, 37, 41–43]. These differences could be caused because of the newest technology used in this study as it is more sensitive to data acquisition than the studies reported by other authors. In our case, our system recorded more points in the space, thus, less intrinsic errors than other technologies are introduced. Moreover, the dimensional reduction proposed in this study using both mean curvature radius and mean asphericity allow to discriminate between healthy and very mild keratoconus from severe stages. However, a deep study is needed to create a function which allow to discriminate all stages.

5 Conclusion

Biconical model is a geometrical reconstruction model able to characterize a variable complexity corneal surface during the keratoconus progression in comparison with other generical modal reconstruction models. The parameters given by the biconical model, both curvature radius and asphericities in the main directions, show a decreasing connection with the disease severity, particularly in the anterior corneal surface. This fact is generated because a lower curvature radius creates a higher curvature, thus the corneal surface is bending with keratoconus progression.

In the posterior corneal surface, a decreasing trend is observed in terms of curvature radius in both main directions. However, asphericity in Y axis presented a high variability with the keratoconus progression.

The dimensionality reduction applied by using mean values for asphericity and curvature radius could discriminate healthy and mild stage from the severe stages. This study allows to validate the biconical model as a useful diagnosis tool in a clinical practice.

References

1. Geng, Z., Bidanda, B.: Review of reverse engineering systems–current state of the art. Virtual Phys. Prototyp. **12**(2), 161–172 (2017)
2. Kainz, W., et al.: Advances in computational human phantoms and their applications in biomedical engineering—a topical review. IEEE Trans. Radiat. Plasma Med. Sci. **3**(1), 1–23 (2018)
3. Lasemi, A., Xue, D., Gu, P.: Recent development in CNC machining of freeform surfaces: a state-of-the-art review. Comput. Aided Des. **42**(7), 641–654 (2010)
4. Lim, S.P., Haron, H.: Surface reconstruction techniques: a review. Artif. Intell. Rev. **42**(1), 59–78 (2012). https://doi.org/10.1007/s10462-012-9329-z
5. Lohfeld, S., Barron, V., McHugh, P.: Biomodels of bone: a review. Ann. Biomed. Eng. **33**(10), 1295–1311 (2005)
6. Schillinger, D., Ruess, M.: The Finite Cell Method: a review in the context of higher-order structural analysis of CAD and image-based geometric models. Arch. Comput. Meth. Eng. **22**(3), 391–455 (2015)
7. Martínez-Finkelshtein, A., et al.: Adaptive cornea modeling from keratometric data. Invest. Ophthalmol. Vis. Sci. **52**(8), 4963–4970 (2011)
8. Ramos-López, D., et al.: Placido-based indices of corneal irregularity. Optom. Vis. Sci. **88**(10), 1220–1231 (2011)
9. Brenner, L.F., et al.: Myopic surface ablation in asymmetrical topographies: refractive results and theoretical corneal elastic response. Am. J. Ophthalmol. **177**, 34–43 (2017)

10. Cavas-Martínez, F., et al.: Study of morpho-geometric variables to improve the diagnosis in keratoconus with mild visual limitation. Symmetry **10**(8), 306 (2018)
11. Cavas-Martínez, F., et al.: Study and characterization of morphogeometric parameters to assist diagnosis of keratoconus. Biomed. Eng. Online **17**(1), 1–18 (2018)
12. Cavas-Martínez, F., et al.: Assessment of pattern and shape symmetry of bilateral normal corneas by Scheimpflug technology. Symmetry **10**(10), 453 (2018)
13. Cavas-Martínez, F., et al.: Keratoconus detection based on a new corneal volumetric analysis. Sci. Rep. **7**(1), 1–10 (2017)
14. Cavas-Martínez, F., et al.: A new approach to keratoconus detection based on corneal morphogeometric analysis. PLoS ONE **12**(9), e0184569 (2017)
15. Alastrué, V., et al.: Biomechanical modeling of refractive corneal surgery. J. Biomech. Eng. **128**(1), 150–160 (2006). https://doi.org/10.1115/1.2132368
16. Calossi, A.: Corneal asphericity and spherical aberration. J. Refract. Surg. **23**(5), 505–514 (2007)
17. Gatinel, D., et al.: Corneal asphericity change after excimer laser hyperopic surgery: theoretical effects on corneal profiles and corresponding Zernike expansions. Invest. Ophthalmol. Vis. Sci. **45**(5), 1349–1359 (2004)
18. Navarro, R., et al.: Lower-and higher-order aberrations predicted by an optomechanical model of arcuate keratotomy for astigmatism. J. Cataract Refract. Surg. **35**(1), 158–165 (2009)
19. Pandolfi, A., Manganiello, F.: A model for the human cornea: constitutive formulation and numerical analysis. Biomech. Model. Mechanobiol. **5**(4), 237–246 (2006)
20. Smadja, D., et al.: Influence of the reference surface shape for discriminating between normal corneas, subclinical keratoconus, and keratoconus. J. Refract. Surg. **29**(4), 274–281 (2013)
21. Martínez-Finkelshtein, A., et al.: Comparative analysis of some modal reconstruction methods of the shape of the cornea from corneal elevation data. Invest. Ophthalmol. Vis. Sci. **50**(12), 5639–5645 (2009)
22. Romero-Jiménez, M., Santodomingo-Rubido, J., Wolffsohn, J.S.: Keratoconus: a review. Cont. Lens Anterior Eye **33**(4), 157–166 (2010)
23. Cavas-Martínez, F., et al.: Corneal topography in keratoconus: state of the art. Eye Vis. **3**(1), 5 (2016)
24. Montalbán, R., et al.: Intrasubject repeatability in keratoconus-eye measurements obtained with a new Scheimpflug photography-based system. J. Cataract. Refract. Surg. **39**(2), 211–218 (2013)
25. Mohan, S., et al.: Parametric NURBS curve interpolators: a review. Int. J. Precis. Eng. Manuf. **9**(2), 84–92 (2008)
26. Espinosa, J., et al.: Optical surface reconstruction technique through combination of zonal and modal fitting. J. Biomed. Opt. **15**(2), 026022 (2010)
27. Rabinowitz, Y.S.: Keratoconus. Surv. Ophthalmol. **42**(4), 297–319 (1998)
28. Wilson, S.E., Lin, D., Klyce, S.D.: Corneal topography of keratoconus. Cornea **10**(1), 2–8 (1991)
29. Rabinowitz, Y.S., McDonnell, P.J.: Computer-assisted corneal topography in keratoconus. Refract. Corneal Surg. **5**(6), 400–408 (1989)
30. Auffarth, G.U., Wang, L., Völcker, H.E.: Keratoconus evaluation using the Orbscan topography system. J. Cataract Refract. Surg. **26**(2), 222–228 (2000)
31. Kennedy, R.H., Bourne, W.M., Dyer, J.A.: A 48-year clinical and epidemiologic study of keratoconus. Am. J. Ophthalmol. **101**(3), 267–273 (1986)
32. Elsheikh, A., et al.: Numerical study of the effect of corneal layered structure on ocular biomechanics. Curr. Eye Res. **34**(1), 26–35 (2009)
33. Gatinel, D., et al.: Corneal elevation topography: best fit sphere, elevation distance, asphericity, toricity and clinical implications. Cornea **30**(5), 508 (2011)

34. Montalbán, R., et al.: Comparative analysis of the relationship between anterior and posterior corneal shape analyzed by Scheimpflug photography in normal and keratoconus eyes. Graefes Arch. Clin. Exp. Ophthalmol. **251**(6), 1547–1555 (2013)
35. Montalbán, R., et al.: Scheimpflug photography-based clinical characterization of the correlation of the corneal shape between the anterior and posterior corneal surfaces in the normal human eye. J. Cataract Refract. Surg. **38**(11), 1925–1933 (2012)
36. Atchison, D.A., et al.: Age-related changes in optical and biometric characteristics of emmetropic eyes. J. Vis. **8**(4), 29 (2008)
37. Piñero, D.P., et al.: Corneal volume, pachymetry, and correlation of anterior and posterior corneal shape in subclinical and different stages of clinical keratoconus. J. Cataract Refract. Surg. **36**(5), 814–825 (2010)
38. Yebra-Pimentel, E., et al.: Asfericidad corneal en una población de adultos jóvenes: Implicaciones clínicas. Arch. Soc. Esp. Oftalmol. **79**(8), 385–391 (2004)
39. González-Méijome, J.M., et al.: Asphericity of the anterior human cornea with different corneal diameters. J. Cataract Refract. Surg. **33**(3), 465–473 (2007)
40. Maldonado, M.J., et al.: Repeatability and reproducibility of posterior corneal curvature measurements by combined scanning-slit and placido-disc topography after LASIK. Ophthalmology **113**(11), 1918–1926 (2006)
41. Ho, J.-D., et al.: Validity of the keratometric index: evaluation by the Pentacam rotating Scheimpflug camera. J. Cataract Refract. Surg. **34**(1), 137–145 (2008)
42. Dubbelman, M., et al.: Radius and asphericity of the posterior corneal surface determined by corrected Scheimpflug photography. Acta Ophthalmol. Scand. **80**(4), 379–383 (2002)
43. Dubbelman, M., Sicam, V., Van der Heijde, G.: The shape of the anterior and posterior surface of the aging human cornea. Vis. Res. **46**(6–7), 993–1001 (2006)

Innovation in Design

Exploring Novel Teaching Methods for Design and Engineering Students in the Field of Nanomaterials

María Isabel Rodríguez-Ferradas(✉), Aitor Cazón-Martín, and Paz Morer-Camo

TECNUN-Universidad de Navarra, 20018 San Sebastián, Spain
mirodriguez@tecnun.es

Abstract. Nowadays new and emerging materials are a key driver for product innovation. However, a gap between the research and development of new materials by universities and research centres and the integration of those materials in the design of products that meet user's needs still exists. This gap delays the adoption of new materials by the industry and consequently reduces the economic return on the investment of Europe in researching more sustainable and with better performance materials. This article presents preliminary findings of the ongoing research project "DATEMATS" funded by the European Commission in the Erasmus + programme. This project aims to develop novel teaching methods for both design and engineering students in the field of Emerging Materials. Tecnun's team participating in this project is focusing on Nanomaterials. We present in this article the activities developed in the first 18 months of this project that included the review of the state of the art regarding teaching methods, and the design and testing of a pre-mobility on-line experience for students. The objective of the pre-mobility was to introduce the participant students in the key concepts related to the nanomaterials and to inspire talks about their potential applications in the field of product design.

Keywords: Emerging materials · Product design · Teaching methods · Nanomaterials

1 Introduction

Nowadays, emerging materials are becoming one of the leading elements of product design practice as a lever to foster innovation and add value to final products. Despite the relevance of these materials for advances in many fields, there is a gap between the design of products that meet the needs of the market, and the research and development of emerging materials to be used in these products. At the heart of this issue, is the problem of the effective transfer of this new knowledge generated in research centers to the companies and professionals who design their products. Designers today need to seek appropriate approaches, strategies, and tools to work with emerging materials [1].

Researchers from Academia have to develop knowledge that can be transferred and used by industrial worlds, to reason on the no-technological dimension as a possibility

F. Cavas Martínez et al. (Eds.): INGEGRAF 2021, LNME, pp. 251–259, 2022.
https://doi.org/10.1007/978-3-030-92426-3_29

to the designer and the company to create new experiences and innovations for specific new and emerging materials and simultaneously to adopt a more entrepreneurial attitude.

Based on these grounds, the project DATEMATS aims at creating a transnational network among Universities, Research Centres and Organizations in contact with SMEs in order to develop and implement interdisciplinary and transdisciplinary methods for emerging materials education in the field of industrial design to boost knowledge transfers from academia and research centres to companies.

The following partners compose the consortium of this research project:

- Universities:
  - Politecnico di Milano – Department of Design (POLIMI) (Italy)
  - Aalto University – Chemarts (AALTO) (Finland)
  - University of Navarra – Faculty of Engineering (TECNUN) (Spain)
  - Copenhagen School of Design and Technology (KEA) – Material Design Lab (Denmark)
- Research Centres about materials:
  - Materially Srl Impresa Sociale (Italy)
  - Fostering Arts and Design (FAD) – Materfad (Spain)
- Business clusters
  - Barcelona Design Center (BCD) (Spain)
  - Centro Italiano per l'Apprendimento Permanente (CIAPE) (Italy)
  - Industrial Development Center West Sweden (IDC) (Sweden)
- Quality assurance
  - Instituto de Soldadura e Qualidade (ISQ) (Portugal)

This project focuses on four specific macro-area of emerging materials:

- ICS for Wearables (by POLIMI): Interactive, Connected, Smart solutions able to establish a two-way exchange of information, to respond contextually and reversibly to external stimuli, linked to another entity or to an external source, and programmable not only through software. These solutions are components of smart electronic devices with micro-controllers embeddable into clothing or worn on the body as implants or accessories.
- Advanced growing materials (by KEA): Materials from a controlled cultivation of organisms (bacteria, yeast, algae, mycelium, etc.) that are directly grown and/or manufactured into their subsequent form, function and performance by tapping into the organisms natural growth behaviour (biofabrication).

- Nanomaterials (by TECNUN): Entirely carbon-based materials, nano-scaled structures or composite blends whose properties are altered by surface and/or substrate doping thereof.
- Wood-based materials (by AALTO): Materials that are processed chemically or mechanically from either trees or other plants for innovative applications. The materials include cellulose fibers, fibrils (microor nanostructured) and derivatives, lignin, bark extractives and novel combinations of these.

This paper presents preliminary findings of this DATEMATS project that started in January 2019 and will finish at the end of 2021 in the field of nanomaterials, which is the research focus of Tecnun's team in this project. Particularly, we will present the results of our activities developed in the first 18 months of this project that included the review of the state of the art regarding teaching methods, and the design and testing of a pre-mobility on-line experience for students. The objective of the premobility tasks developed by our research team of Tecnun was to introduce the participant students in the key concepts related to the nanomaterials and to inspire talks and discussion about their potential applications in the field of product design.

In these times, due to the COVID pandemic, many European universities have been forced to use digital technologies in a massive way in the training of their students. Therefore, the results obtained in this pre-mobility on-line experience can be valuable for understand some keys to the success of on-line learning processes for design and engineering students about emerging materials.

2 State of the Art About Nanomaterials Teaching Methods

Scientists have not unanimously settled on a precise definition of nanomaterials, but agree that they are partially characterized by their tiny size, measured in nanometers. A nanometer is one millionth of a millimeter approximately 100,000 times smaller than the diameter of a human hair [2].

Nano-sized particles exist in nature and can be created from a variety of elements, such as carbon or minerals like silver, but nanomaterials by definition must have at least one dimension that is less than approximately 100 nanometers. Most nanoscale materials are too small to be seen with the naked eye and even with conventional lab microscopes. Materials engineered to such a small scale can take on unique optical, magnetic, electrical, and other properties. These emergent properties have the potential for great impacts in many industrial fields. But, on the other hand, controversial about safety of nanomaterials coming from cosmetic and food sectors negatively affect the perception of these emerging materials [3].

In 2005, the European Centre for the Development of Vocational Training (Cedefop, 2005) in a report entitled "Identification of skill needs in nanotechnology" [4] said that nanotechnology was set to play a key role worldwide in the 21st century since it was a cross-sector technology which was increasingly relevant to economic areas such as chemistry, medical technology, automobile and the food industry. The same study demonstrates new demands for qualification in nanotechnology and the need to improve training at higher education, since the lack of qualified personnel lead become an obstacle in the development and the application of nanotechnologies to develop new products, services and experiences.

The first part of our work in the DATEMATS project was focused on a review of the literature related to nanomaterials didactic methods [5]. It has also been focused on how to transmit that knowledge to more practical levels. At the same time, the search has expanded to other didactic methods used in the transmission of knowledge about new or smart materials. Few references were found from previous research experiences about didactic methods for the training of students about nanomaterials [6–12]. In order to enrich our review, we also included some additional research papers about didactic methods focused in teaching about emerging materials to design students [13–22].

The main issues identified in this literature review related to nanomaterials teaching were the following:

- The difficulty of students to understand something they cannot see. This is mainly related to the lack of proper equipment in Academia due to the expensiveness of special devices needed to work/experience at nano scale. [9–11].
- Difficulties for teachers to comfortably conduct such a multidisciplinary subject that encompasses disciplines such as applied physics, materials science, physical chemistry, polymer science and engineering [11].
- The model of material libraries, so widespread in the field of design training, suffers from the absence of relevant information on this type of material and above all from a multidisciplinary experience. [17].
- Nanomaterials, as emerging materials, are subject to ongoing fluctuations in their availability, processing methods and prices therefore it is very important to transmit to students the sense that these materials are something alive, about which they have to update their knowledge in a continuous way [21].

To conclude, we realised that in order to overcome these issues new teaching strategies are already being tested in the field of nanomaterials. In this sense, some researchers stand that cutting-edge on-line videos and free e-courses make the learning and comprehension of concepts related to nanotechnology faster, deeper and more efficient [8, 12].

3 Pre-mobility Methodology

The following phase of the DATEMATS project aimed at creating the contents of a novel teaching method for design and engineering undergraduate students. Each university involved was responsible for the creation of a training module related to its emerging material addressed to students. The objective of this training module is to introduce the participant students in the basic and key concepts related to emerging materials and to inspire talks on envisioning futures and potential applications of these emerging materials in the fields of design and engineering.

These training modules will be initially tested on partner's students belonging to courses of Design and Engineering on-line. This phase was named pre-mobility, because it would also serve the universities to identify interested students who would be candidates for the mobility activities included in the last phase of the project.

According to the conclusions of their literature review of each emerging material, each university designed a different format for their pre-mobility. This strategy would also make it possible to compare the results of different pre-mobility formats and identify strengths and weaknesses of each format.

Our training module about nanomaterials, which structure and deadlines are shown in Fig. 1, included 2 different types of activities.

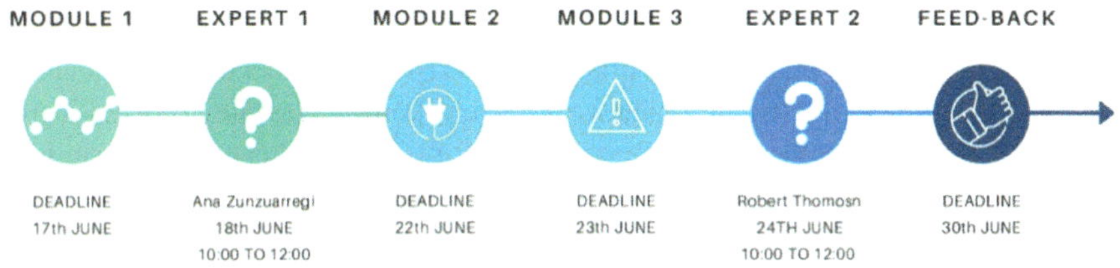

Fig. 1. Structure of the pre-mobility of nanomaterials developed by Tecnun

Firstly, a selection of free videos available in the YouTube, grouped in three modules related to the nanomaterials:

- Module 1: What are nanomaterials and nanotechnology?
- Module 2: Nanomaterials and their superpowers.
- Module 3: Risks of nanomaterials.

Second, two Q&A sessions leaded each of them by an expert in the field, each one with a different profile:

- Expert 1: A scientific expert in the field of basic research in nanomaterials.
- Expert 2: A designer and mechanical engineer working as expert in technological transfer between nanomaterials manufacturers and companies interested in these materials' potential applications.

In order to improve students learning experience, the following measures were taken:

- First, for some of the videos that were a bit long or included some repetitive content, we specified the initial time and finish time with the new contents to watch. But of course, those students interested in delving further into the content could entirely watch the videos.
- The videos were enriched using Kahoot! quizzes about their contents. This strategy looked enhancing the learning experience and motivate students in a more active learning.
- With the aim of testing whether the students showed interest in interacting with the experts or not, the attendance to the Q&A sessions was not compulsory.

A total number of 142 students participated in the complete pre-mobility activity for the 4 EM&T. For the nanomaterial activity, 30 students were involved.

A document with the instructions to develop the nanomaterials pre-mobility was sent to the 30 registered students (9 from Aalto University, 10 from Polimi and 11 from Tecnun). This document included a detailed description of the tasks included in this pre-mobility and also a link to an assessment questionnaire developed by the consortium to collect feedback from students.

4 Results and Discussion

The activities carried out in the framework of the DATEMATS project to date lead us to a better understanding of the strengths and weaknesses of on-line learning processes about emerging materials, specifically in nanomaterials, for design and engineering students.

The assessment of the pre-mobility was based in a brief questionnaire of 7 items that was proposed at the students at the end of the experience. 5 of these items were closed questions measured on a Likert scale with 5 response levels and the last two were open questions. The items of this questionnaire were the following:

1. My overall assessment of the pre-mobility activity (1 = Fair, 2 = Satisfactory, 3 = Good, 4 = Very good, 5 = Excellent)
2. What is your previous knowledge on the emerging material area before the beginning of the activity (1 = None, 2 = Scarce, 3 = Sufficient, 4 = Good, 5 = Expert)
3. The video(s) as a teaching method supported my learning (1 = Strongly disagree, 2 = Disagree, 3 = Neither agree nor disagree, 4 = Agree, 5 = Strongly agree)
4. The discussion(s), exploration(s) and/or assignment(s) method(s) supported my learning (1 = Strongly disagree, 2 = Disagree, 3 = Neither agree nor disagree, 4 = Agree, 5 = Strongly agree)
5. Did the pre-mobility activity meet my expectations? (1 = Strongly disagree, 2 = Disagree, 3 = Neither agree nor disagree, 4 = Agree, 5 = Strongly agree)
6. Why? (open question)
7. General comments and suggestions (on such matters as materials, teaching methods, arrangements, ways of completing the pre-mobility activity and unnecessary overlaps with other course contents, or other matters) (open question).

This assessment provided us information of two types. First, some general results about the teaching methods used in the pre-mobility. Second, some specific results about the specific format designed by our group focused in nanomaterials. In the following paragraphs, we summarize the main results achieved from this information.

The videos were the main source of teaching method. Based on the answers given in the feedback form showed in Fig. 2, the major number of the participants said that the videos supported their learning process (37% strongly agree, 41% agree, 17% Neither agree or disagree, 4% Disagree, 1% Strongly disagree).

Other activities included in the pre-mobility such as Q&A sessions, Kahoot quizzes and suggested readings were also positively evaluated by students to support their learning as shown in Fig. 3. Moreover, thanks to open question we realized that Q&A virtual sessions with experts organized by Tecnun and Aalto, were highlighted by students as places that offered them an opportunity to get to know the people behind the activity

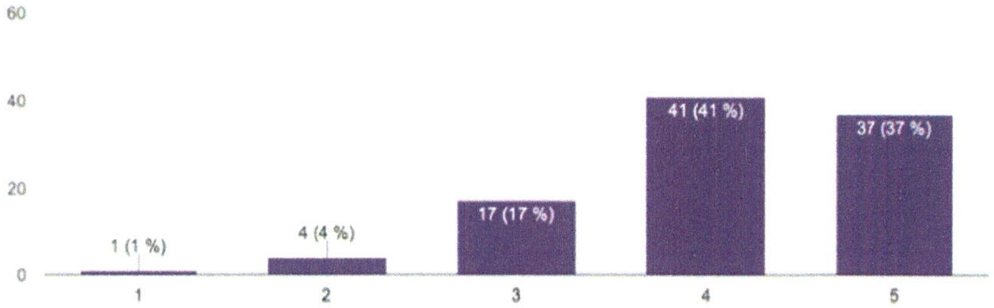

Fig. 2. Assessment by the participants regarding the videos as a teaching method

and to solve questions related to the shared teaching materials and move deeper into the different key concepts. On the other hand, students that participated in the pre-mobility of the other two universities, that did not have this online space in their learning activity, recommended to include one.

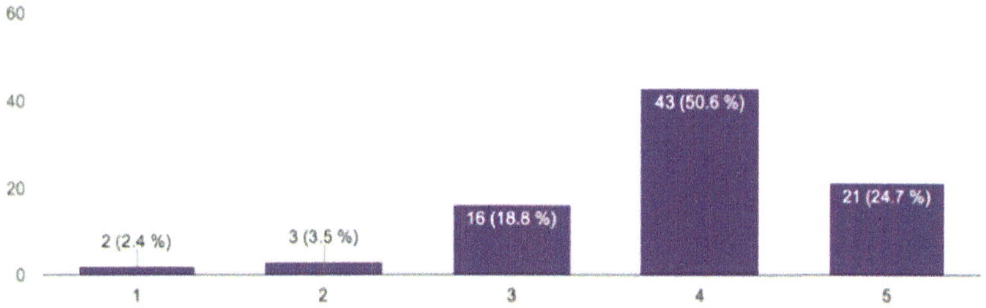

Fig. 3. Assessment by the participants regarding the other teaching activities included in the premobility (Q&A sessions, Kahoot! Quiz, readings, etc.)

Analysing the answers of students to the open questions, students considered the pre-mobility activity informative, clear and inspired them to learn further about the emerging materials and related technologies. The following quote is an example of their answers about nanomaterials pre-mobility "I did learn a lot about nanomaterials and I'm very grateful. It would be even better to learn about the techniques or how you apply these nanomaterials on industrial products. I especially enjoyed the zoom meeting with the expert".

Students also asked for more bibliography, light readings, slides and tutorials to carry some experiments. Another recommendation is to offer various time-slots for the Q&A sessions because sometimes the suggested time may not match with the students' availability.

5 Conclusions

To sum up, even though the activity was mainly developed during the COVID situation and adjustments to the timeline and teaching materials were necessary to make, our objectives set for the pre-mobility activity in the DATEMATS project were met.

The greatest part of the students agreed that the materials developed such as videos and Q&A sessions with experts supported their learning process. There are also several suggestions to improve these materials in order to achieve a better learning experience. The next step in the DATEMAT project will be the mobility activity in summer 2021.

Acknowledgments. The research work reported here was made possible by DATEMATS project (Knowledge & Technology Transfer of Emerging Materials & Technologies through a Design-Driven Approach Agreement Number: 600777-EPP-12018-1-IT-EPPKA2KA) that is co-funded by the Erasmus+ Programme of the European Union.

References

1. Karana, E., Nimkulrat, N., Giaccardi, E., Niedderer, K., Fan, J.N.: Alive. Active. Adaptive: experiential knowledge and emerging materials. Int. J. Des. **13**(2), 1–5 (2019)
2. Miernicki, M., Hofmann, T., Eisenberger, I., von der Kammer, F., Praetorius, A.: Legal and practical challenges in classifying nanomaterials according to regulatory defini tions. Nat. Nanotechnol. **14**(3), 208–216 (2019)
3. Henkler, F., et al.: Risk assessment of nanomaterials in cosmetics: a European Union perspective. Arch. Toxicol. **86**(11), 1641–1646 (2012)
4. Abicht, L., Freikamp, H., Schumann, U., Bulgarelli, A., Lettmayr, C.: Identification of skill needs in nanotechnology. Office for Official Publications of the European Communities, Luxembourg (2006)
5. Ferraro, V., Pasold, A. (Eds.).: Emerging Materials & Technologies: New approaches in Design Teaching Methods on four exemplified areas, pp. 57–81 Franco Angeli, Milano (2020)
6. Irresistible Project Homepage.http://www.irresistible-project.eu/index.php/en/resources/teaching-modules. Accessed 11 May 2021
7. Nanolab Project Homepage. http://www.nanolab.unimore.it/. Accessed 11 May 2021
8. Sebastian, V., Gimenez, M.: Teaching Nanoscience and thinking nano at the macroscale: Nanocapsules of wisdom. Proc. Soc. Behav. Sci. **228**, 489495 (2016)
9. Schulz, M., et al.: Teaching nanoscale engineering to accelerate academic industrial innovation. In: 10th WSEAS International Conference on Engineering Education, vol. 13, pp. 20–22 (2013)
10. Olteanu, R.L., Gorghiu, G., Gorghiu, L.M.: Introducing responsible research and innovation dimensions in non-formal education activities dedicated to industrial application of nanomaterials. J. Sci. Arts **2**, 331–340 (2017)
11. Lisotti, A., De Renzi, V., Goldoni, G.: The NANOLAB project: educational nanoscience at high school. MRS Online Proc. Libr. Arch. **1657**(1), 44–51 (2014)
12. de la Rosa, J.M.: Using nanoHUB.org for teaching and learning nanoelectronic devices in materials engineering. In: Proceedings of the IEEE Global Engineering Education Conference (EDUCON), Marrakech, pp. 1–4 (2012)
13. Karana, E., Pedgley, O., Rognoli, V. (eds.): Materials Experience: Fundamentals of Materials and Design. Butterworth-Heinemann, Oxford (2013)
14. Piselli, A., Dastoli, C., Santi, R., Del Curto, B.: Design tools in materials teaching: bridging the gap between theoretical knowledge and professional practice. In DS 93: Proceedings of the 20th International Conference on Engineering and Product Design Education (E&PDE), Dyson School of Engineering, Imperial College, London. 6th-7th September pp. 193–198 (2018)

15. Veelaert, L., Du Bois, E., Ragaert, K., Hubo, S., Van Kets, K. Bridging design and engineering in terms of materials selection. In: Proceedings of 7th International Polymers & Moulds Innovations Conference, pp. 319–326 (2016)
16. Piselli, A., Baxter, W., Simonato, M., Del Curto, B., Aurisicchio, M.: Development and evaluation of a methodology to integrate technical and sensorial properties in materials selection. Mater. Des. **153**, 259–272 (2018)
17. Bianchi, R., Lucibello, S.: New interdisciplinary perspectives of research on the role of design with new materials: 'open' spaces-methodologies design process. In: 4D Designing Development Developing Design. Conference Proceedings, 28–30 September 2017, Kaunas-Lithuania, pp. 244–255 (2018)
18. Ramalhete, P.S., Senos, A.M.R., Aguiar, C.: Digital tools for material selection in product design. Mater. Des. **31**(5), 2275–2287 (2010)
19. Asbjørn, Ch., Jagtap, S. Warell, A.L: Material Selection in industrial design education-A literature review, DS 83. In: Proceedings of the 18th International Conference on Engineering and Product Design Education (E&PDE16), Design Education: Collaboration and Cross Disciplinarity, Aalborg, Denmark, pp. 708–713 (2016)
20. Ferrara, M., Lucibello, S.: Ensino de materiais para o design. Pesquisa sobre metodologia didática dos materiais em design e desenho industrial. Strat. Des. Res. J. **5**(2), 75–83 (2012)
21. Haug, A.: Acquiring materials knowledge in design education. Int. J. Technol. Des. Educ. **29**(2), 405–420 (2018). https://doi.org/10.1007/s10798-018-9445-4
22. Karana, E., Barati, B., Rognoli, V., Zeeuw van der Laan, A.: Material driven design (MDD): a method to design for material experiences. Int. J. Design **9**(2), 35–54 (2015)

Artisanal Idea Generation as a Creative Method in Higher Education

Mar Melgarejo-Torralba, Dolores Parras-Burgos(✉), and Daniel G. Fernández-Pacheco

Department of Structures, Construction and Graphic Expression, Universidad Politécnica de Cartagena, 30202 Cartagena, Spain
{mar.melgarejo,dolores.parras,daniel.garcia}@upct.es

Abstract. As people's level of education increases, we can see a decrease in their creative abilities. This creativity, although is innate in human beings, can be developed and worked on. The use of new technologies implies in turn an abandonment of manual or craft skills, losing in this area a very relevant possibility of creative development. Most of the methods that are usually used to generate ideas are related to oral and written conceptualization, but manual ideation is practically not contemplated. This study proposes a methodology for generating ideas based on traditional ideation from the physical creation of models as part of the creative process. This methodology has been used with satisfactory results in 38 courses at the Industrial Organisation School (Spain), with 382 people between 18 and 30 years of age participating in the study. It is a fact that companies are showing more and more interest in developing the creative capacity of their employees as a key to innovation. This new approach can also benefit many students, and for this reason it is recommended for use in engineering degrees, where the development of creative skills can have a significant impact on later career development.

Keywords: Creativity · Ideation · Hands-on creation · Innovation · Modeling · Prototyping

1 Introduction

Someone's creative capacities may develop throughout his or her lifetime. According to some studies, a relation exists between creativity and personality whose development in this relation is non-linear, and each age is characterised by a creativity variable: children express their ideas more fluently, youth's thinking is more flexible, and adults come up with more original ideas [1]. In adulthood, people tend to close up to creativity given the premise that everything has been invented or imagined, and creating is an action that is restricted to large enterprises [2].

Nowadays, the education system does not promote all it should in terms of the activities that develop its students' creative thinking, an aspect that might be fundamental for their full academic and professional development. This aspect is more visible in engineering studies where subjects that work this creative aspect are more difficult to find

F. Cavas Martínez et al. (Eds.): INGEGRAF 2021, LNME, pp. 260–267, 2022.
https://doi.org/10.1007/978-3-030-92426-3_30

[3, 4]. Nussbaum [5] sustain that arts, as a way of developing creativity and imagination, have been present during all human beings' periods. However according to Iglesias [6], education does not intend to stimulate either originality or imagination, does not promote creative thinking, but might achieve divergent thinking to a certain point by weakening the trust required for developing a creative personality. To this idea, we can also add a deficiency when developing a creative culture in society [7] insofar as formal education is based on doing and know-how, which leaves abstract thinking to one side, along with modelling and simulation, which are the basis of creative thinking [8]. However, education alone cannot be blamed because strategies need to be sought to lead human beings to creativity by taking the view that something can be accomplished because, as De Bono stated [9], creativity is innate in human beings, and is a capacity that can be developed and trained in.

Creativity as a value is on the rise because of its close relation to problem solving and entrepreneurship. Indeed universities increasingly engage in designing and creating new education methodologies that take students closer to the labour world reality. As enterprises have recently introduced new methodologies to develop their workers' creative thinking, incorporating it into classrooms is an excellent opportunity for enterprises and students [10, 11].

Creativity techniques are procedures that are applied in the idea generation phase. It is important to understand that creativity work is not only based on using these techniques because the creative process is much more complex than only generating ideas. In engineering design work, engineers must be capable of thinking creatively by efficiently alternating between divergent thinking (developing many novel ideas) and convergent thinking (seeking a suitable idea by engineering analyses) [3].

The most representative creative techniques (BrainStorming, SCAMPER, Mental Maps, Synectics, the C-Sketch method, Forced associations, etc.) are mostly based on oral or written ideation to generate ideas. However, techniques whose method is based on artisanal ideation are not generally available; that is, based on manual prototyping as a process to generate ideas. The process followed to stimulate idea generation must, among other things, prioritise a broad observation sense, and one that allows all senses to interact; develop the abstraction capacity to be able to access the essence of problems; develop the capacity to extrapolate conceptual solutions from one knowledge area to another; stimulate the experimentation and creative interpretation of reality [12].

Until quite recently, the aim of prototyping was to help to validate ideas before launching them on the market in an attempt to save time and resources to, thus, avoid mistakes, or to at least avoid making them in a very advanced entrepreneurial project phase. Nonetheless, using prototyping in an early phase, i.e. in the idea generation phase, has not been sufficiently tested. This study reflects the experience of following this method as an idea generator by showing the developed methodology and the results obtained in different academic years to validate this novel technique.

2 Materials and Methods

The participation in the study was offered as a voluntary practice/exercise for the students (ranging in age between 18 and 19 years old) of several specialisation courses to improve their job opportunities, which form part of the training courses given at the Escuela de Organización Industrial (EOI; School of Industrial Organisation). All these students are in a set of 38 different courses (with a mean of 10 students per course) which formed a sample with 380 students. An informed consent was obtained from all subjects of the study.

The students participated in the collective creativity exercise for the last 45 min of a 5-h training session that forms part of a class called "Creativity", which belongs to one of the employability modules that attendees' respective courses include. This allows the same subject to be worked in very different training areas like specialisation courses and programming, logistics or architectural modelling.

The collective creation process with which work was done during these sessions used the so-called creativity elements, as coined by creativity expert Juan Pastor Bustamante [13] on the Neuronilla website (www.neuronilla.com), which didactically included the various creative process phases and allowed them to be easily retained when facing a challenge. It also worked the creativity concept by not only being an act of thinking up and generating ideas, but creativity was also applied to carry out these ideas, to build them and to convert them into reality. This method covered the process using four elements: air, water, earth and fire. The first, **Air**, was the idea generation phase with which to face a challenge. It was associated with the idea of letting oneself go and flying, and was the phase in which many ideas were invented for one same matter. The students worked simultaneously with the BrainWriting technique [14] and with Ideas Association or Random words [15]. Three periods were enacted and each lasted 3 min. Each participant worked his/her own ideas with his/her classmates and many proposals were put forward. The second phase, **Water**, was when the evaluation was made to select a winning idea to face a given challenge from all the ideas worked on. To do so, the students had to vote that idea which met the requirements of being an innovative, different and disruptive idea and one with a real impact on the problem being worked on. In the **Earth** Phase, ideas were built and outlined, and were no longer a concept written on a piece of paper. This was when the students had all types of work materials and 15 min to produce the idea they were working on with their hands. This is very specific prototyping that expects students to make decisions by activating many creative elements that are connected to artisanal performance. Finally, the **Fire** Phase was when the students acted by putting the idea into practice, telling the world about it and activating it. It is pointless having great ideas if they are not carried out. With the selected idea, and with a manufactured element that allowed the idea to be understood, the student group told the other students about its benefits, what it improved and why it was a different creative project.

This methodological framework was devised so that the participants could work on a challenge as groups with four or five members by considering a creative challenge on which to put forward a proposal (Fig. 1).

Fig. 1. The artisanal idea generation process. Selecting different sample groups.

3 Results and Discussion

This method was applied to each student group in the different courses and was able to qualitatively analyse how these participants faced the process of generating ideas and materialising them. It was possible to identify certain learning patterns from observing their conducts and from the way their ideas developed when they worked on them together. Here the main aim was for the **artisanal idea generation process (thinking with hands) to favour greater divergence and to encourage new ideas to be generated.**

The first and most significant pattern was that the method designed for the creative process started with a few principles that had been considerably modified. Although the methodology phases determined a first idea generation part, a second selection part and a third ideas materialisation part that required a divergence process, in this phase it was verified that, by using the capacity to generate by hands, most of the student groups continued to generate ideas, improve them and build a true innovating project. From the initial idea taken from the ideation process to the ideas "building" process, the group rebuilt the initial idea, improved and processed it, which meant making much qualitative progress in defining this idea. This systematically took place in all the analysed cases (the 38 courses worked on), mostly because an idea was developed with specific materials, collectively and by using manual work to once again activate the group's generation capacity.

Table 1 details and describes the qualitative leaps with nine examples taken from these courses. The first columns describe the course and the number of students with which these examples were worked, as well as the creative challenge they faced. This matter in some examples is similar and is the basis with which the group started the ideation process. From the idea generation process, the participants selected and voted (in the Air and Water Phases), and arrived at their first concept, which they later developed in the Earth Phase, that of materialising ideas. During the next period, the students worked with this idea using the materials that they were provided with and, as the examples show, the initial idea substantially changed in many cases.

Table 1. A qualitative leap in the idea generation process using artisanal generation.

Course	Initial idea chosen	Idea developed during the materialisation process	Picture
BIM architectural modelling (October 2019) 13 students in two groups **Challenge:** How could we improve mobility in our city?	(1) Experience mobility up in the air	**Flying among buildings** The group started out with the idea of being transported over a zero level. However, when they begin materialising their idea using their hands, this idea becomes an element that could be built between buildings and be related to air	 Flying between buildings  Under the sea
	(2) Improve maritime transport	**Beneath the sea** The participants started work on the idea of working at sea. During the materialisation process, they designed a public submarine project by converting the proposal into a new and different tourist proposal.	
360 marketing technician (May 2019) 10 students in two groups **Challenge:** How could we improve mobility in our city?	(3) Flying buses	**A bus with legs** The processing during which the idea is materialised led the students to reconstruct the main process towards transport in two ways: flying public transport and individual vehicles at another level	 Bus with legs  Experiential bus
	(4) Musical background on buses	**Experiential buses** The initial idea began by converting public transport into a placed with personalised music. When building the idea, the students developed this concept much more by working with an experiential space by transforming walls, smells, colours, etc.	
Full Stack programming (February 2019) 16 students in three groups **Challenge:** How could we improve mobility in our city?	(5) Incentives for using public transport	**Transport app** They began by undertaking a public transport incentives project. Materialising the idea helped them to build many more options for incentives. They developed an interface for a mobile application with awards.	 Transport app  ilights  Teletransport in a magnetic field
	(6) Improve pedestrian's lives in cities	**ilights** Based on the idea of improving pedestrians' walking on footpaths, the students began designing an urban space. While materialising the idea, they ended up building a smart traffic lights project, which were red or green depending on users.	
	(7) Being quickly transported	**Teletransport in a magnetic field** The group began with the idea of speeding up people transport. This idea was modelled during the materialisation process using highly innovative ideas that proposed a magnetic field to help move people around it immediately. This proposal involved an attempt to make progress, and the group members set no technical limit to imagine feasible solutions	
Cobol programming (April 2019) 11 students in two groups **Challenge:** How could you improve local commerce services?	(8) Digitising fashion shops	**Interactive suit** The students began by working the concept of digitising local commerce. When they faced the challenge of its material nature, they decided to leap forward to really apply digitisation to a dress. This prototype enabled to collect users' data and to improve their direct purchasing experience	 Interactive suit 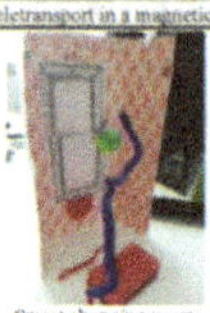 Smart changing room
	(9) Improving shopping users' experience	**Smart changing rooms** The students worked the concept of improving shopping experience by starting with the material nature and designing a physical shop. However during the building work, they redesigned the model to create a smart changing room which, by means of a screen, provided customers with information about how the same garment in another size would suit customers, seek possible colours and models of the same garment, etc.	

One specific example of this process can be described with the work done by the student group described in row 7 of Table 1. This group was made up of four students from the "Full Stack Programming" course, who worked on the creative challenge known as "How could we improve mobility around our city?".

In the first phase, **idea generation (the Air phase)**, 98 ideas were generated in 9 min thanks to the BrainWriting and Word Association techniques. The aim of this first approach was to generate lots of ideas in an attempt to obtain many ideas and to follow Linus Pauling's spirit, who stated that it is necessary to have lots of ideas to have great ideas [16]. This was, therefore, the right time to let oneself go and write down the conceptualised ideas that had not been evaluated at the time. Most of the 98 ideas were probably ruled out, but this led to a time of collective creation as an attempt to find a critical number of concepts with which to start working. Moreover, these initial ideas were not well-developed and only allowed some words to be seen that can be further worked on later.

From this first idea generation approach, the group moved on to the second phase when it had to select which of these ideas it would continue to work with: **evaluation (the Water Phase).** To do so, group members had to vote which ideas best met the requirement of being innovative and having a strong impact. The students were motivated to contemplate the possibility of these ideas being "impossible" ones or the students not being blocked because they were technically unaware as to how to solve them because the objective here was to attempt to build different and groundbreaking proposals. The most voted ideas in this group in the Water Phase were: automatic street lamp lights; X-ray glasses; Waste-cleaning medals; Growing vegetables in parks; Personalising background music on public transport; Individual passenger rooms on buses; Automatic baggage transport platform; BlaBlaCar car-sharing collection points; people-propulsion jetpacks; Party buses.

This list of the first selected ideas was made up of the phrases with the first concepts that were not well-established. For all these phrases, an endless number of ways could be followed. With the selected method, ideas were still generated through artisanal building but, as we previously described, many methods were available to continue defining an idea with prototypes. The present study chose the material nature and building of ideas because it is not a widespread method, but one that allows the learning performed in this creativity technique to be observed and retrieved.

Finally, the group decided to work on the "*people-propulsion jetpacks*" concept. This idea was the most voted one because they thought that it was very groundbreaking and, even though they did not know how to technically materialise it, they considered it to be an innovative option. After selecting this concept, the participants started materialising the idea using materials and their own hands **in the Earth Phase: artisanal idea generation.** The process that next took place is a clear example of how working with one's hands can bring about true qualitative changes to an idea. The students began working on the people-propulsion jetpacks idea, and started selecting different materials to begin the materialisation work. They took a drill to cut paper in the shape of someone small to obtain a scale for their model. Quite by chance, they found a paper marked with crosses, which suddenly led the group to generate a new idea: *"we don't know how to make jetpacks, but this paper reminds me of a magnetic field"*. This was when the qualitative leap in their work appeared because the decorated paper inspired them with a new concept that completely changed their initial idea. So now people would not wear jetpacks, but would climb onto a platform that would travel around run different points in space by means of magnetism. The students were very excited and went to the other groups to tell them about their work in the **Fire Phase: activating their idea**. They were quite clear about their idea being technically difficult to materialise and to understand in this phase, but acknowledged that it could be an inspiring idea towards sustainable transport that would allow people to quickly travel while, for instance, moving around their cities. By describing this work example, this idea generation process can be recognised through the material nature and the initial concepts, which were the main feature and were rebuilt as more groundbreaking and differentiating ideas.

Collectively Building an Idea Encourages Collaborative Group Work and Allows the Process to Flow Better

Another of the patterns that came to light when working this creative process was the advantage of the collaborative work allowing ideas to be artisanally materialised. Unlike a process in which ideas are generated through words, with a Post-it or written on a panel, hands would be employed because manual work allows more participants' engaging. They all worked on a format that allowed them to visualise what they wished to do, improve the idea and make decisions together. It also permitted them to start debates and to talk in visual language, which allows much more fluent communication.

Manual Work Experience Offers Many Positive Moments that Promote Participants' Creativity

The majority of the participants in this study told their teachers that they had not worked with this type of materials since Primary or Secondary education. Offering an experience of this type meant that students took a very positive attitude to begin this work. While we observed the groups, we noticed many moments of fun, laughter, highly constructive remarks and a relaxed atmosphere to work in, which students do not feel during other learning processes. This attitude towards a creative process leads to very effective results, as many other studies into this aspect have indicated [17, 18]. This finding also falls in line with the Botín Foundation [19], which demonstrates that the people in whom a sense of joy is induced participate in more flexible and original thinking.

4 Conclusions

Indeed relevant findings were obtained during the teaching-learning process in this study after applying the proposed methodology during several creativity sessions and with different student groups. Generating ideas with hands exponentially activates creativity: the provided materials influenced the result of the designed or proposed object (very open elements allowing the creating idea capacity to develop, where a ball can be a person, a wheel, a button off a suit, a tree, etc.). These techniques need to be used in class because students no longer train in working with their own hands. Using these techniques in other teaching contexts or anywhere where design takes place is missing. The herein obtained results revealed a new approach that can benefit many students at all levels whose use in engineering degrees is particularly recommended. Students' developing their creative capacity, which is highly demanded by companies today, can notably affect their future professional development.

References

1. Sánchez, N.S.: Relaciones entre creatividad y personalidad en varones y mujeres de educación primaria, secundaria, universitaria y adultos. Int. J. Dev. Educ. Psychol. **1**, 119–134 (2010). https://www.redalyc.org/articulo.oa?id=349832324013
2. Inglehart, R.: Modernización y posmodernización: el cambio cultural, económico y político en 43 sociedades. Siglo XXI, Madrid (1997)
3. Hirshfield, L.J., Koretsky, M.D.: Cultivating creative thinking in engineering student teams: can a computer-mediated virtual laboratory help? J. Comput. Assist. Learn. **37**(2), 587–601 (2020). https://doi.org/10.1111/jcal.12509
4. Badran, I.: Enhancing creativity and innovation in engineering education. Eur. J. Eng. Educ. **32**, 573–585 (2007). https://doi.org/10.1080/03043790701433061
5. Nussbaum, M.C.: Sin fines de lucro. Por qué la democracia necesita de las humanidades. Katz editores, Madrid (2010)
6. Iglesias Casal, M.I.: La creatividad en el proceso de enseñanza-aprendizaje de ELE: caracterización y aplicaciones. In: Nuevas perspectivas en la enseñanza del español como lengua extranjera: Actas del X Congreso Internacional de ASELE, Universidad de Cádiz, Servicio de Publicaciones, Cádiz (2000)

7. Guilford, J.P., Lagemann, J.K., Eisner, E.W.: Creatividad y educación. Paidós, Barcelona (1983)
8. Samaja, J.: Sobre el lugar de la abducción y la analogía en la creación cognitiva. F. Psicologia. UBA.: Buenos Aires (1996)
9. De Bono, E., Castellote, R.M.: Creatividad: 62 ejercicios para desarrollar la mente. Paidós, Barcelona (2008)
10. Jordán-Fisas, A., Diestra-Espino, G.E.: La metodología design thinking y su implementación en las aulas universitarias: Análisis del proyecto B-SMART en UIC Barcelona durante el curso 2019/20. Emprendimiento y Negocios Internacionales **5**(1), 18–23 (2020). https://doi.org/10.20420/eni.2020.328
11. Chen, H.B., Tao, T.Q., Zhou, C.F.: Fostering creative young engineers in Chinese universities. Int. J. Eng. Educ. **34**, 329–339 (2018)
12. Vanden Broeck, F.: El diseño de la naturaleza o de la naturaleza del diseño. Universidad Autónoma Metropolitana, Unidad Azcapotzalco, División de Ciencias y Artes para el Diseño, Departamento de Medio Ambiente (2000)
13. Bustamante, J.P.: C x I. Creatividad e innovación: Factores clave para la gestión e internacionalización. ICEX España Exportación e Inversiones (2013)
14. BrainWriting: https://xn--designthinkingespaa-d4b.com/brainwriting. Accessed 15 Mar 2021
15. Guilera, L.: Anatomía de la creatividad. Marge Books, Barcelona (2020)
16. Kirwan, J.: It's good to have lots of bad ideas. Nature **548**(7668), 491–491 (2017). https://doi.org/10.1038/nj7668-491a
17. Amabile, T.M.: Creativity in Context: Update to the Social Psychology of Creativity. Routledge, New York (2018)
18. Kumar, M., Roy, S., Sameer, A.: Influence of creative thinking and playfulness on creative styles of the individuals. In: Chakrabarti, A., Chakrabarti, D. (eds.) Research into Design for Communities, Volume 2, pp. 465–474. Springer Singapore, Singapore (2017). https://doi.org/10.1007/978-981-10-3521-0_40
19. Clouder, C.: Artes y emociones que potencian la creatividad. Informe Fundación Botín 2014. Fundación Botín: Santander (Spain) (2014)

State of the Art of the Impact of Emerging Technologies in Product Design

Laura Diago Ferrer(✉)

Department of Design and Manufacturing Engineering, Maria Luna 3, 50018 Zaragoza, Spain
lauradf@unizar.es

Abstract. For a few years we have been witnessing a process in which information and communication technologies are changing our lives and our professions. The way we communicate and interact with other people, the way we enjoy content and even our way of seeing the world is transforming. Artificial Intelligence as well as Augmented Reality or Virtual Reality open new fields of exploration for technicians, which allow us to be more efficient in our work, adapt and customize the designs to different users as well as offer a much more attractive and interactive products or services. One of the conclusions of the study conducted by the World Economic Forum in collaboration with McKinsey & Company on the impact of emerging technologies (Artificial Intelligence, Augmented Reality, Virtual Reality and Blockchain) in the creative economy, is that they are changing value chains for creative content from start to finish as well as transform storytelling and the way content is experience. That is why, in this study, we analyze the impact that some of these emerging technologies may have as creative tools in Product Design engineering. These new creation tools open very interesting research fields between Technology, Design and Society.

Keywords: Artificial Intelligence · Design product · Innovation · Creativity

1 Introduction

Today, we are immersed in the digital age. Our environments are filled with sensors, intelligent detection and management systems, algorithms, etc. We are no longer able to imagine our life without Smartphones, Tablets, without internet, without GPS navigation systems, etc. [1]. We produce thousands and millions of data daily that are stored in the cloud and managed with Big Data systems. There is no sector that can escape this new reality. In fact, data analysis begins to gain such importance, and to be so essential when tackling a professional job or an investigation, that recently Google has made available to users 25 million free datasets that they can be searched with the Dataset Search [2].

Augmented Reality, Virtual Reality and Artificial Intelligence are changing the value chains of the creative content process from production to distribution and consumption of products [3]. And this reality is so much for the way in which content is generated that will be consumed as well as for the way in which it conceptualizes, processes, distributes and consumes an industrial design product. This paradigm shift can have

F. Cavas Martínez et al. (Eds.): INGEGRAF 2021, LNME, pp. 268–274, 2022.
https://doi.org/10.1007/978-3-030-92426-3_31

positive implications. It will allow creative content to be adapted more and more to user needs as well as offering new experiences. Moreover, It will also help creators to be more creative or help to create their own content more easily.

Kai Fu Lee [4] talks about two axes, one would correspond to the variation of creativity and the other one to the variation of emotions (or feelings such as love, compassion or empathy). For him, as for many other theorists, these are the qualities that characterize us with respect to machines and that make us different. Thus, AI (Artificial Intelligence) could be an ally in the design processes. According to Lee "a serendipity that comes to free us from routine work and remind us what makes us human." AI will take over routine jobs and humans will have a greater presence in jobs that require greater empathy or compassion. AI will become a great tool for creators helping them to be more creative and productive.

And it is that creativity based on digital technologies is promoting the democratization of art and the socialization of design. The original and the copy take on new meanings. Creative projects based on computer systems awaken a new formula in the viewer, producing a more participatory communicative experience both on a physical level and on a conceptual level [5].

1.1 Augmented Reality and Artificial Intelligence in Creative Activities

Although Augmented Reality and Virtual Reality applied to product design give us the opportunity to live environments and designed products, easing us communication with the client, Artificial Intelligence is the technology that get more possibilities and have more impact in creative activities. And that is why its application covers the entire value chain, from conceptualization to the sale and consumption of the product.

Thus, computational creativity or artificial creativity, emerges as a novel concept that combines fields as Artificial Intelligence, with others as cognitive psychology, philosophy and the arts. This computational creativity could be define as "the study and simulation of the natural and artificial creative behavior of human beings through computers and computer programs" [6] and allows computers to make autonomous decisions about creative designs.

Without going into deep disquisition about whether AI can have the ability to be creative in itself or not, we can point out that computer systems are already playing very significant roles in creative processes such as design, music, fine arts and science and that Big Data will be the creative canvas for artists of the future [7].

In this regard, some scopes such as literature, journalism, art and music have already taken the first steps. We have examples like FlowMachines, an AI software managed by a human that is able of composing songs; the robot Paul, created by Patrick Tresset in 2011, capable of autonomously producing works of art imitating his sketching style; the 3D printer that has replicated Rembrandt's pictorial style by creating a painting that seems painted 3 centuries ago after applying Machine Learning techniques or even the generation of books and articles written by software with AI such as the three-page chapter "Harry Potter and the portrait of what looked like a large pile of ash"created by Botnik Studios.

Likewise, scopes as design, engineering and architecture, Artificial Intelligence is becoming increasingly stronger to generate content. As evidence of it, the exponential increase in AI and engineering publications that have been taking place in scientific journals since 2010 [8] (Fig. 1). The automation of calculations, the optimization of the designs both in initial phases and in production phases, or the creative exploration of different design options are just some of the doors that open the application of this technology to technicians.

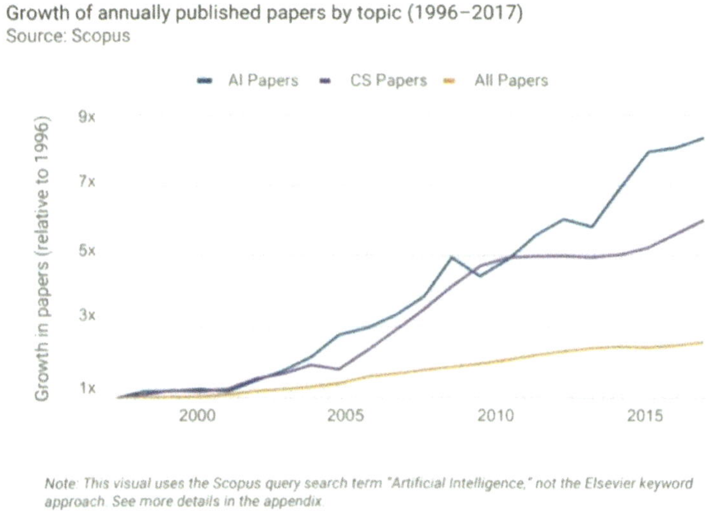

Fig. 1. Growth of annually published papers by topic. Source: AI Index 2018 Annual Report

2 Artificial Intelligence in the Scope of Design

Although there are some publications about technologies' applications like Augmented Reality or Virtual Reality to design, especially in museum and teaching scopes; its volume and interest is not comparable with the Artificial Intelligence publications applied to the field of product design engineering. In general, the global existing literature could be classified into three groups. A first group would be constituted by those studies whose purpose is the improvement of security; a second group would correspond to those investigations whose designs improve functionality and efficiency products, and finally, we would find studies that focus on design to improve user´s comfort. Moreover, those publications referring to the use of parametric design applications with AI tools should also be mentioned.

However, the use of AI technology in social design purposes is still lacking. The social responsibility of the designer is already something very assumed, however that is not reflected at the moment in which new technologies such as AI come into play. Ethics in AI is a hot topic in recent years and governments and institutions are beginning to publish rules to achieve a more ethical and responsible AI. Designing products and services with data-biased technology can have unintended consequences for the designer.

2.1 Artificial Intelligence Applied to Graphic Programs

From years, parametric design has altered the conception of engineering designs. Thus, it has gone from using the computer within a mechanical paradigm to a vision of digital technology as more than a representation tool. A fundamental advantage of the new design programs, is the ability they have to directly generate information about the manufacturing processes from the design information together with the possibility of transferring it to the new processes and techniques of digital design and production [9].

One of the applications of Artificial Intelligence to parametric design is generative design. Generative design is a process that use algorithms for finding ways that mimics the evolutionary approach of nature, so that, these evolutionary algorithms allow you to quickly generate a wide spectrum of possible solutions to a given problem. Among them, we can highlight the project "Dreamcatcher" of Autodesk. This project allows designers to create a definition of the problem by means of objectives and limitations, in order to generate alternative design solutions that respond to the proposed parameters. Another program, also from Autodesk, is the "Netfabb 2018" [10]. Netfabb allows even predicting and altering 3D designs to compensate for the distortion that the piece will have in the printing process.

On the other hand, we also find the "Finch" algorithm, developed by BOX Bygg and Wallgren Arkitekter and which is now a Grasshopper's complement. This algorithm generates different spatial configurations (according to predetermined parameters) as the total area of the design space changes [11] and helps to define zones in the initial stages of the project. Later, the space could be refined according to the specific requirements of the order. Some of this application are found in the experimental research project, "Evolving Floor Plans", by the programmer Joel Simon, which explores designs of speculative and optimized plants using a generative design. Interested in the intersection of computer science, biology and design, Joel organized rooms and the expected flow of people through a genetic algorithm, to minimize walking time, the use of hallways, etc. Its objective was to approach the design of the plant only from the perspective of optimization and without taking into account the convention or the construction capacity. Similarly, Zaha Hadid, in the "The Winton Gallery" for the Science Museum in London, takes into account parameters such as the distance between objects, the number of objects, the maximum and minimum rotation between them, the iteration or their size or the number of story lines for the generation of the definitive form of the building [12]. In other areas we find the "Cityscope" project of the MIT City Science research group that is developing tangible and digital platforms, an urban planimetry formed with pieces like lego bricks that allow drawing and visualizing in three dimensions on time what the designer is projecting; or the "M-Table" designed by Gramazio & Kohler (Fig. 2). This table is personalized by each customer with their mobile. Thus decides the roughness or shape of your product based on the combination of seven forces that are preset. After being designed, it is produced by direct manufacturing. In web design we highlight "Wix Adi", a tool that leverages AI to design websites, making developers have easier to work; "Colormind" that use neural networks to help designers to generate color palettes automatically based on the color combinations that have already worked or "Fontjoy" which, like colormind, finds combinations of harmonious fonts.

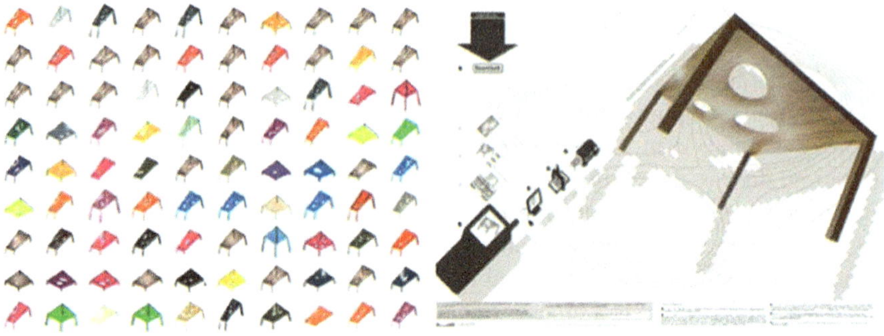

Fig. 2. mTable. Source: Grazio Kohler Architects

Also, another of the interesting applications of AI is the combination of programming languages like Python or Dynamo to BIM programs such as Revit, Rhino, etc., which allows automating calculations or processes of design work documentation. This conjunction allows to increase work efficiency, reduce errors, eliminate repetitive tasks without added value and improve the quality control of products, among other things.

2.2 Artificial Intelligence to Improve Efficiency, Comfort and Safety

Another applications of Artificial Intelligence to design, which is being used most because its ability to analyze large amounts of data and learn from users, are the tools that improve the safety, functionality, comfort or efficiency of products both in the design and in the production phase.

Security has a range from the most technical to the most functional. This allows us, from the project phase, to study the safety and stability of designs with simulations of use, in order to anticipate problems or control safety during the production or construction phase in the case of architecture (which is one of those more favorable occupational accidents). Some concrete examples of this would be the "BIM 360 Project IQ" of Autodesk that collects data of the work (audio, video and construction management documents) and with machine learning, predicts and prioritizes for the contractor high risk or risk problems that subcontractors can generate [13]. Similarly, the "Smartvid.io" or "Smartconstruction" of Komatsu (2015), help construction teams to observe, monitor and predict risks in areas of safety, productivity and project quality [14], through an industrial management platform photos and videos that include a "SmartTag" engine that can tag photos and videos of the work site automatically by taking advantage of neural networks and deep machine learning, voice recognition and image.

In this sense we also find "Doxel AI". "Doxel" is a Silicon Valley born company with advanced work monitoring. It uses robots and automated drones that are equipped with cameras and sensors to periodically scan a work from excavation to completion details. This information is processed and contrasted with the project and the planning to calculate the exact progress of the project and of each of its components. "Doxel" also detects deviations from planning or any alteration of the project preventing the general work schedule from being affected. Thus, the director and the construction manager receive at all times the information they need to direct the construction and adjust the work plan [15].

On the other hand we find research projects such as Matarazzo, Santi, Pakzad, Carter, Ratti, Moaveni, Osgood and Jacob that predict the collapse of the structure of an infrastructure [16], in this case a bridge, by analyzing the data that they receive from the mobile phones of the users that pass through it or software with reinforced learning that use algorithms to carry out trial-error studies of structures in the design phase, in order to execute a risk-free action, or "GenMEP", a Revit add-on that focuses on mechanical and plumbing design (MEP), ensuring that 3D models of MEP systems do not conflict with the building design. These softwares use machine learning to explore all possible design alternatives and tests until you find which design is the best. Regarding efficiency [17, 18], we highlight the role of IBM Watson software applied to product optimization. Thus, from the sustainable design or eco design, the energy efficiency of the products is improved by learning from the user use experiences. In terms of comfort we find the improvement of the user experience with chatbots or, with the creation of intelligent products such as intelligent prostheses, which improve the quality of life of people thanks to the joint work between AI and product design.

3 Conclusions

In conclusion, we can say that if according to a study by McKinsey & Company the emerging technologies (AI, Augmented Reality, Virtual Reality and Blockchain) are transforming the value chains of the creative content process in all its phases, it is true that, Artificial Intelligence will be the technology who have the greatest impact in the engineering.

The prediction that AI will be established as an indispensable working tool for creative activity in the coming years, extending and generalizing its incorporation as a tandem in all professional areas of creative fields, including the design of product. This "cooperation between humans and machines will allow both to improve" as stated in an interview Jeremiah Harmsen, responsible for artificial intelligence applied to Google in Europe [19]. Some companies like Alphabet are already working to demonstrate that machine learning can be used to unlock and improve the creative potential of all people. Thus, "Magenta" is a Google open source research project that explores the role of machine learning as a tool in the creative process [20].

Today, AI enables the creation of more open and participatory experiences, so that the original and the copy acquire new meanings. Creative projects based on computer systems provoke a new formula in the viewer, producing a more collaborative communicative experience both at a physical level and at a conceptual level.

References

1. Joseph, A., Paradiso: El cerebro sensorial aumentado. Como conectarán los humanos con el internet de las cosas. In: El próximo paso. La vida exponencial. (2016). https://www.bbvaopenmind.com/libros/el-proximo-paso-la-vida-exponencial/

2. https://towardsdatascience.com/google-just-published-25-million-free-datasetsd83940e 24284
3. World Economic Forum, McKinsey & Company: Creative Disruption: The impact of emerging technologies on the creative economy, White Paper. (2018). https://www.weforum.org/whitepapers/creative-disruption-the-impact-of-emergingtechnologies-on-the-creative-economy
4. https://www.ted.com/talks/kai_fu_lee_how_ai_can_save_our_humanity/up-nextl
5. Blanco Rodriguez, M.: Emoción y creatividad en Inteligencia Artificial. Tesis doctoral de la complutense de Madrid (2015)
6. https://bisite.usal.es/es/investigacion/lineas-investigacion/creatividad-computacional
7. BBVA: Elpróximopaso: Lavida exponencial (2017). www.bbvaopenmind.com/libros
8. Shoham, Y., et al.: The AI index 2018 annual report. In: AI Index Steering Committee, Human-Centered AI Initiative, Stanford University, Stanford, CA, December 2018. https://aiindex.org/
9. Kolarevic, B. (ed.): Architecture in the Digital Age. Design and Manufacturing. Spoon Press/Taylor & Francis, Nueva York (2003)
10. http://www.3dcadportal.com/el-poder-del-diseno-generativo-estara-en-autodesk-netfab b2018.html
11. http://finch3d.com
12. Oosterhuis, K.: Programmable Architecture, Milano: l'Arca Edizioni (2002)
13. https://www.youtube.com/watch?time_continue=15&v=nuWlpqevfIk&feature=emb_logo
14. https://www.smartvid.io/
15. https://www.doxel.ai/
16. Matarazzo, T.J., et al.: Crowdsensing framework for monitoring bridge vibrations using moving smartphones. Actas del IEEE **106**(4), 577–593 (2018)
17. Souza, P., et al.: A Nairobi experiment in using low cost air quality monitors. Clean Air J. **27**(2), 12–42 (2017)
18. Aowabin, R., Vivek, S., Amanda, D.S.: Predicting electricity consumption for commercial and residential buildings using deep recurrent neural networks. Elsevier **212**, 372–385 (2018)
19. https://retina.elpais.com/retina/2017/04/09/tendencias/1491755541_042969.html
20. https://magenta.tensorflow.org/

Teaching – Learning in Graphic Engineering

Spatial Skills Training Proposal in Virtual Reality Learning Environments

Javier Salgado Fernández(✉), Francisco José Soto Lara, and Manuel Damián Marín Granados

Departamento de Expresión Gráfica, Diseño y Proyectos Escuela de Ingenierías Industriales, University of Málaga, c/Dr. Ortiz Ramos s/n. Campus de Teatinos, 29071 Málaga, Spain
{javier.salgado,fsoto,mdmarin}@uma.es

Abstract. In the last decade, Virtual Reality (VR) technologies have become a driving force for Information and Communication Technologies (ICT) educational change. Several authors and Computers and Education publications state that immersive experiences provide better assimilation of the concepts studied. The educational goal of ICT such as VR is to deliver tools that enable the learning process as it could happen in the classroom, through three-dimensional simulations where information is organized and presented to the user who interacts directly with it. These simulations are referred as Virtual Reality Learning Environments (VRLE).

This paper proposal is an immersive VRLE where the user performs tasks through direct interaction with 3D objects to train and enhance spatial skills demanded for Engineering undergraduates in the Graphic Expression area, including identification of 3D object projections, translation, rotation and cutting plane operations. This VR based method aims at optimizing time dedication to learn traditional Computer-Aided Design (CAD) software for training spatial skills set and motivate students to participate in a program to improve and evaluate competences required in engineering occupations.

This methodology allows telematic learning via VR head-mounted displays with connectivity, that has become more relevant in the light of lockdown scenarios experienced during pandemic in the year 2020.

Keywords: Graphic expression · Spatial skills · Virtual reality · ICT and education · Collaborative engineering

1 Introduction

Traditional higher education models are based on the direct knowledge transmission found in the classroom teacher-pupil relation. In front of this, current model set by European Higher Education Area (EHEA) is student-centered and the skills and competences required to contribute to academic, professional and social needs. Competence framework was established as main criteria to evaluate the higher educational activity success [1]. In technical degrees programs, spatial visualization and relation acquisition is defined as a basic competence to be acquired in first degree year of the Graphic Expression knowledge area [2].

F. Cavas Martínez et al. (Eds.): INGEGRAF 2021, LNME, pp. 277–283, 2022.
https://doi.org/10.1007/978-3-030-92426-3_32

Since the 1980s, program studies have progressively reduced the number of hours dedicated to technical drawing and Graphic Expression subjects in Engineering Schools as illustrated in [3, Fig. 1]. This change was motivated by the reduction and summary of the contents, as well as the inclusion of new Computer-Aided Design (CAD) software [4]. Ten years later, new CAD subjects were firstly introduced as optional courses, which educational goals overlapped with core technical drawing studies. As a result, proposed activities were solved in less time, allowing a reduction in the student's dedication. Direct three-dimensional interface workflow, with less or even no level of abstraction needed by the user and low-complexity interactions in contrast with traditional methods, made the work with graphic projections more accessible and immediate. The student will nevertheless face the learning curve of standard CAD software to address required skills and knowledge of these subjects in the same period of time [5].

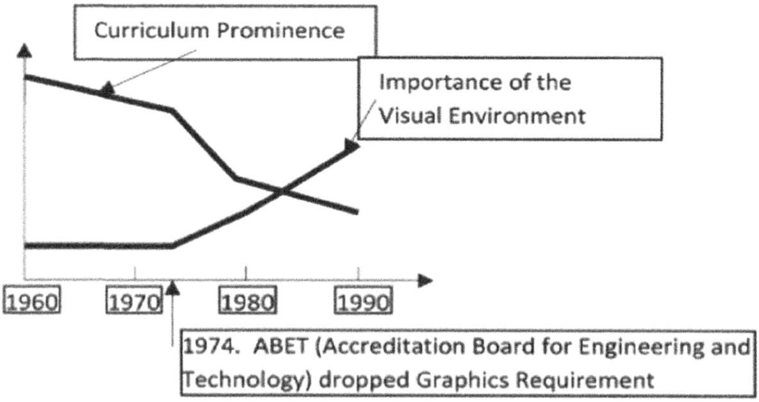

Fig. 1. Graphic Illustration prominence against Visual Environment importance as shown in high education studies. Source: M.H. Pleck [3]

Computer graphics introduction to technical studies produced a significant distribution in dedication between hand and computer guided practices, in order to acquire spatial skills competences. In this context, teachers provide best learning environment for student's familiarization with fundaments and theory behind the concepts described, without focusing on the tool used for that purpose.

Educators' role is to create learning environments and validate the information provided as to cultivate learners with valid criteria needed for future professional background.

There is an extensive bibliography on training and spatial skills development, as well as debate whether is it possible to enhance them. As part of intellectual capacities [6], spatial skill can be measured and, with the appropriate means such as standardized tests [7], CAD software usage [8] or other interaction and representation systems [9], they can be improved and adapted to engineering context [10].

In the last decade, Virtual Reality (VR) technologies have become a relevant alternative as Information and Communication Technologies (ICT) specifically for technical educational courses [11]. Its application to educational scenarios has been partially favored by technologic evolution and development cost reduction. Not only referred as a set of technologies that produce immersive experiences, VR is also defined under broader concept Extended Reality (XR) that involves any technology combining biological and virtual elements in visual representations [12]. Inside this conception, Augmented Reality (AR) and Mixed Reality (MR) are included. Inside a VR simulation a fully documented workspace is presented, 2D and 3D models are presented along with audiovisual resources so the student can put into practice knowledge acquired or solve proposed problems.

2 Objectives

2.1 Research Goal

A Virtual Reality Learning Environment (VRLE) is proposed as a bridge connecting theory concept of Graphic Expression subjects and virtual interface of CAD software. This tool grant access to immersive experience where tasks solving and skill training exercises are presented to be completed without previous knowledge in the mentioned tool due to the direct interaction between user and VR representation.

2.2 Key Indicators

The proposal is oriented to motivate higher technical programs students to conduct practices and exercises for Graphic Expression subjects. Immersive three-dimensional environment offers a series of graphic models and resources which facilitates acquisition and retention of concepts discussed in the classroom, inviting to perform different nature activities, in contrast to standard questionnaires. This way, independently on the familiarity level of the user with ICT tools, time dedication is optimized by offering a method that does not rely on the individual CAD expertise, enabling more time to acquire spatial competences required and measure results obtained from this proposal as to improve the methodology for future research.

3 Material and Methods

3.1 Development Engine

A reliable and active source of information was required to develop a functional VRLE, given the lack of previous documentation in the area of knowledge. After review of platform availability, Unity engine is chosen due to its extensive use in game development in C# programming language, documentation [13] low learning curve in relation to other platforms and active support from online community. These factors are key to introduce an environment where materials, physics and other operations common to CAD software are present. To implement mechanics such as movement or object grabbing in the VR application, Extended Reality Toolkit asset from Unity digital store is used ensuring compatibility with VR headset usage.

3.2 Head-Mounted Display: Oculus Quest 2

Head-Mounted Display (HMD) market is growing exponentially with numerous solutions with requisites to run proposed VR software. Based in use characteristics such as simple controllers, inside-out tracking, wifi connectivity and no need for wired connection to a computer from Oculus Quest 2™ headset, this HMD launched in 2020 is the hardware chosen to work with.

3.3 VR Design Principles

Still in early adoption stages, experience design for educational VR is a key discipline, as per specialized media to provide immersive environments adapted to learning based on interaction. Based on the research of [14] and [15], following design principles are identified to build the proposed solution:

User Movement: Continuous movement is enabled for user, similar to walking inside the 3D space. Instant teleportation is also an option to move quickly and exactly to chosen destination.

Basic Object Interaction: Inside-out tracking of the head and hands, it is possible to grab objects, rotate or scale them. Contextual menus are also present so user can choose between options available. Proposed exercises are based around simple actions so user can solve them efficiently.

Realistic Surroundings: A replication of a real-world like environment is required for better location and orientation. Including elements, for example doors or tables increases usability of a VR space since their function is already known.

Immediate Response: Auditive, haptic or text stimuli are to be produce in real time to assist interaction.

User Generated Content: Inclusion of 2D and 3D content created prior in CAD software is a possibility to broaden usage of the tool.

Instructions and Tutorial: Through explicit instructions or step by step tutorial, instructions are provided inside the VR experience. These instructions can be written, Heard or provided by another user connected to the network.

Knowledge Test: Questionnaires or time trials.

Rewards: As recognition for completing all proposed tasks successfully. Educators can judge if academic recognition may be also applicable.

3.4 Information Research

For the proposal development, a search for systematic VR studies in educational fields was conducted in Scopus and Web of Science, with special interest in publications between 2010 and 2020. Keywords used were: *education, virtual reality design, spatial skills, graphic expression, collaborative engineering*.

4 Results

Virtual Reality Learning Environment proposed for spatial skills development presents a workspace where students can perform operations such as rotation, assembly or sections with cut planes. Operation and 3D model can be selected, using a catalogue created intentionally for this development or imported CAD models converted from external software and compatible with Unity.

As per rotation, user is provided with a series of models to be examined through direct interaction. After exploring the geometry, four projections will be presented from which the user will have to select the ones corresponding with models shown previously. Additionally, plane cut option is available for insertion to produce different sections. Alternatively, workflow can be inverted for each exercise so the projections are demonstrated at first and the user is shown different three-dimensional models to choose from as illustrated in Fig. 2.

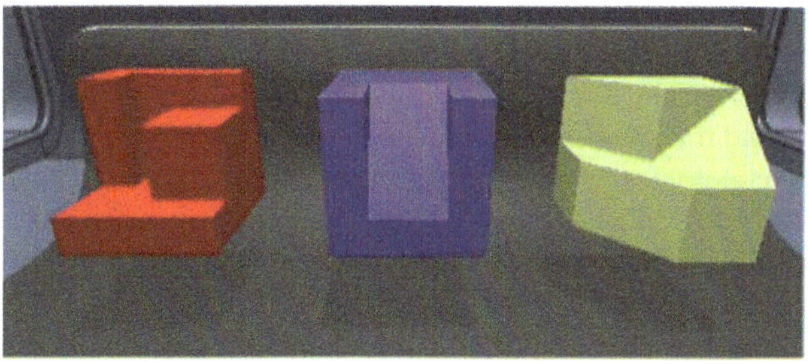

Fig. 2. Imported 3D models from external CAD software. Source: authors

Currently work in progress is the tool's capability to create three dimensional models using learning procedural generation algorithms. This method can create models based on the decisions made by the user in previous tasks performed. If certain complex geometries are identified as challenging for the user, the application will add them for future exercises to facilitate detection and improve understanding of 3D models and projections.

Eventually, assembly tasks are presented as a series of parts that belong to an assembly set, see Fig. 3. User will explore and with the tools mentioned before, execute the assembly with unambiguous solution.

Fig. 3. Left: Component assembly. Right: Object grabbing with VR controller. Source: authors.

5 Conclusions

Virtual Reality experiences and applications had been main topic for scientific research, but none have provided a learning theory perspective in field studies yet. In this paper, an innovative methodology is proposed for spatial skills enhancement with application of a Virtual Reality Learning Environment for Graphic Expression subjects in higher technical education programs. The main goal is to establish the fundaments for VR active learning strategies, promoting student's creativity through immersion and interaction. In recent times, ICT technologies grant access to extensive volume of information may overwhelm those in search for specific information. With proposed method, an alternative to classic measure systems as seen in standardized spatial test [1, 6, 17, 18] is presented.

The proposal discussed in this document is not to replace descriptive drawing practice or CAD learning in academic scenarios but to address the gap between theory understanding and skills acquisition for graphic expression students. Graphic Engineering has experienced massive digitalization in the 21st century, being more accessible to educators and undergraduates, allowing conclusion collaborative projects valuable for the professional background of the future engineers. The latter is a noteworthy event as part of Concurrent Engineering, where all parties involved in a project use graphic language to communicate and transmit information between areas [19].

Immersive technologies postulate as useful link between professional requirements and academic areas. Graduates often face challenges when real world tasks are presented. This problematic can be halved with the usage of immersive technologies. Recent events of SARS Covid-2 pandemic have highlighted the relevance of telework and non-presential educational strategies, specially when field work cannot take place in the same location. Extended Reality hardware is at the brink of massive adoption, still some barriers are to be surpassed for example high pricing for HMD and lack content quality. Primarily, developers and stakeholders are the main drivers for improving these circumstances. Until then, the main target is to define and adapt educational strategies which include immersive technologies which strengthen competence acquisition.

Acknowledgements. The research results included in this paper are part of the final project "Cooperative Learning with Virtual Reality" from Master's Degree Project in "Representación Gráfica en Ingeniería y Arquitectura" and research work performed at Escuela de Ingenierías Industriales of Malaga University.

References

1. MEC. Borrador de propuesta Directrices para la Elaboración de Títulos Universitarios de Grado y Máster, Madrid (2006)
2. Blazquez-Parra, E.B., Ladrón-de-Guevara-López, I., Marín-Granados, M.D.: Comparativa y análisis de la Ingeniería Gráfica en el Grado en Ingeniería en Tecnologías Industriales. Escuela de Ingenierías Industriales de Málaga (2019)
3. Pleck, M.H., et al.: Factors affecting the engineering design graphics curriculum: past, present, future. In: Proceedings of the NSF Symposium on Modernization of the Engineering Design Graphics Curriculum, Austin, Texas (1990)

4. Farrerons Vidal, Ó., Olmedo Torre, N.: Las TIC y la Ingeniería Gráfica, Barcelona. OmniaScience (2016)
5. Almirall, G.: Proyecto de información de la docencia de técnicas de representación gráfica (TRG) en la escuela técnica superior de ingenieros industriales de Barcelona (ETSEIB). In: Actas del XI Congreso Internacional de Ingeniería Gráfica (INGEGRAF), Logroño-Pamplona (1999)
6. Gardner, H.: Frames of Mind: The Theory of Multiple Intelligences. Basic Books, New York (1983)
7. Gómez Hermosa, F., Miravet Garret, L., Ortíz Zamora, F.J., Ladrón de Guevara Muñoz, C., Nieto Reigal, F.: Relationship between spatial reasoning skills in first year industrial engineering students. ICERI2018 Proc. **1**, 2253–2258 (2018). https://doi.org/10.21125/iceri.2018.1497
8. Sorby, S.A.: Educational research in developing 3D spatial skills for engineering students. Int. J. Sci. Educ. **1**(3), 459–480 (2009)
9. Marín-Granados, M.-D., Gómez-Hermosa, F., Ladrón-de-Guevara-López, M.-D.-C., Miravet-Garret, L., Soto-Lara, F.-J.: Analysis of spatial comprehension through a retrospective study of its effects among first year electrical engineering students. In: Cavas-Martínez, F., Sanz-Adan, F., Camo, P.M., Lorza, R.L., Peña, J.S. (eds.) Advances in Design Engineering: Proceedings of the XXIX International Congress INGEGRAF, 20-21 June 2019, Logroño, Spain, pp. 339–346. Springer International Publishing, Cham (2020). https://doi.org/10.1007/978-3-030-41200-5_37
10. Olkun, S.: Making connections: improving spatial abilities with engineering drawing activities. In: International Journal of Mathematics Teaching and Learning (2003)
11. Abulrub, A.-H.G., Attridge, A.N., Williams, M.A.: Virtual reality in engineering education: the future of creative learning. In: IEEE Global Engineering Education Conference (EDUCON), pp. 751–757 (2011)
12. Milgram, P., Kishino, F.: A taxonomy of mixed reality visual displays. In: IEICE Transactions on, n° 12, pp. 1321–1329 (1994)
13. Unity User Manual - 2019.4 LTS (2020). https://docs.unity3d.com/Manual/
14. Radianti, J., Majchrzak, T.A., Fromm, J., Wohlgenannt, I.: A systematic review of immersive virtual reality applications for higher education: design elements, lessons learned, and research Agenda. Computers and Education (2019)
15. Jensen, L., Konradsen, F.: A review of the use of virtual reality head-mounted displays in education and training. In: Education and Information Technologies (2017)
16. Vanderberg, S.G., Kuse, A.R.: Mental rotations, a group test of three-dimensional spatial visualization. Percept. Mot. Skills **47**, 599–604 (1978)
17. Guay, R.: Purdue Spatial Visualization Tests. West Lafayette (1977)
18. Bennett, G.K., Seashore, H.G., Wesman, A.G.: Diferential Aptitude Tests - Spatial Relations Subset (1947)
19. Bertoline, G.R., Wiebe, E.N., Miller, C.L., Mohler, J.L.: Dibujo en Ingeniería y Comunicación Gráfica. McGraw-Hill (1999)

Automatic Evaluation of Facility Layouts Through Graph Matching

Lucía Díaz-Vilariño(✉), José Luis González-Cespón, José Antonio Alonso-Rodríguez, and Antonio Fernández-Álvarez

Department of Engineering Design, School of Industrial Engineering, Universidade de Vigo, Campus Universitario, 36310 Vigo, Spain
lucia@uvigo.es

Abstract. This work presents a method for determining whether an industrial layout configuration is compliant with design specifications and constraints. Functional specifications are modelled by an undirected graph. The nodes of the graph represent those spaces which are considered essential for the industry to carry out its activity. An edge connecting two nodes indicates that such activity involves frequent movements of people or materials between the corresponding spaces, and therefore they must be located close to each other for the sake of efficiency. Manual verification of compliance with these specifications is time consuming and error prone. In order to automatically check if a facility layout includes all the essential spaces and they are arranged so that all proximity requirements are met, we developed a software which converts a layout drawing into its graph representation and compares the obtained graph with the target graph. The proposed system comes in handy in an engineering teaching setting, as it saves the instructor from manually verifying that the layouts created by the students comply with the project specifications. To demonstrate the usefulness of this approach we evaluated different solutions to a facility layout problem with the developed software. The results show that our system reliably detects any non-compliance with specifications.

Keywords: Facility layout problem · Layout planning · Automation · Undirected graph

1 Introduction

The problem of facility placement within a plant, generally referred to as Facility Layout Problem (FLP), is a well-known optimization problem commonly dealing with the spatial arrangement of the facilities required in an organization while satisfying a set of preferences and constraints. Layout problems range in scale from urban planning by the assignment of activities to cities, sites or buildings, to the location of equipment and personnel on a single floor, or even to the location of components in printed circuit boards. Evidence of the huge applicability of FLP in real life situations is proved by the use in layout design for manufacturing such as in case of cell designs [1], logistics

F. Cavas Martínez et al. (Eds.): INGEGRAF 2021, LNME, pp. 284–293, 2022.
https://doi.org/10.1007/978-3-030-92426-3_33

systems [2], chemical processes [3], hospitals considering transit of patients, personnel and materials [4], multiple construction stages with safety considerations [5], space utilization in warehouses [6], etc.

The layout problem is known to have a noteworthy impact on work processes, manufacturing costs, productivity, and lead times [7]. Material handling cost represents 20–50% of the total operating expenses in manufacturing environments [8]. And these costs can be considerably reduced if an efficient layout configuration is designed. Consequently, the main attention of the research in this area has been focused on the manufacturing applications. Accordingly, minimizing the material handling cost has been considered the most common objective for optimization in the literature, but account was also taken of reducing work in progress, lead times, utilizing existing space more effectively, making plant adaptive to future changes, providing health, convenient and safe environment for employees [9].

The FLP has been extensively researched in the last decades and most of efforts have been aimed to develop algorithms to find optimal or suboptimal solutions to complex problems. Although there is no doubt regarding to the effectiveness and potentiality of algorithms in solving complex layout problems, the solution is highly dependent on how the problem is modelled. Computer-based design algorithms do not always capture the qualitative and intelligence aspects of layout design, and it has been noted that in most cases they could not replace human judgment and experience [10]. Consequently, the manual creation of layout designs is still being used especially for solving simple layout problems.

In this work we propose a method for the evaluation of facility layout designs. Our system reads a layout drawing created using CAD software and automatically generates a graph model of the design. In this model the different spaces that make up the layout are represented as nodes, whereas connections between pairs of spaces are represented by edges. Then the obtained graph is compared with a target graph which defines the functional specifications, namely all the essential spaces for the industry to carry out its activity as well as certain proximity relationships between spaces that must be fulfilled in order to minimize costs. We applied the proposed method to check whether facility layouts designed by fourth year engineering students comply with the project specifications provided by the instructor. It clearly emerges from the tests performed that our software accurately detects any non-compliance with specifications. The usefulness of such a CAD automation tool is twofold: on the one hand it saves the instructor from manually checking that students' projects are compliant with specifications, and on the other hand, it allows students to self-check that their projects are compliant with specifications prior to final submission.

The remainder of this article is organized as follows. Section 2 reviews different aspects of the facility layout problem. Section 3 introduces the proposed evaluation method and describes some relevant implementation details. In Sect. 4, we test our system with a set of real plant layouts and analyze the obtained results. And finally, Sect. 5 presents some concluding remarks and discuss promising research directions for future work.

2 The Facility Layout Problem

In the last decades, a great deal of research effort has been put into the facility layout problem because of the impact upon manufacturing costs and productivity, and because of the complexity of layout optimization, generally considered as NP-hard. As a consequence, extensive reviews and classifications of optimization algorithms and systems have been published [7, 11–14]. This section provides the reader with a brief overview of the related research, with a main focus on the modelling problem and the resolution approaches. The interdependence between modelling and resolution approaches makes it difficult to design methods for the quantitative evaluation of manually created layouts.

2.1 Modelling the Layout Design

The modelling problem refers to the structural characteristics that are considered to represent and formulate the problem and which, in turn, influence the resolution approach. Those structural characteristics involve aspects related to the configuration of the manufacturing system, the objective function, and the space representation.

The configuration of the manufacturing system generally depends on the products variety and on the production volumes. Accordingly, four key organizations have been commonly accepted in literature: *fixed product layout* in which the resources move around the product, *process layout* in which facilities are grouped together according to functionality in order to address a wide variety of products, *product layout* used when there is a high production volumes and low variety of products, *and cellular layout* in which all these distributions are combined into cells [15]. In addition, the configuration of the problem can be considered dynamic or static depending on whether the flow between spaces change over the time or not, respectively. When addressing dynamic configurations, both the cost related to rearrangement of layouts and the costs incurred due to loss of production time should be taken into account [16, 17].

The objective function is a function whose value we want to maximize or minimize. Although real-world problems include many objectives which may not be expressed in terms of cost, the majority of literature has considered a single objective function, being the minimization of material handling cost the most popular [11]. The objective function can be expressed quantitatively (by distance-based or cost-based measures), qualitatively (by security factors or proximity relationships), or as a combination of both [18].

The space can be represented in a wide variety of ways through parameters such as shape (regular, irregular), area and dimensions (fixed dimensions, aspect ratio). Moreover, although most of approaches deal with one-floor problems, some solutions have been presented to solve the multi-floor problem [9, 19]. In this case, horizontal and vertical flows need to be considered.

Finally, the plant layout can be represented by discrete units or in a continuous way. In the first case, the plant layout can be subdivided into rectangular blocks with same size and area, and each block is assigned to a facility. If facilities should have different areas, they can be assigned different numbers of blocks [20]. When the plant layout representation is continuous rather than discrete, facilities can be located anywhere – their position could be defined, for example, by the centroid coordinates – with the only constraint they do not overlap [21]. The continuous representation is more suitable for addressing certain constraints such as localization, orientation, or security distance [7].

2.2 Resolution Approaches

The resolution approaches refer to those methods implemented to solve the optimization problem given the structural characteristics considered during the modelling phase. The approaches can be categorized into exact and heuristic methods. While exact methods eventually provide an optimal solution, heuristic methods are not guarantee to find an optimal solution. However, as FLP is NP-hard, exact methods are often impractical for large-scale size problems. As a consequence, heuristic and metaheuristic approaches have become very popular for solving complex models.

Despite high computational complexity, some authors have attempted to solve FLP optimally. Most of the exact methods proposed in the literature are based on branch and bound algorithms [7], although other methods are based on decomposing the problem into some simpler problems which are solved individually using exact approaches [11]. The combinatorial complexity of the FLP formulations has boosted the development of numerous efficient heuristic and metaheuristic methods, including *genetic algorithms, simulated annealing, tabu search, random search, naive evolution, ant colony*, etc., to name a few successful examples [7, 11].

Intelligence and knowledge-based methods, mainly *Decision Support Systems (DSS)* and *Expert Systems (ES),* represent a class of computer-based systems that use the knowledge about a specific application domain to assist decision makers by providing appropriate strategies and actions. While DSSs are based on summarizing and analysing relevant data, ESs are based on applying reasoning and inference methods built from a large repository of knowledge. It is worth mentioning that these systems have also been applied in combination with heuristic algorithms to the resolution of the FLP [9].

2.3 Evaluation of Plant Layouts

As said in the previous paragraphs, resolution methods are intended to find a solution for a specific optimization problem. The obtained solution, which can be optimal or suboptimal, strongly depends on how the problem is modelled, i.e. what "features" are fed to the resolution method (shape and/or size of the spaces, adjacency relations, types of process, etc.). As a consequence, a straightforward comparison of the goodness of layout designs generated through different modelling approaches can be tricky and usually leads to erroneous conclusions. This intrinsic difficulty explains why most of the research effort is focussed on evaluating modelling and resolution approaches jointly and is the underlying reason why only a few papers address layout design evaluation regardless of the modelling approach [7].

The resolution approaches just described are powerful methods which make it possible to solve complex facility layout problems. However, some authors consider that human expertise and know-how are invaluable, and they claim that a seasoned designer can outperform sophisticated computer-based layout design algorithms [10]. In fact, plant layouts are still often created manually, especially when it comes to simple layouts.

3 Methodology

3.1 Overview

In this work we address the problem of evaluating manually created plant layouts in a teaching setting. The evaluation of layouts designed by engineering students is usually carried out manually by the instructor according to the criteria defined in a rubric. One important assessment criterion is the compliance with project specifications which can include a wide variety of requirements and constraints. This paper focuses on simple cases, in which the overall process to be carried out in the facility is decomposed into a set of subprocesses or activities, each of which has to be run in a specific space. In addition, students are asked to place certain spaces close to each other to minimise the costs associated with the movement of people or handling of goods. Manually verifying that a particular layout contains all the specified spaces and pairwise connections require a lot of scrutiny and is prone to mistakes. Therefore, it would be very convenient if those tedious, cumbersome checks could be performed automatically. That is the principal motivation of our work.

The proposed method to achieve that goal consists in representing a layout through a graph, in which the nodes play the role of the different facility spaces, and the adjacency relations between pairs of spaces are modelled by edges. The resulting graph is then compared to a target graph that includes all the necessary edges and nodes. In the upcoming paragraphs we show that this graph comparison is an adequate way to highlight any non-compliance of the corresponding layout with the project specifications.

It is important to note that in order to be able to automatically convert a layout into a graph the drawings must meet certain requirements, for example, different elements of the layout such as inner walls, outer walls, inner doors, outer doors, etc., have to be drawn in separate layers, and these layers must have specific names. Besides, the drawings must be accurate for the automatic detection of pairwise connections to work. Concretely, the boundary of each space has to be perfectly closed as if the boundary leaks two spaces that are not connected in the layout could be linked by and edge in the generated graph. The complete set of requirements that a layout drawing must adhere to is described in [22].

3.2 Implementation

To automatically verify that a layout conforms to specifications we developed a program in Python which works as follows. First, we launch AutoCAD and open a drawing. All the entities that make up the drawing, such as walls, doors, borders, space IDs, etc., are retrieved. We start iterating through all the connectors, namely, inner doors and borders, and we hide one connector at a time. For each connector we run a nested loop. In this case we iterate through pairs of spaces until we find that two text entities (the space IDs) are inside the same closed boundary. When this happens, it means that those spaces are connected through the hidden connector. The nested loops are executed until all the connectors and pairs of spaces are exhausted. In our program we make extensive use of a component of the `pywin32` package [23], concretely the `win32com.client` module, which makes it possible to create and use COM objects and automate Windows applications with Python. Unfortunately, the Autocad COM API is poorly documented, and as a consequence, interacting with instances of the `IAcadDocument` class can be very challenging. The class interface is particularly obscure in those aspects regarding exception management and attribute access. To work around this issue, we had to implement part of the program in AutoLISP, a dialect of the programming language Lisp built specifically for use with AutoCAD. The hack we developed to implement the unsupported functionality consisted in creating AutoLISP snippets as Python strings and passing them in as an argument to the. Send- Command() method.

The second stage of the program deals with graph processing. Although the graph generated in the first stage provides a full representation of the layout, this complete graph cannot be directly compared with the target graph due to two major reasons. The first one is that some essential spaces can be split into two or more subspaces. For example, it may well happen that space X is distributed across subspaces X.1, X.2 and X.3. To cope with that situation, we need to merge these subspaces into one single space and modify the graph edges accordingly. The second limitation of the complete graph has to do with corridors or aisles. So far, we have considered that two spaces are connected if they are separated by a single door or border in the layout. However, there is a third possibility that should be taken into account. If two spaces, say X and Y, are connected to Z and Z is a corridor then we assume that X and Y are connected. Whenever this situation occurs, we need to replace the edges XZ and YZ together with node Z by an edge XY. This replacement must be done carefully in order not to break other connections. In summary, to pass from the complete graph to a simplified representation that can be directly compared with the target graph we carry out fusion and pruning operations. Most of the functionality necessary to simplify the complete graph is available in the `networkx` library [24].

For reproducible research purposes, we have made all code and data available in a Github repository [25].

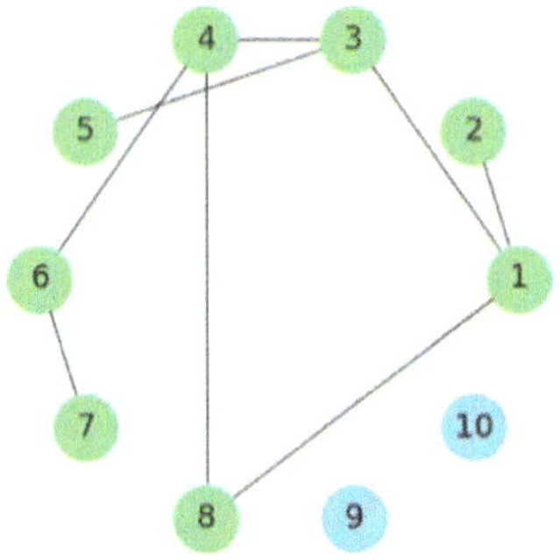

Fig. 1. Graph depicting the essential spaces (green) and the required spaces (blue) of the motorhome conversion center, as well as the necessary connections between pair of essential spaces. The meaning of the numbers is explained in Table 1.

4 Experiments

To assess the effectiveness of the system just described we tested it on a set of 20 different plant layouts created by fourth year engineering students enrolled in a course on layout planning. The students were asked to design in AutoCAD a motorhome conversion center, i.e. a workshop specialized in converting vans into campervans. The students were informed that their designs had to meet two major constraints. The first one is that the proposed layouts had to include eight essential spaces, two required spaces and a variable number of optional spaces (see Table 1). Further, a set of pairwise connections between the essential spaces was also imposed. These requirements can be conveniently represented through the target graph (Fig. 1). The second constraint is geometric rather than functional and is that the floor plan outline is defined in advance and cannot be modified. Figure 2 shows a sample layout together with its corresponding graphs (complete and simplified).

After processing the 20 plant layouts that make up this study case, we thoroughly checked the automatically generated graphs to assess the accuracy of the method. We found out that all layout drawings were converted to graphs correctly, resulting in a 100% success rate.

Table 1. Keys and designations of the spaces used in the case study.

Essential spaces	Required spaces	Optional spaces
1: Production area	9: Toilet	11: Staff lounge
2: Warehouse	10: Changing room	12: Project room
3: Parking area		13: Meeting room
4: Vehicle reception		14: Entrance hall
5: Delivery area		15: Showroom
6: Administrative office		16: Broom closet
7: Manager office		17: Cargo unloading
8: Waste store		18: Garage
		19: Finish area
		20: Corridor
		21: Server room
		99: Outside

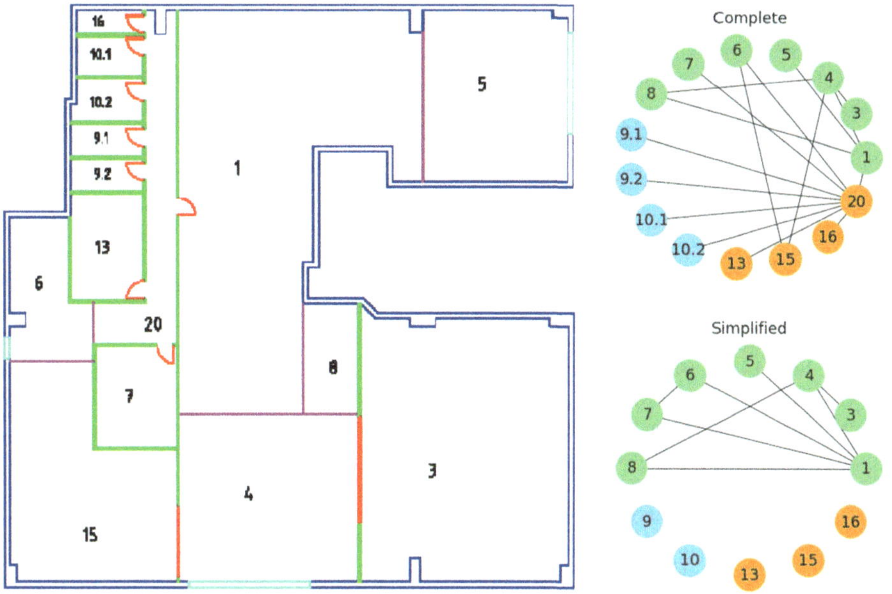

Fig. 2. Sample layout and resulting graphs.

5 Conclusions and Future Work

We have presented an automation tool for the evaluation of plant layouts. Our software converts a CAD drawing in `.dwg` format into a graph which is then compared with the target graph, i.e. the graph that models the functional specifications the facility layout must comply with. The system generates a report which flags missing spaces or connections, if any; otherwise, it indicates fulfillment of the specifications. The proposed system proves handy for both students and instructors in an engineering teaching setting. On the one hand, it saves instructors from manually checking that students' submissions are compliant with the specifications of an assignment. Experimental results show that this time-consuming, error prone task can be reliably performed in an autonomous manner. The proposed system therefore constitutes a valuable aid for teachers when grading students' projects through rubrics. On the other hand, students can also benefit from our software, as it allows them to self-check that their projects meet the specifications before they submit the final version. This would likely result in improved overall grades and a better learning experience.

In the future, we plan to take our software one step further so that it not only verifies that requirements and constraints are met, but also automatically scores student-designed solutions for a facility layout problem. To that end, we need to come up with a custom graph similarity measure that takes into account all the assessment criteria included in the grading rubric. For the value of this measure to be in accordance with the score that a human grader would assign manually, we will have to adjust some parameters through machine learning techniques.

Acknowledgements. The authors thank José Carlos Troncoso for helpful suggestions and advice on AutoLISP programming. The authors also thank to Xunta de Galicia for human resources grant (ED481D 2019/020) and competitive reference groups (ED431C 2020/01), and to the Ministerio de Ciencia e Innovación - Agencia Estatal de Investigación/https://doi.org/10.13039/501100011033 (Proyecto PID2019-105221RB-C43).

References

1. Sadeghi, A., Suer, G., Sinaki, R.Y., Wilson, D.: Cellular manufacturing design and replenishment strategy in a capacitated supply chain system: a simulation-based analysis. Comput. Ind. Eng. **141**, 106282 (2020). https://doi.org/10.1016/j.cie.2020.106282
2. Grobelny, J., Michalski, R.: A concept of a flexible approach to the facilities layout problems in logistics systems. In: Borzemski, L., Grzech, A., Świątek, J., Wilimowska, Z. (eds.) Information Systems Architecture and Technology: Proceedings of 36th International Conference on Information Systems Architecture and Technology – ISAT 2015 – Part I. AISC, vol. 429, pp. 171–181. Springer, Cham (2016). https://doi.org/10.1007/978-3-319-28555-9_15
3. Xu, S., Wang, Y., Feng, X.: Plant layout optimization for chemical industry considering inner frame structure design. Sustainability **12**(6), 1–19 (2020). https://doi.org/10.3390/su12062476
4. Helber, S., Böhme, D., Oucherif, F., Lagershausen, S., Kasper, S.: A hierarchical facility layout planning approach for large and complex hospitals. Flex. Serv. Manuf. J. **28**(1–2), 5–29 (2015). https://doi.org/10.1007/s10696-015-9214-6
5. Huang, C., Wong, C.: Optimisation of site layout planning for multiple construction stages with safety considerations and requirements. Autom. Constr. **53**, 58–68 (2015). https://doi.org/10.1016/j.autcon.2015.03.005
6. Derhami, S., Smith, J.S., Gue, K.R.: Optimising space utilisation in block stacking warehouses. Int. J. Prod. Res. **55**(17), 1–17 (2016). https://doi.org/10.1080/00207543.2016.1154216
7. Drira, A., Pierreval, H., Hajri-Gabouj, S.: Facility layout problems: a survey. Annu. Rev. Contr. **31**(2), 255–267 (2007). https://doi.org/10.1016/j.arcontrol.2007.04.001
8. Francis, R.L., McGinnis, L.F., White, J.A.: Facility Layout and Location: An Analytical Approach. Pearson College Division (1992)
9. Ahmad, A.-R., Basir, O., Hassanein, K., Azam, S.: An intelligent expert systems' approach to layout decision analysis and design under uncertainty. In: Phillips-Wren, G., Ichalkaranje, N., Jain, L.C. (eds.) Intelligent Decision Making: An AI-Based Approach, pp. 321–364. Springer, Heidelberg (2008). https://doi.org/10.1007/978-3-540-76829-6_12
10. Tompkins, J.A., White, J.A., Bozer, Y.A., Tanchoco, J.M.A.: Facilities Planning, 3rd edn. Wiley, Reading (2002)
11. Ahmadi, A., Saman Pishvaee, M., Reza Akbari Jokar, M.: A survey on multi-floor facility layout problems. Comput. Indust. Eng. **107**, 158–170 (2017). https://doi.org/10.1016/j.cie.2017.03.015
12. Kusiak, A., Heragu, S.S.: The facility layout problem. Eur. J. Oper. Res. **29**(3), 229–251 (1987). https://doi.org/10.1016/0377-2217(87)90238-4
13. Liggett, R.S.: Automated facilities layout: past, present and future. Autom. Constr. **9**(2), 197–215 (2000). https://doi.org/10.1016/S0926-5805(99)00005-9
14. Meiler, R.D., Gau, K.-Y.: The facility layout problem: recent and emerging trends and perspectives. J. Manuf. Syst. **15**(5), 351–366 (1996). https://doi.org/10.1016/0278-6125(96)84198-7

15. Mohammadi, M., Forghani, K.: Designing cellular manufacturing systems considering S-shaped layout. Comput. Indust. Eng. **98**, 221–236 (2016). https://doi.org/10.1016/j.cie.2016.05.041
16. Balakrishnan, J., Cheng, C.H.: Dynamic layout algorithms: a state-of-the-art survey. Omega **26**(4), 507–521 (1998). https://doi.org/10.1016/S0305-0483(97)00078-9
17. Braglia, M., Zanoni, S., Zavanella, L.: Layout design in dynamic environments: strategies and quantitative indices. Int. J. Prod. Res. **41**(5), 995–1016 (2003). https://doi.org/10.1080/00207540210162983
18. Muther, R.: Systematic layout planning, Cahners Books (1984)
19. Park, K., Koo, J., Shin, D., Lee, C.J., Yoon, E.S.: Optimal multi-floor plant layout with consideration of safety distance based on mathematical programming and modified consequence analysis. Korean J. Chem. Eng. **28**, 1009–1018 (2011). https://doi.org/10.1007/s11814-010-0470-6
20. Wang, M.J., Hu, M.H., Ku, M.Y.: A solution to the unequal area facilities layout problem by genetic algorithm. Comput. Ind. **56**(2), 207–220 (2005). https://doi.org/10.1016/j.compind.2004.06.003
21. Dunker, T., Radonsb, G., Westkämpera, E.: Combining evolutionary computation and dynamic programming for solving a dynamic facility layout problem. Eur. J. Oper. Res. **165**(1), 55–69 (2005). https://doi.org/10.1016/j.ejor.2003.01.002
22. Díaz-Vilariño, L., González-Cespón, J.L., Alonso-Rodríguez, J.A., Fernández-Álvarez, A.: Una propuesta para la autoevaluación de layouts en la asignatura de Oficina Técnica y Proyectos. In: VII Jornadas Iberoamericanas de Innovación Educativa en el Ámbito de las TIC y las TAC. Las Palmas de Gran Canaria, 19 y 20 de noviembre de 2020. https://accedacris.ulpgc.es/handle/10553/76539
23. Python for Win32 extensions (pywin32). https://github.com/mhammond/pywin32
24. Network Analysis in Python (NetworkX). https://networkx.org/
25. layout2graph. https://github.com/tonechas/layout2graph

Complete and Automated Generation of Configurable Virtual Prototypes of Products Based on Parameterization Tools and Rules. Application to a Case Study

Virgilio Véliz Vega[1], Francisco Albert Gil[2(✉)], and Nuria Aleixos Borrás[2]

[1] Universitat Politècnica de Valencia, Valencia, Spain
virveve@etsii.upv.es

[2] Departamento de Ingeniería Gráfica, Universitat Politècnica de València, Valencia, Spain
{fraalgil,naleixos}@dig.upv.es

Abstract. In engineering, the 3D model of a product is essential. A 3D model allows making modifications by editing the characteristic modeling functions, which implies knowing in detail the process followed in the modeling. In addition, the flexibility in the configuration is limited, since the modifications are made on geometry and parameters that are dependent on each other. In this work, a parametric modeling approach allows the generation of 3D models with different specifications by modifying a reduced number of parameters. To demonstrate the functionality, an application has been developed, using Autodesk Inventor iLogic, for the modeling of an engine with V-cylinder arrangement. Taking as input key parameters (number of cylinders,...), it can generate virtual prototypes with different configurations, facilitating the selection of the best product configuration by allowing to evaluate different alternatives.

Keywords: CAD · 3D models · Parameterization · Product configuration · Rules

1 Introduction

CAD tools are a basic element in engineering, facilitating the design and management of products and industrial facilities in all phases of their development, as well as communication between the people involved in the process [1].

For engineering students, CAD tools bring them closer to the professional world [2], simulating real devices with resources within their reach (personal computers and educational licenses of CAD tools provided by universities), increasing their motivation and learning [3] by relating theoretical concepts and practical application, improving the understanding of different mechanisms.

In the first years of engineering degrees, students receive technical drawing courses to improve their spatial vision and become familiar with the language of engineering drawings. These courses usually focus on the use of pencil and paper representation

F. Cavas Martínez et al. (Eds.): INGEGRAF 2021, LNME, pp. 294–301, 2022.
https://doi.org/10.1007/978-3-030-92426-3_34

systems and 2D CAD tools. In later courses it is common to be introduced to 3D modeling tools [4] and even simulation [5].

In order to optimize the use of 3D CAD applications and facilitate the understanding of real mechanisms, the 3D models used must allow changes in their geometry without having to restart the project again. The modeling strategy if fundamental for this [5]. There are many ways to model the same component, but to create optimal 3D models it is necessary to perform a parametric design that allows controlling, with a few parameters, the geometry of the component [6].

The objective of this work is to establish a parametric modeling methodology, applied in this case to combustion engines (although it can be extrapolated to other engineering mechanisms) that allows the automatic generation of 3D models with different specifications by modifying a few parameters, which allows students to immediately have the 3D model of a new engine for their study, without having to perform the modeling again from scratch.

In this work (the result of a TFM supervised by professors of the Department of Graphic Engineering of the UPV for the qualifying master's degree in Industrial Engineering) the iLogic tool of Autodesk Inventor Professional has been used. iLogic allows the implementation of rules applicable to the parameters that govern the functionality of the engine and affect the modeling operations.

Section 2 presents the set of parameters that define the specifications and how the geometrical parameters are derived from them, Sect. 3 shows some of the relationships and rules implemented, Sect. 4 shows the results and Sect. 5 presents the conclusions.

2 Main Parameters and Geometrical Parameters

First, the main parameters (those that, within a certain range of values, will allow different configurations available) will be determined. From them, the geometric parameters used in the modeling will be derived. Table 1 shows the main parameters used with the ranges of values considered. With the different levels available for selection, there are more than 7000 possible configurations.

Table 1. Main engine parameters.

Parameter	Description	Units	Range of values
VT	Total cylinder capacity	Cubic centimeters (cc)	[1000, 7000]
z	Number of cylinders	–	[4, 12] (even number)
cd	Stroke ratio	–	[0.7, 1.4]
α	V angle	Sexagesimal degrees (°)	60°, 90° and 120°

Figure 1 shows an engine with a specific configuration. To reduce complexity, only the essential components have been modeled and the details have been simplified.

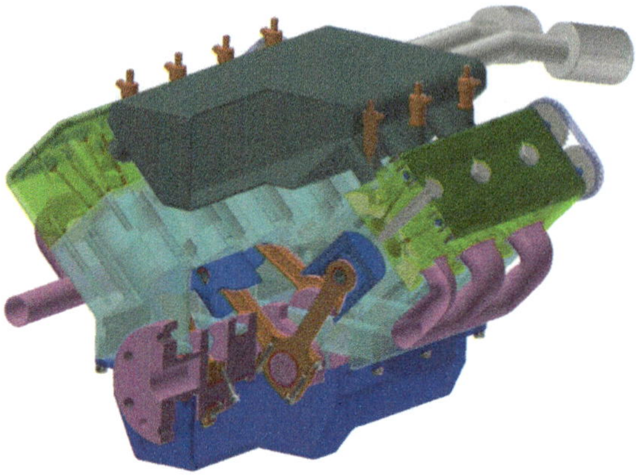

Fig. 1. Engine with 8-cylinder V configuration at 90° with 4000 cc and stroke ratio 1.

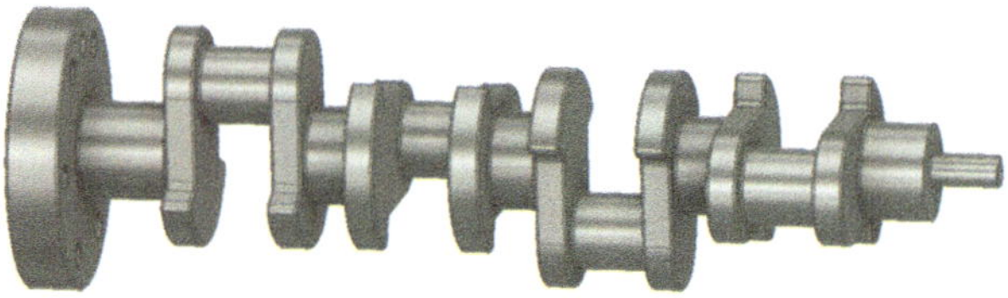

Fig. 2. Engine crankshaft in Fig. 1.

Figure 2 shows the crankshaft modeled with the specifications corresponding to the engine in Fig. 1. Together with the engine block, this is the component whose modeling is more dependent on the main parameters selected.

The geometrical parameters of the engine (some of them are shown in Table 2 and Fig. 3) depend on the main ones and are obtained from them with equations that are implemented by means of Autodesk Inventor iLogic rules.

Table 2. Most important geometrical parameters of the engine.

Parameter	Description	Units
VD	Unitary capacity	Cubic centimeters (cc)
S	Stroke	Millimeters (mm)
D	Piston diameter	Millimeters (mm)
Ap	Piston section	Square Millimeters (mm^2)
a	Crank distance	Millimeters (mm)
L	Connecting rod length	Millimeters (mm)
γ	Crank/rod ratio	–

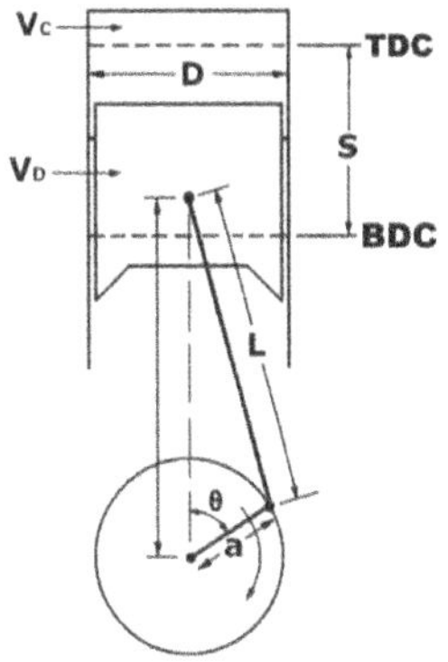

Fig. 3. Relationship between geometric parameters.

The unitary capacity (1) is the volume displaced by the piston. It is obtained from the total cylinder capacity and the number of cylinders.

$$V_D = V_T/Z \tag{1}$$

The stroke (2) is the distance from top dead center (TDC) to bottom dead center (BDC). It is obtained from the unitary capacity and the piston cross-section.

$$S = V_D/A_p \tag{2}$$

The piston diameter (3) is obtained from the stroke and the stroke ratio.

$$D = S/cd \tag{3}$$

The piston cross-section (4) is obtained from the diameter.

$$A_p = \left(\pi \cdot D^2\right)/4 \tag{4}$$

From Eqs. (2), (3) and (4) the diameter (5) and the stroke (6) are obtained.

$$D = \left(\frac{V_D \cdot 4}{\pi \cdot cd}\right)^{1/3} \tag{5}$$

$$S = D \cdot cd \tag{6}$$

The crank (7) transforms the translational motion of the piston into the rotational motion of the crankshaft. It is the distance between the crank pin shaft (where the connecting rod heads are housed on the crankshaft) and the crankshaft axis.

$$a = S/2 \tag{7}$$

The connecting rod length (8) is the distance between the crank pin axis and the central axis of the connecting rod foot. It is calculated from the parameter γ and the crank. γ is set between 0.2 and 0.4 and in this work has been used $\gamma = 0.25$.

$$L = a/\lambda = (S/2)/0.25 = 2 \cdot S \tag{8}$$

Other geometrical parameters are obtained directly from the above. Table 3 shows some of the piston parameters derived from the diameter.

Table 3. Secondary geometrical parameters of the piston.

Parameter	Usual range	Selected value
Piston height	(0.6–1.5) D	1.2 D
Distance from pin to piston head	(0.3–0.6) D	0.6 D
Segment height	(0.02–0.025) D	0.025 D
Piston pin diameter	(0.2–0.4) D	0.26 D

3 Relationships and Rules

All the components modeled in this work have been parameterized according to the main parameters, but only a few of the relationships and rules implemented will be briefly described.

The dimensions of the main engine block sketch (Fig. 4 left) are directly related to the piston, connecting rod and crankshaft. The dimensions preceded by "fx:" have been obtained by means of an algebraic relation or equation.

In Fig. 4, the framed distance is the length from the crankshaft axis to the face of the piston head when at the top dead center position (9) and it is obtained by adding the lengths of the crank (7), connecting rod (8) and the distance from the pin to the piston head (Table 3).

$$(S/2) + (2 \cdot S) + (0.6 \cdot D) = 2.5 \cdot S + 0.6 \cdot D \tag{9}$$

Figure 5 shows the parameters of a part and the dependencies implemented by means of the previous equations.

Figure 6 shows the engine configuration interface (left) and part of the source code of the rule (right) which, depending on the number of cylinders, defines the position of

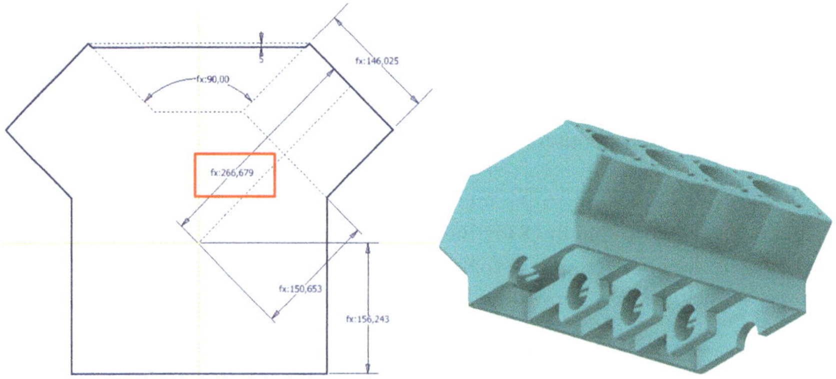

Fig. 4. 2D main sketch and full 3D model of the engine block.

Parameters	Used by	Units/Type	Equation
User parameters			
rel_cd	carr, diam	su	1 su
num_cil	cilin_u	su	8 su
mili	diam	m	1 m
diam	carr	mm	(((4 su * (cilin_u / 1000000 su) / (PI * rel_cd))) ^
cilin_u	diam	su	cilin_t / num_cil
cilin_t	cilin_u	su	4000 su
carr		mm	diam * rel_cd
ang		gr	90 gr
Model parameters			

Fig. 5. Parameters of a part obtained by equations.

the connecting rods, inserts them and adds the bearing matching constraints. The result of its application can be seen in the Fig. 7.

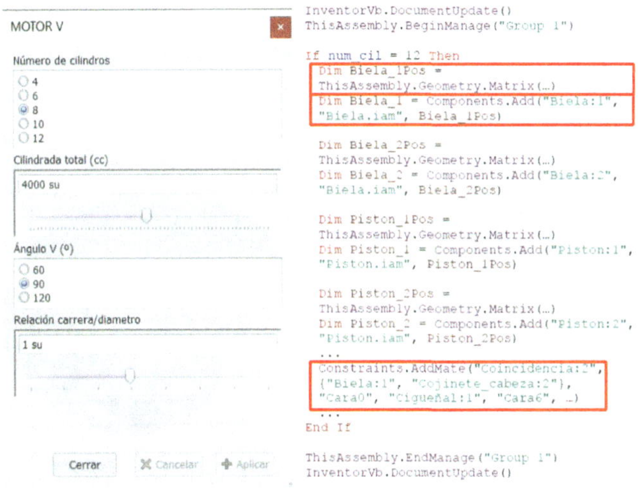

Fig. 6. User interface and source code of a rule.

4 Results

Figure 8 shows the result of two different configurations: the geometry, size and number of parts, as well as the assembly, are generated automatically according to the values selected in the application interface (Fig. 6 left). In a few seconds, two different configurations are generated.

Thanks to the time reduction, different configurations can be compared immediately and the best option can be chosen. Figure 9 shows (by dynamic simulation with Autodesk Inventor), for an 8-cylinder engine, using the same crankshaft and the same average effective cylinder pressure, how the 90° configuration presents less vibration and achieves a higher rpm.

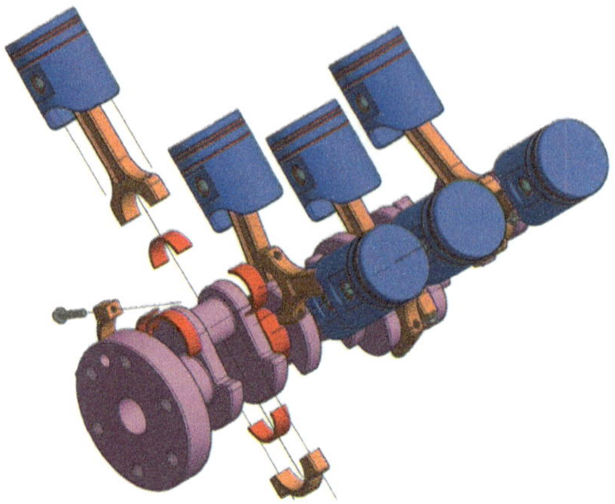

Fig. 7. Generation, location and constraint of the first connecting rod with the rule of the Fig. 6.

Fig. 8. Two different V-engine configurations (6-cylinder at 30° and 4-cylinder at 90°).

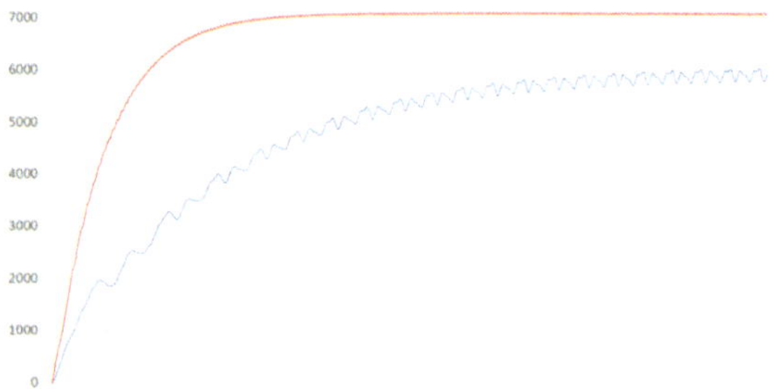

Fig. 9. Speed evolution of an 8-cylinder V-cylinder engine at 90° (red) and at 60° (blue).

5 Conclusions

This paper presents a modeling strategy that, using constraints and rules, allows to automatically generating virtual models with different configurations in order to compare them and choose the optimal one. This strategy needs a previous deeper study, but it allows creating virtual prototypes in a more efficient way without the need of repeating the modeling and assembly tasks.

As an example, a virtual prototype of a V-cylinder engine has been developed, which allows different configurations by modifying the main parameters.

The work presented shows the viability of this strategy and its potential in the field of engineering education.

References

1. Khanzode, A., Fisher, M., Reed, D.: Challenges and benefits of implementing virtual design and construction technologies for coordination of mechanical, electrical and plumbing systems on large healthcare project. In: Proceedings of CIB 24th W78 Conference, Maribor, pp. 205–212 (2007)
2. Richards, L.: Incorporating 3D modeling and visualization in the first year engineering curriculum. In: Proceedings Frontiers in Education 25th Engineering Education for the 21st Century, Atlanta, vol. 2, pp. 3c5.15–3c5.20 (1995). https://doi.org/10.1109/FIE.1995.483156
3. Hsi, S., Linn, M., Bell, J.: The role of spatial reasoning in engineering and the design of spatial instruction. J. Eng. Educ. **86**, 151–158 (1997). https://doi.org/10.1002/j.2168-9830.1997.tb00278.x
4. Barbero, B.R., Pedrosa, C.M., Samperio, R.Z.: Learning CAD at university through summaries of the rules of design intent. Int. J. Technol. Des. Educ. **27**(3), 481–498 (2016). https://doi.org/10.1007/s10798-016-9358-z
5. Buckley, J., Seery, N., Canty, D.: Heuristics and CAD modelling: an examination of student behaviour during problem solving episodes within CAD modelling activities. Int. J. Technol. Des. Educ. **28**(4), 939–956 (2017). https://doi.org/10.1007/s10798-017-9423-2
6. Salehi, V., McMahon, C.: Development of a generic integrated approach for parametric associative CAD systems. In: Proceedings of the 17th International Conference on Engineering Design (ICED 2009), pp. 145–156 (2009)

Competences Assessment and Gamification Strategies to Incentive Students

Faust Séculi, Fernando Julián(✉), F. Xavier Espinach, and Manel Alcalà

Escola Politècnica Superior, Universitat de Girona, 17003 Girona, Spain
fernando.julian@udg.edu

Abstract. Competency-based education is already adopted by higher education institutions around the world. This study presents the results of an investigation carried out with engineering students. The objective is focused on exploring aspects of innovations in teaching and learning through a competence assessment procedure and some aspects of gamification. The goal is to improve content learning, as well as the academic performance of the students. Cycloid is the ICT tool used in previous years for the assessment of competencies. The study showed some results detected as weaknesses by the student. As a prevention measure, it was decided to introduce gamification as the strategy in the subject of technical drawing in the Degree in Engineering in Industrial Technologies at the University of Girona. This contribution describes a practical experience through the use of certain aspects of gamification during the course 2019–2020, student's motivation, and acquisition of competencies. We worked with a total of 78 students. In conclusion, comparing with previous studies, the academic results obtained were positive and the students were more enthusiastic about this subject. They also collaborated to continue team working, creating a good atmosphere. They also faced the change in weaknesses observed in the assessment of competencies, positively.

Keywords: ICT · Competences · Gamification · Assessment

1 Introduction

For several years now, we have been moving from learning knowledge to learning competences, with it becoming important to be able to assess these competences. Higher education, specifically in engineering [1], shows that it is feasible to do so by basing their study on a proprietary tracking model. Other cases can be found [2].

Universities usually include various blocks of competences, divided as follows: singular, transversal, specific, and subject-specific. Boyazis [3] establishes another classification and divides them into cognitive, emotional intelligence, and social intelligence competences. Therefore, future engineers need to acquire cognitive, emotional intelligence, and social intelligence competences, which are encompassed in what universities call transversal and specialty-specific competences.

Some of our students tell us about the difficulties they have in our Graphic Expression subjects. In some cases, this difficulty is concerning spatial vision; in other cases, the

F. Cavas Martínez et al. (Eds.): INGEGRAF 2021, LNME, pp. 302–308, 2022.
https://doi.org/10.1007/978-3-030-92426-3_35

difficulty lies in sketching, or in understanding the operation of manufacturing processes of certain parts. Faced with this situation, we decided to analyze the competences of our students to be able to act based on the data obtained.

To carry out the study we worked with the Evolute platform, an ICT tool that can be found in Liikamaa [4]. Recent examples of use can be found in Koskialho et al. [5] and Imran and Kantola [6]. The Evolute platform has different applications, one of them being Cycloid, with which we can assess competences. Self-evaluation is considered an efficient method for personal development and for committing to the objectives of a project [7].

The Cycloid competence model developed and described in detail in Liikamaa [4], groups competences into two broad groups: personal competences (self-awareness, self-control, cognitive ability, and self-motivation) and social competences (empathy and social skills). It works with 30 transversal competences that have an important connection with those chosen by the Universitat de Girona, from hereon in UdG. Once the results were obtained, we wanted to motivate them to improve those competences that they themselves had identified as problematic in their training.

One way to motivate them was to introduce game-based elements to make the students feel more comfortable and encouraged. Werbach and Hunter [8] talk about the importance of play and that it should be considered in three fundamental aspects: engagement, experimentation, and results.

The use of games in a teaching context is not a new practice. The term gamification has been used frequently for many years now. According to Laja [9], 'gamification is the concept of using game design elements in non-game applications to make them more fun and engaging. The article points out that it applies to both business and education. There is a strong concept behind gamification: the main thing about gamification is that the user has a feeling of living a game experience.

Gamification in education is understood as game mechanics applied to non-game environments to achieve educational objectives. Classrooms are a great place to put this into practice.

2 Objective

The main objective of this experience was to improve learning the content as well as students' academic performance. To this end, the process employed consisted of exploring certain aspects of teaching and learning through a competence assessment procedure and, based on the data obtained, applying solutions, in this case focusing on certain aspects of gamification in the subject. To incentivize them and make this process easier for the students, it was decided to introduce certain elements that have rarely or not been used before in an attempt to increase student motivation, thus facilitating the acquisition of competences.

The process, which included the introduction of some gamification aspects, consisted of trying to improve the learning methodologies of the subject content and the idea of gamification. Use of the gamified experience aimed to motivate our students. Possible secondary objectives could be to improve the classroom climate, improve academic performance, and promote cooperative work.

Our research group carried out a teaching experience in gamification work during the last academic year, considering the basic ingredients defined by Werbach and Hunter [8] to carry out the gamification process, and changing the assessment methodology, such as a game. The game could bring excitement, enthusiasm, and optimism. The main changes introduced in this case were cooperative work and gamification.

The main idea was always to favor the learning of subject content, working in groups to make the work more cooperative. It was important to have a good classroom climate and to get our students to improve in the different competences, analyzed in the creative tension obtained from the Cycloid ICT application.

3 Innovation Development

The present study was conducted with first-year Industrial Engineering students in the 2019–2020 academic year, during the first semester. 78 students were participating in the study, divided into two classroom groups. None of the groups was considered as a control because there was no previous data.

The self-assessment tool for this research consisted of 120 questions that were evaluated based on a fuzzy logic procedure, on the current level and the target level. For each competence, the learner measured their current level and the level they think they should reach. The difference between these two levels (the necessary or desired, and the real) is the creative tension [10].

In this proposal, we raised the use of gamification. For its implementation, we started by looking at the experiences of some authors such as Lee Sheldon [11]. The interesting point of Sheldon's work lies in the fact that he explains his experience over years of teaching and recounts both successes and failures.

Concerning the game elements that can be applied to a gamified experience, there are two that stand out for being the most widely used by various authors. The first approach is that proposed by Werbach and Hunter [8] who organize game elements into three categories: dynamics, mechanics, and components. The second approach is described in Hunicke, Leblanc, and Zubek [12], who define three types: dynamics, mechanics, and aesthetics. This study is based on this second approach, as it is considered to be more comprehensive since it defines the concepts of mechanics and dynamics in a broader way and without limiting them to a closed list, as in Werbach and Hunter's proposal.

Aesthetics-narrative: This refers to the graphics, music, and story that is explained in the game. It is everything that the player perceives and that gets them hooked to the game. Dynamics: This consists of defining how the game will work; how the player behaves during the game. What they can do based on what the game mechanics will allow them to do. Mechanics: This refers to the various control actions offered to the player within a game context. These are the basic rules of the game. It defines how the game will be developed in terms of parameters such as scoring, ranking, or levels.

Afterward, a questionnaire was given to the students to assess various aspects of gamification. The questionnaire contained several questions with various items that could be rated on a scale. 1- Did you find it useful? 2- What did you value most about the experience? 3- What would you improve? A total of 216 responses were received from the 78 students who responded to the questionnaire. This figure, which is not very high, makes the results difficult to translate.

4 Results

Based on the data collected, we proceeded with an analysis (Fig. 1). Regarding the two main groups of competences divided into personal and social competences, the greatest number of responses for strengths corresponded to the group of social competences. As for weaknesses, the greatest number corresponded to the competences considered to be personal.

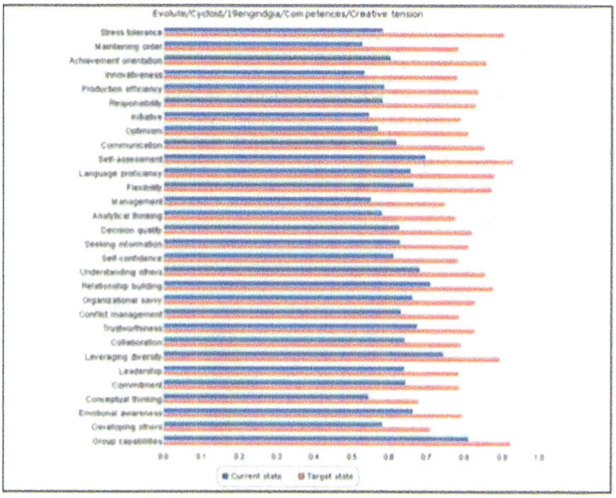

Fig. 1. Example of the result obtained with one of the groups. Source: Cycloid-Evolute.

In terms of creative tension, the biggest gap was also in personal competences. Regarding the sub-classification of personal competences (self-awareness, self-control, cognitive ability, self-motivation) and social competences (empathy, social skills) into six groupings, the most outstanding ones (those we consider strengths) were in Empathy and Social Skills, mainly within the group of competences considered to be social, as pointed out above. As for weaknesses, and with the greatest creative tension, the data we obtained was in the competences of Self-control, followed by Self-motivation, both within the group of competences considered to be personal. The proposal focuses on competences related to motivation, due to the results observed. We proceeded to implement some of the concepts of gamification within the subject of Graphic Expression within the studies of the Industrial Engineering Degree. During the 2016–2017 course, the researchers made a similar study adding gamification aspects [13]. The first step was to explain to the students the different parts of the game.

Narrative-Aesthetics: The students were to be engineers working in an international company. The ultimate goal was to create a robot that could research the planets. Some would have to design certain parts and the rest, others. At no point is the robot model shown, as it is a secret project and they cannot know what the other groups are doing. The professor indicates the requirements of each phase. As the project progresses, the level of difficulty of the following sections increases. They are told that they will start the ideas with sketches, which will then undergo a review process by the supervisor (in this case the professor), to continue with the proposal in the Solid-Works program.

Dynamics: This is explained after the theoretical concepts, in this case drawing of assemblies and exploded views. It sets out how the teams are to be set up, the phases involved, the delivery dates, and the assessment systems. The professor tells each group the object they have to design and then attach to the robot. They are provided with concrete specifications. The initial difficulty faced by the students focuses on proposing imaginative solutions and representing the different projections of the proposed mechanical parts. Thanks to the professor's in-class tutorials, this can be solved. The problem of group synergy lies with the leaders, who will be assisted by the professor in case of conflict.

Mechanics: The game begins. They now know the deadlines and the additional points they can earn in the final assessment of the subject. Two exercises were proposed to the whole class. On the one hand, different parts with threads are presented, and they are asked to put together the assembly. An assembly drawing of a simple valve is also proposed and the exploded view of each part is requested.

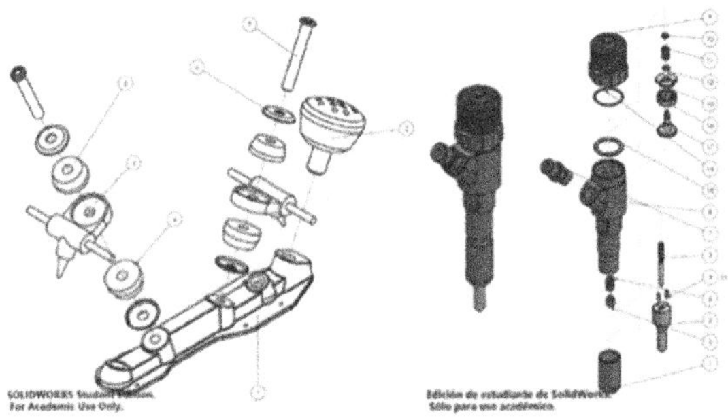

Fig. 2. Proposals of two teams Source: UdG students.

From here comes the first evaluation to define the group leaders. Each classroom group has 39 students. Eight groups of five students each are chosen. The five top-performing students per class become group leaders. The leader is the head of the technical team and the highest authority and shall ensure the smooth running of the team. They also have the last word. From this moment on, leaders have an additional point in their final marks of the course. They will have to hand in proposed solutions on a designated day, which are then discussed. The professor assesses the results of each group. Teams that have correctly fulfilled the requirements will be compensated with half a point more for each component in their final marks of the course. As for the second deadline, there is no reward in points (Fig. 2).

The professor gives each team new requirements, this time more complex. When the professor approves the newly submitted proposals, they request that the different groups change cards, so that each team receives the proposal of another group. The exchange is carried out by the professor. Now, each team must make their proposal adapt to the other team's proposal and follow a certain logic, making different parts of the components fit together. These proposals are shown once they have been defined by the other group. Both groups must give their approval or improve them. From this moment on, leaders have an additional point in their final marks of the course.

In a first analysis of the data received, certain relevant trends can be discerned, which are described in the following section. The game began by grouping the students, with a leader for each group. Four students joined each group. We observed students who did not know where they should be, partly due to being in their first year and not knowing each other well enough. From the proposed experience some of the students made new friendships and can be seen together. On the other hand, competition between the different groups was important. They proved to be competitive. The project lasted four weeks. The questionnaire was then given to the students to assess the experience.

Regarding the question, 'What would you improve?' the answers focused on the request for more time; to be able to visualize the robot at the beginning, and to be able to see the robot with the parts they had designed. They also told us that, in the beginning, they lacked the software classes to carry out the proposals they wanted, and that these were appropriate to the difficulty of handling the program.

5 Conclusions

The Cycloid ICT tool shows the results that our students detect as weaknesses in their transversal competences, as well as creative tension: the competences of Self-control, followed by Self-motivation. As an improvement measure, it was decided to introduce gamification as a way of boosting the motivation of our students in the subject of Graphic Expression. The study presented has pursued a twofold objective: Firstly, to improve learning the content as well as students' academic performance. Secondly, to collect and analyze the opinions of the students who have experienced the project to determine whether gamification favors student motivation.

The results obtained allow us to affirm that, with the limitations offered by the sample, the first objective has been somewhat fulfilled, since the results of the qualifications have been improved concerning other academic years. Regarding the second objective, we can affirm that it has been fulfilled and, therefore, that the use of certain aspects of gamification in a subject offers benefits for student learning.

The research opens interesting future lines of research like the temporal value of collected data and its resilience against cultural change.

References

1. Vargas, H., et al.: Automated assessment and monitoring support for competency-based courses. IEEE Access **7**, 41043–41051 (2019)
2. Plua, J.J.L., Betancourt, L.A.A., Díaz, R.T.: Estrategia de evaluación de competencias profesionales en estudiantes de Ingeniería en Sistemas. REFCalE: Rev. Electrón. Form. Calid. Educ. **5**(2), 17–32 (2017). ISSN 1390-9010
3. Boyazis, R.: Competencies in the 21th century. J. Manag. Develop. **27**(1), 5–12 (2008)
4. Liikamaa, K.: Piilevä tieto ja projektipäällikön kompetenssit (Tacit Knowledge and Project Manager's Competences), PhD Thesis, Tampere University of Technology, Publication 628 (2006)
5. Koskialho, J., Einolander, J., Vanharanta, H.: Commitment and motivation in professional organization. In: Kantola, J.I., Barath, T., Nazir, S., Andre, T. (eds.) Advances in Human Factors, Business Management, Training and Education. AISC, vol. 498, pp. 47–58. Springer, Cham (2017). https://doi.org/10.1007/978-3-319-42070-7_5

6. Imran, F., Kantola, J.: A co-evolute approach to analyze the competencies of sales personnel of banking sector of Pakistan. In: Kantola, J., Barath, T., Nazir, S. (eds). Advances in Intelligent Systems and Computing, vol. 594. Springer, Cham (2018). https://doi.org/10.1007/978-3-319-45991-2
7. Bikfalvi, A., Rafart, A., Mancebo, N.: The assessment of competence in University adult education: a dynamic approach. Int. J. Manag. Knowl. Learn. **2**(1), 3–23 (2013)
8. Werbach, K., Hunter, D.: The Gamification Toolkit: Dynamics, Mechanics, and Components for the Win. Prensa Digital Wharton (2015). http://blogs.cccb.org/lab/article_gamificar-vol-dir-fer-jugar/
9. Laja, P.: How to use gamification for better business results (2012). https://blog.kissmetrics.com/gamification-for-better-results/. Accessed 21 Feb 2012
10. Senge, P.M., Scharmer, C.O., Jaworski, J., Flowers, B.S.: Presence: an exploration of profound change in people, organizations, and society. Currency (2005)
11. Sheldon, L.: The Multiplayer Classroom: Designing Coursework as a Game. Course Technology/Cengage Learning, Boston (2012)
12. Hunicke, R., Leblanc, M., Zubek, R.: MdA: a formal approach to game design and game Research (2004). http://www.aaai.org/Papers/Workshops/2004/WS-04-04/WS04-04-001.pdf#page=1&zoom=auto,-65,798
13. Espinach, F.X., Aguilar, M.D., Tarrés, J., Vilavella, M.A., Pérez, F.J.: Gamification as a methodology to incentive students. In: Edulearn 18. 10th International Conference on Education and New Learning Technology: Palma, 2nd–4th of July, 2018 (2018)

Learning 3D Design Applied to the Recovery of Historical Aeronautical Heritage

Laura García-Ruesgas[1(✉)], Francisco Molero-Garrido[2], and Ignacio Rosales-Silván[1]

[1] Graphical Engineering Department, University of Seville, 41092 Seville, Spain
lauragr@us.es
[2] Ryanair, 41020 Seville, Spain

Abstract. The learning of industrial design programs allows multiple applications in practice. For more than 15 years, one of the work lines of the Department of Graphical Engineering of the University of Seville is to deepen the 3D design by modelling and virtual recreation of all types of old industrial machinery, as well as aircraft, thus digitally recovering the industrial and aeronautical historical heritage. The objective is to achieve a final product starting from very scarce information such as sketches, photographs or a very limited number of plans. Thus, students need to study the industrial and historical situation of a specific period of time from which they must obtain all kind of information. Sometimes being in contact with museums is fundamental and in those cases where they work with post-industrial revolution mechanisms, it is an obligatory step to visit companies or retired professionals. They are invited to propose the machine or aircraft and to review the material and tools used in the selected era. The final products usually include photorealistic images or improvements. More than 60 projects have been submitted. Two of them are presented in this article: Sopwith triplane and Mil v-12 helicopter.

Keywords: Industrial graphic design · Interactive 3D modelling · Antique historical aircrafts · CATIA

1 Introduction

Computer-aided design (CAD) allows designers and engineers, among others, to create virtual models that attempt to faithfully reproduce the main characteristics of a product, such as the size or shape of each of its components, while also making it possible to simulate virtual images. On the other hand, it facilitates the designer the task of modifying the initial design with simplicity as the development of the final product progresses.

In this way, CAD/CAM systems help to improve the design, development and manufacture of products, as well as significantly reducing costs and increasing productivity [1].

These systems began to be used in the early 1970s in very specific sectors such as the automotive or aeronautical industries, due to the economic unfeasibility of a

F. Cavas Martínez et al. (Eds.): INGEGRAF 2021, LNME, pp. 309–317, 2022.
https://doi.org/10.1007/978-3-030-92426-3_36

high percentage of companies to invest in this type of technology. Currently, its use is widespread throughout the industry in general and as a result of its extraordinary importance in the business environment, many universities have opted to introduce the learning of this technology in the curriculum of some of their degrees, considering this training essential for the professional future of their students. At the Escuela Técnica Superior de Ingeniería de Sevilla, Computer Aided Design and Manufacturing is taught in its multiple degrees, among which is the degree of Aerospace Engineering [2]. A more in-depth study of the contents of this subject is carried out in the Final Degree and Master's Degree Projects.

In particular, CATIA, standard software in the aeronautical industry in Europe, the United States and Canada [3], is used in Aerospace Engineering studies. Since 2002, the Department of Graphical Engineering of the ETSI in Seville has 41 CATIA licenses. In other degrees, there is an unlimited number of campus licenses of Solid Edge and Siemens NX programs, among others. Currently, CATIA is considered one of the most powerful tools for CAD file processing, providing support from design conception to production and product analysis, being used by some of the most important engineering companies in the world such as Boeing, Airbus, Bombardier, Embraer or Dassault Aviation.

Our students have access to complete modules that integrate all the necessary tools for the design, analysis, simulation, presentation, manufacturing, production and maintenance of products, being possible to combine different modules. It is common to start using those modules that facilitate the design of components in 3D to continue with those that allow their assembly in 3D or make it possible to obtain 2D drawings of the designed components. Subsequently, they learn to handle the modules that analyze and simulate the machining of parts, those in which a finite element analysis or electrical, electronic or piping designs are performed, and even the ergonomic analysis required in certain cases.

Another important feature of this software is the extensive library of materials available, from wood or stone to composite materials, which are increasingly used in the aeronautical industry.

This strong economic investment by our School in the acquisition of specific software, coupled with a strong commitment to project-based learning, is rewarded by the high level of training of our students, which is confirmed by the satisfaction shown in the aeronautical industry and its ancillary companies [4].

In undergraduate studies, learning how this software works is limited due to lack of time and there are many students who demand a more exhaustive knowledge. This is the main reason why we are starting to propose the development of a type of projects in which the use of a greater number of modules of this software is encouraged, thus offering students new design challenges that allow them to enhance their creativity.

These projects must be able to be executed in a limited time and must depend solely on the students and their corresponding tutors.

As a result of this learning, students acquire more than enough skills to perform virtual modelling and recreations [5], not only oriented to their future and immediate professional activities, but also to others related to the rescue and cataloguing of the historical aeronautical heritage [6], which is currently very neglected, since it does not represent a remunerated professional activity for a group of engineers whose professional future is guaranteed in the aeronautical production industry.

For the reasons mentioned above, the aeronautical heritage, unlike others such as architectural or industrial heritage, has not been historically addressed. In particular, in Seville, the only Spanish city where aeronautical production activity has been uninterruptedly developed during the last century [7], it is striking that there is not even a museum related to this subject.

During the last years, several professors of the Department of Graphic Engineering of the University of Seville have tutored projects related to the recovery of the historical aeronautical heritage by means of 3D modelling and virtual re-creation [8], being the total number of projects carried out by the authors of this work close to 15, out of a total of about 30 in the whole department. Most of these projects can be consulted in the library of the University of Seville, since they must have obtained a high score to be housed in the repository. All types of old aircraft have been modelled, finding a large majority belonging to the First and Second World War.

2 Methodology

The methodology used is shown graphically in Fig. 1. The first step is to perform an "Information Search" for an old aircraft for which no digitized information is available. The time range is not too wide, since it can start with the beginning of the first known aircraft and end in the early 1970s, when CAD was introduced in the industry.

The period of invention of a given aircraft is the one that defines the objectives of the project. On the one hand, when working with older aircraft, the starting information is usually very scarce, starting with simple sketches and some images. In these cases, the student must develop his creativity and adapt to the materials and tools of the moment. In those cases, where the student works with aircraft closer to the end of the first half of the twentieth century (40s and 50s), it is possible to obtain some support plane on which to measure and from which it is possible to select the most appropriate materials, although the number of elements in these cases is greater.

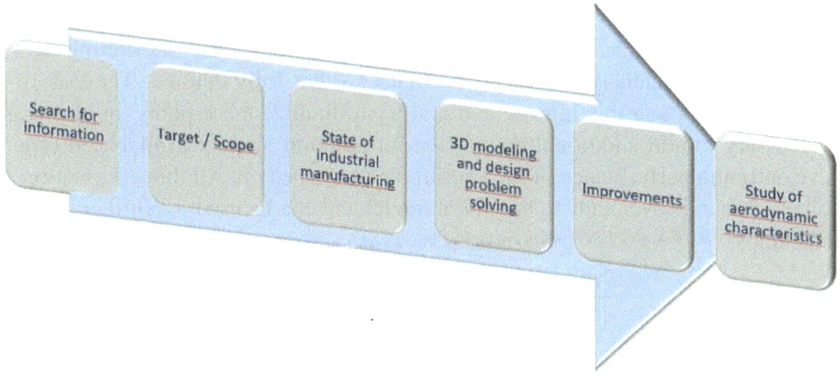

Fig. 1. Methodology.

Therefore, it is necessary to study the mechanical designs, the available manufacturing machines, as well as the materials. The next step is the design and three-dimensional virtual recreation, as well as detecting possible design errors that do not appear in the sketches, plans, images or any other type of information available, such as replicas in museums. Students should be able to analyze and correct these errors. Sometimes, the corrections imply improvements in the definition of the aircraft.

2.1 Search for Information

The biggest problem faced by students when tackling a project of this type is the difficulty in finding useful information. Usually, the available information is very scarce and, except in extraordinary occasions in which the students go to a museum or a company and are lucky enough to get information about the aircraft to be modeled, the usual thing to do is to carry out exhaustive searches on the net, in forums, go to private collections or even old books. There are those who, due to the impossibility of obtaining any plane of certain quality, opt to buy models of the aircraft in question, and work from that information.

2.2 Aim/Scope/State of Industrial Manufacturing

The scope of the projects depends on a large extent on the information available for the particular aircraft. If the information available is only images, the objective will be to obtain a three-dimensional model as faithful as possible to those images. However, if the source of information also includes drawings with measurements, other types of studies can be carried out by the student, such as the study of aerodynamic characteristics by means of computational fluid simulations. Static and dynamic interference analysis tools are also usually studied.

On the other hand, obtaining a model with acceptable reliability requires the correct selection of the most appropriate materials and the precise application of non-standard measurements in specific mechanisms such as gears or screws.

2.3 3D Model and Design Problem Solving. Enhancements

All the aircraft studied lack complete information, either because the starting data are only drawings or sketches or because the plans are not fully defined. The task of the students is to complete this information and try to finalize these plans. In most cases, it is necessary to add additional data to the information already available in order to achieve optimum performance of the aircraft being worked on. All this information will be collected as improvements to the initial model and can be used for further studies of various kinds.

3 Results

This section includes, as an example, two of the virtual recreations that have been carried out and tutored by the authors of this work.

3.1 Example 1: Mil Mi-12 Helicopter

The Mi-12 project, a helicopter which was built at the height of the Cold War, was initiated in 1965 with the objective of producing a vertical take-off aircraft capable of transporting heavy nuclear missiles and resources from Siberia, presenting a clear technological superiority over other aircraft [9, 10].

Due to its large dimensions, its two engines are located on the sides and consist of twin intakes, driving five-bladed metal rotors that ensure better reliability, fatigue life and stability. Interestingly, the use of counter-rotating twin main rotors eliminated the requirement for a tail rotor, present on most helicopters as it can be seen in Fig. 2. Of note is the presence of a plethora of rods attached to the landing gear [11]. This helicopter received a total of eight world records in 1969, mostly related to the ability to lift a certain weight to a certain height.

The project began with an exhaustive search for information on the web, where two plans were found, with more than acceptable information regarding the general outside structure, but not at the detail level, making it necessary to apply proportionality in some dimensions [11, 12].

The design of this helicopter was divided into four sections which were the fuselage, nose, tail and stabilizers in the third division and finally the wing assembly, engines and any other details of the structure, such as fuel tanks or rods. First, the external structure was modeled followed by the internal structure and finally the assembly of each of the parts.

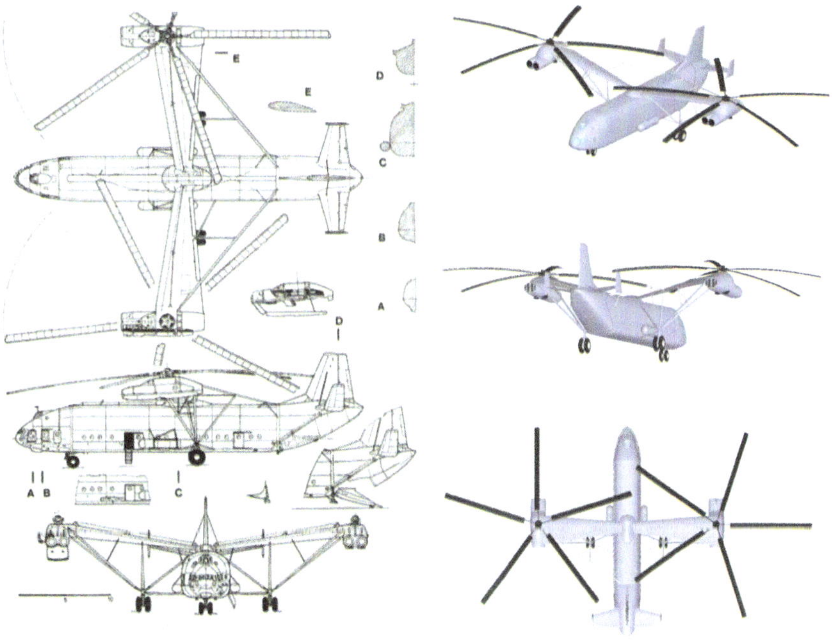

Fig. 2. Drawings of aircraft's outside (left) and final modeling (right).

The information about the inside of the aircraft was minimal, so all the designs related to these areas were made based on the observation in photographs [13], by approximations and estimations, being the images shown in the upper part of Fig. 3 the two main ones. Nevertheless, the results obtained were very accurate in the vast majority of cases.

The final result is shown in the lower part of Fig. 3.

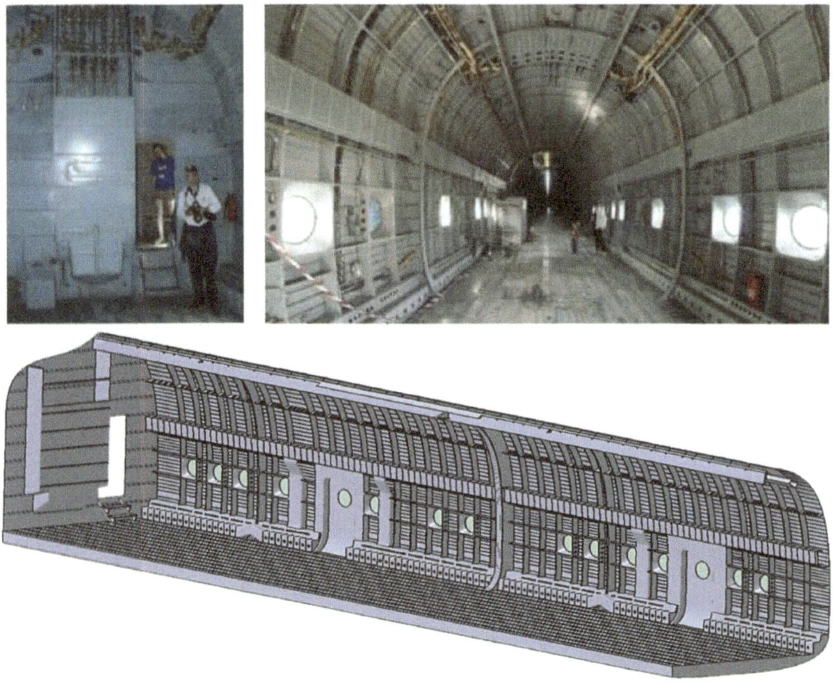

Fig. 3. Images of aircraft's inside (top) and final modeling (bottom).

Fig. 4. Rendered image of final model.

Finally, several rendered illustrations were generated to simulate the actual helicopter in flight (Fig. 4).

3.2 Example 2: Sopwith Triplane Aircraft

In this second example, the British Sopwith triplane fighter was designed as it was one of the most important aircraft during the First World War. The Sopwith triplane succeeded in granting war superiority to the Royal Air Force due to its fast climb and turn rates [14]. A surface area three times larger than that of all other fighters made the aerodynamic efficiency of the tri-plane noticeably greater than that of its rivals, compensating for the excess weight of the third wing. It is corroborated that the vertical speed of the triplane corresponding to horizontal speed of 60 mph is much higher than the fighters of the time [15]. On the other hand, its turning speed is not the highest, but considering the combination of climb and turning speed, the Sopwith triplane is superior to its contemporary fighters.

Few drawings of this aircraft exist, probably due to the fact that it was the first aircraft to be built with three wings. After a long and complex research process, two plans were obtained that were quite similar to the existing photographic material of this aircraft. These plans could not be downloaded in original format, so it was necessary to obtain their scale (Fig. 5). Regarding the photographic images, it was possible to find good quality material. Its specification table was also available. The modelling process followed the following order: main structure, depth and steering rudder, uprights with slats and ribs, wings, joints between structural parts, assembly and, finally, fuselage and covering of the aerodynamic surfaces. The latter were made of stiffened fabric with

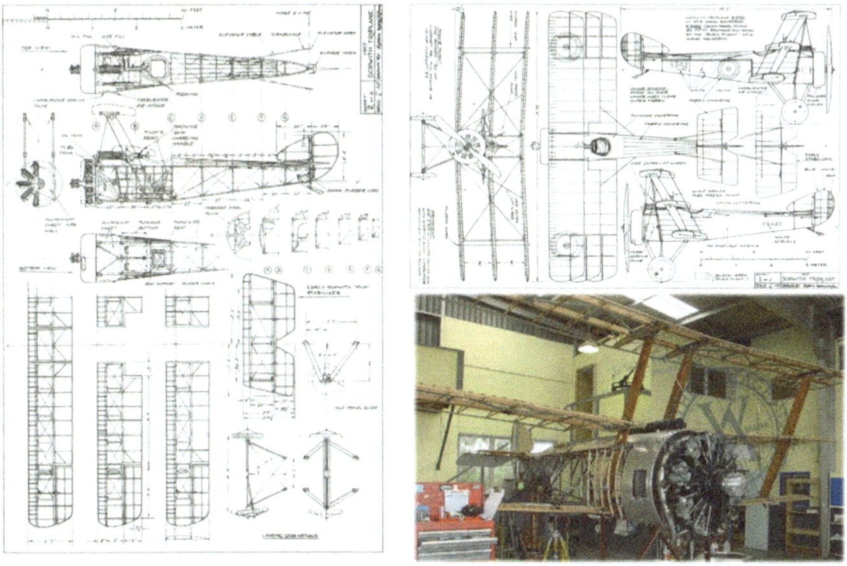

Fig. 5. Drawings and simple image available.

varnish, to which a strong layer of paint was applied, so that the approximate thickness of this "composite material" was one millimetre.

The materials used in the manufacture of this aircraft were steel and wood, specifically birch, which gives the fuselage habitability and good aerodynamics, with continuous, smooth and fuselated shapes (Fig. 6).

Fig. 6. Final modelling with materials (left) and fuselage (right)

Once the modelling was completed, the flight ceiling of the Sopwith was determined at an altitude of 2000 m and the aerodynamic characteristics of the triplane were studied by performing a flight at this altitude (Fig. 7). The two mechanisms of lift generation inherent to aerodynamics (Bernoulli's Principle and Newton's 3rd law) were verified. Also, it was corroborated that the Sopwith has an aerodynamic efficiency within the typical range of the fighters of the time.

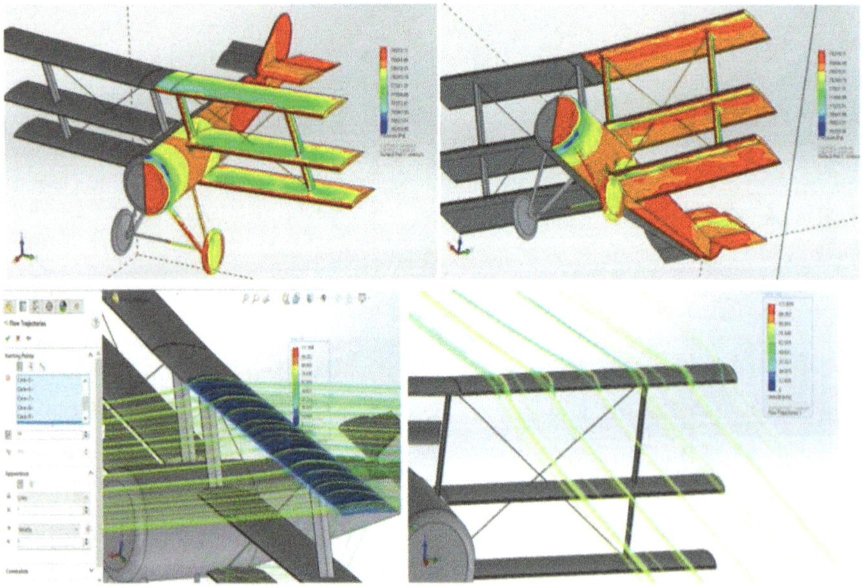

Fig. 7. Aerodynamic characterization of pressures (top) and velocities (bottom).

4 Conclusions

Since 2002, more than 30 students have chosen this type of project as their final project to complete their studies. The challenge of designing aircraft that have never been manufactured or that, if they have, do not exist or are incomplete, is enormously attractive, as well as carrying out simulations to check the flight behavior of the modeled aircraft, thus obtaining an exhaustive aerodynamic description. Motivation is a fundamental pillar in education and this type of projects linked to 3D design, with a large self-learning component, are clear examples.

References

1. Lorenzo-Yustos, H., et al.: Towards complete product development teaching employing combined CAD - CAM - CAE technologies. Comput. Appl. Eng. Educ. **18**(4), 661–668 (2010)
2. Morales-Palma, D., Martínez-Donaire, A.J., Centeno, G., Vallecano, C.: Teaching experience for the virtualization of machine tools and simulation of manufacturing operations. J. Mater. Sci. Forum. **853**, 79–84 (2016)
3. Hürlimann, F., Kelm, R., Dugas, M., Oltmann, K., Kress, G.: Mass estimation of transport aircraft wingbox structures with a CAD/CAE-based multidisciplinary process. J. Aerosp. Sci. Technol. **15**, 323–333 (2011). https://doi.org/10.1016/j.ast.2010.08.005
4. Aguilera, C.M., Castañeda, A., Guerrero, F.: Past, present and future of the Andalusan aeronautical cluster. J. Netw. Centr. Collab. Supp. Framew. **224**, 583–590 (2016)
5. Ghiţescu, I.-M., Scutaru, M.L., Ghiţescu, M., Borza, P.N., Marin, M.: New command mechanism of flaps and wings of a light sport aircraft. Symmetry **13**(2), 221 (2021). https://doi.org/10.3390/sym13020221
6. Fix, P.D.: From sky to sea: the case for aeronautical archaeology. The Oxford Handbook of Maritime Archaeology (2012). https://doi.org/10.1093/oxfordhb/9780199336005.013.0043
7. Martín-Caballero, J.: La industria aeronáutica en Sevilla en el siglo XX: el caso de industrias subsidiarias de aviación (1938–1966) (2018). https://idus.us.es/handle/11441/92802
8. Ross, W.A., Aukstakalnis, S.: Virtual reality: potential for research in engineering education. J. Eng. Educ. **83**(3), 287–291 (1994)
9. http://www.aviastar.org/helicopters_eng/mi-12.php
10. http://www.aerohispanoblog.com/leviatanes-rusos-iv-mil-mi-12-homer
11. http://fdra.blogspot.com/2015/02/helicoptero-pesado-la-mole-homero-urss.html
12. https://caza-y-ataque.blogspot.com/2015/09/mil-mi-12-homer.html
13. http://pilotspost.com/arn0001996
14. Frank, N.: Sopwith Pup Aces of World War 1. Osprey Publishing, USA (2005)
15. http://www.aviation-history.com/sopwith/triplane.htm

Graphic Interpretation of Surfaces with the Support of Augmented Reality as a Training Complement in Engineering Studies

Dolores Parras-Burgos(✉), Mar Melgarejo-Torralba, Francisco J. F. Cañavate, and Daniel G. Fernández-Pacheco

Department of Structures, Construction and Graphic Expression, Universidad Politécnica de Cartagena, 30202 Cartagena, Spain
{dolores.parras,mar.melgarejo,francisco.canavate, daniel.garcia}@upct.es

Abstract. Currently, some difficulty has been detected in engineering students when interpreting graphical representations of surfaces. The combined use of new technological tools together with mobile devices involves a very interesting option to support academic training in drawing subjects. This paper presents the ARTID (Augmented Reality for Teaching, Innovation and Design) tool used for the representation of mechanical components in engineering studies, which has allowed the development of various teaching materials for the representation of surfaces in the Dihedral system. For this purpose, QR codes are used to accompany the drawings of different surface exercises, increasing students' interest by using mobile devices to visualize in 3D the models studied. The response of the students to the use of these complementary contents has been very satisfactory in the Industrial Design II course of the Degree in Mechanical Engineering at the Technical University of Cartagena. The evaluations have been collected in a questionnaire to address the conclusions of the experience.

Keywords: Surfaces · 3D modeling · Spatial vision · Dihedral system · Augmented reality

1 Introduction

In the teaching field, augmented reality is acquiring special relevance in different subjects and educational levels [1–5]. The visualization of 3D objects in space is one of the main challenges faced by engineering students in their academic training. The rapid evolution of technology, together with the growing popularity and easy access to smart mobile devices, make possible the development of new opportunities that can serve to improve the quality of teaching [6, 7], achieving greater involvement and learning levels in students [8] and high motivation in teachers who use them [9]. In this sense, it has been proven that the use of this type of technology in the classroom and outside can

F. Cavas Martínez et al. (Eds.): INGEGRAF 2021, LNME, pp. 318–326, 2022.
https://doi.org/10.1007/978-3-030-92426-3_37

increase the understanding and motivation of students in technical drawing subjects in engineering degrees [10–12].

The advantage of augmented reality is to allow students to zoom in, zoom out and rotate the 3D models, perceiving the content in a more attractive way and motivating them to learn. From the resources analyzed, it is worth mentioning those that use augmented reality by scanning QR codes with mobile phones [13], such as the work performed in an aquarium museum where users could express, create and interact with their creativity through coloring activities [14], or the development of board games to enhance and support the learning of social sciences and mathematics in primary education [15].

Within the field of Graphic Expression, spatial vision is an ability that in most cases needs to be trained. Previous studies have shown that the use of various technologies, including augmented reality, can favor this skill in STEM (Science, Technology, Engineering and Mathematics) studies [16–21]. In order to facilitate the understanding of graphical representations of dihedral surfaces, a mobile application called ARTID (Augmented Reality for Teaching, Innovation and Design) has been developed to visualize 3D virtual models using augmented reality, which has been used successfully in fields such as agronomy [21]. The aim is to use this type of tools inside and outside the classroom, to promote understanding and motivation of students in different subjects within these degrees. The experience shown in this study has been carried out in the Industrial Design II subject of the Degree in Mechanical Engineering at the Technical University of Cartagena during the 2020/2021 academic year.

2 Materials and Methods

2.1 ARTID Application

For the development of the ARTID application, different software have been used to obtain an easy-to-use and open access tool (Fig. 1). Its development process can be divided into five characteristic phases:

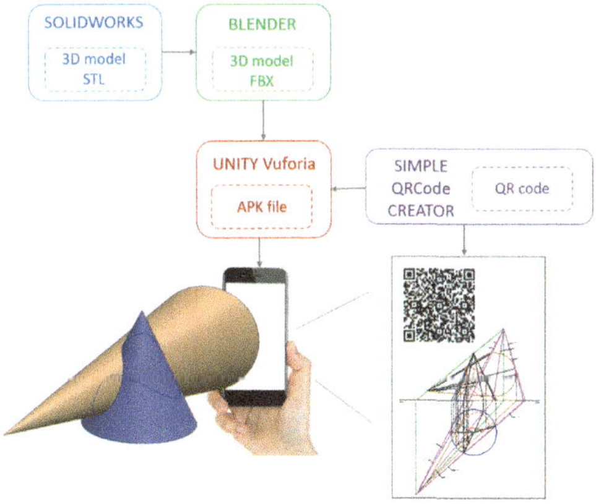

Fig. 1. Equation formatting.

a) **Creation of QR stamps to scan and reproduce the surfaces**: the stamp corresponds to the image to be scanned through the mobile phone to project the piece in augmented reality. For the creation of the QR code, different options can be used, as for instance web pages or applications. In this work, the *Simple QR Code Creator* application was used for this task. It is a free and easy-to-use application obtained directly from the Microsoft Store. To use it, it was only necessary to enter a text and the encryption level, which was set to high.
b) **3D modelling of the parts**: in a second stage, the 3D modelling of the parts is performed with the aid of the 3D Design and Modelling software SolidWorks v2019 (Dassault Systèmes).
c) **Exporting the 3D model to FBX format**: in order to import the 3D models into Unity software (Unity Technologies, licensed software that can be used for educational and non-profit purposes), they must be in FBX format. To do this, the 3D models are exported to STL format from SolidWorks and then imported into Blender (free multiplatform software). Finally, they are exported to FBX files, which are necessary to work with augmented reality. If models are directly drawn in Blender, the first conversion is not be necessary.
d) **Joining the stamps and the pieces within the Unity platform**: Vuforia is an extension of the Unity program that allows working with augmented reality. In this stage there are two work phases: first, the stamps created in the previous stage are uploaded, using the Vuforia website (https://developer.vuforia.com), generating a file that must be added to Unity; and, secondly, the file generated in Vuforia must be run with Unity executed and adding the stamps that have been chosen. Then the *ARCamera* option is selected through the *GameObjects* tab, and the stamps (*ImageTarget*) are added and, later, the 3D models.
e) **Generating the application**: a series of procedures are necessary to generate the application. Initially, the platform on which it will be executed. In this case, the Android operating system was chosen due to its accessibility to a greater number of users. Once the platform is selected, the settings for the mobile device are modified in the *Player settings*, activating *Vuforia Augment Reality* and indicating the name of the application. Once all these steps have been completed, the *Build* button must be clicked to generate the ".apk" file that will be later installed on the mobile device.

2.2 Contents Used in the Experience

The theoretical-practical contents of the Industrial Design II course of the Degree in Mechanical Engineering are based on the analysis of technical surfaces. The study of incidences, tangencies, plane sections and intersection of surfaces can be highlighted among the exercises developed in class. After explaining in class the different procedures for each of the processes, the students are provided with templates to solve each of the exercises. Figures 2 and 3 show some examples of exercises accompanied by the QR code for their visualization with the ARTID tool. This code is available to students to visualize, if necessary, the 3D model of the elements of the plan, and thus facilitate their understanding in order to solve it.

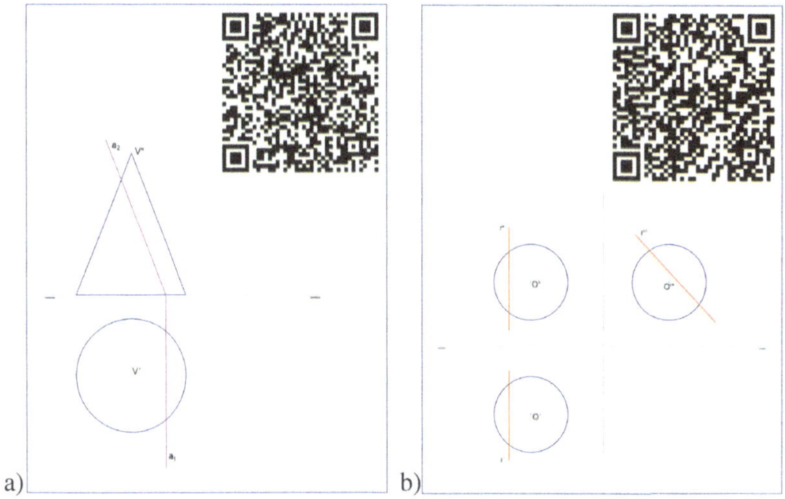

Fig. 2. (a) Exercise to obtain the plane section produced to a cone by a projecting plane. (b) Exercise to obtain the points of incidence between a straight line and a spherical surface.

Fig. 3. (a) Exercise to obtain the conditional planes tangent to a cylindrical surface (b) Exercise to determine the intersection curve between two conic surfaces

2.3 Participants

The experience was developed with the aid of 76 students of the Industrial Design II course of the Degree in Mechanical Engineering at the Technical University of Cartagena during the academic year 2020/2021. The participation in the study was offered as a voluntary practice/exercise for the students and an informed consent was obtained from all subjects of the study. The experiment was developed through the TEAMS platform, interacting with the students via videoconference to explain the process of downloading and installing the ARTID tool, as well as how to work with the proposed exercises to visualize the 3D models. The students were able to download the application from the website www.artid.es where it is intended to be a resource for students on augmented reality and technical drawing.

3 Results

Figure 4 shows the process of visualization and resolution of one of the exercises proposed in the experience carried out. Once the application has been installed on the mobile device, it is enough to open it and focus on the QR codes using the camera. Once the application has recognized the QR code, the related 3D model is displayed, which can be viewed from different angles to obtain diverse perspectives. Figures 5 and 6 show other examples of visualization of dihedral exercises with the help of Augmented Reality.

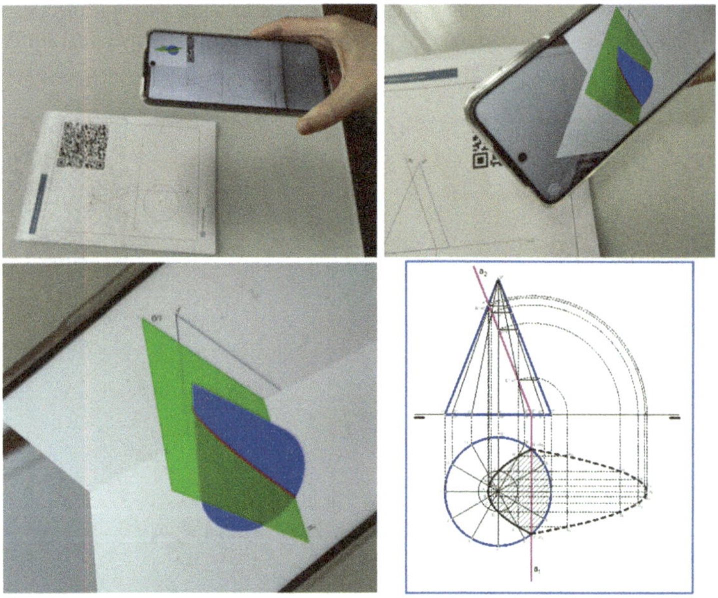

Fig. 4. Visualization process and resolution of the cone plane section exercise.

In order to obtain information about the acceptance of this type of tool among students and their assessment as a training complement to the Industrial Design II subject, 67 students who are studying this subject at the Technical University of Cartagena during the 2020/2021 academic year filled an anonymous questionnaire through Google Forms. As the tool is only available for Android devices, approximately 10% of the students who participated in the experience could not use their own device because they were running the IOS operating system. However, they were able to see through TEAMS the images of the different visualizations that could be obtained with the tool.

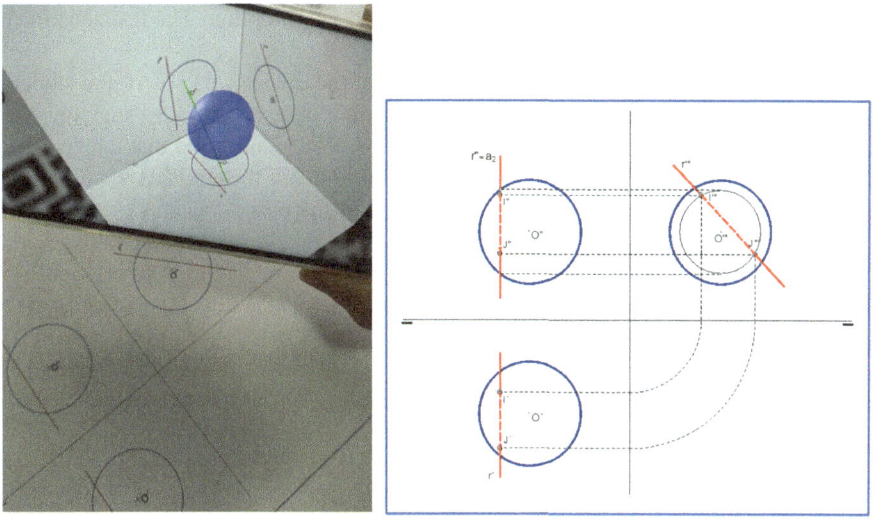

Fig. 5. Process of visualization and resolution of the exercise of incidence between straight line and sphere.

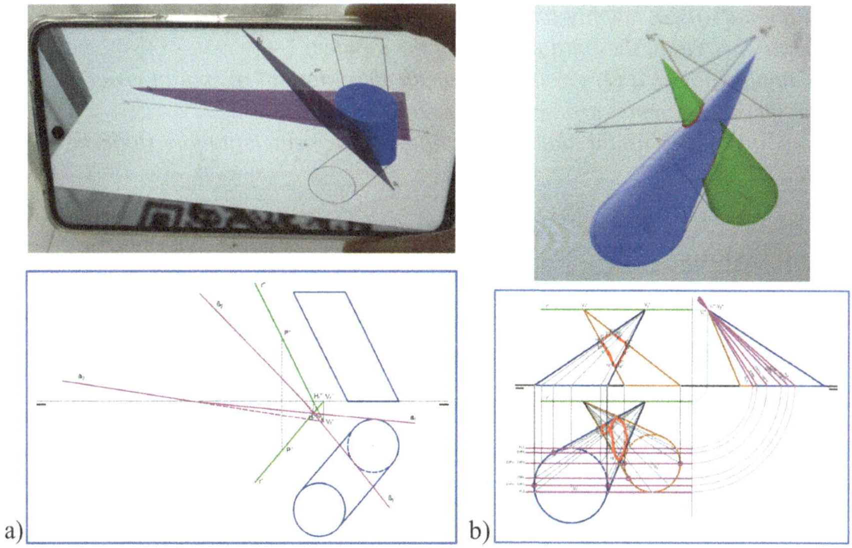

Fig. 6. (a) Process of visualization and resolution of the exercise of conditioned planes tangent to a cylindrical surface, (b) Determination of the intersection curve between two conical surfaces.

The questionnaire consisted of five questions that included technical aspects of the tool's operation and evaluation of the contents shown during the experience. A Likert scale with a score from 1 to 4 was used for four of the questions in the questionnaire [22]. The four response options had a specific definition associated with them, with two answers in favor and two against, thus avoiding the neutral option. The last question of the questionnaire only offered two answer options: yes or no.

Regarding the functioning of the application on the devices, 51.9% of the students indicated that the application worked very well on their device and 33.3% that it worked quite well. Regarding the reading of the QR stamps, 77.8% indicated that they read the stamps very well and 22.2% quite well. On the other hand, the students were asked if they considered that this tool favored the learning of the basic concepts of the subject and 85.2% said that it favored learning very much. In relation to spatial vision, 81.5% considered that the use of this tool greatly favored their spatial vision. As a last question, the students were asked if they would like this type of tools and contents to be part of the drawing subjects, and 100% answered yes.

The overall assessment of the experience was very positive. The students enjoyed the experience of using their mobile devices to visualize each of the proposed exercises and highlighted their interest in using this type of tool as a training complement in drawing subjects. Some of the comments or suggestions made by the students were:

- *"I think it is very useful, especially for students who have not studied technical drawing at baccalaureate level"*.
- *"Develop when possible in IOS"*.
- *"Very good and very interesting initiative"*.
- *"It is a good way of helping to visualize the elements."*
- *"It is very helpful in the spatial vision of the problem"*.
- *"The application is a very good tool for the development of spatial vision, which is one of the most complicated points to learn in dihedral"*.
- *"I think it is a very useful tool and should be implemented in more subjects, such as graphic expression"*.

4 Conclusions

This study describes the development of an open access tool for mobile devices that allows the visualization, through augmented reality, of different types of surfaces represented in dihedral through plans that incorporate QR codes. 3D visualization of objects in space is a challenge among engineering students and a skill that can be trained using this type of technology. The development of this type of application opens up a wide range of possibilities in teaching, allowing students to interact with different graphic representations that complement the contents of the subjects. Through this interactive form of teaching, learning is motivated in a more effective way, especially with those graphic representations that can be more complex with a more conventional type of visualization.

References

1. Arora, R., Kazi, R.H., Grossman, T., Fitzmaurice, G., Singh, K.: SymbiosisSketch: combining 2D & 3D sketching for designing detailed 3D objects in situ. In: Conference on Human Factors in Computing Systems - Proceedings (2018)

2. Atrio Cerezo, S., Guardado Moreno, E.: La realidad aumentada y su presencia en un modelo docente tecnológico para la didáctica de la Química en Bachillerato. Educación y tecnología (2012)
3. Chang, H.-Y., Hsu, Y.-S., Wu, H.K., Tsai, C.-C.: Students' development of socio-scientific reasoning in a mobile augmented reality learning environment. Int. J. Sci. Educ. **40**(12), 1410–1431 (2018). https://doi.org/10.1080/09500693.2018.1480075
4. Ibáñez, M.B., Delgado-Kloos, C.: Augmented reality for STEM learning: a systematic review. Comput. Educ. **123**, 109–123 (2018). https://doi.org/10.1016/j.compedu.2018.05.002
5. Suh, A., Prophet, J.: The state of immersive technology research: a literature analysis. Comput. Hum. Behav. **86**, 77–90 (2018). https://doi.org/10.1016/j.chb.2018.04.019
6. Ferrer Torregrosa, J., Jiménez Rodríguez, M.Á., Barcia González, J.M., Torralba Estellés, J.: La realidad aumentada en la docencia de ciencias de la salud. Nuevos caminos para la comprensión (2014)
7. López Lineros, M., Sánchez Jiménez, F.J., Mateo Carballo, F., Vázquez Tatay, C.: Realidad Aumentada como herramienta docente en la asignatura de Construcción en las enseñanzas de Ingeniería (2016). https://idus.us.es/bitstream/handle/11441/48743/24CUIEET_paper_23.pdf?sequence=1&isAllowed=y
8. Redondo, B., Cózar-Gutiérrez, R., González-Calero, J.A., Sánchez Ruiz, R.: Integration of augmented reality in the teaching of english as a foreign language in early childhood education. Early Childhood Educ. J. **48**(2), 147–155 (2019). https://doi.org/10.1007/s10643-019-00999-5
9. Rodriguez-Garcia, A.M., Hinojo-Lucena, F.J., Agreda-Montoro, M.: Design and implementation of an educational project on interculturality in early childhood education using augmented reality and QR codes. Educar **55**, 59–77 (2019). https://doi.org/10.5565/rev/educar.966
10. Juan, C., YuLin, W., Tjondronegoro Dian, W., Wei, S.: Construction of interactive teaching system for course of mechanical drawing based on mobile augmented reality technology. Int. J. Emerg. Technol. Learn. **13**, 126–139 (2018). https://doi.org/10.3991/ijet.v13i02.7847
11. Martín-Gutiérrez, J., Saorín, J.L., Contero, M., Alcañiz, M., Pérez-López, D.C., Ortega, M.J.C.: Design and validation of an augmented book for spatial abilities development in engineering students. Comput. Graph. **34**, 77–91 (2010). https://doi.org/10.1016/j.cag.2009.11.003
12. Polhmann, T., Parras-Burgos, D., Cavas-Martínez, F., Cañavate, F.J.F., Nieto, J., Fernández-Pacheco, D.G.: Augmented reality and mobile devices as tools to enhance spatial vision in graphic representations. In: Cavas-Martínez, F., Sanz-Adan, F., Camo, P.M., Lorza, R.L., Peña, J.S. (eds.) Advances in Design Engineering: Proceedings of the XXIX International Congress INGEGRAF, 20-21 June 2019, Logroño, Spain, pp. 420–427. Springer International Publishing, Cham (2020). https://doi.org/10.1007/978-3-030-41200-5_46
13. Nguyen, M., Le, H., Lai, P.M., Yan, W.Q.: A web-based augmented reality plat-form using pictorial QR code for educational purposes and beyond. In: Proceedings of the ACM Symposium on Virtual Reality Software and Technology, VRST (2019)
14. Sing, A.L.L., Ibrahim, A.A.A., Weng, N.G., Hamzah, M., Yung, W.C.: Design and development of multimedia and multi-marker detection techniques in interactive augmented reality colouring book. In: Alfred, R., Lim, Y., Haviluddin, H., On, C.K. (eds.) Computational Science and Technology. LNEE, vol. 603, pp. 605–616. Springer, Singapore (2020). https://doi.org/10.1007/978-981-15-0058-9_58
15. Fracchia, C.C., Boas, A.M.V.: Augmented reality in the primary education: design of table games for the social sciences and mathematics area. Rev. Electron. Investig. Docencia Creat.-DOCREA **2017**, 89–104 (2017)
16. Bairaktarova, D., van Den Einde, L., Bell, J.E.: Using digital sketching and augmented reality mobile apps to improve spatial visualization in a freshmen engineering course. In: ASEE Annual Conference and Exposition, Conference Proceedings (2019)

17. Durlach, N., et al.: Virtual environments and the enhancement of spatial behavior: towards a comprehensive research agenda. Presence Teleoper. Virt. Environ. **9**(6), 593–615 (2000). https://doi.org/10.1162/105474600300040402
18. Hartman, N.W., Connolly, P.E., Gilger, J.W., Bertoline, G.R., Heisler, J.: Virtual reality-based spatial skills assessment and its role in computer graphics education. In: ACM SIGGRAPH 2006 Educators program, pp 46-es (2006)
19. Kaufmann, H., Steinbügl, K., Dünser, A., Glück, J.: General training of spatial abilities by geometry education in augmented reality. Annu. Rev. Cyb. Therap. Telemed. Decade VR **3**, 65–76 (2005)
20. Kim, J., Irizarry, J.: Evaluating the use of augmented reality technology to improve construction management student's spatial skills. Int. J. Constr. Educ. Res. (2020). https://doi.org/10.1080/15578771.2020.1717680
21. Parras-Burgos, D., Fernández-Pacheco, D.G., Barbosa, T.P., Soler-Méndez, M., Molina-Martínez, J.M.: An augmented reality tool for teaching application in the agronomy domain. Appl. Sci. **10**(10), 3632 (2020). https://doi.org/10.3390/app10103632
22. Borgers, N., Hox, J., Sikkel, D.: Response effects in surveys on children and adolescents: the effect of number of response options, negative wording, and neutral mid-point. Qual. Quant. **38**, 17–33 (2004). https://doi.org/10.1023/B:QUQU.0000013236.29205.a6

The Role of Graphic Design in Perceived of Enjoyment in Online Education

Benyamin Soleimani(✉) and Larisa Dunai

Department of Graphic Engineering, Universitat Politècnica de València, Camino de Vera s/n, 5L, 46022 Valencia, Spain
bsoleim@doctor.upv.es, ladu@upv.es

Abstract. Covid-19 is a pandemic disease and as the evidence shows, the world is not ready for such a disease like this. Therefore currently, online education is used to prevent the limitations of learning. This paper seeks a way to expand and develop the joy of perception and encourage more teachers and students to use the online environment to conduct classes. The results of this article show how the use of graphic design material help to online educational be more interactive and motivate students. In this study, we try to use a package of General Standards for using colors, text, Illustrations, and pictures, video clips, using soundtrack, and music. But the most important thing is when we convert the whole chapters of classes to short movies like comic video or cartoons the participation rate increase significantly. Also, we convert a small video of the next class or theme in order to motive the students for the next session. In this case, using technology in graphic design education will create a sense of pleasure in learning and teaching for both categories of teachers and students.

Keywords: Graphic design · Enjoyment · Online · Education

1 Introduction

Early spring 2020 as a result of Covid19 pandemic, most universities decided to use online educational system by the approach of regular and virtual education. After a short time, millions of faculty members started to lecture in front of their computer screens and the students made to stay at home and pass online courses. As education is necessary demand, the previous educational system has been replaced by this system since then. Although online learning system was exist previously, it is currently demanded more than ever as the education cannot be stopped.

After emerge of modern technologies, educational institutes in different countries tried to employ such technologies in their programs through internet to help students and teachers in occurrence of special conditions as pandemic or similar occasions. The main focus of this study is on higher education in universities, increasing efficiency of education and creating satisfaction and pleasure in online education. Our main purposes are discovering for more attraction of students of graphic design by using general standard package which includes color, text, images, video clips, soundtrack and music. This new

F. Cavas Martínez et al. (Eds.): INGEGRAF 2021, LNME, pp. 327–336, 2022.
https://doi.org/10.1007/978-3-030-92426-3_38

idea also transforms every session of a class to a complete chapter for more cooperation of students and create pleasure in perception. The other key point is using maximum time for increasing students' cooperation in online virtual classes.

Nowadays, cooperation in educational system has considerably decreased comparing the past which is one of the reasons of replacing regular education to virtual education. Non-attendance of students in practical online courses like art courses which require practical works and execution of discussed subjects at same time is more than other courses. Moreover, the studies indicated that long quarantine period enforced by governments caused people to stay at home and continue their work, study and education through online courses. Therefore, creating advantageous educational environment for students and teachers is required more an ever. Moreover, volume of provided information shall not confused students but provide occasion for increasing their effectiveness. This procedure will result in pleasure in virtual classes [12].

This reality that technology has main role in digital era comparing previous generations, resulted in more technological literacy of current generation. Increase of literacy and recent developments of technology had resulted in technological development in education. Moreover, merger of continuous grading, early feedbacks, obvious goals, bonuses, challenges and improving positive points are new nature of digital education. Most importantly, measures for improving technology in classes, increasing usage of social media, providing learning resources and increasing learning and data-based evaluation are mandatory in this occasion [4].

Online art education has been prospered by development of technology in providing demands of current educational system but there are still various obstacles and issues in this type of education [9]. Nevertheless, online learning had solved important difficulties of traditional educational system. Performance of online education may be considered by tracing students' efforts, time of work, grades and other scientific information [5]. Teachers can also employ tools in various learning systems that are in proportionate with educational curriculum and useful. Many schools and organizations had invested large amount in learning management systems [3].

While popularity of online educational systems is increasing continuously, many teacher, managers, parents and students are involved with traditional education [11]. As Novak said, most of online courses designers are increasingly involved with technology for education. Before employing new technology, we are inquired to ask ourselves that learning technology is attractive and meaningful for students or not? How this technology resulted in deep learning and how we can modify our technologic programs to serve our population information? [3].

Novak believes that in early 2000s, two-third of faculty members had not accepted online education as a scientific method of education. Nevertheless, our perception had changed since then. During 12 years, members of online courses increased from 1.6 million to 8 million people and although the studies revealed that online learning may have positive aspects, some of the faculty members feel isolated and evaluation of student's interest is difficult [10].

Generally, graphic design education requires computer environment as "Graphic is training present design on the basis of technology". Nevertheless, it was revealed that not studying in analogue and computer environments that are important in increasing creativity of graphic students is considered as a gap in literature review of this study [2]. Development of technology had considerable impact on traditional form of graphic design. Therefore, this resulted in development of graphic design. Although traditional concept of graphic design may have not required attraction for students, digital graphic design with magnificent visual aspects may satisfy students. Therefore, new media science is very important in graphic design education [7].

Although there are various challenges for students and teachers in online education, the good news is that holding online classes was an efficient measure that governments had done in minimizing students' activity loss because there are few replacements for online classes. Meanwhile, students and professors shall be motivated and consider this procedure as a challenge. The students shall remember that they are main applicant of this procedure and shall have motivation to employ their maximum efforts in online classes [1]. Moreover, more concentration on optimizing infrastructures, educating teachers and creating motivation in students is highly required to increase effectiveness of online education. In accordance with WHO notice, COVID19 will continue for two or three years. Therefore, online education can be one of the most important educational systems [6].

Online education in the field of graphic design provided considerable occasion for professors and students to employ technology in creating more effective condition and create motivation in students. Currently, planning class and questions, helping communication, performing new exercises, helping students' expression and improving general development and learning of students are the factors in aiding students to leave boring conditions of the classes and encourage them in their lectures. The most considerable factor about this educational environment is that online education can be attractive, entertaining and appropriate for every person.

2 Methodology

The main purpose of this study is employing graphic design standards in online education. Symbol of the study is creating a program for executing a complete research study. Our goal is increasing students' cooperation level, creating motivation and finally causing satisfaction and pleasure in learning. We tried to provide efficient solution by reviewing current status of online classes to reach appropriate quality for virtual teaching and fill the gap between current condition and demanded quality by such solution.

2.1 Material

Photoshop course of bachelor's degree students of Pars University was inspected in this study. Photoshop software is a specialized course in the field of graphics in curriculum of bachelor's degree student which is usually instructed in two three-credit courses naming basic Photoshop and Advanced Photoshop. Graphic standards divided into two parts of aesthetics (theoretical issues) and main elements of graphic design i.e. principles, level recognition, volume, texture, color, text, type and sound were employed in this study. Moreover, AIDA model, questionnaire and opinion forms were employed in this study.

Educational environment was university system of the university which was connected to Adobe Connect software. Employed software was Photoshop.

2.2 Method

As a result of non-attendance of teachers and students in classrooms and lack of eye contact, class environment is different from its ordinary and verbal form. We studies all new methods to prevent boringness of long dialogues. Advanced Photoshop class divided into 16 sessions for bachelor's degree students. The main purpose is holding class with more satisfaction. Therefore, we are looking for a solution to change a class to more active one. A special planning was employed in this study for every session of education. Therefore, each session divided into two parts. In first part, current techniques were instructed under the purpose of education. On the other hand, different items were added to convince students and create more motivation among them. Accordingly, introducing software tools, filters, text, merging images and also items and issues that can be notified by students in increasing the speed of software usage are recommended as a common method. In next method, other items are provided such as playing music during rest time, playing music during solving exercises by teacher, providing new exercise on the basis of soundtrack, playing short movies on special times as series, representing output of the work before and after the work by employing 3D tools and also transforming next sessions to a short movie that students watch at the end of each session to recognize subjects of next sessions. Therefore, the method of this study was on the basis of comparing 2 sessions of a lecture. AIDA model which is one the most common marketing models was used in this study. This model is employed by many marketing authorities intentionally or unintentionally. If you had purchased a product being motivated by advertisements or filled opinion form or send SMS, be sure that your action affected by AIDA model.

AIDA is modern marketing theory which is on the basis of market movement and advertisements caused by clients' perception. This model was firstly developed by E. St. Elmo Lewis in 1898. Each character in AIDA refers to a special level. Attention, Interest, Desire and Action. AIDA is abbreviation used in marketing and advertisements and is describing list of common events that may occur in encounter of consumers and advertisement (Fig. 1) [8].

This model stands for four words as follows: A - Attention: to attract customer attention I - Interest: to create interest and desire, D - Desire: to create a desire to buy and have a product, A - Action: to buy a product.

This model is revealing one process and system for motivating people to purchase from you. This means that: firstly, you are required to attract client's attention to your product and yourself. Secondly, you should attract client's interest by describing product's specification. Thirdly, considering their interest on product, you should create desire for your product and fourthly you should direct them in final decision making. Now, if we replace this model on the basis of online classes study, the first stage is creating attention on online classes. It means that we proposed recommendation to attract students' attention in virtual classes. Second stage is creating interest and desire to attend in such classes. It means that class procedure shall develop people's interest in participating in classes. The next stage is creating desire among students in class. That is class shall be considered as an important and basic demand. The last stage is creating the sense of efficiency and satisfaction from participation in online classes. AIDA model is represented in following figure.

Fig. 1. AIDA model

As mentioned, attracting attention is first stage which is transforming a class to center of attention. Therefore, each session is divided into two sections. The first section is fulfilled by representing common subjects and simultaneous exercises. In second section, new elements and exercises are being used to attract students' attention and focus on class. It means that we behaved as the pattern of using modern entertainments. According to this definition, if we say something with high value and importance at the beginning, the remaining process will attract attention and you should repeat previous sentences for a while. But after following this pattern, important and main part may be revealed at the middle of session. At the end of each session, a short movie will be played that is presenting next session and its subjects. Such as Marvel movies that summary of next episode of movie or series is being introduced at the end of each episode. This factor may be used to increase student's attention to the class. You can suffice to considered questions during the conversation in order to evaluate this stage's reliability. For example, students are randomly asked during the class about explained subjects and the student were considerably able to answer the questions and they had adequate attention to the lecture and class as indicated in figure number 2 (Fig. 2).

Second stage is creating interest. This means that the addressees shall feel at the end of class that issues of previous lecture were attractive and interesting for them and encourage them to perform exercises at the end of each session. Therefore, new exercises are considered for students that were not fulfilled previously and encourage them to perform such exercises. In cases that they were asked to perform only one task for next session, they fulfilled the task because of its attraction which indicates that the exercise was not mandatory and interest in performing it was more than before. Performing tasks were considered as home works which is designing new exercises. It was decided to consider exercises as home works in accordance with similar issues from discussed subjects. For example, in designing Farsi logos, always traditional and regular

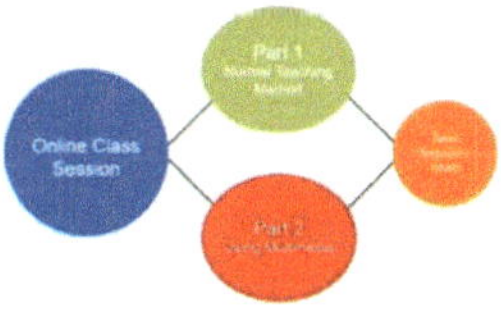

Fig. 2. Online class method

methods were considered. Traditionally, we always use pencil and paper to create one type of logo that sometimes are mixed with some shapes and then scan the shape or insert the picture in relating software and transform the design to digital file. We have to consider coordination and combination regulations in this case. Considering previous method, we want to fulfil a task with more interest. Therefore, we are directly employing computer without any sketch. Accordingly, all letters shall be typed in English and then we should select our required letters on the basis of our typography and after some changes, create our logo type in other language. This means that you can fulfil a task without using your alphabet. Therefore, after regular exercises, the students were asked to design an English text by employing English letters for a coherent complex. Figure 3 is a sample of new exercises on the basis of typography.

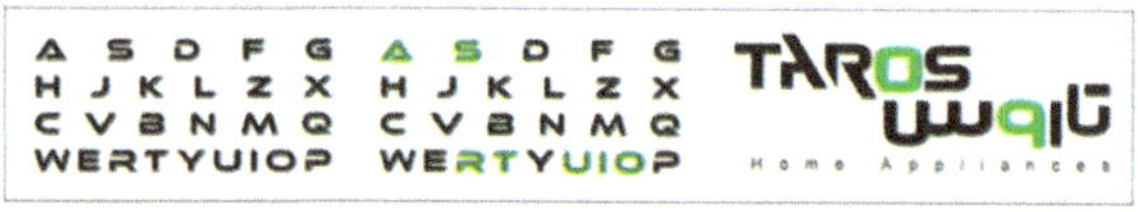

Fig. 3. Typography exercies based on English letters

The next stage is creating desire among students which is providing an environment to have desire for such trainings. As you know, online education is very important regarding its availability, in a manner that every person may have his/her specific time and planning. Sometimes it is possible to minimize cost of travel, stress of reaching destination and some other cases. Therefore, when an addressee is in appropriate educational route and sessions with new directions and entertaining trainings are designed for them, class will be gradually changed from a mandatory form to session with students' demand. In cases that academic routine are progressed correctly, desire sense will be achieved in shorter time. For example, there are students that are asking the instructor how they could perform a task on the basis of their needs.

The final stage is creating the sense of effectiveness which result in pleasure. It means that after passing a virtual course, a student feels that his/her desires are provided and had achieved his/her educational demands. In other word, in the progress of study and learning, he/she had successfully passed the course and feel the internal sense of pleasure that did not exist previously and may cause more motivation to participate in similar classes. Students' satisfaction is the most important item in this case. In other word, although the course was a mandatory course to be graduated, are they feel satisfied with the course or not? Results of this polling revealed that although the course and relating class had finished, students were interested to participate in similar classes outside of the university and learn to work with other software.

3 Result

In this section we want to compare results from analysis of online classes. Moreover, the tables are designed to divide each session to two parts and details of Sects. 1 and 2 of the class are provided in Table 1. Following table is comparing two different parts of each course. These data are gathered by using questionnaire and student's cooperation in each section. As you can see, number of notices has significantly increases in second column of this table as a result of basic language learners' movement from traditional education to modern one and employing new methods and techniques.

Table 1. Summary of student status in online class after questionnaire

Variable	P1 N (part 1)	P2 N (part 2)
Number of students	30	30
Online technology experience	30	30
Online learning experience	30	30
Class sections	16	16
Interest about learn (from 1–10)	5	9
Participation in class (from 1–10)	5	9
Items (graphic design requirement)	10	10
Number of lessons	8	8
Number of class exercise	10	10
Number of homework	10	10
New training	2	10
Class enjoyment (from 1–10)	5	9
Satisfaction (from 1–10)	5	9

As it was indicated, value of interest and cooperation as a number is 5 which is half of total score. Therefore, interest and cooperation in entities is equal to 5.

Results of polling were inspected to make sure about accuracy of the results of 4 stages of AIDA model. It is noteworthy that this polling was performed before end of the course and relating final test. Therefore, this was done before the final exam to prevent concordance of students' score and results of polling. Accordingly, the students were asked to fill online questionnaire. In the next stage, the results were compared to previous semesters that education was normal and verbal. As indicated in Table 2, all 4 stages of AIDA model were increased comparing previous classes that this amount of students' cooperation and performing tasks at home may be considered as the reason for developing online education environments. It is noteworthy that scores of students in this semester was considerably increased comparing previous semesters. Usually, there were one or two absentees at exam sessions and many of students couldn't archive passing grade. As indicated, not only all 30 students of an online class achieved passing grade, but also their grade were increased one level comparing previous classes which is an obvious reason for effectiveness of this course. Moreover, it was asked at the end of semester that are the students interested to participate in other classes of specialized software or not? The results revealed that 80 percent of them are interested to participate in future classes.

Table 2. Questionnaire form for photoshop online course

Online course	Present class (percent of 100)	Online class (percent of 100)
Create attention	60	80
Create interest	65	95
Scene of need	60	80
Scene of satisfaction	65	85
Timely attendance at class	60	75
Participation	50	70
Homework	40	65
Make contact after class	40	75
Class exercise	50	75
Quality of final work	65	85
Average grade	B	A
Participation for other software class	50	80

For additional analysis, Table 3 is indicating every subject discussed at class. As you can see, various parts are being shared and discussed in two different parts (sessions) of the class. Most of primary hours were used to introduce general issues, software knowledge and most of theoretical issues. But second part were used to introduce fast methods of fulfillment of tasks, using movies, 3D image, etc.

Table 3. Different section of teaching (part 1 and part 2 for each lesson)

General items	Presentation in part 1	Presentation in part 2
Basic speech about workspace	Yes	Yes
Using new tools of software	Yes	Yes
New methods for design	No	Yes
Teaching how to work faster	No	Yes
Short cuts	No	Yes
Basic graphic items	Yes	Yes
Mixing shapes	Yes	Yes
Using color for particular purpose	No	Yes
Using text in different styles	No	Yes
New method for designing logo	No	Yes
Using 3D photos	No	Yes
Video clip	No	Yes
Sound and Music	No	Yes
Converting sessions like a series	No	Yes

As per Table 3 and relating results, it can be concluded that value of cooperation, discuss and opinion sharing is highly increased in second part of the class. However, value of interest and satisfaction in second part was increased comparing first part. Therefore, analyzing tables' data revealed that we can increase attention, interests and participation of students by entering new items and employing multimedia devices.

4 Discussion

Very important elements are effective in online education success on the basis of causing satisfaction. Main specifications are internet speed, type of computer system, using media for instructing issues, etc., but important note in the field of artificial education is analyzing current academic system and finding appropriate method such as ability of providing a design in accordance with facilities of studied communication to provide condition for best action in this case. It can be concluded that results of this semester were considerably different from other semesters. Maybe the reason is transferring educational media. Analysis of results table indicated that when new issues are instructed in new methods accompanying new exercises, people's cooperation is being considerably increased. Moreover, transforming these items to a coherent series was effective in increasing students' interest. In fact, changing educational environment with new exercises is inviting students to a new challenge to prevent repetition of previous exercises and guide them in this endless route. Moreover, significant increase of students' grades revealed that although this course is mandatory for graduation, comparing it with previous semesters indicate that online instruction with new techniques is very effective in attention and interest of students. Although it is very regular to analyze each course at the end of semester by relating exam, method of examining in specialized courses was different as the exam was on the basis of theoretical issues. Students are obliged to send their tasks to the instructor on specific days same as previous classes. But an important issue in this case is method of presenting tasks and their high quality. What was misrecognized in previous semesters. For example, we were observed 25 final work with high quality from aesthetics and accuracy viewpoint in a 30-person statistical society.

Although Online system has advantages and disadvantages same as traditional educational systems. Some of its advantages are its availability in every time and place, ability of more concentration on ideas, decreasing cost of travel and availability to previous sessions' archive. On the other hand, considering current conditions that is learning in Covide19 occasion, cause demand to such environments more than ever and such academic environments may be updated by removing difficulties such as internet speed and technology issues and change the systems to a powerful one or even replace with previous systems.

5 Conclusion

Considering the main purpose of this study which is providing appropriate condition for creating motivation among students and finally causing interest among them, the results of this study indicated that how using graphic design items will help students to have more cooperation. For creating more interest in this study, we tried to encourage students by introducing main elements of graphic design and aesthetics issues in online environment by the purpose of increasing students' cooperation and performance in educational conditions. Comparing results of online classes revealed that although virtual learning environment had emerged in many countries, it may considerably help students of graphic design in increasing educational quality. Therefore, transforming sessions to a type of entertainment, using new exercises and employing them for creating condition for learning had been studies simultaneously. Moreover, if we provide technology facilities in educational environments, it may be hoped to have increased public cooperation in this academic process comparing the past.

References

1. Alam, A.: Challenges and Possibilities of Online Education During Covid-19 (2020)
2. Alhajri, S.: The effectiveness of teaching methods used in graphic design pedagogy in both analogue and digital education systems. Univers. J. Educ. Res. **4**(2), 422–425 (2016)
3. Comas-Quinn, A.: Learning to teach online or learning to become an online teacher: an exploration of teachers' experiences in a blended learning course. ReCALL **23**, 218−232 (2011). https://doi.org/10.1017/S0958344011000152
4. Hashim, H.: Application of technology in the digital era education. Int. J. Res. Couns. Educ. **2**(1), 1–5 (2018)
5. Hubbard, R.: The Really Useful e-Learning Instruction Manual: Your Toolkit for Putting e-Learning into Practice. Wiley, New York (2013)
6. Khati, K., Bhatta, K.: Challenges of online education during COVID-19 pandemic in Nepal. Int. J. Entrepreneurship Econ. Issues **4**(1), 45–49 (2020)
7. Liu, Y.: Research on the development trend and application of digital media art in graphic design education. Int. J. Electr. Eng. Educ. 0020720920984310 (2021)
8. Mackey, A.: The Practice of Advertising, 5th edn. Elsevier Butterworth, Sydney, Australia (2005)
9. Michinnov, N.: Procrastination, participation, and performance in online learning environments. Comput. Educ. **56**(1), 243–252 (2011)
10. Novak, K.: Udl Now! A teacher's guide to applying universal design for learning in today's classrooms. Professional Publishing, Wakefield, MA (2016)
11. Patel, F.: Online learning: an educational development perspective. Nova (2014). https://doi.org/10.1007/978-1-f63321-088-2
12. Stolbova, I., Nosov, K.: About Innovation Practice Within Geometric-Graphic Training (2020)

Development of an Application for the Automatic Evaluation of the Quality of 3D CAD Models

Inmaculada Pou Schmidt, Alejandro Rodríguez Ortega, Francisco Albert Gil, and Nuria Aleixos Borrás(✉)

Departamento de Ingeniería Gráfica, Universitat Politècnica de València, València, Spain
naleixos@dig.upv.es

Abstract. In the 3D modeling of products, the use of an adequate methodology that ensures the capture of the design intention is very important. The sequence of operations is key, like for instance, the sketches have to be completely restricted and the references of the modelling functions have to be correctly chosen without generating unwanted dependency relationships, among others. In the best of cases, the team leader dictates best practice manuals and then supervises the design work, ensuring that quality, which will facilitate future modifications or new designs based on existing models. However, this is not an established process, causing multiple failures in cascade when modifying or reusing the models is approached. This work has consisted of the development of an application that allows the automation of the quality analysis process in the models and has been developed for the Autodesk Inventor application using its iLogic tool. This work is the result of a Master's Thesis, where for the evaluation of the developed application, the examination models of the students of the subject of Graphic Engineering of the 4th year of the Degree in Engineering in Industrial Technologies of the Universitat Politècnica de València have been used.

Keywords: Design · Quality evaluation · 3D CAD models

1 Introduction

More and more companies are using 3D modeling tools. However, how do companies know that their designers make proper use of these tools? It would be necessary to ensure that the methodology followed is adequate through some type of control [1, 2]. Are those controls carried out efficiently? Let us see two examples of this in the industry to help focus the problem that has been tried to solve in this Master's Thesis (Project from now on). The first example company manufactures radiators for the automotive industry where some of their parts are very problematic, because further modifications are tough to work (see Fig. 1a). In order to solve or minimize this process, the company collaborated with the local University through a national research project (Ref.: 1FD97-0784) to identify the key aspects of the design process and developed Best Practices manuals for their designers to follow when modeling these products.

F. Cavas Martínez et al. (Eds.): INGEGRAF 2021, LNME, pp. 337–344, 2022.
https://doi.org/10.1007/978-3-030-92426-3_39

The second company manufactures molds for the automotive industry, where the first author of this work did her traineeship. In order to verify that its designers meet the design requirements set by the company, the person in charge reviews each model manually using a 75-point checklist (see Fig. 1b). This process is very laborious and represents one of the bottlenecks of the technical office department of the company. In the checklist, things such as collisions, diameters, and some general characteristics of the model are checked over. Despite this supervising control, failures in the molds are still found, which implies remaking the molds from the scratch, representing many hours of work.

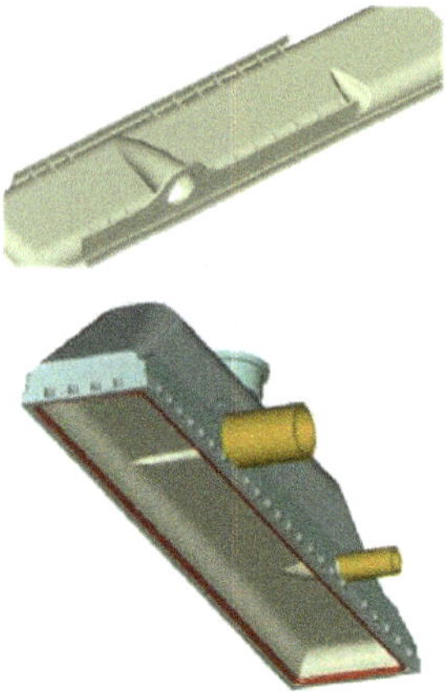

(a)

1	USE COLLISION FILTER. NON UNJUSTIFIED COLLISIONS SHOULD APPEAR
2	USE DRILL ALIGNMENT FILTER. NON UNJUSTIFIED DRILL SHOULD APPEAR
3	CHECK THAT NO PART OF THE MODEL TREE IS ON ALERT.
4	CHECK THE BELT ELEVATION [H=46mm – STRIP-EXTRACT] (H=331 mm BELD-BENCH)
5	CHECK THE FASTENER OF THE BELT ELEVATION.
6	CHECK FOR FALLING OFF REMNANTS (INTERFERENCE), SCRAPPING.
7	GLUED STOPS [SPLIT BETWEEN TOP AND BOTTOM].
8	DESING THE PART EJECTOR.
9	CHECK THE ADVANCE. MUST BE EQUAL TO OR LESS THAN QUOTED (89 mm)
10	CHECK THE BOBBIN WIDTH. MUST BE EQUAL TO OR LESS THAN QUOTED. (171 mm)
11	ADVANCE SENSORS. THE FINAL SENSOR IS PRIORITISED AGAINST THE NON-RETURN SENSOR, IF POSSIBLE.
12	CHECK THAT THE OUTPUT OF PARTS IS SUITABLE FOR THE PRESS. MANUAL
13	CHECK BELT DRIVE.
14	CHECK THE SAFETY ROUTE OF NITROGEN GASES (10%). RV 170-15 (C-10)
15	CHECK LIQUID EVACUATION AREAS IN NITROGEN GAS ACCOMMODATION.
16	CHECK CHAMFERS OF PRESSPLATE, KEYHOLE TOOL, CUTTER-SHAPER, FOR SHEET METAL DEFORMATION IN CAP MOLDS.
17	CHECK THAT ALL INTERNAL ELEMENTS HAVE AIR OUTLETS.
18	CHECK DIRECTION OF CUTTING BURRS.
19	CHECK TOTAL HEIGHT OF CLOSED MOLDS AND VERIFY WITH MACHINE MOTION. 344mm CLOSED DIE – MOTION ~130
20	CHECK FLANGED GROOVES ARE IN ACCORDANCE WITH THE STANDARD AND PRESS. IT IS FLANGED WITH FLANGES.
21	CHECK THAT COMPONENT CHANGES ARE APPROPIATE, SAFE AND QUICK.
22	CHECK THAT THE COMPONENT CHANGES HAVE 2 M6 THREADS FOR EASY REMOVAL.
23	CHECK CRITICAL TOLERANCES TO ACHIEVE PLANE' TOLERANCES.
24	CHECK PENETRATION OF THE PUNCHES INTO THE CUTTING PLATE. 6mm
25	CHECK THE PROTRUSION OF THE DIE-HOLDERS WITH RESPECT TO THE PUNCHES. 2mm
26	CHECK THE PATH OF THE DIE-HOLDERS IN RELATION TO THE PUNCHES (TRAIL TRM). 10 mm
27	CHECK THE SHAPER GAP BETWEEN THE MALE CUTTER-SHAPER AND THE CUTTER PLATE (CAPS MOLD)
28	CHECK THE HEIGHT DIFFERENCE BETWEEN THE DIE-HOLDER AND THE CUTTER-SHAPER PLATE (CAPS)
29	CHECK THE NUMBER AND DIAMETER OF THE KEYHOLE TOOL IN RELATION TO THE WORKPIECE PLANE.
30	CHECK THAT THE INNER EXTRACTOR IS ABOVE THE OUTER EXTRACTOR AND THE MALE CUTTER-SHAPER.
31	CHECK PART EXTRACTION. Manual

… (up to 75 key-points)

(b)

Fig. 1. a) Radiator components subject to good practice manuals; b) Key-points checklist during the mold design process.

Thus, despite best practices' manuals have been developed and applied or some type of control is done during the process of design, normally the evaluation of the CAD model is still subjective and done manually, which hinders the reuse of the models or makes their modification a laborious task [3, 4]. The work presented here is the Project carried out in the Master's Thesis of the first author of this paper and has consisted on the development of an application that allows automatically assessing the quality of the geometric 3D CAD models [5]. More specifically, the main objective was to develop an application aimed to assess the exams of the students of the 4th course subject Graphic Engineering of the Degree in Engineering in Industrial Technologies of the UPV (Universitat Politècnica de València). Attempting in turn to reduce the correction time and to suppress the variability among the professors since, due to the high number of students in the subject, this task is commonly carried out by several professors.

2 Materials and Methods

In this section, the materials and methods used for the development of this Project are briefly described.

2.1 CAD Tools

The Autodesk Inventor application has been chosen since it is the modeling application used in the laboratory classes in the subject Graphical Engineering of the 4th course of the Engineering in Industrial Technologies degree, and also in the 2nd course of the Industrial Engineering Master degree, in the subject of Industrial Applications of Solid Modeling. The Autodesk Inventor iLogic programming tool is the platform where the application has been developed (Fig. 2), which uses the VisualBasic.NET as programming language, allowing the development of specific applications using rules and forms to be executed inside Inventor.

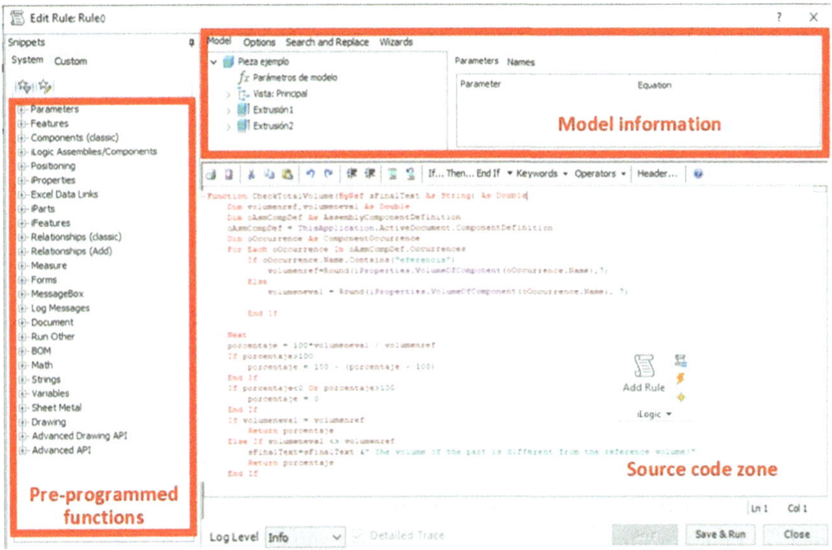

Fig. 2. Autodesk inventor iLogic tool.

The Forms allow interacting with the user to enter parameters or a specific configuration, whereas Rules are functions that are created with a programming language to perform specific tasks. The iLogic tool already provides basic pre-programmed functions contained in libraries, that can directly access to the internal database of the model [6].

2.2 Models

The 3D CAD models used for this work are those used in the subjects previously mentioned. Figure 3 shows some examples of the parts modeled as part of the laboratory exams of the students.

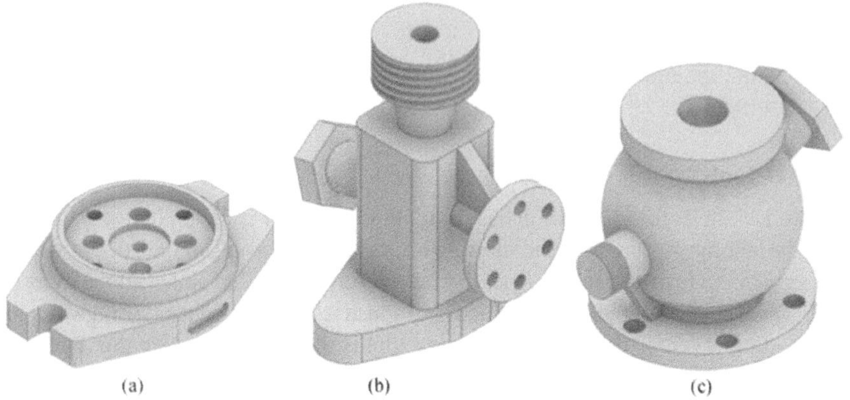

Fig. 3. Examples of 3D CAD models used in the exams.

The functions/features that can be found in the model tree of the 3D models used in this project are extrusions, revolutions, ribs, holes, patterns, symmetries, threads, fillets, chamfers and work features.

2.3 Quality Specifications Under Evaluation

Although the quality of the 3D CAD models depends on many aspects and it is also very dependent on the model itself, these aspects have been organized into the following three differentiated blocks: (1) general considerations, (2) functions/features used, and (3) sketches; which usually are always taken into account in more or less depth, when quality is evaluated. More specifically, the aspects checked inside each of the three main blocks are:

- Block-1. General considerations: orientation and model sequence, what means that main features have to be at the beginning of the model tree and the secondary features have to be located at the end of the model tree.
- Block-2. Functions/features: the use of specific suitable features instead of "any" similar feature.
- Block-3. Sketches: should be fully constrained with no crosses between geometric entities or superimposed entities.

For this first approach developed in this Project, the application operates comparing two models, the reference model consisting on the model that the professor creates as the solution of the exam, and the student model. Regarding the general considerations, the orientation and location of the two models (reference and student model) have to match, including the total volume. Additionally, the main features are checked to be in the upper half of the model tree and secondary features (like fillets, chamfers, threads, and ribs) are checked to be in the bottom half of the model tree. Regarding the use of suitable features to create specific geometry, the students were asked to rename some functions with a certain name, so the application can identify them in the student model and check the type of feature used and its parameters. With regards to the sketches, the items checked out are the number of dimensional restrictions, the number of some geometric constraints (like tangencies), and also if the contained profile is unique (no crosses or superimposed geometric entities) and if it is fully constrained.

3 Results and Discussion

In order to study the performance of the application developed, named EVALUATE, it was tested on several student models. More specifically, the exam chosen for the tests was the one of the first partial exam (December 2020) of the 4th course of the subject Graphic Engineering of the Degree in Industrial Technologies (Figs. 4 and 5).

Fig. 4. 3D model resulting of the exam statement in Fig. 5.

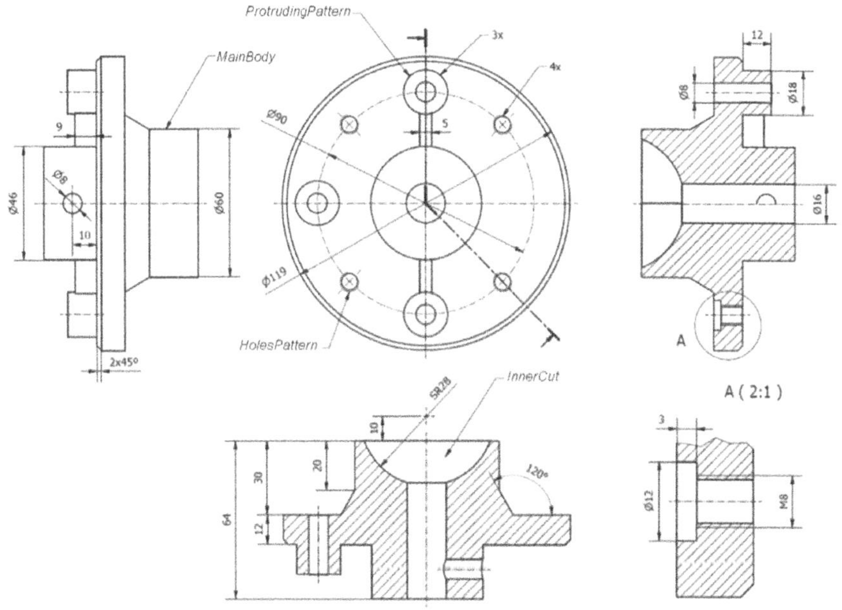

Fig. 5. Statement of the exam used for the tests (first partial exam December 2020).

In general, it is a good practice and highly recommended to label the features with logical names to find them easily by designers when modeling or modifying. In the statement of the exam, students were asked to label some features with a certain name in order to find them and later inspect them properly.

3.1 Application Developed

Figure 6 shows the general flowchart of the application developed. EVALUATE loads the reference model from the professor and interacts with the user to weight the items considered in each of the three blocks to assess the student models. Later on, EVALUATE dumps a global numeric mark and some comments in an Excel file. On the other side, two professors marked each of the student models attending to a standard criteria set for the exams of the subject, which served as reference.

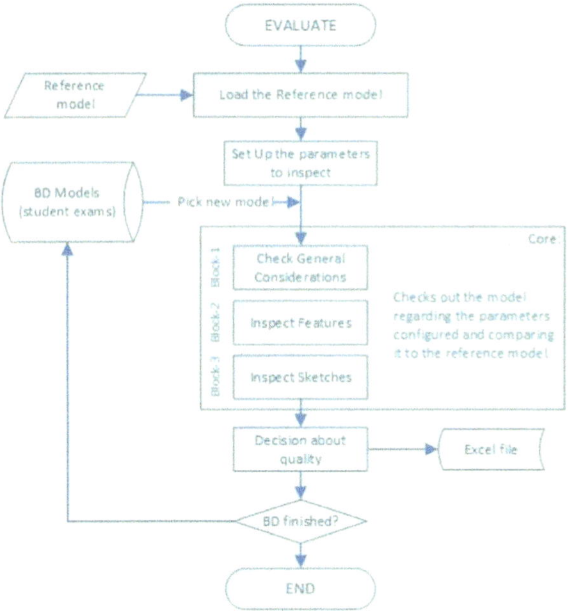

Fig. 6. General flowchart of the application EVALUATE.

3.2 Validation

As we can see in Figure 7, the differences between the professors' marks are similar to those obtained between the average of both professors and the result of the application developed.

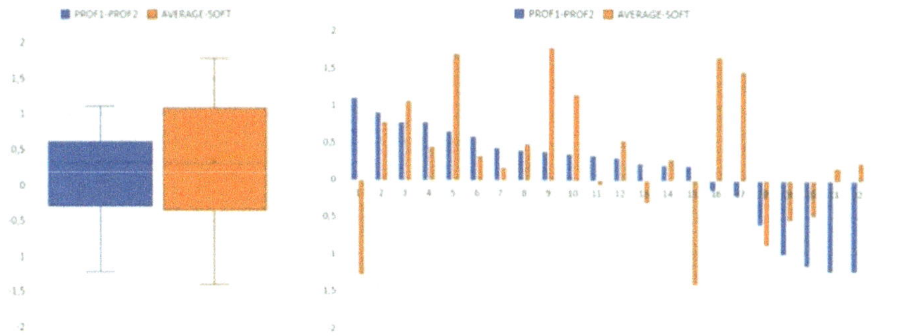

a) b)

Fig. 7. For each student exam: a) differences between global marks of the two professors (blue); and differences between the professors' average and the application mark (orange). (Color figure online)

However, we can also see that there is a greater deviation between the results of the application and the average of the professors (Fig. 7a), which may be because the criteria used by the professors are much more detailed or rigid than those configured in this first approach of EVALUATE. It also should be noted that, in those cases where the differences between professors are minimal, the system penalizes more, and when the difference between professors is more dissimilar, the result obtained by the application is close to the average between professors, which can be taken as a goodness (Fig. 7b).

4 Conclusions

As this first approach to automate the assessment of the quality CAD models demonstrates, it is possible to create automated strategies to check whether student exams meet minimum quality standards. Additionally, the application developed has shortened the correction time considerably, also suppressing the subjectivity due to the long correction time (fatigue) or to the variability among professors. Besides, the results obtained have proved a more objective and unified result for all students.

Reference

1. Andrews, P.T.J., Shahin, T.M.M., Sivaloganathan, S.: Design reuse in a CAD environment – four case studies. Comput. Ind. Eng. **37**(1), 105–109 (1999)
2. Amadori, K., Tarkian, M., Ölvander, J., Krus, P.: Flexible and robust CAD models for design automation. Adv. Eng. Informat. **26**(2), 180–195 (2012)
3. Company, P., Contero, M., Piquer, A., Aleixos, N., Conesa, J., Naya, F.: Educational software for teaching drawing-based conceptual design skills. Comput. Appl. Eng. Educ. **12**(4), 257–268 (2004)
4. Contero, M., Company, P., Aleixos, N., Vila, C.: Metodología de modelado con herramientas CAD/CAM avanzadas. In: XII Congreso Internacional de Ingeniería Gráfica. ISBN 84-8448-008-9 (2000)

5. Hekman, K.A., Gordon, M.T.: Automated grading of first year student CAD work. In: ASEE Annual Conference and Exposition (2013)
6. Autodesk: Glossary for the Engineer's Handbook, Inventor 2021, Autodesk Knowledge Network (n.d.). https://knowledge.autodesk.com/support/inventor/learn-explore/caas/CloudHelp/cloudhelp/2018/ENU/Inventor-Help/files/GUID-3E0794C5-75B4-4425-8163-45FCCF8597B3-htm.html. Accessed 12 May 2021

Application of Convergent Technologies in Teaching: Flipped Classroom and Augmented Reality

Irene Sentana-Gadea(✉), Juan Llorca-Schenk, and Mª Carmen Díaz-Ivorra

Alicante University, Carretera San Vicente del Raspeig s/n, 03690 San Vicente del Raspeig, Alicante, Spain
Irene.sentana@ua.es

Abstract. Teaching methodologies and learning processes have varied over time depending on historical milestones and technological development. Nowadays, there are pedagogical models that transfer certain learning processes out of the classroom. In this way, classroom time is dedicated especially to the practice of previously acquired knowledge and to resolve doubts or to expand concepts. In this teaching research, a combined methodology that includes flipped classroom and augmented reality is exposed. The objective is looking for tools that allow engineering degrees students improve spatial perception by facilitating autonomous study. This technique initially includes the flipped classroom, in which students must watch some videos where basic concepts are explained. Then they work with basic exercises, which include augmented reality tools. This article shows the results carried out by two groups of students who are student chemical and robotic engineering degrees. It has been proven to be a valid and positive tool to increase the spatial perception for both, lectures and students.

Keywords: Flipped classroom · Augmented reality · Graphic engineering · Third view

1 Introduction

Society worldwide is in constant transformation, in particular, these huge changes that have been taking place in the last decades, affect the productive model and the economic model. Experts indicate that we are in the fourth industrial revolution [1, 2]. We must be aware that all these changes will affect the workplace, generating new types of jobs that in one way or another will be modified by the new technologies. Therefore, workers will be forced to change, and stable jobs in specific locations will be relegated to the past, developing into more changing jobs with a wider geographical coverage.

These changes directly affect education, a changing society must be trained to be able to adapt to these challenges, hence in June 1999 the Bologna Declaration was signed [3]. The main objective of this document was to undertake a profound reform of the European University System, in order to improve the quality of teaching and to establish quality criteria that would put the different degrees in the European Union on an equal

F. Cavas Martínez et al. (Eds.): INGEGRAF 2021, LNME, pp. 345–357, 2022.
https://doi.org/10.1007/978-3-030-92426-3_40

level and, therefore, promote employment mobility with certain standards. A European Credit Transfer System (ECTS) was established, which corresponds to a total of 25 h of dedication. These hours of dedication are not all face-to-face, but involve 10 face-to-face hours and 15 non-face-to-face hours [4].

These changes have not only affected a new structuring of bachelor's and master's degrees, which was perhaps the most evident change that was made in the eyes of society. Since then, with these new ways of working, lectures must make an effort to change methodologies, with the main premise that the change must be centered on the focus. If before the focus was centered on teaching, now the focus must be oriented on student learning, where training must be guided so that professional competencies, both specific and transversal, are developed. The student now has an active role and must be the protagonist of his or her learning, and the teachers must move away from giving information in a passive way, instead there must be an active involvement and we are responsible for promoting it [5].

For this change to be effective, one of the options we have is to integrate active methodologies in our teaching, that is, strategies that allow and encourage student involvement in their own learning, we must be able to accompany and encourage students to acquire the skills they will need in the future.

There is a wide range of active methodologies that can be applied at different educational levels, among others we find: Case studies, Flipped Classroom, project-based learning, Analysis Method, simulations, workshops, seminars, Serious games, learning through information and communication technology (ICT), cooperative-collaborative learning, design thinking, inquire based learning.... [6–9]. The choice of one or several methodology depends on many factors, and should always be adapted and transformed according to the group and the objectives proposed for it.

In this case, we will focus on the use of two technologies, on the one hand, the flipped classroom and the use of ICT, in our case augmented reality.

Figure 1 shows the differences in application between the traditional method and the Flipped Classroom (FC) method [10]. The FC model reverses the access to knowledge, the first experience is an individual work outside the classroom, where the student uses various materials, usually videos where the basic concepts and approaches to be worked with later are exposed. This is followed by a subsequent assimilation of the instructions given. Finally, it will be in the classroom, and usually in a group, where knowledge is consolidated, clarified and fixed, and where the lecturer will support the learning process consolidation and expansion.

In addition, an information and communication technology, augmented reality (AR), has been used in this research. AR is a technology that allows the combination of digital information and physical information in real time with the use of technological devices such as phones, glasses, tablets. AR allows virtual information to be added to physical information [11].

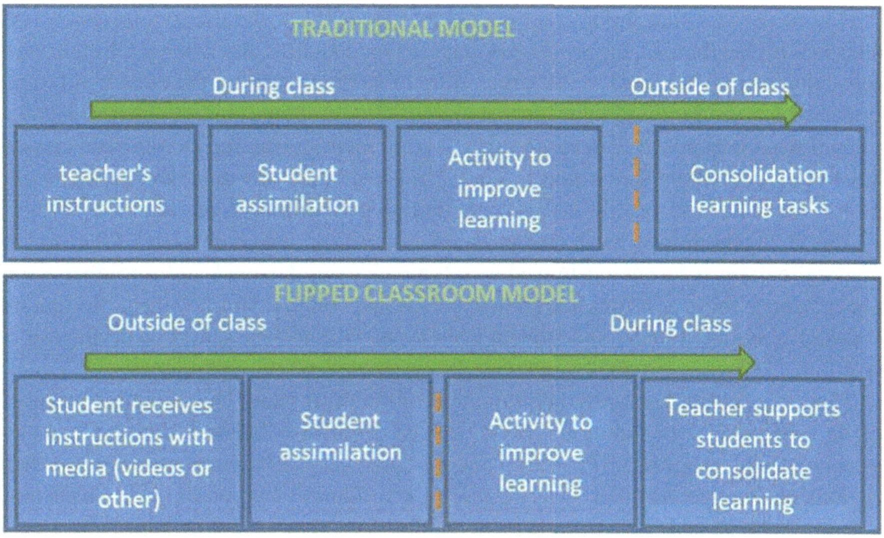

Fig. 1. Structure of the traditional model versus the flipped classroom

2 Objectives

In the process of adapting the teaching staff to the new pedagogical approach, and in the area of Graphic Expression in Engineering, the need to improve the students' interest in learning was raised. For this purpose, a group of professors who teach in Graphic Expression subjects decided to apply a double active methodology with several groups of students, Flipped Classroom and augmented reality.

This methodology has been developed and applied to two undergraduate degrees, Robotics Engineering and Chemical Engineering at the University of Alicante, and specifically within the subject Graphic Expression/Graphic Engineering. The topic selected for the application of this methodology was in these subjects, but specifically in the topic of "Determination of the third view" for the improvement of spatial perception.

The reason why this need arises is based on the difficulty that the professors had in managing the theory classes with more than 75 students in the classroom in terms of transmission of content, given both the number of students and the poor layout of the classrooms that made a simple lecture class not viable.

The objective, therefore, is focused on exposing the methodological development carried out by the group and the results of the experience of applying FC and AR, as an improvement of the teaching quality to these groups of students.

3 Innovation Development

As mentioned above, the two degrees selected are Robotics Engineering and Chemical Engineering, where the basic subjects taught are: Graphic Expression (first year) and Graphic Engineering (second year) respectively, both of which include the subject of "determining the third view" in their programmes.

Due to the fact that the average number of students in the theory classes of both subjects is higher than 75, the professors detected special difficulty when teaching certain topics, due on the one hand to the large group of students, and on the other hand to the poor classroom layout, which means that practically 50% of the students do not visualize in detail what is exposed with technological equipment during the face-to-face classes. This led us to the conclusion that the face-to-face classes had to change substantially and for this reason we opted for the FC methodology mentioned above.

The FC methodology has a first phase of instruction, which consists of the student receiving certain information that allows him/her to understand and introduce some concepts, once these concepts are known, he/she goes on to assimilate them through the proposed activities. Finally, the student goes to the face-to-face session with the rest of the group.

The topic selected was the determination of the third view. This topic was selected not only because it is common to both subjects, but also because the way students understood and progressed in basic concepts was very different depending on their previous knowledge, and there was a phase of assimilation that required very little time for some students and more for others, depending to a great extent on their capacity for spatial perception.

3.1 Instruction Phase

To start the instruction phase, different alternatives were analysed to provide them with the study concepts (written bibliographic material, web pages, public videos...), finally, since none of these materials met the requirements that combined the set criteria of concreteness and simplicity, it was decided to develop our own material in the form of a video.

To address this topic, three videos were developed which students could access through the Moodle platform.

3.2 Assimilation Phase

Although it was relatively easy for us to control how we provided the initial information to the students, we found that, in the assimilation phase, when a series of tasks are proposed to the students, they had difficulties in knowing how far they had progressed in their learning. How to communicate with the students in this phase and how to ensure that the student has a minimum of feedback to know their progress, this is where augmented reality came into play.

Initially, the different types of classification of augmented reality were analyzed, for which it was decided to start from a physical component of level 3, this level corresponds to a "2D" image, and depending on the virtual component, it was assessed that it would be based on both a 3D model and a video.

Different augmented reality software were analyzed.

- Unity 3D [12]
- Augment [13]
- Aurasma [14]
- Blippar [15]

Finally, due to the fact that it is free, easy to use and can be applied with a mobile phone, it was decided to use the Blippar program.

3.3 Real Implementation and Conclusions

Once all the material had been prepared, a pilot test was conducted during the 2020–2021 academic year with the two selected groups. Once the pilot test had been carried out, a questionnaire was developed to determine the usefulness of the material developed during the instruction phase, as well as the material developed during the assimilation phase.

4 Results

4.1 Instruction Phase

Three videos were produced. Figure 2 shows three images captured from them [16].

- Third view. Parallel surfaces.

  o Video time 9.30 min

- Third view. Inclined surfaces

  o Video time 7:20 min

- Third view. Curved surfaces

  o Video time 9:10 min

The objective of the videos was to explain and guide the students in the different techniques with which they should approach the resolution of the third view, as well as the approaches they should take into consideration.

In this instruction phase, the student was also provided with a pdf file with the exercise description, so that while watching the video, he/she could simultaneously or subsequently develop the exercise (Fig. 3).

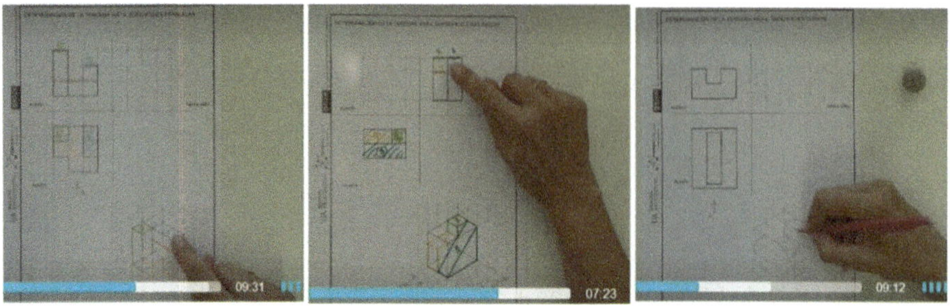

Fig. 2. Models of the three videos developed [16].

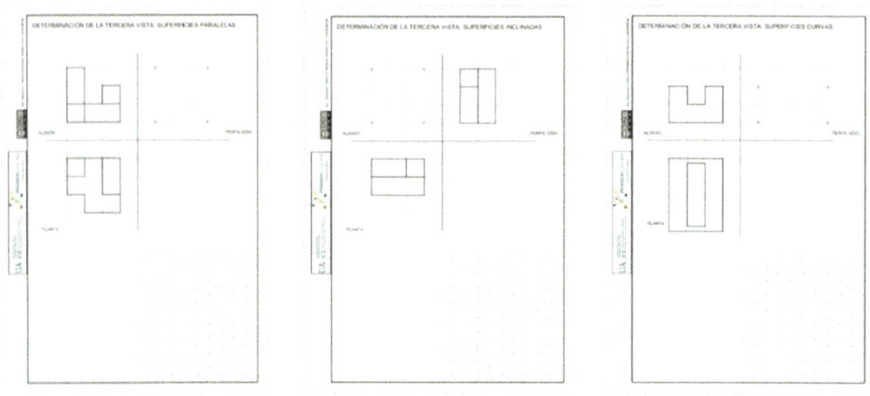

Fig. 3. Practical material provided to follow the concepts in the video.

4.2 Assimilation Phase

The assimilation phase is carried out through the proposal of simple practical cases that the student must perform before the classroom session. The aim of these exercises is that the student, starting from two views, tries to imagine, draw and capture the design of that part, that is, to improve his ability to visualize 3D objects given by means of their representation in 2D.

For this work, the student was provided with a statement similar to the practical material with which he/she had to work previously.

The difficulty detected with these exercises is the problems that students have in verifying whether their final solutions are correct or not. The aim is to improve the students' spatial vision, and therefore, the fact of directly showing them a solution means that, as soon as they see the exercise solved, the learning process is broken. If they have not worked sufficiently on the problem and the solution is incorrect, even if they reproduce it correctly, the required improvement in spatial perception will not be achieved.

Therefore, the idea of including AR arises. With AR, the aim is: If the solution is correct, the student is shown an image and a video of the piece for comparison. If the solution is not correct, the student is shown a small preview and an idea of the solution, but not the whole piece, so that he/she can continue thinking and looking for a valid solution.

As an example, one of the models made using the Blippar application is shown in Figs. 4 and 5.

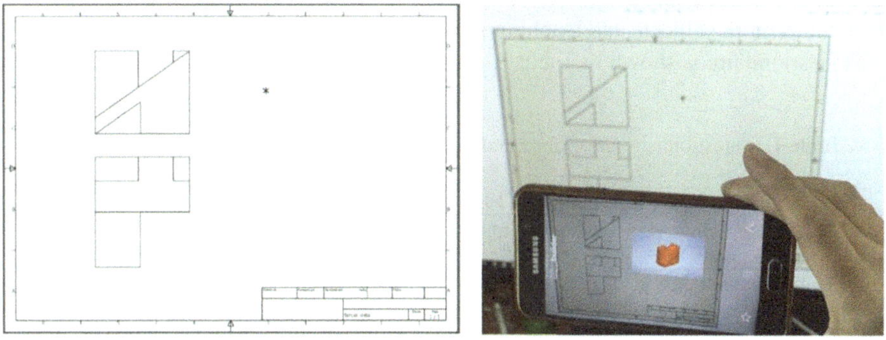

Fig. 4. Example of exercise and partial result with the Blippar mobile application.

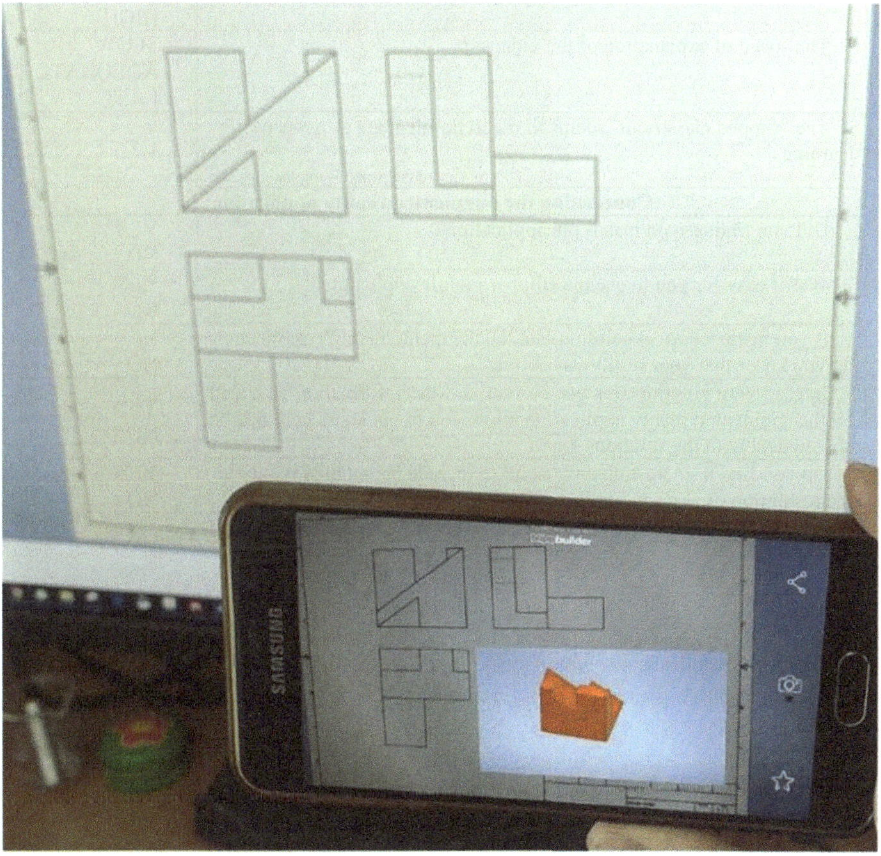

Fig. 5. Example of the final result with the Blippar mobile application.

4.3 Evaluation of Teaching Research

In order to analyze if the proposed methodology is effective, and therefore to be able to extend it to more subjects and teaching groups, a student questionnaire was conducted. This questionnaire is shown in Table 1.

Table 1. Questionnaire developed and handed out to students through google forms.

Concerning the Flipped classroom and videos	
Did you previously know what a "flipped classroom" was?	YES NO
Did you watch the videos before the classroom lessons as the lecturer instructed you?	ALL SOME NONE
For me, the level of the videos is:	LOW
	ADEQUATE HIGH
The speed of explanation of the videos is:	SLOW ADEQUATE FAST
The "flipped classroom" seems to me to be adequate as a means of learning. is inadequate and 5 is totally adequate	1 2 3 4 5
Concerning the augmented reality application	
Did you find easy to install the application?	YES NO
Was it easy for you to use the Blippar mobile application?	YES NO
If you made a correct solution, did the augmented reality application let you know that your result was correct?	YES NO
In case your solution was not correct, did the information provided by the augmented reality application allow you to get some help and be able to deal with the solution	YES NO
Do you find it interesting and useful to include augmented reality in the resolution of case studies?	YES NO
What did you like the most and the least about the use of this technology?	

Achieved Results

The results obtained are shown in Figs. 6, 7, 8, 9, 10, 11, 12 and 13. The results are presented for the two groups of robotic engineering and chemical engineering students.

Concerning the Flipped Classroom and Videos

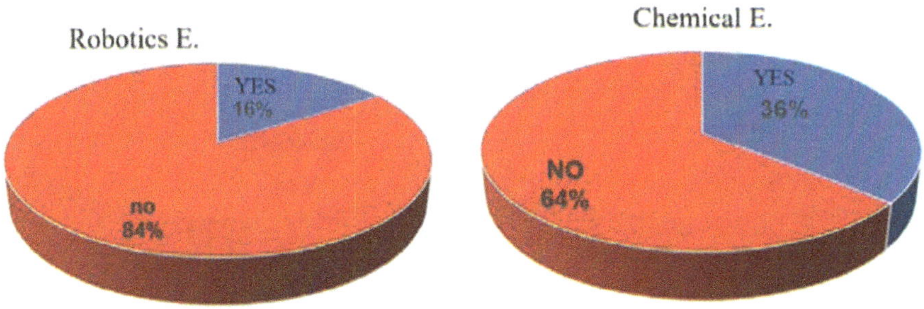

Fig. 6. Did you previously know what a "flipped classroom" was?

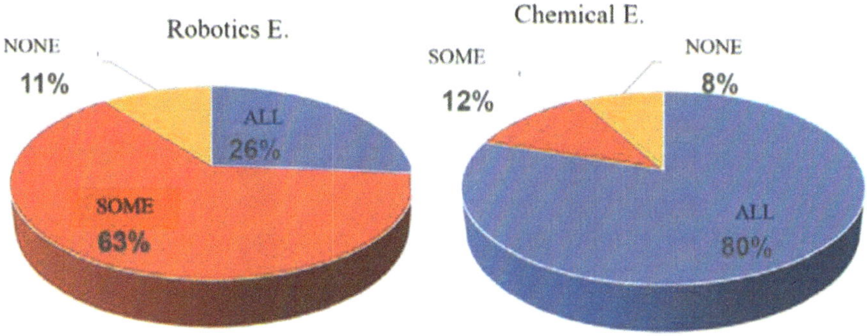

Fig. 7. Did you watch the videos before the classroom lessons as the lecturer instructed you?

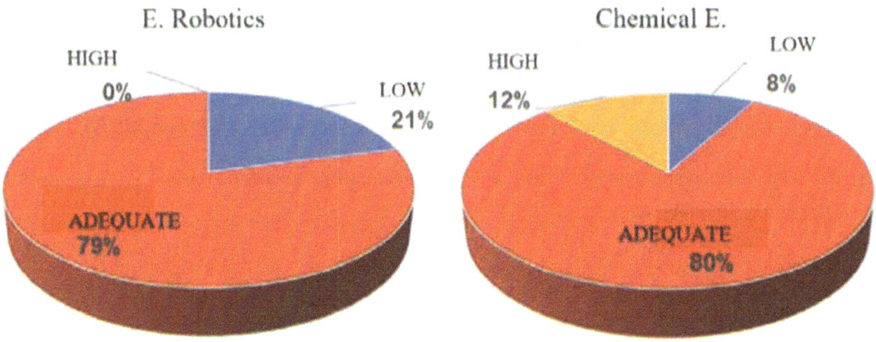

Fig. 8. For me, the level of the videos is: (high, adequate, low)

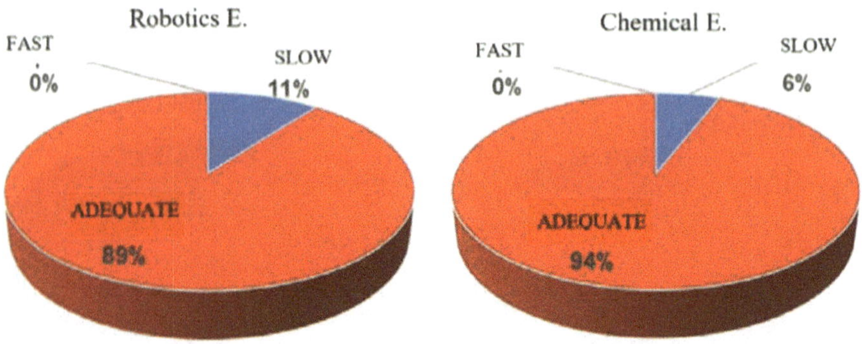

Fig. 9. The speed of explanation of the videos is: (fast, adequate, slow)

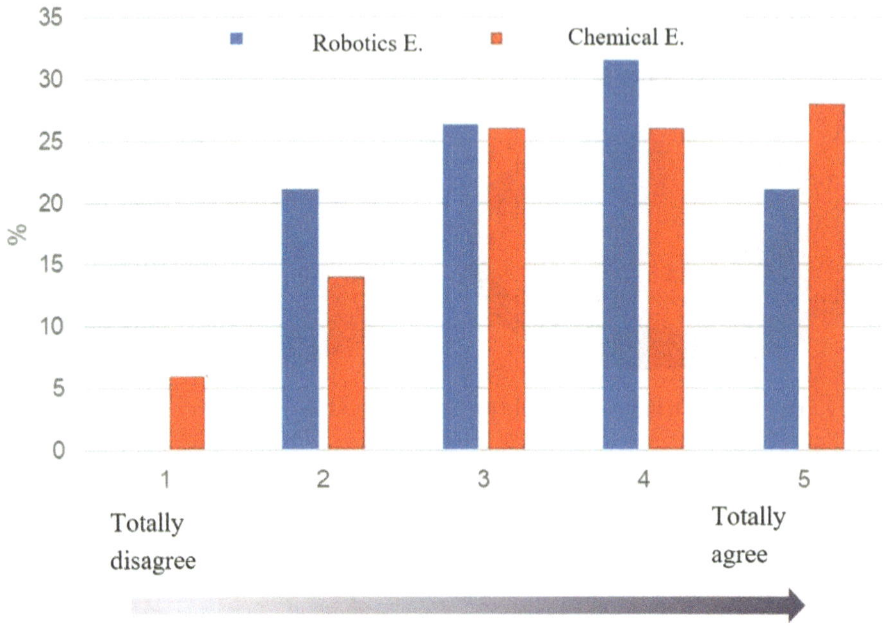

Fig. 10. The "flipped classroom" seems to me to be adequate as a learning medium. 1 is inadequate and 5 is totally adequate.

Concerning the Augmented Reality Application

The question about: Did you find it easy to install the application? 100% of the students of the two degrees answered yes.

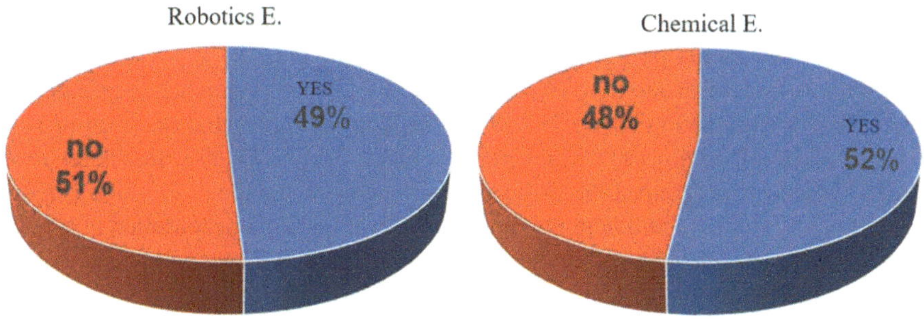

Fig. 11. Was it easy for you to use the Blippar mobile application?

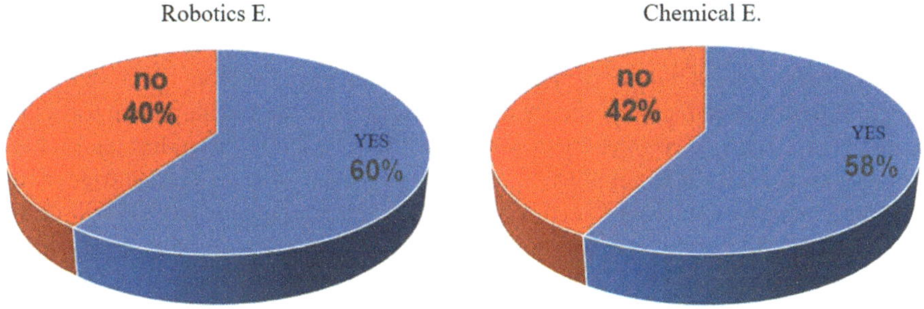

Fig. 12. If you made a correct solution, did the augmented reality application let you know that your result was correct?

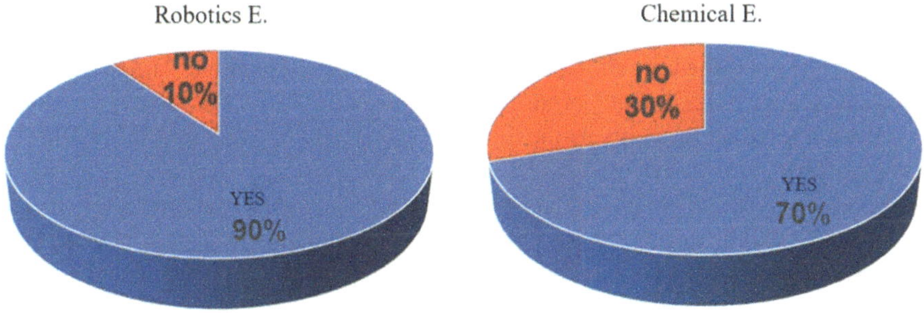

Fig. 13. In case your solution was not the correct one, the information provided by the augmented reality application has allowed you to get some help and be able to cope with the solution.

Concerning the Flipped classroom and videos: in general, the evaluation of the methodology is good. It is surprising that most of the students had not worked with this methodology before in schools or high schools and therefore did not know it. Although we initially believed that this methodology was more established in schools and high school, we have found that this is not the case.

In general, most of the students watched the videos, although there is a small percentage of about 10% who did not watch them beforehand, it would be necessary to think about how to improve this percentage, probably by including in the videos some exercise that they should perform or give feedback to the teacher.

In general, for both groups, the level and speed of the videos seems adequate, although in the robotics engineering group 21% of the students consider that the level of some videos is low, while in no case do they indicate that it is high. On the other hand, in the chemical engineering group, 12% of the students consider that the level of the videos is high. These differences are due to the fact that 90% of the students enrolled in robotics engineering have taken technical drawing in high school, while in chemical engineering only 10% have taken this subject.

In general, they do consider this teaching methodology to be adequate to favor their learning. In other comments, they indicate and suggest that this methodology be extended to other subjects of the course, and they emphasise that the fact of being able to stop and watch the videos several times has been useful. As improvements, they suggest that it should also be extended to the resolution of more complex exercises.

Concerning the evaluation of the augmented reality part, although the installation of the program seemed simple, the specific application of the mobile phone for its use has an average rating. Specifically, they highlight that this application has the problem of trigger focus, since it is not always easy to detect, even with the correct trigger, and sometimes it takes a long time to read. In general, the AR has helped them to know if their solution was correct and if not, they have found a little help to solve it by them- selves. As suggestions they indicate the possibility of including different levels of resolution in case the solution they present is not correct. They also suggest testing other applications that give less trigger reading failures.

5 Conclusions

In general, the combined teaching experience is considered favorable. The two methodologies have improved student learning and have been useful for the proposed purpose. The students value more positively the flipped classroom with video than the augmented reality material.

Therefore, as future goals, lectures will gradually include this methodology in other subjects or expand the level of these subjects. Also in future applications of the methodology, we intend to compare the academic and knowledge evaluation results of students who have used this new methodology versus a group of "control" students who have used traditional methodologies. In this way, it is intended to justify comparatively and quantitatively the benefits of the new methodology.

As for augmented reality, it is proposed to make a study of other basic and free programs that allow to improve the interpretation of triggers with the mobile, so that errors due to technology are minimized and therefore it can be a much more useful tool.

Acknowledgments. We would like to thank the research group "Diseño de Ingeniería y desarrollo tenol6gico" (DIDET), as well as the program "Redes de investigaci6n docente of the University of Alicante" for the support and aid provided for the technological and teaching improvement.

References

1. Schwab, K.: Cuatro principios de liderazgo de la Cuarta Revolución Industrial. World Econ. Forum. (2016)
2. Alam, G.M., Forhad, A.R., Ismail, I.A.: Can education as an 'International Commodity' be the backbone or cane of a nation in the era of fourth industrial revolution? - a comparative study. Technol. Forecast. Soc. Chang. **159**, 120184 (2020). https://doi.org/10.1016/J.TECHFORE.2020.120184
3. Ministerio de Educación.: Declaración de Bolonia de 1999: Declaración conjunta de los ministros europeos de educación. Dispon. en la Dir. web http//www.Educ.gob.es/boloniaen secundaria/img/Declaracion_Bolonia.pdf (consultada en noviembre 2011) (1999)
4. de Wit, H.: European integration in higher education: the bologna process towards a European higher education area. Int. Handb. High. Educ. 461–482 (2007). https://doi.org/10.1007/978-1-4020-4012-2_23
5. McCarthy, J.P., Anderson, L.: Active learning techniques versus traditional teaching styles: two experiments from history and political science. Innov. High. Educ. **244**(24), 279–294 (2000). https://doi.org/10.1023/B:IHIE.0000047415.48495.05
6. De Gloria, A., Bellotti, F., Berta, R.: Serious games for education and training. Int. J. Serious Games (2014). https://doi.org/10.17083/ijsg.v1i1.11
7. Cobo, G., Valdivia, S.: El Aprendizaje Basado En Proyectos. Rev. Panor (2018)
8. Schönhals, A., Hepp, T., Gipp, B.: Design thinking using the blockchain. In: CRYBLOCK 2018 - Proceedings of the 1st Workshop on Cryptocurrencies and Blockchains for Distributed Systems, Part of MobiSys 2018 (2018). https://doi.org/10.1145/3211933.3211952
9. Lam, P., Lau, C.K.M., Chan, C.H.: Flipped classroom assessment: a learning process approach. In: Multi Conference on Computer Science and Information Systems, MCCSIS 2019 - Proceedings of the International Conference on e-Learning 2019 (2019). https://doi.org/10.33965/el2019_201909f016
10. WordPress.com site: Education Blogger
11. Cabero-Almenara, J., Vázquez-Cano, E., López-Meneses, E.: Uso de la Realidad Aumentada como Recurso Didáctico en la Enseñanza Universitaria. Form. Univ. (2018). https://doi.org/10.4067/s0718-50062018000100025
12. Unity Technology: Unity 3D. https://doi.org/10.3969/j.issn.1002-6673.2014.04.028
13. Augment: Virtual Reality vs. Augmented Reality (Augment News)
14. Holzschuh, V.R.B., Bogoni, T.N.: Aurasma: a tool for education. In: Proceedings - 19th Symposium on Virtual and Augmented Reality, SVR 2017 (2017). https://doi.org/10.1109/SVR.2017.40
15. The Blippar picture. Eng. Technol. (2013). https://doi.org/10.1049/et.2013.0433
16. Díaz Ivorra, M. del C.: Vistas (2019). http://rua.ua.es/dspace/handle/10045/97693

BIM Workflows in the Classroom: A Topographical and Earthworks Experience with Autodesk Revit® and AutoCAD Civil3D®

Jacinto Santamaría-Peña(✉), Sergio Rojo-Vea, and Félix Sanz-Adán

Department of Mechanical Engineering, University of La Rioja, 26004 Logroño, (La Rioja), Spain
jacinto.santamaria@unirioja.es

Abstract. Interoperability is one of the fundamental characteristics required of any software that aspires to be used successfully within the BIM methodology.

And the multiple strategies that architecture and engineering projects require today to be developed successfully make it difficult to face any company with a single pro- gram, no matter how commercially successful it is. On the other hand, the development of digital skills within BIM flows makes it essential for students to routinely assume continuous interface changes.

Autodesk Revit and AutoCAD Civil3D are just two of the many softwares that make up the Autodesk AEC Collection (Architecture, Engineering and Construction Collection). The fact that they are part of the same family facilitates their interoperability, and therefore opens up a didactic experience with an optimal learning curve.

This article will discuss how to approach the learning of the information flow between Revit and AutoCAD Civil3D, through a simple topographical exercise, with analysis of levellings, embankments and earthworks.

Keywords: Interoperability · Workflows · BIM · Revit · AutoCAD Civil3D · Topography · Earth works

1 Introduction

In this paper, two different workflows will be analyzed when dealing with the teaching and learning of typical exercises for surveying and earthmoving work. As a starting material, there is a point cloud of a plot, obtained by topographic surveying using GNSS. It is proposed to generate a grading at a determined elevation with its corresponding slopes.

It is intended to discover and compare the advantages and disadvantages of two methodologies, or to follow a BIM process using Autodesk Revit®, or to use CAD3D software for topographic orientation such as AutoCAD Civil3D®. This will make it possible to reach conclusions regarding which of the two methodologies offers the best results at the teaching level.

F. Cavas Martínez et al. (Eds.): INGEGRAF 2021, LNME, pp. 358–365, 2022.
https://doi.org/10.1007/978-3-030-92426-3_41

2 Phases of the Process of Creating a Grade Using 3D CAD Software

2.1 Phase 1: Creation of the Surface. Assignment of Points. Triangulation

In general, the first task to be carried out is the creation of the digital surface that defines the natural terrain, either through a point cloud from a topographic survey or through previously obtained contour lines. In this first phase, a specific style of visualization of its components (plot boundary, contour lines, triangles,…) and a rendering material are usually associated with cited digital surface. The surface, in general, will be configured as a TIN (Triangulated Irregular Network) surface, starting from cited points, forming a network of triangles whose vertices correspond to cited points in their position X, Y, Z (Fig. 1).

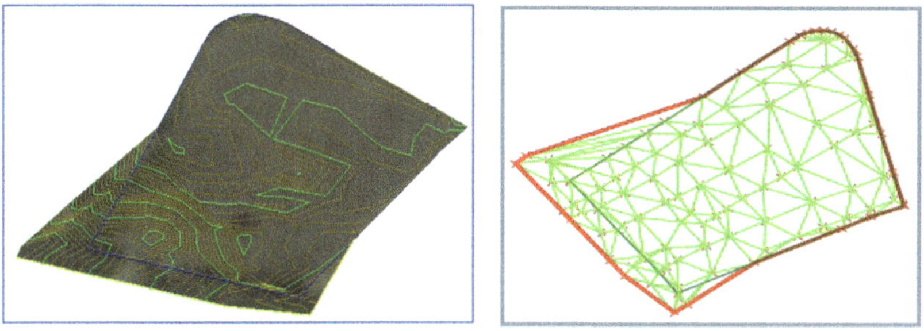

Fig. 1. TIN surface creation and surface style assignment.

The default interpolation method to perform the triangulation is the Delaunay method, based on the fact that the circumscribed circumference of each triangle must not contain any vertex of another triangle, obtaining homogeneous triangles without excessively sharp angles (Fig. 2).

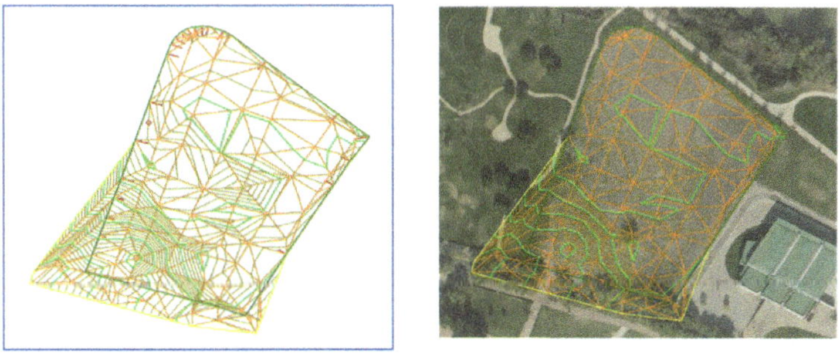

Fig. 2. Triangulation, plot boundary and contour line of the terrain.

2.2 Phase 2: Definition of the Parcel Perimeter and Grading Slopes

The projected grading will start from a perimeter line defined in the project and from which the cut or fill slopes will emerge with the assigned slopes. In our example, we will create a grading that starts from the perimeter line (parcel boundary) and generates slopes with a slope of 2H:1V towards its interior, except on one of the access sides to the parcel, which will have a slope of 5H:1V and These slopes will all ascend until reaching a height equal to 423 m. In this way, the generated esplanade will be slightly elevated above the grade of its access sidewalk (Fig. 3).

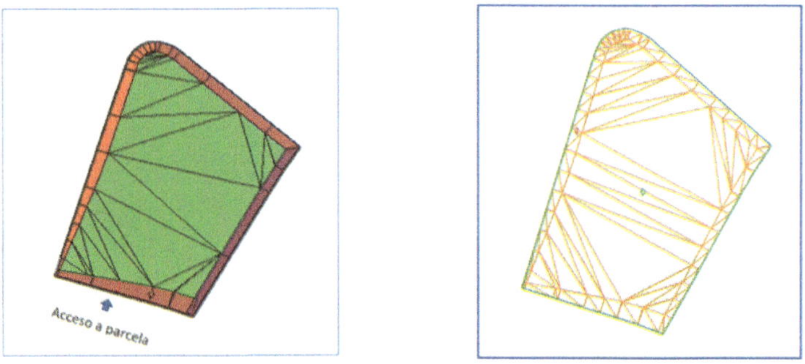

Fig. 3. Generation of slopes and grading surface.

2.3 Phase 3: Generation of Grading Surfaces and Land Volume Calculation

Once the slopes and the platform are defined at elevation 423 m, the grading and platform surfaces are generated, so that these surfaces can be compared with the original surface of the natural terrain. The volume of this ground movement determines the need to carry out a 38,128 m^3 embankment (Figs. 4 and 5).

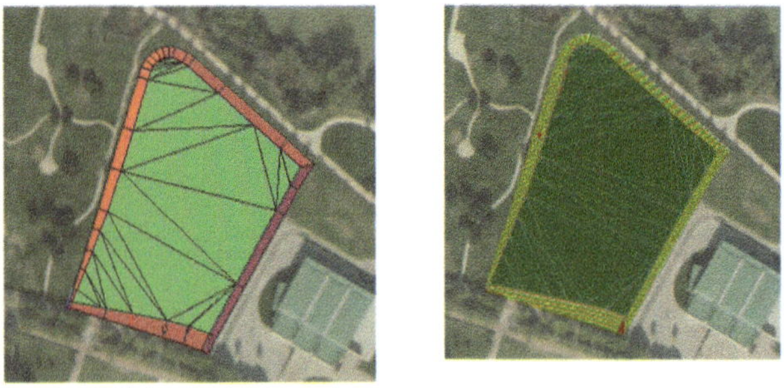

Fig. 4. Target surface model.

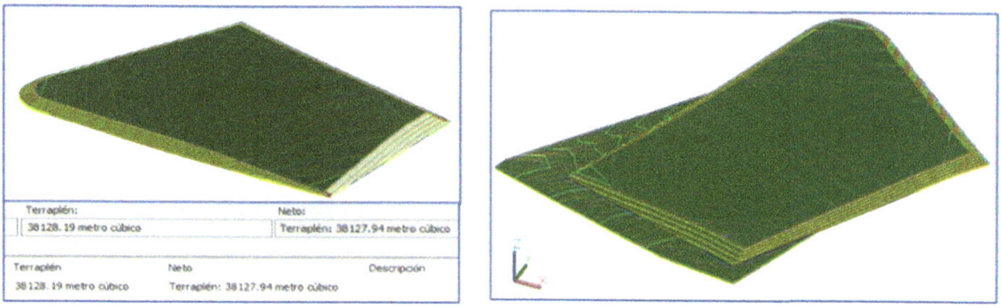

Fig. 5. Original and definitive Surface. Earthworks volume.

3 Phases of the Process of Creating an Earth Grading Using Autodesk Revit® 2022 Software

3.1 Phase 1: Creation of the Surface. Point Cloud Import

In the project browser, to work in an orderly manner, three phases are created (Manage/Phasing) and assigned to three new views that have the same name. These phases would be Existing (as we obtain the plot from the point cloud), Earthworks (the phase in which we operate) and New Construction (where the final phase appears). Thus, the existing view corresponds to the existing phase and so on. Tool panel, Manage/Process by phases tab.

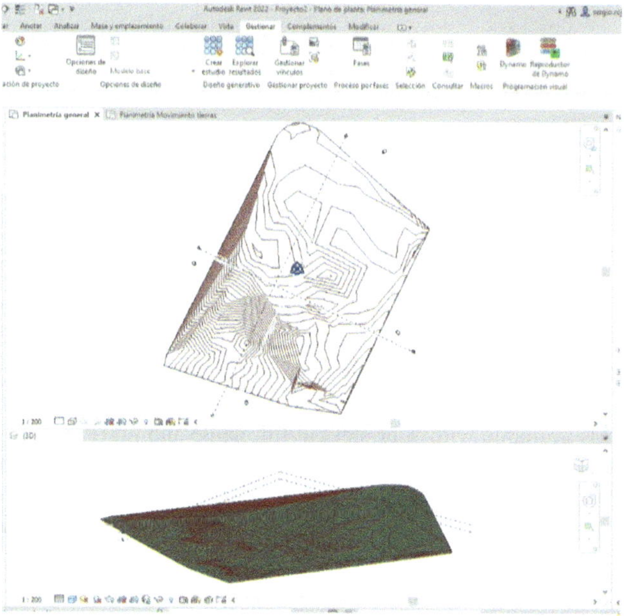

Fig. 6. Creating the topographic surface from a file *.txt. Contour lines at 0.10 m.

We start from a point cloud in * .txt format. On the Massing & Site/Toposurface/Create from Import/Specify Points File toolbar.

Visualizing the data that the point cloud gives us, we can assign to the levels 1 and 2 that come by default in the Revit template, the lowest and highest dimensions respectively. The lowest elevation would be 419 for level 1 and the elevation of our platform would be 423 for level 2 (Fig. 6).

3.2 Phase 2: Ground Movement

We pass to the Ground Movement phase, in the view of the same name. To define the platform limits, whose only data is its altimetry, we model embankments with the given slope (2H:1V towards the inside, except on one of the access sides to the plot, which will have a 5H:1V slope). And we interrupt them at their intersection with the +423 elevation. In this way, we obtain the perimeter of the platform (Figs. 7 and 8).

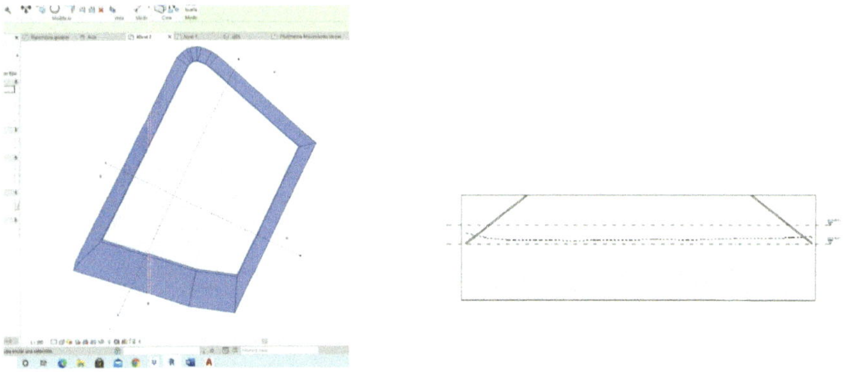

Fig. 7. Creation of auxiliary embankments, with the command "roof" from the architecture toolbar. This allows us to specify the slope and its intersection with the altimetric plane +423.

When obtaining the perimeter, we are placing points at the elevation +423 coinciding with that limit.

This involves a point-to-point editing of the altimetry, which is tedious. In addition, we do not apply an exact slope to the embankment, since, by hand, the contour lines are placed in parallel. We compose a planning table with the data that interest us in relation to topography. Analyze/Schedule tab. In our case, we will select the corresponding topography fields available with: Cut, net cut/fill, name and fill (Fig. 9).

It can be seen how, in our case, the cut volume is 0 m^3, and the fill volume is 37,638.95 m^3, similar to 38,128 m^3 obtained following the previous methodology. Specifically, 1.2% less, which is still a good approximation. Therefore, one of the first conclusions may be this, that obtaining results is acceptable. However, we must bear in mind that the tools that have been used to calculate the fill are inside the building bar, in the architecture group. This is undoubtedly a disadvantage with respect to AutoCAD Civil3D, since the embankments have been calculated as if they were skirting, taking as a reference a crowning platform that arises from the intersection of a horizontal plane at

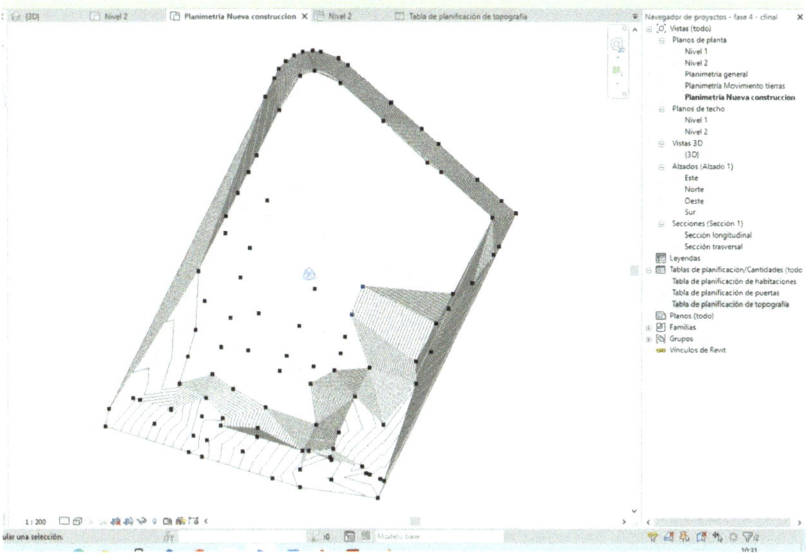

Fig. 8. Creation of a platform, its perimeter already known at level +423.00. The points of that height are placed on the perimeter.

<Topography schedule>			
A	B	C	D
Name	Cut	Fill	Net cut/fill
Existent plot	0.00 m³	0.00 m³	0.00 m³
Earthworks	0.00 m³	37638.95 m³	37638.95 m³

Fig. 9. Topographic planning table of the worked model.

elevation +423.00 with some slopes with a given slope whose origin coincides with the entire perimeter, and is located at the elevation +419.00. This would imply an inaccuracy, since the program does not monitor the altimetric situation of the perimeter of the surface, but rather extrapolates the altimetric data of the same to a situation in which all of it is at the same height, exactly as if it were an eaves of a roof (Fig. 10).

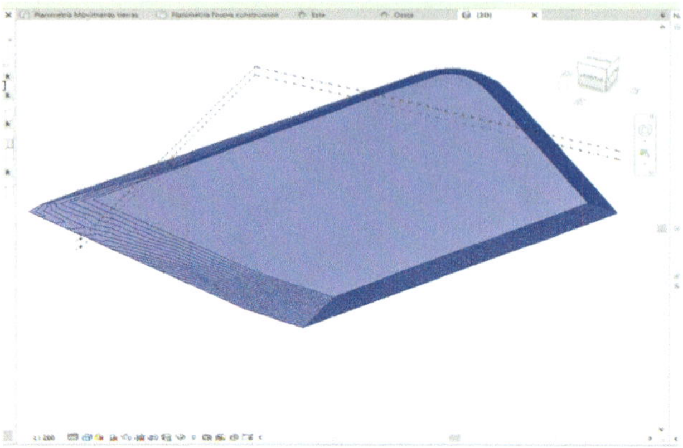

Fig. 10. Final surface.

4 Flowcharts

In Figs. 11 and 12, the different phases of the process of creating the surface model, grading and measuring the earthworks, using AutoCAD Civil3D® and Revit®, are schematized.

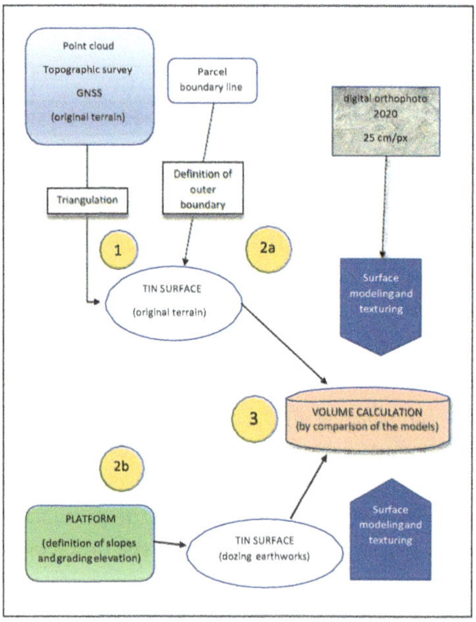

Fig. 11. Flow chart using Civil3D®.

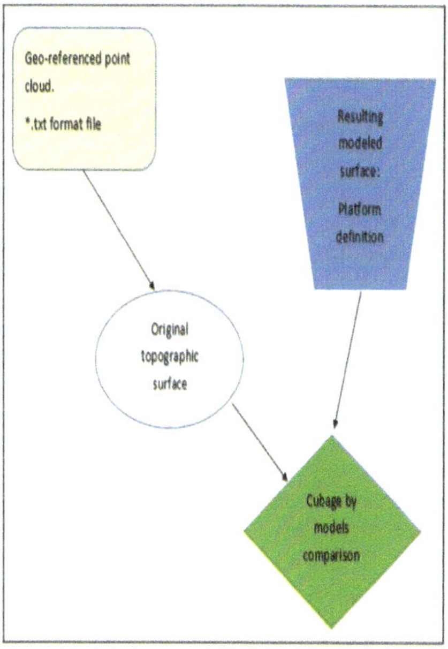

Fig. 12. Flow chart using Revit®.

5 Conclusions

After the description of both methodologies and the practical development of the different processes, it is concluded that the AutoCAD Civil3D® software is better oriented to the execution and calculation of ground movements in topographic works, with specific tools that greatly facilitate student learning. The BIM environment with Autodesk Revit 2022®, although it allows working with digital terrain models, uses solutions that are not very intuitive and offer little versatility to the user, standing out in the design and management of construction elements and installations for post- earthworks execution.

References

1. Reyes, A.M.: Diseño y documentación de un edificio industrial. Anaya, Spain (2020)
2. Martínez, D.: Civil3D. Anaya, Spain (2019)
3. Lv, S.: Study on BIM modeling method of bridge railway integration based on Revit and Civil3D. In: 2018 7th International Conference on Energy and Environmental Protection (ICEEP 2018), pp. 214–220. Atlantis Press, Amsterdam (2018)
4. Petschek, P.: LandscapingSMART, digital terrain modeling and BIM. In: Grading, pp. 163–184. Birkhäuser, Berlin, Boston (2019). https://doi.org/10.1515/9783035620375-009
5. Menéndez Cárdenas, Margarita: Incorporación de Metodología BIM en la Gestión Inte- grada de Proyectos. Máster Universitario en Gestión de Proyectos de Edificación (2016)

Topographic Levelling from the Point of View of Active Learning Using the Flipped Classroom Model

Elidia Beatriz Blázquez-Parra(✉), Daniel López Granero, Francisca J. Castillo Rueda, Laia Miravet-Garret, M. Carmen Ladrón de Guevara-Muñoz, and Francisco David Trujillo-Aguilera

University of Malaga, C/Doctor Ortiz Ramos, s/n, Malaga, Spain
ebeatriz@uma.es

Abstract. University education has experienced an increase in the use of ICT-related tools, to some extent driven by the pandemic situation that occurred during the past academic years 2019/20 and 2020/21. Face-to-face small group practices that required the use of specific equipment by students have been relegated to other types of activities that allow keeping an adequate distance between individuals. At this point, active methodologies acquire a fundamental role for large group virtual theoretical classes, when having to keep focussed for one hour and a half on the traditional masterful method, thus becoming a true challenge. This paper presents the work carried out to complement the teaching of the subject Graphic Engineering and Topography, and more precisely, in the area related to topographic levelling to incorporate the flipped classroom model as a pilot experience in some topics. The follow-up of this experience has been performed by means of activity reports and a satisfaction survey.

Keywords: Active methodology · Teaching-learning process · Flipped classroom model · Graphic engineering · Levelling

1 Introduction

The incorporation of technology and other computer resources in all educational systems has recently undergone a revolution following our social and professional habits [1, 2]. Among the potentialities of Information and Communication Technologies (ICT) in education, the active profile attained by students while the teacher becomes a mere guide in their learning process, stands out [3, 4]. This framework boosts an improvement in students' motivation, autonomy, participation, and attitude and, ultimately, an increase in the quality of their training [5, 6]. However, during the Covid19 pandemic, the use of ICT has become essential support for the educational community [7], leading to a profound transformation of very high-quality teaching-learning processes based on active learning [8].

Although it is true that in university education, students already have a considerable capacity for autonomous and self-regulated study [9] and even greater effectiveness in

F. Cavas Martínez et al. (Eds.): INGEGRAF 2021, LNME, pp. 366–374, 2022.
https://doi.org/10.1007/978-3-030-92426-3_42

the virtual system [10], the change to the online system has not been chosen voluntarily, which may have had a negative impact on students' motivation [11]. This condition has also required higher levels of autonomy since, besides having had to face external stressors due to the exceptional situation, they have had to manage the resulting reduction of social interaction, the lack of close support from their classmates and teachers, as well as having had to cope with possible distractions taking place at home [12, 13].

In a face-to-face environment, teachers can act based on the manifest behaviour of students as evidence of their attention or interest. Nevertheless, in the online environment, the teacher may hopefully see a few students' faces, greatly hindering their ability to capture students' attention. Therefore, one disadvantage of synchronous virtual classes is that they may find it challenging to remain engaged for long periods of time. Two aspects should be borne in mind to establish strategies aimed at increasing student participation in both face-to-face and online courses: a) the synchronous learning format is helpful for discussing complex ideas while b) asynchronous learning or, in this case, offline, can be used to carry out self-learning activities, these being recommended to be relatively short and having questionnaires or short essays associated, such as suggested videos on the fundamental concepts of the studied subject [14].

Here is where the flipped classroom model is presented as a potentially vital tool for teaching-learning processes that have already proved to provide a more effective model at the university level than methodologies based on expository lectures without any digital resource employment [15]. One of the main contributions is, as stated before, the increase in student motivation and commitment to the task [16].

The flipped model allows students to view digital contents at their own pace and needs anytime and anywhere. During the online synchronous session, they will be able to deepen into their self-acquired knowledge through practice.

Topography field practices and topographic levelling are taught in the course called 'Graphic Engineering and Topography', which corresponds to the second year of the 'Industrial Technologies' degree at the University of Malaga. They are developed in the open air in any way comparable to future professional work. They give students the chance to practice theoretical concepts through the actual handling of topographic instruments, which students especially welcome. The cancellation of these practices due to the impossibility of guaranteeing the Covid security conditions entails a double challenge to the teachers of this course: to compensate most faithfully for the learning derived from these open-air practices and reinforce students' motivation.

This manuscript is structured as follows: the introduction is outlined in Sect. 1, materials and methods are explained in Sect. 2, the results and discussion are displayed in Sect. 3 and conclusions and future lines of research are presented in Sect. 4.

2 Material and Methods

The flipped classroom model is carried out taking into account the following phases:

0. *Preparation of the material* by the teaching staff.
1. Work to be done before the class. The student carries out a personal study with the help of didactic material. Hence, he/she has to receive explicit information with

updated material following the competencies that the student is expected to acquire, to facilitate the resource and understanding of Bloom's cognitive dimension. In addition, there is a process of assimilation of the information, aided by the completion of questionnaires.
2. *Work during class.* To improve the analysis of what has been learnt through exercises and complementing knowledge.
3. *After class.* The student works on mechanization and completes his or her understanding.

Among the different methodology phases, this article focuses on the first two (preparation of material and work before class).

This section includes the materials and methods used in the development of the flipped classroom model. In the development of this model, students are a fundamental part of its development. Thus, this pilot experience in engineering was carried out with students in the second year of the Degree in Industrial Technologies Engineering, precisely in Graphical Engineering and Topography of the School of Industrial Engineering of the University of Malaga, which is taught in the first semester. The data refer to the 2020–21 academic year. Although the total number of students was 134, distributed in 3 large groups and five subgroups, a sample subgroup of 25 students was chosen.

3 Development of the Content of Topographic Levelling

The subject of Engineering Graphics and Topography consists of 16 topics, with topic 13 on levelling, profiles and earthworks being the subject of study for this work. In total, a dynamic work strategy is established with the definition of the scripts of each video according to the competencies that students wish to acquire: the use of software for the creation of material, the design of a questionnaire for a better understanding of the subject and the revision and validation of each content.

Five different videos of less than 5 min in length were chosen from among the videos produced.

3.1 Script

After the analysis of the main topics that make up the syllabus of the subject, it was decided to make a video on the definition of levelling and the type of instrumentation. This way, students could learn the basics of topographic levelling. Another video was made on geometric levelling, another on trigonometric levelling, the fourth video on barometric levelling and finally a video and examples on the calculation of a levelling notebook, Fig. 1.

Each of the videos is studied independently, first establishing a strategic plan that takes into account what it has to contain, who it is aimed at, how the information is to appear and what format is to be used. Also, a briefing was made with the strengths and weaknesses, problems and resources needed in order to be as objective as possible. A synopsis was written including the treatment of different scenes and defining what is to

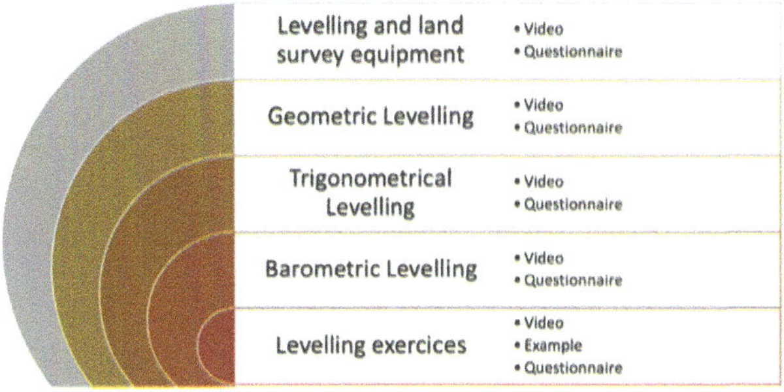

Fig. 1. Content of flipped classroom on topographic levelling.

appear in each photo. It is also important to determine timing and correct sequencing of the video to ensure continuous learning.

For the validation of the video and its use by students, and based on the analysis of the fundamental aspects of the video, a questionnaire on the contents was carried out, which is used for the first phase of the flipped classroom model.

3.2 Software Used

Inkscape, a program that works with vector images, was used to create the images, while the OpenShot Video Editor was used to edit the videos (Fig. 2). Google Forms was chosen for the questionnaires and Microsoft Excel for the exercises.

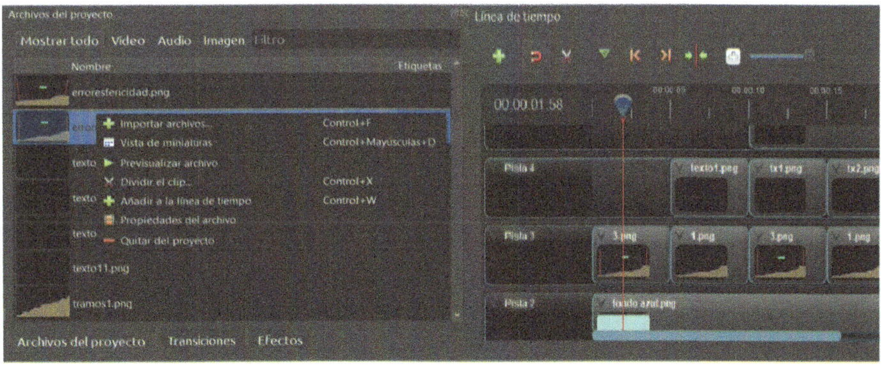

Fig. 2. View of the video editing software OpenShot Video Editor.

3.3 Videos and Quizzes

Using the chosen software and based on the script, the five videos indicated above were made. For a better understanding of the subject, as reinforcement and feedback for the teaching staff, each of the videos is accompanied by a questionnaire in order to reinforce the knowledge gained from the video, Fig. 3.

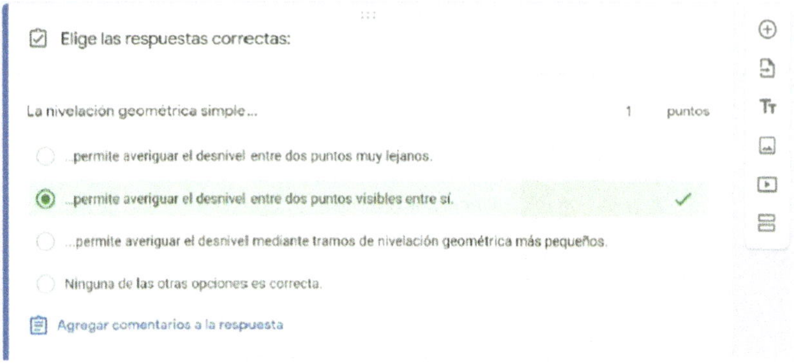

Fig. 3. Sample questionnaire question in Google Forms.

Finally, to check the material generated, it was viewed numerous times in search of any type of error to see if the video time was adequate and if the questions were correctly formulated. Once the material had been validated in the first phase, it was launched to the study group to set up a pilot experience.

4 Results

The material generated for levelling has a tremendous educational value in university teaching. They are simple videos that explain the necessary concepts in a straightforward way (Fig. 4).

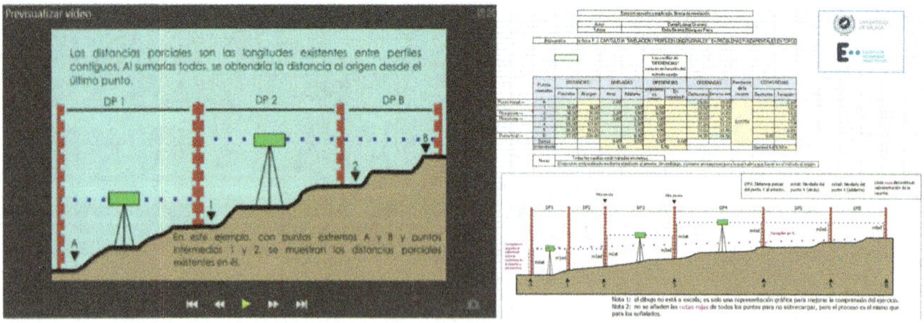

Fig. 4. Example of a video frame and levelling notebook.

These videos generated to be able to carry out the work in front of the class studying the subject were uploaded to the Moodle platform used at the University of Malaga, which is available to all students of this subject (Fig. 5).

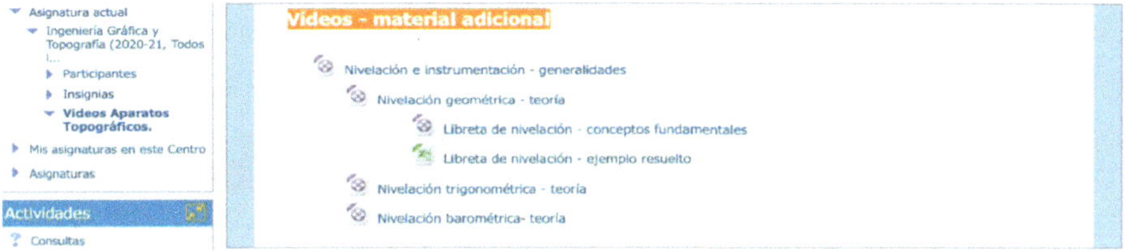

Fig. 5. Implementation in the subject of Engineering Graphics and Topography in the subject.

During the course, the reports generated by the virtual campus platform allow us to check the level of use of the material produced (Fig. 6).

Nivelación e instrumentación - generalidades	38 por 25 usuarios	viernes, 23 de abril de 2021, 18:29 (30 días 4 horas)
Nivelación geométrica - teoría	34 por 26 usuarios	lunes, 1 de febrero de 2021, 06:18 (111 días 15 horas)
Libreta de nivelación - conceptos fundamentales	63 por 31 usuarios	viernes, 26 de febrero de 2021, 13:43 (86 días 8 horas)
Libreta de nivelación - ejemplo resuelto	62 por 38 usuarios	jueves, 22 de abril de 2021, 20:14 (31 días 2 horas)
Nivelación trigonométrica - teoría	21 por 16 usuarios	jueves, 22 de abril de 2021, 20:17 (31 días 2 horas)
Nivelación barométrica- teoría	16 por 14 usuarios	jueves, 22 de abril de 2021, 20:20 (31 días 2 horas)
Encuesta vídeos de nivelación	20 por 15 usuarios	domingo, 23 de mayo de 2021, 19:35 (3 horas 23 minutos)

Fig. 6. Report on the documentation generated by Engineering Graphics and Topography.

Figure 6 shows that, although the target audience consisted of 25 students, in some cases, such as the leveling notebook, there were 38 users with a total of 62 views of the material. The second most viewed video corresponds to general concepts followed by trigonometric levelling. Nevertheless, it can be seen that students have consulted the material, but participation in the surveys has been low.

In addition to the questionnaires on each topic, a survey was carried out in order to assess the usefulness of the material generated. Questions were asked about the usefulness of each video generated. The result is shown in Fig. 7(a), where it can be seen that participation considering gender was lower for women (33%) than for men (67%). However, if we take into account the rate of students in the subgroup, the participation of women was relatively higher. Whether the material has helped to improve understanding of the subject, 75% gave it a value of 5 out of 5 on the Likert scale. Figure 7(b) shows that the evaluation of the project was very positive.

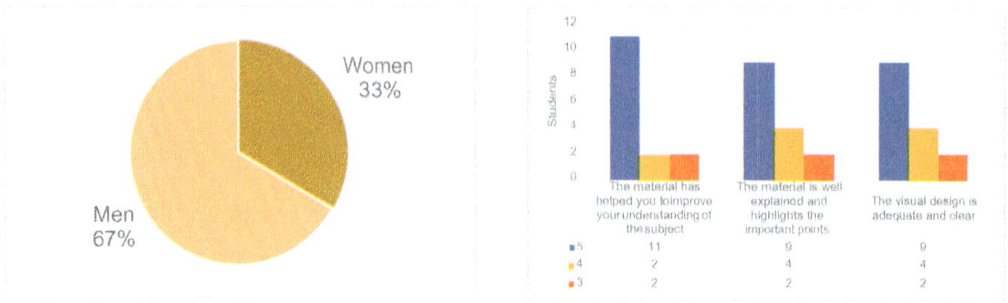

(a) (b)

Fig. 7. (a) Percentage of participation by gender, (b) General assessment of the videos produced. Report on the documentation generated by Engineering Graphics and Topography,

Figure 8 shows the evaluation with respect to the material studied, separating them by videos. It can be deduced that, in general, students' evaluation of the material generated is high, being more concentrated in values 4 and 5.

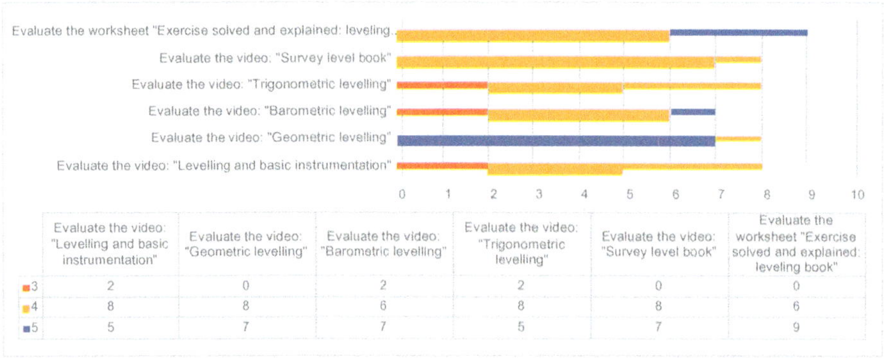

Fig. 8. Individual assessment of the material produced in the course.

On the other hand, of all the videos, the practical example of the levelling notebook is the one that has had the greatest acceptance with 62% of those surveyed, giving it the highest score, and, as explained in the previous section, this same video is the one that has been viewed most often.

5 Conclusion and Future Lines of Research

The flipped classroom model is viable for its application to active methodologies in the university teaching of Engineering Graphics. It improves the understanding of the syllabus by the agents involved in the teaching/learning process.

This project has generated a series of digital material available to all students of the subject. Although it has been carried out through a study group of 25 students, the idea is to disseminate it to other students, taking into account the pilot experience carried out.

The use of free software has become a requirement for obtaining material. It is available to any user, regardless of whether the university has a licence for other types of software. Therefore, to create quality material, it is not necessary to pay a high financial outlay. What is more, what is needed is good planning when creating the material.

The pandemic situation during this academic year has made it possible for students to follow the subject virtually. In addition to this advantage, there are others such as personal commitment to active learning, it is considered universally accessible in that it can be accessed from anywhere with a suitable connection, it facilitates understanding, analysis and critical thinking, it is timeless in the fact that the useful life of the contents can be extended over time, it is a fundamental part of the digital transformation of society. And it has become a new channel for the dissemination of this subject.

As for future lines of investigation, an improvement in the comprehension questionnaires is proposed, extrapolating it to other degrees that are directly related to these planned subjects. The development of new material for different issues in the subject with a standard design format, in such a way that there is unification and standardization, is also proposed. This extrapolation would mean that around 535 students from the School of Industrial Engineering could benefit from the material generated.

Another future research line would be creating a universally accessible channel so that students enrolled in the subject can access it.

Finally, related to the previous line, is the internationalization of contents so that they appear in different languages and subtitles, allowing accessibility to contents so that they become inclusive.

Acknowledgements. The research work reported here was made possible by the collaboration between Graphical and Survey Engineering students and professors. It has also been supported by Education Innovation Project PIE159, University of Malaga (Spain).

References

1. Capilla, M.M., Torres, J.M.T., Sánchez, F.R.: Percepciones acerca de la integración de las TIC en el proceso de enseñanza-aprendizaje de la universidad. Píxel-Bit, Revista de Medios y Educación **46**, 103–117 (2014). https://doi.org/10.12795/pixelbit.2015.i46.07
2. Rodríguez Correa, M.: La integración de las TIC en los centros universitarios: la visión de los directivos. Edutec. Revista Electrónica de Tecnología Educativa, **64**(386), 41–50 (2018). https://doi.org/10.21556/edutec.2018.64.975
3. Figueroa Rojas, R.: Rol del docente universitario en el proceso de enseñanza y aprendizaje de las Ciencias Naturales a través de internet. Educere, **16**(1316–4970), 37–42 (2012). http://www.redalyc.org/articulo.oa?id=35623538005
4. López-Quintero, J.L., Pontes-Pedrajas, A., Varo-Martínez, M., Luis López-Quintero, J., Pontes-Pedrajas, A., Varo-Martínez, M.: Las TIC en la enseñanza científico-técnica hispanoamericana: Una revisión bibliográfica. Dig. Educ. Rev. **35**, 229–243 (2019). https://doi.org/10.1344/der.2019.35.229-243
5. Graells, P.M.: Impacto de las TIC en la enseñanza Universitaria. In DIM: Didáctica, Innovación y Multimedia (Issue 11) (2008). http://dewey.uab.es/pmarques/ticuniv.htm
6. Ocaña-Fernández, Y., Valenzuela-Fernández, L., Morillo-Flores, J.: La competencia digital en el docente universitario. Propósitos y Representaciones **8**(1), 455 (2020). https://doi.org/10.20511/pyr2020.v8n1.455

7. Ieng Lei, S., Siu Ian So, A.: Online teaching and learning experiences during the COVID-19 pandemic-a comparison of teacher and student perceptions (2021). https://doi.org/10.1080/10963758.2021.1907196
8. Latorre-Cosculluela, C., Suárez, C., Quiroga, S., Sobradiel-Sierra, N., Lozano-Blasco, R., Rodríguez-Martínez, A.: Flipped Classroom model before and during COVID19: using technology to develop 21st century skills. Interact. Technol. Smart Educ. **18**, 189–204 (2021). https://doi.org/10.1108/ITSE-08-2020-0137
9. Zimmerman, B.J.: Becoming a self-regulated learner: an overview. Theory Into Pract. **41**(2), 64–70 (2002). https://doi.org/10.1207/s15430421tip4102_2
10. Cerezo, R., Bernardo, A., Esteban, M., Sánchez, M., Tuero, E.: Programas para la promoción de la autorregulación en educación superior: un estudio de la satisfacción diferencial entre metodología presencial y virtual. Europ. J. Educ. Psychol. **8**(1), 30–36 (2015). https://doi.org/10.1016/j.ejeps.2015.10.004
11. Hsu, H.C.K., Wang, C.V., Levesque-Bristol, C.: Reexamining the impact of self-determination theory on learning outcomes in the online learning environment. Educ. Inf. Technol. **24**(3), 2159–2174 (2019). https://doi.org/10.1007/s10639-01909863-w
12. Quintiliani, L., Sisto, A., Vicinanza, F., Curcio, G., Tambone, V.: Resilience and psychological impact on Italian university students during COVID-19 pandemic. Distance learning and health. Psychol. Health Med. (2021). https://doi.org/10.1080/13548506.2021.1891266
13. Son, C., Hegde, S., Smith, A., Wang, X., Sasangohar, F.: Effects of COVID-19 on college students' mental health in the United States: Interview survey study. J. Med. Inter. Res. **22**(9), e21279 (2020). https://doi.org/10.2196/21279
14. APA: Managing attention and distractibility in online learning. In American Psychological Association. https://www.apa.org/topics/covid-19/managing-attention-distractibility-online-learning (2020)
15. Martínez, T.S., Díaz, I.A., Rodríguez, J.M.R., Rodríguez-García, A.-M.: Eficacia del método flipped classroom en la universidad: meta-análisis de la producción científica de impacto. REICE. Revista Iberoamericana sobre Calidad, Eficacia y Cambio en Educación **17**(1), 25 (2018). https://doi.org/10.15366/reice2019.17.1.002
16. Blázquez-Parra, E. B., Miravet-Garret, L.: El modelo de Clase Invertida en la ingeniería y en el área de la Expresión Gráfica. In: Santamaría-Peña, J., Sanz-Adán, F., (Eds.), La Transformación Digital en la Ingeniería Gráfica: Experiencias metodológicas para una enseñanza/aprendizaje innovadora de la Ingeniería Gráfica, pp. 79–93) (2019). https://dialnet.unirioja.es/servlet/articulo?codigo=7151261

Didactic Strategy Based on Experimental Process and Physical Manipulation for a Meaningful Learning of the Dihedral System

Jon Mikel Cabezas Escaño[1], Itziar Gonzalez Gurrutxaga[1], María Lozano Chico[1], Jose Antonio Oriozabala Brit[1], and Lorena Ugarte Soraluce[2](✉)

[1] Department of Graphical Design and Engineering Projects, Faculty of Engineering, Gipuzkoa, Universidad del País Vasco/Euskal Herriko Unibertsitatea, Plaza Europa 1, 20018 Donostia-San Sebastian, Spain

[2] Department of Graphical Design and Engineering Projects, Faculty of Engineering, Gipuzkoa-Eibar Section, Universidad del País Vasco/Euskal Herriko Unibertsitatea, Avenida Otaola 29, 20600 Eibar, Spain

lorena.ugarte@ehu.eus

Abstract. A didactic strategy, which aims to get a meaningful learning of the theoretical basics in descriptive geometry in the subject Graphical Expression, is presented. As starting hypothesis, it is considered that the experimental processes based on the interaction with physical objects help the development of spatial vision capacity and hence, the comprehension of graphical representation techniques. To materialize this hypothesis, a didactic strategy based on experimentation-deduction processes was designed and applied. The strategy consisted on two pillars. On the one side, Cooperative Activities were developed to work on theoretical basics by means of experimentation-deduction processes of interactive character. On the other hand, didactic resources were created. These resources allowed students to interact with three-dimensional reality and work in the proposed experimentation-deduction processes. Results showed an improvement on the qualifications of students. In addition, the 70% of student body considered that the strategy was helpful for the comprehension of theoretical basics. As future lines, it is proposed to extend the strategy to the practical part of the subject.

Keywords: Descriptive geometry · Spatial vision capacity · Didactic strategy · Experimentation-deduction processes

1 Introduction

Spatial vision capacity is the main competence of the subject Graphical Expression, taught on the first course of Industrial Engineering grades at Gipuzkoa Engineering School (University of the Basque Country). According to the European Higher Education Area (EHEA), students of engineering grades should acquire spatial vision capacity and

F. Cavas Martínez et al. (Eds.): INGEGRAF 2021, LNME, pp. 375–383, 2022.
https://doi.org/10.1007/978-3-030-92426-3_43

knowledge of graphical representation techniques, both by traditional metric geometry techniques and by computer aided design applications.

According to Saorín, [1] spatial vision capacity can be defined as the ability to mentally manipulate the objects in a two-dimensional and three-dimensional space. Specifically, the ability of performing rotations of three-dimensional cubes and the ability to recognize three-dimensional parts by folding and unfolding of their faces are mentioned in his work. Keller et al. [2] emphasize that the spatial visualization implies the comprehension of the movements performed by twoand three-dimensional objects in the space. With relation to this, the spatial knowledge may occur when the visualization and the relational thinking are applied simultaneously [3]. In this scenario, Sorby [4] points out that providing students with a high spatial vision capacity may ease the comprehension of theoretical basics in Graphical Expression and, furthermore, they might manage better in the graphical representation of three-dimensional objects in two-dimensions and vice versa.

The faculty is aware of the poor grades in Dihedral System part of Graphical Expression course. According to the previously mentioned, the main reason could be the poor development of spatial vision capacity. It is to mention that Blasko et al. [5] observed that the educational antecedents, the motivation and the performance on spatial and verbal reasoning might affect the performance and dropout rate of first course engineering students. In other work, Blasko et al., [6] identified that secondary school students showed deficiencies in spatial capacity. Moreover, it is also to mention that only a part of the studenthood in Graphical Expression course have received previous instruction in Dihedral System, creating differences on the starting point.

The faculty also identifies a lack of didactic resources to accompany students in the training of spatial vision capacity. Many authors believe that the interaction with objects is the most effective way to develop spatial vision capacity [7–9]. According to Tristando et al., [10], in the starting phases of spatial vision capacity development, the interaction with physical objects is the most appropriate. In the middle phases, however, interaction by means of CAD software is proposed, since this software allows rapid design of exercises and increases the motivation of students. With traditional teaching methods, although not intentionally, students are pushed towards memory learning [11]. However, this learning method implied a very low conscience of real situations. Since all the spatial situations and exercises set out in the subject are presented in different contexts, a high conscience of real situations is required and, in consequence, a deep comprehension of theoretical basics is essential.

In this scenario, the faculty has developed a didactic strategy based on the interaction with physical objects. The foundations of the strategy align with the pedagogical theory of constructivism. According to Jean Piaget, the knowledge is acquired as a result of person-object interaction [12] and, as Seymour Papert stated, the learning is more significant when it is acquired as a result of an experiment based on the manipulation or construction of an object [13]. To apply the strategy, different didactic resources were created. Firstly, 3D cut-out models of objects were designed to carry out physical interactions. Secondly, an element simulating the projection planes in the space, *Fronton*, was created. *Fronton* facilitates the transition from three-dimensional space to two-dimensional representation. Once the element is projected in the spatial projection planes, it can be unfolded to get the projections in the plane. Finally, Cooperative Activities were designed. These activities meet two criterions. First, they have experimentation-deduction nature. They are based in an experimentation-deduction process carried out by both physical manipulation and three-dimensional and two-dimensional representation. Second, they are collaborative in order to promote student-to-student interaction.

2 Purpose and Hypotheses

The purpose of the work was to help engineering first course undergraduates to achieve a more meaningful learning of the Dihedral System, by means of a new didactic strategy.

To carry out the research, the following hypotheses were defined:

1. The proposed didactic strategy helps in the development of spatial vision capacity.
2. The deduction work and the interactions with 3D models carried out by the experimentation processes proposed in Cooperative Activities, contribute to a correct identification and interpretation of the geometrical relationships and positions between the elements involved. In consequence, students achieve a more meaningful learning of theoretical basics.
3. The reconstruction of the real situations proposed by Cooperative Activities in 3D models contributes to a more meaningful understanding of the situations proposed in the exercises. In consequence, students may choose the correct solution method.

These hypotheses would reflect in the following indicators:

1. Improvement of the qualifications in the Dihedral System exam of Graphical Expression course.
2. Positive answering on the questionnaire created to ascertain the agreement of students with the hypotheses.

3 Methods

This study aims to implement a didactic strategy for a particular group of students to verify if these didactic resources are beneficial to developed their spatial vision capacity.

The following describes the characteristics of the participants and the procedure for the implementation of the didactic strategy. The didactic resources and instruments for data collection are also defined.

3.1 Participants

The participants were 149 students in the first course of Industrial Engineering grades at Gipuzkoa Engineering School. A control group consisting of 52 students and an experimental group consisting of 97 students were defined. The students were assigned to each group at random. As a consequence in every group there were both students with previous knowledge of dihedral system and students with no previous instruction in the subject (Table 1).

Table 1. Sample data.

Participants	Knowledge of dihedral system	Without knowledge of dihedral system	Total group
Experimental group	84	13	97
Control group	38	14	52

3.2 Procedure

The research consisted of three stages. In the first stage, theoretical work was carried out and the didactic resources were created in order to improve the spatial vision capacity development. In the second stage, the implementation of the didactic strategy was implemented in the experimental group. Finally, on the third stage, the indicators were collected and analyzed.

3.3 Didactic Resources

Three teaching resources were developed. The first two teaching resources, the manipulative 3D cut-out models and *Fronton*, were described in the *Introduction* and shown in Fig. 1. The third teaching resource are the Cooperative Activities. These activities are the medium to work in the concepts involved in Dihedral System, both individually and in group, giving place to reflective thinking.

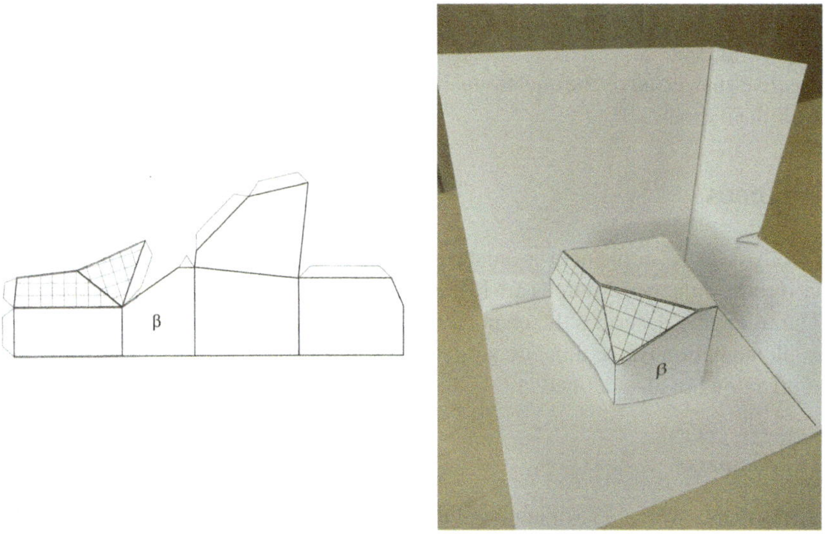

Fig. 1. Unfolded 3D cut-out model (left) and *Fronton* containing the 3D cut-out model (right)

The structure of Cooperative Activities is composed by three sections. In the first one, the learning objectives are defined. The second section outlines the experimentation deduction process and finally, in the third section, the discussion of the conclusions is promoted.

In the section concerning the experimentation-deduction process, students work on the theoretical basics. They individually rebuild the spatial situation proposed in the Cooperative Activity by using the 3D cut-out models and the *Fronton*. The manipulation of the physical objects in this stage helps the interpretation of geometrical position between the elements by means of spatial vision. Next, students are asked to represent by dihedral projections the situation reconstructed in three dimensions. Continuing with the experimentation-deduction process, students undergo a structured deduction process working in pairs or in groups. This deduction process aims a meaningful understanding of implied theoretical basics. Finally, in the third section, the activity finishes by sharing the key concepts deduced from the experimentation deduction process.

Nine Cooperative Activities were designed. Six of them (CA1-CA5, CA7) work on the alphabet of Dihedral System since knowing the favorable positions of the elements is considered a key for a meaningful understanding of the theoretical basics. Moreover, CA6, CA8 and CA9 are designed to work on the theorems and geometrical relations upon which the different solution methods are based.

The designed Cooperative Activities are listed hereafter.

- CA1. Point, straight and plane.
- CA2. Parallel straights to projection planes.
- CA3. Perpendicular straights to projection planes.
- CA4. Perpendicular planes to projection planes.
- CA5. Parallel planes to projection planes.
- CA6. Auxiliary views.
- CA7. Straight of maximum slope (s.m.s.) and straight of maximum inclination (s.m.i.)
- CA8. Real magnitudes of planes through plane changes.
- CA9. Intersections of 3D objects by auxiliary plane method.

3.4 Instruments for Data Collection

Two instruments were defined to collect the indicators: the Dihedral System exam and a questionnaire.

The indicator qualification was collected from Dihedral System exam. The second indicator was collected through a questionnaire in the experimental group. The purpose of this instrument was to check whether students corroborate the hypotheses on which the didactic strategy is supported. The questionnaire was composed by the following questions:

1. Do you think that the spatial vision capacity is an essential key to master the subject?
2. Do you think that the teaching methodology helped you in the development of spatial vision capacity?
3. Do you think that the reconstruction of the real situations proposed in Cooperative Activities helped you in the correct identification of the geometrical position between involved elements?

4. Do you think that carrying out the experimentation processes by means of manipulation in Cooperative Activities helped you understanding of theoretical basics?
5. Do you think that Cooperative Activities help you achieving a more meaningful learning of theoretical basics?
6. Do you think that the correct identification of the geometrical position between elements helped you in choosing the correct solution method?

4 Results and Discussion

Concerning the success rate of students, defined as the percentage of passed exams over the total; it is to mention that a value of 43.6% was achieved in the experimental group. In contrast, in the control group, the value decreased until the 16%. Considering that previous training may influence the success rate, a comparison between the success rates for students with previous training and with no previous training was carried out, both in the experimental and in the control groups. The obtained results are shown in Fig. 2. It can be demonstrated that previous training influenced the success rate. In the experimental group, the success rate reduced to a half approximately, from 48.2% to 23.1%. In the control group, it is to mention that all students who passed the exam had previous training in Dihedral System. In fact, the 21% of students with previous training passed the exam in the control group. In view of that, the influence of the new teaching strategy was analyzed by comparing the success rate of students with same level of previous training.

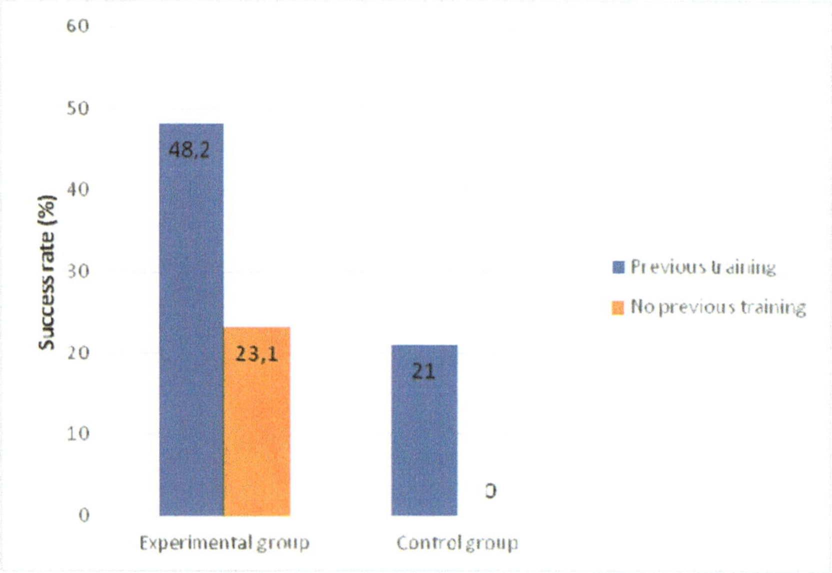

Fig. 2. Comparison of success rate between the experimental group and the control group depending on previous training on Dihedral System.

Considering students with previous training in Dihedral System, the success rate was 48.2% in the experimental group while it decreased to 21% in the control group. This may evidence the efficacy of the new didactic strategy. Considering students with no previous training, the success rate was 23.1% in the experimental group and 0% in the control group. This may suggest that the new didactic strategy is more effective than the traditional teaching methods also with those students that are in the very beginning stages of spatial vision development.

The results collected from the questionnaires are shown in Fig. 3. First of all, it is to mention that 80% of the students consider that spatial vision is a key ability in dihedral system (question 1). This is aligned with the idea around which the teaching methodology was developed. Moreover, according to question 2, it is remarkable that 70% of students believe that the new teaching strategy helped them in the development of spatial vision capacity, in good agreement with the hypothesis 1. Students point of view about hypothesis 2 was assessed by questions 3-5 and the obtained results were very similar in all of them. Around 70% of students consider that the simulation of real situations, the manipulation of objects and the deductionexperimentation processes carried out by the methodology are helpful for the identification of geometrical position, for the understanding of theoretical basics and thus to get a more meaningful comprehension. Finally, with relation to hypothesis 3, 80% of students considered that the correct interpretation of geometrical position, as a consequence of the reconstruction of real situation, helped them in choosing the correct solution method (question 6).

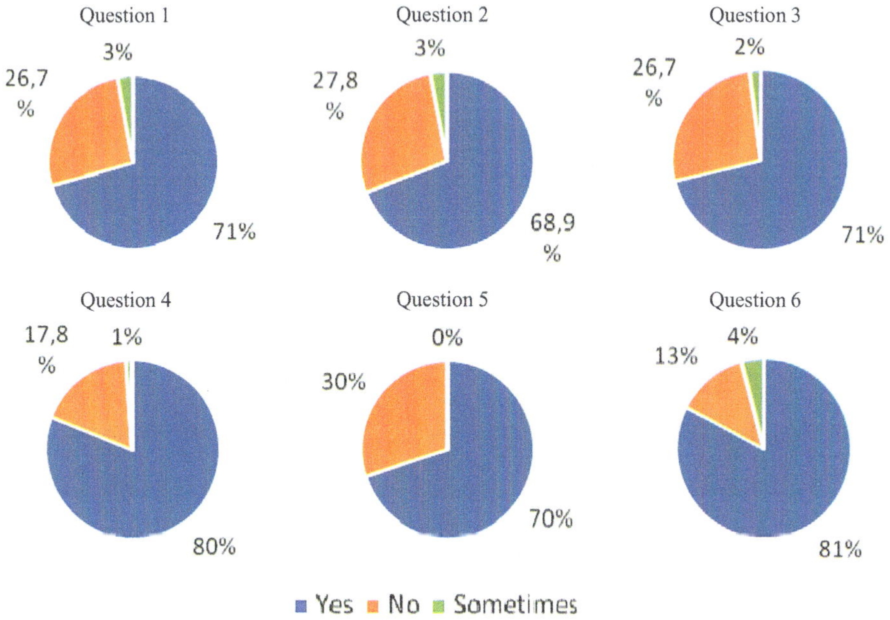

Fig. 3. Results of the questionnaire carried out on the experimental group.

5 Conclusions and Future Lines

It was verified that the implementation of the didactic strategy improved the qualifications of students. Although previous training in Dihedral System influenced the results, it was observed that the number of students passing the exam was higher in the experimental group, both in the case of students with previous training and students with no previous training. These results suggested that the didactic strategy would fit to both students profile.

As commented previously, the spatial vision capacity implies the ability of conducting the transition from three-dimensional space to two-dimensional representation in a plane. The *Fronton* allows carrying out this transition gradually, giving place to a better understanding of the process. This might be the key to develop the ability of performing this transition mentally, gaining spatial vision capacity. Moreover, Cooperative Activities not only train in the representation of objects by the dihedral system: since the theoretical basics are worked through an experimentation-deduction process, a more meaningful understanding is achieved.

As future lines, in view of the positive embracement of students, the developed teaching strategy will be extended to the practical part of the subject. Moreover, it is proposed to use 3D models of industrial objects in order to put students closer to their professional field.

Acknowledgments. Financial support from the University of the Basque Country (UPV/EHU) in the frame of Proyectos de Innovación Educativa/Hezkuntza Berrikuntza Proiektuak (PIE/HBP 2018–2020: Adierazpen grafikoaren irakaskuntza, ikaskuntza esperimentalaren bidez) is gratefully acknowledged. Authors deeply thank the involvement of student body of the first course of Industrial Engineering grade at Faculty of Engineering, Gipuzkoa (UPV/EHU).

References

1. Saorín, J.: Estudio del efecto de la aplicación de tecnologías multimedia y del modelado basado en bocetos en el desarrollo de las habilidades espaciales. Doctoral Thesis. Polytechnic University of Valencia, Spain (2006). https://doi.org/10.4995/Thesis/10251/1881
2. Keller, B., Wasburn-Moses, J., Hart, E.: Improving students' spatial visualization skills and teachers' pedagogical content knowledge by using on-line curriculum-embedded applets: overview of a research and development project (2002)
3. Lowrie, T.: Developing Talented Children's Mathematical Ability through Visual and Spatial learning Tasks. Australian Association for Research in Education (1994)
4. Sorby, S.: Educational research in developing 3-D spatial skills for engineering students. Int. J. Sci. **31**(3), 459–480 (2009). https://doi.org/10.1080/09500690802595839
5. Blasko, D., Holliday-Darr, K.: Engineering and Psychology: Joint Project on Visualization and Engineering. In: meeting of the American Society for Engineering Educators North Central Section. Erie, PA (1999)
6. Blasko, D., Holliday-Darr, K., Mace, D., Blasko-Drabik, H.: VIZ: The visualization assessment and training Web site. Behav. Res. Methods Instrum. Comput. **36**(2), 256–260 (2004)

7. Gutierrez, J.: Estudio y evaluación de contenidos didácticos en el desarrollo de las habilidades espaciales en el ámbito de la ingeniería. Doctoral Thesis. Polytechnic University of Valencia, Spain (2010)
8. Melgosa, C.: Diseño y eficacia de un gestor web interactivo de aprendizaje en ingeniería gráfica para el desarrollo de la capacidad de visión espacial. Doctoral Thesis. University of Burgos, Spain (2012)
9. McGee, M.G.: Human spatial abilities: Psychometric studies and environmental, genetic, hormonal, and neurological influences. Psychol. Bull. **86**, 889–918 (1979)
10. Tristancho Ortiz, J.A., Contreras Bravo, L.E., Vargas Tamayo, L.F.: Evaluación de técnicas tradicionales y TIC para el desarrollo de habilidades espaciales en estudiantes de primer semestre de ingeniería industrial. Dig. J. North Catholic Univ. Chile **43**, 34–50 (2014)
11. Jacques, J., Teixeira, F.: Development of CAD software for descriptive geometry practice. Volume 16. In: 12th International Conference on Geometry and Graphics. August 6–10, Salvador, Brazil (2006)
12. Piaget, J.: The role of Action in the Development of Thinking. In Knowledge and Development. Springer US (1977). https://doi.org/10.1007/978-1-4684-2547-5_2
13. Papert, S.: Harel Idit. Situating Constructionism. Ablex Publishing Corporation, Norwood, NJ (1991)

Assembling a Reducer Using VR

Óscar D. de-Cózar-Macías(✉), Fernando Gómez-Hermosa, José Luis Martínez-Torres, and Jorge Pérez-García

University of Málaga, 29071 Málaga, Spain
odecozar@uma.es

Abstract. This project develops a virtual reality (VR) application for training in the assembly and disassembly of a mechanical speed reducer. Another important concept to work on is the idea of enhancing and improving spatial skills in terms of spatial vision and relations by rotating figures in space. This assembly procedure, together with other training activities included in the same application, will allow students to become familiar with the process of assembling mechanical units. In this case, the procedure was developed in a virtual environment in which the assembly of a speed reducer can be carried out step by step, so that the student is able to understand the process to be followed and put knowledge into practice. In addition, the application will help the user to understand how the mechanism works once it is assembled. To achieve all this, Unity was used, along with various design and 3D modelling tools.

Keywords: Gearbox · Assembly · Virtual reality · Training · Unity 3D

1 Introduction

The Department of Graphic Expression, Design, and Projects of the School of Industrial Engineering at the University of Málaga is researching the educational benefits offered by Virtual Reality (VR) [1] and Augmented Reality (AR) technologies. Along these lines, a proposal was made to create a VR application that would help students to understand mechanical assembly procedures of a certain degree of difficulty. From here, we decided to work on the assembly of a speed reducer in a virtual environment. Starting with the 3D model of the assembly set, it was necessary to design a work environment and to program the actions and procedures that the user can carry out during the assembly and disassembly operations of the unit.

As teaching materials, VR applications offer a multitude of advantages when compared with other traditional methods. Firstly, these applications allow for the manipulation of the components of the mechanical unit with the safety offered by a digital environment, i.e. the student is not subject to the risks that can be caused by the manipulation of physical elements, such as cuts, falling objects, and blows. Secondly, there is no wear and tear from manipulating the virtual components, nor does it require maintenance or replacement of elements. However, in our opinion, the main advantage offered by VR is the skills that the student acquires through the use of the application and the motivation with which he or she undertakes them.

F. Cavas Martínez et al. (Eds.): INGEGRAF 2021, LNME, pp. 384–389, 2022.
https://doi.org/10.1007/978-3-030-92426-3_44

2 State of the Technique

The advantages of using VR as an educational [2] and psychological [3] resource have received supported from a number of researchers. One author [4] conducted a study to demonstrate how the use of VR can influence engineering students being trained to use a lathe. The findings revealed that those students who were new to the subject benefited from using virtual reality before acquiring any ideas about the subject, since they could get a general idea of what was being studied. However, students who already had some idea of the subject were able to better assimilate the concepts after having acquired the basic notions of what was being worked on.

One of the areas of engineering in which VR has the greatest influence is in the training associated with production processes and assembly lines. According to [5], the use of these virtual environments also allows for the detection of possible problems in the assembly line. This means that any problems that may arise can be corrected without the need to invest a considerable amount of resources.

The application of VR in the field of machine assembly has been carried out at the University of the Armed Forces (ESPE) in Sangolquí, Ecuador [6]. Their project consisted of creating a virtual reality application that offers an environment similar to that of an automobile workshop, in which one can get a clear view of the process involved in assembling the different parts that make up a vehicle.

The course of action in the previous case was the same as that followed in our project. First, 3D modelling was carried out before moving to the "Unity 3D" application [7], where it was given its corresponding utility.

2.1 Speed Reducer

The first part of the work consisted of generating the 3D model of the chosen reducer, formed, as can be seen in Figs. 1 and 2, by a set of mechanical elements such as the gear train. This will provide the transmission of the movement from the entry point to the exit point. Axis 1 is the input point with connection to the motor and Axis 4 is the output element. In addition to the gear wheels, the assembly is made up of a series of elements such as shafts, bearings, seals, cotter pins, and retaining rings that allow the reducer to fulfil its function.

The reducer has a speed ratio between Axis 1 and 4, as indicated in Eq. (1), so for an input speed of 3000 rpm the output speed would be 20 times lower, i.e. 150 rpm. It is essential to know the speed ratio in order to later make the corresponding animation in the operation simulation.

$$\frac{\omega_1}{\omega_4} = \frac{z_2 * z_4 * z_6}{z_1 * z_3 * z_5} = \frac{25 * 80 * 80 \text{ teeth}}{20 * 20 * 20 \text{ teeth}} = 20 \qquad (1)$$

Where ω_n is the speed of the n axis and z_n is the number of teeth of the n wheel. The assembly and disassembly process, as well as the animation of the operation, was implemented in the process of programming the application with Unity.

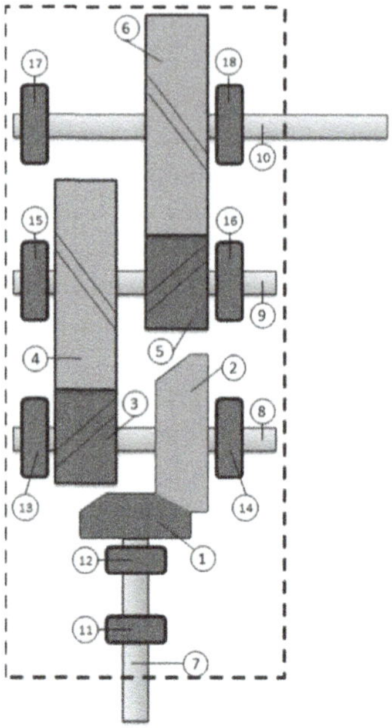

1	Gear1	10	Axis4
2	Gear2	11	Bearing1
3	Gear3	12	Bearing2
4	Gear4	13	Bearing3
5	Gear5	14	Bearing4
6	Gear6	15	Bearing5
7	Axis1	16	Bearing6
8	Axis2	17	Bearing7
9	Axis3	18	Bearing8

Fig. 1. Arrangement of the speed reducer axes.

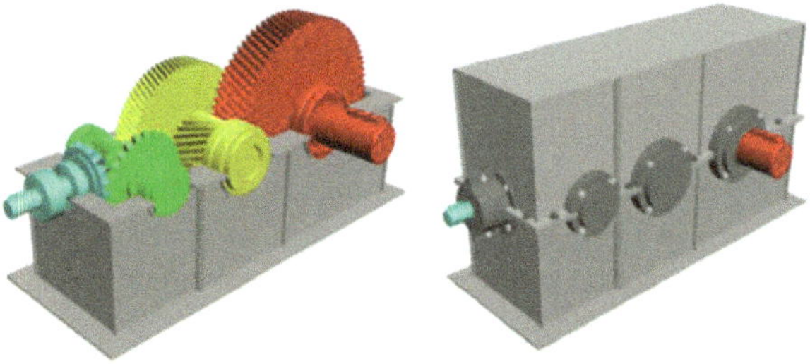

Fig. 2. Virtual representation of the reducer.

2.2 Development with Unity

Modelling was carried out using the Solid Edge (.u3d) and AutoCAD (.dwg) applications, so that the elements are structured in an orderly way, both for the reducer and the scene where the interactive practice of assembly and operation would take place. Once the modelling was completed and ordered, it was necessary to generate a model of exportable quality. To do this, the 3DStudio Max application was used as an intermediate step for its export to Unity (.fbx).

The virtual workroom (Fig. 3) was configured and arranged so that no long trips were required for assembly or the implementation of additional functions. The dimensions are 5.5 × 5 m.

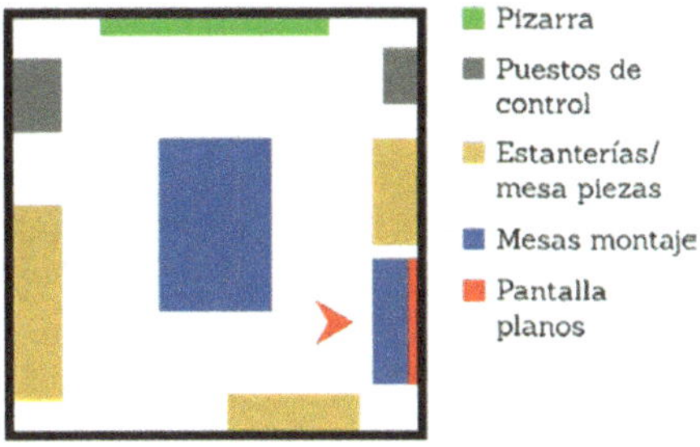

Fig. 3. Diagram of the workroom layout.

The shelves are used to distribute the different elements of the unit throughout the room. The work desks are used to carry out the guided or autonomous assembly, an option that is permitted by the application. There are two control stations, one for interacting and for executing actions in the application whilst the second has a single button to control the reducer disassembly function. And finally, the screens are used to display information about the actions and operation of the reducer. In addition to the functional elements of the application, objects are included for decorative purposes such as a blackboard, a panel, and various lights.

The application will have a set of restrictions and actions that are allowed for the user, which will need to be programmed using Unity 3D [8, 9] and Visual Studio. These include the collision and fixing function of physical elements (Fig. 4), the simulation of the force of gravity and the launch speed of a body and the area enabled for mobility in the room. Other components that bring a sense of realism to the scene include the assignment of materials and lighting that generate shadows. The assembly must be carried out in an orderly and sequential manner. The relative position and the joint movement between elements must be adequate, so it will be necessary to program the application to avoid undesired situations.

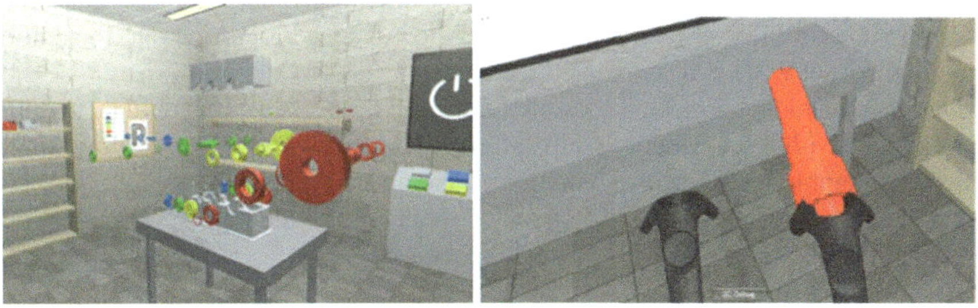

Fig. 4. Application restrictions on collision and fixing function.

The control station allows for activating and deactivating the assembly application and creating different configurations of the reducer operation by playing with the input and output speeds. The screen above this station will display the information required to work on the reducer (Fig. 5).

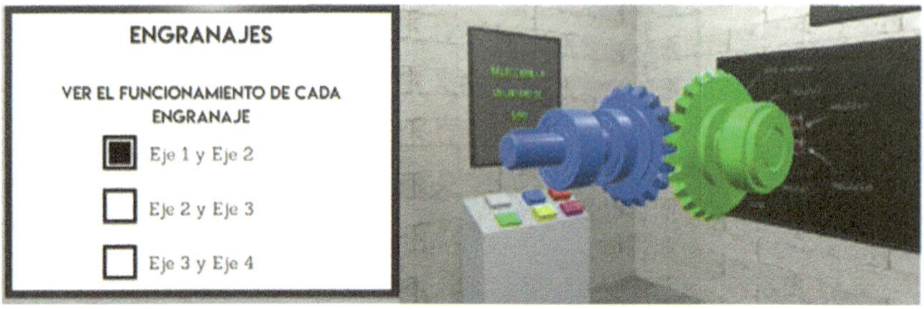

Fig. 5. Control panel display and action options.

3 Conclusions

VR is a very useful tool for students to put their knowledge into practice. This requires that the work carried out to develop the application is thorough in order to prevent any flaws from degrading the user experience.

One of the main advantages observed during the development of the project was the infinite possibilities that this technology offers to the field of engineering. Virtual laboratories for the implementation and training of practices would allow for a better understanding and comprehension of knowledge in different subjects. It must be noted that the initial investment in the equipment of this technology can lead to the rejection of its use, but this would be completely amortized due to the physical technology that could be virtualized.

Teaching or training with these methods has also been found to be more effective because of the ability to isolate the user within the application, eliminating most external distractions. It is therefore essential that the virtual working environment created is as realistic as possible.

In this case, the assembly process followed the same guidelines as those that would be used if the procedure were carried out in reality. It has been made possible for the entire assembly process to be guided. As far as the operation of the reducer is concerned, it was possible for the user to see how transmission of speed is carried out. In addition, the assembled set will be disassembled step-by-step.

On the basis of this project, a possible future line of investigation could be to implement the tool within the Moodle platform and thus assess students according to their ability to assemble the reducer in addition to the questionnaires that are administered during the process. Thus, the work conducted by the user within the application would be another factor to consider when evaluating its use.

References

1. Santana, A., Lira, A., Lara, G., Peña, A.: Evolution of virtual reality's interaction devices. In: 2018 7th International Conference on Software Process Improvement (CIMPS), Guadalajara, México, pp. 119–123. IEEE (2018). https://doi.org/10.1109/CIMPS.2018.8625634

2. Freina, L., Ott, M.: A literature review on immersive virtual reality in education: state of the art and perspectives. In The International Scientific Conference eLearning and Software for Education, pp. 133–141 (2015). Accessed 22 Jan. 2020. https://ppm.itd.cnr.it/download/eLSE%202015%20Freina%20Ott%20Paper.pdf
3. Botella, C., García-Palacios, A., Baños, R.M., Quero, S.: Realidad virtual y tratamientos psicológicos. Cuadernos de medicina psicosomática y psiquiatría de enlace **82**, 17–31 (2007)
4. Antonietti, A., Imperio, E., Rasi, C., Sacco, M.: Virtual reality and hypermedia in learning to use a turning lathe. J. Comput. Assist. Learn. **17**, 142–155 (2001)
5. Dwivedi, P. C.: Manual assembly training in virtual environments. In: IEEE 18th International Conference on Advanced Learning Technologies, 395–399 (2018). https://doi.org/10.1109/icalt.2018.00100
6. Quevedo, W.X., et al.: Virtual reality system for training in automotive mechanics. In: De Paolis, L.T., Bourdot, P., Mongelli, A. (eds.) AVR 2017. LNCS, vol. 10324, pp. 185–198. Springer, Cham (2017). https://doi.org/10.1007/978-3-319-60922-5_14
7. Linowes, J.: Unity Virtual Reality projects: learn Virtual Reality by Developing More Than 10 Engaging Projects with Unity 2018 (2ª ed.). Packt Publishing Ltd (2015)
8. Unity Technologies: The Game view Unity Manual (2019). Accessed 22 Jan. 2020. https://docs.unity3d.com/Manual/index.html
9. Unity Technologies: Scripting Unity Manual (2019). Accessed 22 Jan. 2020. https://docs.unity3d.com/Manual/index.html

Design Learning Adaptation of Sheet Metal Elbows to Covid 19 Pandemic

María Gloria Del Río-Cidoncha(✉), Rafael Ortiz-Marín, Alfonso Martínez-Del Río, and Juan Martínez-Palacios

University of Seville, Seville, Spain
cidoncha@us.es

Abstract. Universities have been gradually adapting themselves to the framework set by the European Higher Education Area (EHEA). Information and communication technologies (ICT) are offering tools for this adaptation, such as videos, mobile learning (mlearning) and QR codes (Quick Response). Initially, a teaching-learning methodology of the design and development for cylindrical and conical elbows applied in the 2018/19 academic year is presented. Two technologies are mixed, first as basis for the construction and development of the elbows are shown in an audio-visual way and the other, a two-dimensional QR barcode that allows access to these fundamentals from anywhere just being connected to the internet. The advantage of this methodology is that the student can access the information at any time (a video or an Adobe 3D file) by any device (smartphone, tablet, laptop, etc.). Then compares this alternative to the one used when adapted to the situation created by the Covid 19 on the course 2019/20.

Keywords: Sheet metal · Cylindrical and conical elbows · QR codes · Mobile devices · Videos · M-learning · E-learning · Covid 19

1 Introduction

According to experts from the Future Trends Forum [1], thanks to the unstoppable globalization, internet technology and mobile devices, which imply free access to training and information, the concept of teaching and learning is becoming more and more important, as well as the didactic strategies and methodologies to carry it out. The student must learn how to revise and the teacher becomes the facilitator of the process, evolving [2] from the concept of teaching by the professor, to active learning by the student alone.

Today's students will be tomorrow's professionals and, as recommended by the European Commission [3], they must be capable, throughout their lives, of designing their own training curriculum (Life Long Learning) that will allow them to acquire competences and skills adapted to the changes that society and the job market demand at such a dizzying pace. In this situation, information and communication technologies (ICT) are used to improve the quality of the teaching and learning process, supporting both students and teachers [4].

F. Cavas Martínez et al. (Eds.): INGEGRAF 2021, LNME, pp. 390–397, 2022.
https://doi.org/10.1007/978-3-030-92426-3_45

This paper presents the implementation of a learning and teaching strategy based on these resources in the academic year 2018/19 and its comparison using an e-learning platform in the academic year 2019/20 with the situation generated by Covid-19.

We present the implementation of a learning and teaching strategy, based on ICT resources such as QR codes, online video screencasts, mobile devices and e-learning for the design and development of cylindrical and conical sheet metal bows. As this content has somewhat complex constructions and developments from the instrumental point of view, it was decided to reinforce it after its conceptual explanation in the classroom. The QR codes are provided to the students in the classroom on a piece of paper, as part of the practical exercises statement to be carried out in this subject. The student can capture the image of the QR code with their mobile phone or tablet at any time and from anywhere and go directly to the website where the teaching videos of the subject are uploaded without the need to type the long and complex URL address. Everything is immediate and convenient. The following course was implemented using the resources of the paid e-learning platform Blackboard Collaborate. Therefore, both experiences and their results are presented.

2 Materials and Methods

2.1 Aim

The main objective is to evaluate whether the proposed initial methodology improves the teaching-learning process of the students and consequently generates a satisfactory response and to compare it, with the results obtained after the use of an e-learning platform.

2.2 Videos

YouTube, a website dedicated to video sharing, is one of the most visited spaces by these digital natives on the Internet. That is why Whatley and Ahmad [5] point out that the use of video in learning environments, both in face-to-face and distance learning, is a particularly attractive and a motivating practice for students, helping them to understand and improve their retention of information. Screencasts is a video where only the action that occurs on a computer screen is recorded and edited with specific software. It can include audio or subtitles to clarify the content. Scholars [6, 7] find screencasting beneficial as it helps to reduce cognitive load and also have been found to have a positive correlation with student exam performance.

The software used to draw 2D both construction and elbow developments is Solid Edge [8] while to produce the screencast videos, the software Camtasia was chosen [9].

2.3 QR Codes

The term QR stands for Quick Response, meaning the immediate response they provide after their reading, is an evolution of the conventional barcode but with greater storage capacity. It is a system of alphanumeric information encoded in a twodimensional matrix. It was developed by the Japanese company Denso Wave in 1994 to track the manufacturing process of Toyota vehicles.

The QR code is an open source of international standard (ISO/IEC 18004) as Denso Wave has chosen to make it. The graphic symbol or image of the QR code is easily identifiable by its square shape and by the three squares in the top and bottom left corners. And, by the series of scattered squares encoding the alignment and synchronization. There are websites that allow QR codes to be created online, offering very intuitive interfaces.

Mobile devices (smartphones, tablets, etc.) are capable of capturing images decoding the code by downloading a free APP and lead directly to links to a website, files, email, etc. Hence their use in many different areas of society. According to Traxler [10], its application to the field of education, although it has just happened, has changed the meaning and importance of learning, as Cacheiro [11] states. By incorporating QR codes into teaching materials, an attractive design of these materials is achieved, and the student obtains immediate accessibility to the content both inside and outside the classroom, especially now that, according to Cubillo et al. [12], it is necessary to continue learning not only during the educational period but throughout life.

QR codes can be created using free tools, such as websites or PC applications. In this experience, the CODES-QR website [13] has been used to create them online and insert inside them the links to the virtual hard disk where the videos have been placed. The creation consists of encoding the information to be inserted inside, by generating an image in a "png" format of the QR code.

Table 1 shows the information provided in the class with the contents and the QR codes needed to access the videos with detailed explanations of the construction and development of these codes. Access to the virtual hard disk website was not restricted by a password, although this was possible.

2.4 Mobile Devices

The use of mobile technology, connected to the network, in education is part of the new educational model adapted to the user [14] which, according to [15], constitutes a challenge in the educational methodological approaches [16] attaching its attractiveness to its ease of use and immediacy of results.

In this project, mobile phones, Smartphones and Tablets have been used as they are the most used by students. Some technical problems were detected in relation to the resources that the students employ to download the applications for reading the QR codes, viewing videos, etc.

Some problems arose when downloading the apps for reading QR codes, depending on the one chosen, on the characteristics of the mobile device or on the most suitable browser for the mobile. For example, one of the most common problems was using the QR code reader app on the mobile and clicking on search, when in some cases the URL displayed indicated that the page did not exist; this was solved by copying the link and entering it manually in Google or changing the type of QR reader.

Table 1. Contents and QR codes

TYPE CODE	CONTENTS	QR CODES
Link to virtual disk	Solution cylindrical elbow	
	Solution cylindrical elbow (I)	
	Solution development cylindrical elbow	
	Solution top view cylindrical elbow	
	Solution conical elbow	
	Solution development conical elbow	
	Solution top view conical elbow	

2.5 E-learning Platforms

An educational platform, online or virtual learning platform encompasses different types of computer applications installed on a server whose function is to facilitate learning over the Internet [17]. They can be used to comprehensively manage distance learning or as a complement to the face-to-face teaching. They can be classified into: Commercial: for which an installation and/or maintenance fee has to be paid, which usually varies according to the number of users and has to be renewed from time to time. These include Blackboard, Almagesto, Educativa, Saba, etc.

Free software: also known as open source platforms, they are designed to be distributed and used freely. They are part of the public domain. Moodle, Atutor, Chamilo, etc. stand out.

Pay-per-use in the cloud: this is an alternative that is currently being used nowadays. In this platform the subscriber pays according to the use, although it is considered that its main use is to provide support for the face-to-face class. Examples include Edmodo, Udemy, Schoology.

3 Results and Discussion

The main study adopts a quantitative research design, with a survey used as an instrument for analysis. The process followed in its development is described below.

The experience was carried out with 56 students in the first year of the Degree in Engineering of Industrial Technologies at the University of Seville in the subject of Graphic Expression.

The survey given to the students at the end of the experience for their evaluation consisted of 4 questions measured on a Likert-type scale with 5 levels of response (1 Strongly disagree, 2 Disagree, 3 Indifferent, 4 Agree and 5 Strongly agree).

The questions asked were:

Q1. Did the videos help you to understand the construction of cylindrical and conical bends?
Q2. With these videos would you be able to do self-learning on the construction of bends?
Q3. Do you find the videos useful as a learning tool using mobile devices and QR codes?
Q4. Would you find these tools useful for learning other subjects?
Of the 56 students in the sample, only 34 responded to the survey, with the results shown in Figs. 1 and 2.

In Fig. 1 the strongly agree rating (5) is predominant in the survey. In questions P1 and P2 it exceeds or equals 50% of the responses. Around 25% in all the question has a response in agreement (4). The least Likert-rated responses (1, 2 and 3) total 21% in questions P1 and P2, while in P3 and P4 they reach 30%. Regarding videos as an educational aid, in both questions P1 and P2, 80% of the students responded with scores of 4 and 5; and in P3 70%. This confirms the students' positive assessment of the use of videos as a learning tool.

Figure 2 shows that the student´s answers to the 4 questions have the same Mode with a value of 5. The answers to questions P3 and P4 have the same Median, a value of 4, while questions P1 and P2 have 5 and 4.5 values, which are also very close. With the mean, it is observed that the student´s evaluation in questions P3 and P4 is close to the value 4, while in questions P1 and P2 it is slightly higher.

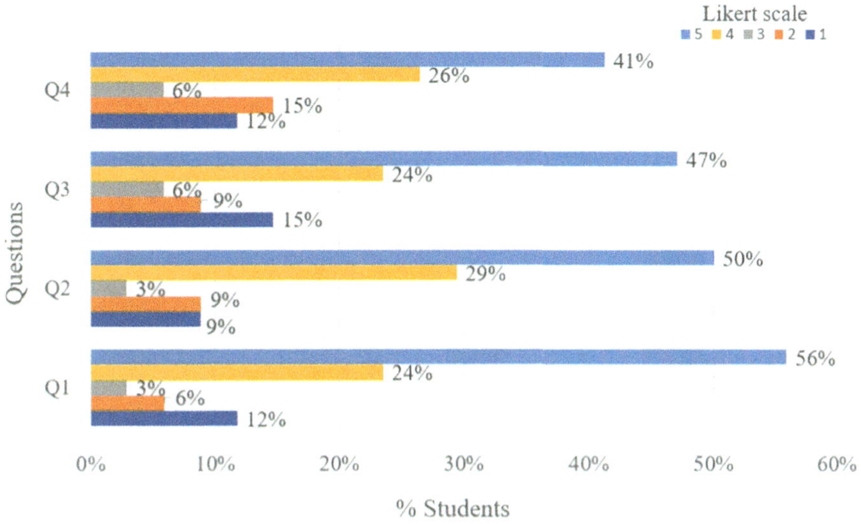

Fig. 1. Survey results.

The data to analyses the teaching and learning experience carried out through the virtual teaching platform Blackboard were obtained from the same one. Each time students access the platform, the platform keeps track of the activity carried out on it, indicating date, duration and options used.

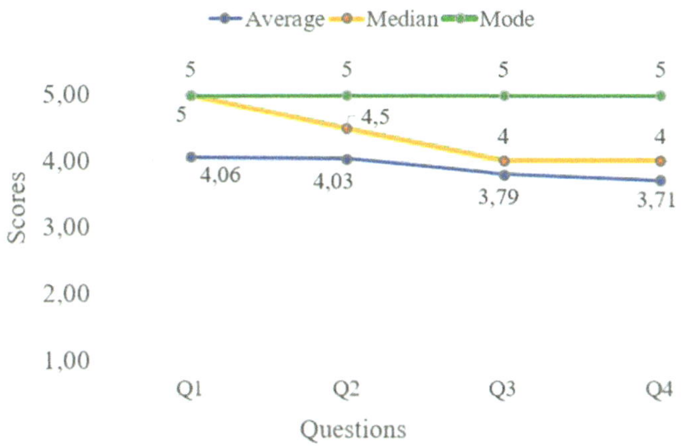

Fig. 2. Average, median and mode of the answers

As it can be seen in Fig. 3, students have a preference for both theoretical and practical content uploaded to the platform in video format, followed by the exercises that already had a solution. The least use was made of theoretical content and unsolved practical exercises.

Since students pay more attention to video sequences, it could mean that they are looking for a quick approaching, without really focusing into the acquisition of knowledge in depth, considering only the easiest way to obtain it with a minimum effort.

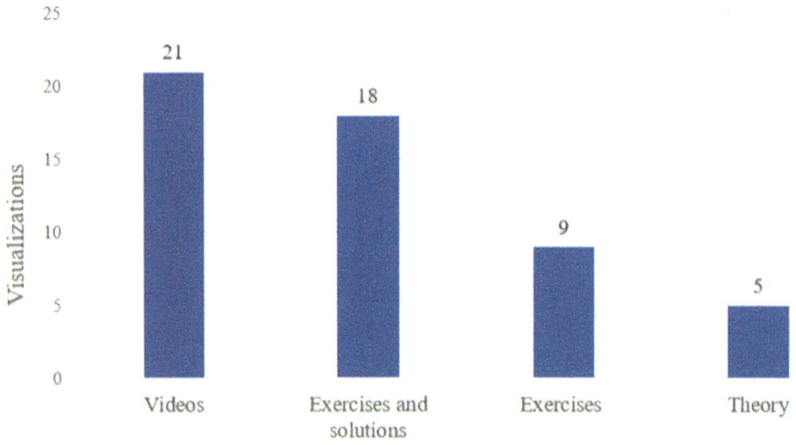

Fig. 3. Average uses of students' visits to the blackboard platform

4 Conclusions

Considering the effectiveness of the integration of the QR code and online video as an m-learning tool, and analysing the results of the surveys, the grade of satisfaction on the part of the students has been high. In addition, it has helped students to take responsibility for their own learning by reinforcing the content themselves. This invites us to make teaching videos on new topics around the subject or even to use other ICT resources.

But the good results obtained on the platform, and the already known, interest of students for to the quickest and easiest search for the minimum knowledge required to pass the assessment, raise doubts as to whether this first-year students are sufficiently motivated and autonomous to make use of new teaching approaches using only elearning tools when faced with new unforeseen situations such as Covid or if they will continue to make a right use of these tools once we came back to the traditional education system. The combination of such tools as an accompaniment to the face-to-face teaching system is considered to be more productive for the student.

References

1. Future Trends Forum 2011: La educación del siglo XXI. Una apuesta de futuro. Fundación de la Innovación Bankinter, https://www.fundacionbankinter.org/, Accessed 10 Dec. 2020
2. Martín, E., Moreno, A.: Competencia para aprender a aprender. Alianza Editorial, Madrid (2007).978-84-206-8410-9

3. Comisión Europea.: Competencias clave para el aprendizaje permanente https://www.educacionyfp.gob.es/dctm/ministerio/educacion/mecu/movilidadeuropa/competenciasclave.pdf?documentId=0901e72b80685fb1. Accessed 18 Dec. 2020
4. Mendoza, L., Zermeño, M., Zermeño, R.: Desarrollo de habilidades cognitivas y tecnológicas con aprendizaje móvil. Revista de Investigación Educativa de la Escuela de Graduados en Educación **3**(6), 30–39 (2013). 2007-3003
5. Whatley, J. Ahmad, A.: Using video to record summary lectures to aid students' revision. Interdis. J. Knowl. Learn. Obj. **3**, 185–196 (2007). http://ijklo.org/Volume3/IJKLOv3p185-196Whatley367.pdf, Accessed 19 Feb. 2020
6. Tisdell, C., Loch, B.: How useful are closed captions for learning mathematics via online video? Int. J. Math. Educ. Sci. Technol. **48**(2), 229–243 (2017). https://doi.org/10.1080/0020739X.2016.1238518
7. Grabe, M., Christopherson, K.: Optional student use of online lecture resources: resource preferences, performance and lecture attendance. J. Comput. Assist. Learn. **24**, 1 (2008). https://doi.org/10.1111/j.1365-2729.2007.00228.x
8. Solid Edge ® software Homepage. https://solidedge.siemens.com/es/. Accessed 15 Nov. 2020
9. Camtasia ® software Homepage. https://www.TechSmith.es. Accessed 17 Oct. 2020
10. Traxler, J.: Learning in a mobile age. Int. J. Mob. Blend. Learn. **1**(1), 1–12 (2009). https://doi.org/10.4018/jmbl.2009010101
11. Cacheiro González, M.L.: Recursos educativos TIC de información, colaboración y aprendizaje. Pixel-Bit. Revista de Medios y Educación **39**, 69–81 (2011). 11338482
12. Cubillo, J., Martín, S., Castro, M.: New technologies applied in the educational process. In: IEEE Global Engineering Education Conference (Educon 2011), pp. 575–584, IEEE, Amman, Jordan (2011). ISBN 978-1-61284-643-9
13. Codigos-QR ® software Homepage. https://www.codigos-qr.com/. Accessed 15 Nov. 2020
14. Wu, Y. (ed.): Advanced technology in teaching - proceedings of the 2009 3rd international conference on teaching and computational science (WTCS 2009). AISC, vol. 116. Springer, Heidelberg (2012). https://doi.org/10.1007/978-3-642-11276-8
15. Wang, M., Shen, R., Novak, D., Pan, X.: The impact of mobile learning on students' learning behaviours and performance: Report from a large blended classroom. Br. J. Edu. Technol. **40**(4), 673–695 (2009). https://doi.org/10.1111/j.14678535.2008.00846.x
16. Chong, J.L., Chong, A.Y., Ooi, K.B., Lin, B.: An empirical analysis of the adoption of mlearning in Malaysia. Int. J. Mobile Commun. **9**(1), 1–18 (2011). https://doi.org/10.1504/IJMC.2011.037952
17. Sanchez Rodríguez, J.: Plataformas de Enseñanza Virtual para entornos educativos. PixelBit. Revista de Medios y Educación **34**(1), 217–233 (2009). 1133-8482

Peer-Assessment for Student Learning and Empowerment in the Subject "Design Workshop I"

Laura Diago Ferrer(✉), Jorge Sierra Pérez, and Eduardo Manchado Pérez

Department of Design and Manufacturing Engineering, Maria Luna 3, 50018 Zaragoza, Spain
lauradf@unizar.es

Abstract. Traditionally, the evaluation of teaching has been closely linked to qualification. This has been verified throughout the years of teaching experience in the subject of Design Workshop I, in Industrial Engineering Design and Product Development Degree at the University of Zaragoza. It has been observed that students present some difficulties in the development of applicable critical thinking in the context of project-based learning (PBL). On the other hand, there are experiences that have obtained satisfactory results when applying the learning-oriented assessment methodology (EOA) in similar cases. The objective of this work is to present the results of the empirical studies carried out during the 2020–2021 academic year framed in the PIIDUZ 2019_20 Teaching Innovation Project, for the subject of Design Workshop I: Fundamentals and Product Communication. New tools will be used for peer-to-peer evaluation and self-evaluation, both aimed at learning and empowering students. Through the application of rubrics in the evaluation processes, evaluation has been approached as a process and not as an end in itself. The results of this study allow us to validate the methodology used.

Keywords: Teaching innovation · Peer-assessment · Evaluation for empowerment · Evaluation for learning

1 Introduction

Traditionally, evaluation in the teaching field has been closely linked to homework, exam, and the grade [1]. The evaluation process has been confused with the qualification process even though they have nothing to do with it. Different educational spheres and especially from the university have tried to alleviate this in recent years, focusing on working on the essential competencies or capacities to be developed in a transversal way. We could say that we have gone from learning assessment to assessment for learning [2].

However, the student, specially in the first years degree, who generally has not yet had this type of teaching reference, is still closely linked to the exam-qualification process. For this reason, it is difficult for him to dissociate that a constructive criticism or a feedback is not an absolute and definitive rating, but is part of a learning process and awareness of improvement based on the strengths and weaknesses of a work process.

F. Cavas Martínez et al. (Eds.): INGEGRAF 2021, LNME, pp. 398–403, 2022.
https://doi.org/10.1007/978-3-030-92426-3_46

Evaluation as a learning process, as well as evaluation for empowerment articulated by Fetterman in 1993 are novel approaches from a social perspective [3, 4]. In them, the value of the evaluation does not reside in the qualification but in the awareness of individual progress within a broader global context, which allows the use of technique, findings for improvement and self-determination. Making the student more critical and resilient in the face of external opinions, fostering reflection on their own learning and self-criticism, has been pointed out by many authors as a main objective to prepare future professionals in a society as changing as the current one [5, 6].

The Design Workshop I subject is part of the first year of the Degree in Industrial Design Engineering and Product Development, a compulsory subject that takes place in the second semester. After taking this course, the student, through experimentation, should be better able to tackle a design project from a conceptually appropriate approach and based on a basic work methodology that will be enriched in later years. And specifically, it will allow you to develop skills such as teamwork, critical thinking or the communication of messages through visual language. For this, the subject is divided into two parts: a theoretical part in the form of lectures and a practical part whose learning is based on the resolution of practical cases.

The objective of the teaching innovation project presented is the implementation of an evaluation process among the students of some of the works developed in the subject, carrying out an analysis and evaluation of the work of their classmates based on some rubrics that are provided to them.

2 Methods

The case study presented began to be launched in the 2019–2020 academic year and although it had to be interrupted by the COVID-19 pandemic, it has been resumed during the second four-month period of this academic year 2020–21. It is framed of a PIIDUZ teaching innovation project from the University of Zaragoza. The scope of application has been during the second and third exercise of the course. In both cases, the exercise consisted of creating a structurally resistant form using spaghetti or foam cardboard as material. The sculpture created had to be capable of communicating a previously selected abstract concept of a current news story.

Throughout the semester, students performed six exercises with a variable and progressive workload. All of them are structured in three phases. In the first week or first phase (information) the work to be carried out is introduced, focusing on the learning objectives, as well as the correction criteria. As part of this phase, the evaluation rubric is explained to them and the necessary doubts are resolved. In the second week or phase, they analyze what they have been working on individually, discussing between all constructively weak and strong points, as well as how they could be improved. The role of the teacher is such a coach who facilitates the debate between them, the exchange of opinions, etc. The third and last session is in which the students orally present their exercise. After presenting, they discuss the exercises as a team based on the judgments made.

A total of 75 students belonging to five class groups participated in the experiment. All students evaluated only their peers in their own group both orally and in writing. Of these, a total of 25 students voluntarily and anonymously answered a questionnaire about the experience.

The objective was, that through a discussion among equals and led by the teaching staff, each group would discuss the success of taking advantage of the potential of the material, the symbolic value, the form-meaning relationship, the quality of the graphic and constructive representation of the form, as well as the quality of the verbal defense of the exercise.

For this, once the work had been presented orally in the third session, each student received a red post-it and two green ones. A quantitative grade was not sought, but rather that the students express which works were, in their opinion, the ones that had come closest to achieving the objectives of the proposed exercise, using the green post-its. The red post-it indicated which was the job that was furthest from the objectives initially set (Fig. 1). For this, they had been provided with a rubric and a brief with the strategic points.

Fig. 1. Example of the structures developed and their evaluation

3 Results

The data extracted from five groups of both the 2019–20 and 2020–21 courses have been analyzed, obtaining very similar results in each of them and extrapolated from one course to another. For this reason, only the graphs of the results of two groups have been included in this work (Figs. 1 and 2). These figures show each students in the group, expressed as numbers, indicating to which partner their positive (green) and negative (red) evaluations have gone. Furthermore, the reasons expressed during the evaluation have been coded as: a) Good concept representation, b) Good structure execution, c) Bad concept representation, d) Bad structure execution, and e) By comparison with others.

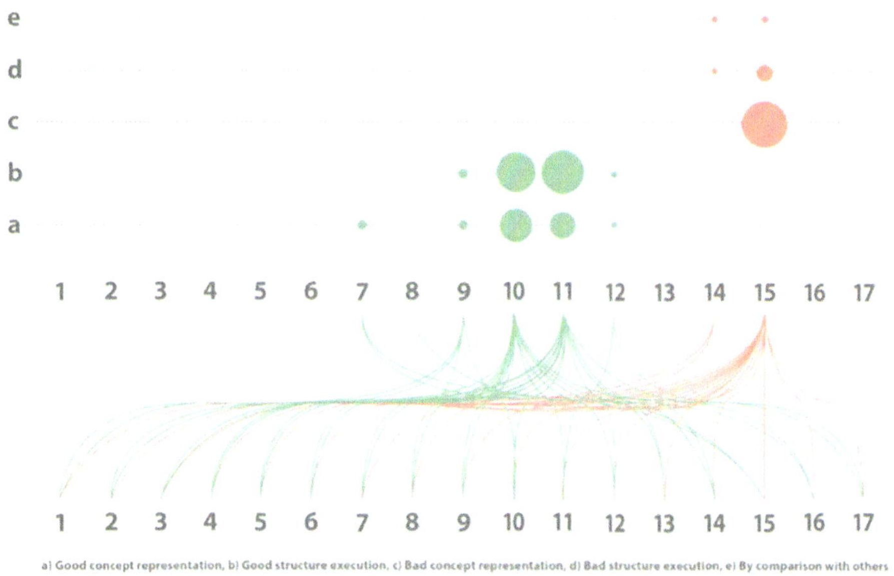

Fig. 2. Group 1 results analysis

Figure 2 shows how the exercises developed by students 10 and 11 concentrate the majority of positive evaluations, both for the correct representation of their concepts, and for the good execution of the structure. On the other hand, exercise 15 concentrates the vast majority of negative votes, especially due to the low correspondence of the concept to be represented.

Figure 3 shows that the positive evaluations have been more dispersed, although two thirds of the group gave a positive vote to exercise 8. The second positive vote of the rest of the group was dispersed in other exercises. On the other hand, exercise 13 had almost unanimously the negative vote of the group, due to the poor execution of the structure exercise.

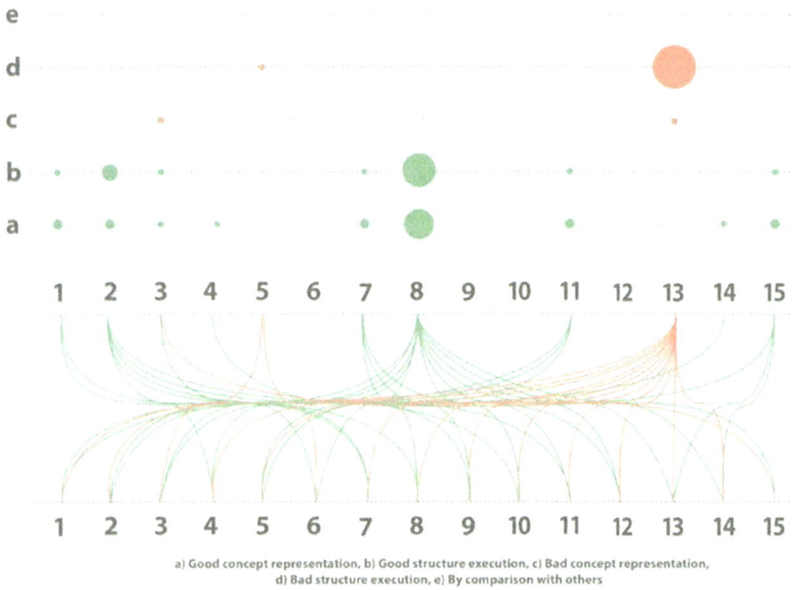

Fig. 3. Group 2 results analysis

When a negative assessment has to be made, it has been observed that in many cases students are uncomfortable when assessing the work of their classmates, so they resort to apologizing, to counteract the opinion with a positive assessment or even to contribute a suggestion for improvement. This fact has also been observed by other authors who consider it to be one of the possible drawbacks of this type of evaluation [7, 8].

At the end of the evaluation process, students were surveyed to find out their opinion about this peer evaluation process, as well as whether this process had seemed appropriate for their learning process. The results of the survey can be seen in Fig. 4. As can be seen, the students have valued this process positively, especially noting that the students consider that the evaluation and opinion of their colleagues about their work is useful to them.

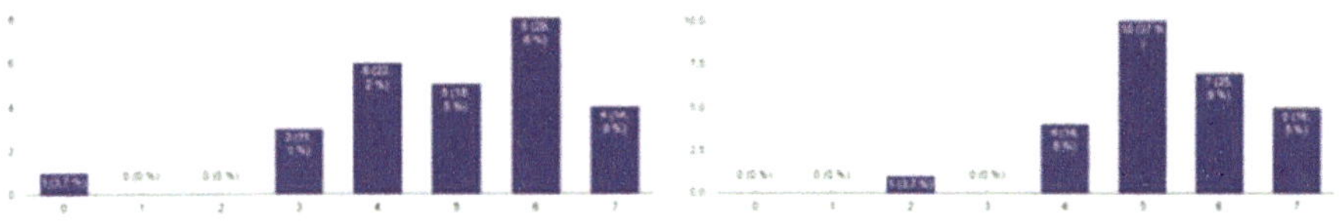

Fig. 4. Student satisfaction survey result.

4 Conclusions

The study presents the results of a peer evaluation process in the Design Workshop I subject, an eminently practical subject where students begin to acquire the basic skills of an industrial designer.

The evaluation process has been defended both orally and in writing, following a rubric proposed by teaching staff. It is worth highlighting the involvement of the students in the entire evaluation process, helping peer evaluation to individual awareness and express criticism based on objective criteria.

The learning process has been satisfactory for both students and teachers according to the results of the surveys as well as for having satisfactorily achieved the results obtained in each exercise.

Acknowledgments. The research work reported here was made possible by the participating students and teachers from different subjects of the 2nd semester of 1st. Degree course in Industrial Design Engineering and Product Development.

References

1. Sanmartí, N.: 10 ideas clave: Evaluar para aprender. Colecci6n ideas clave (España) (2007)
2. Sambell, K., McDowell, L., Montgomery, C.: Assessment for Learning in Higher Education. Londres, Routledge (2013)
3. Fetterman, D.M.: Foundations of Empowerment Evaluation, Sage Publication (2001)
4. Fetterman, D.M., Kaftarian, S.J., Kaftarian, S.J. Wandersman, A.: Empowerment Evaluation Knowledge and Tools for Self-Assesmente & Accountabilit. Sage Publications (1996)
5. Drew, S., et al.: Formative observation of teaching: focusing peer assistance on teachers' developmental goals. Assess. Eval. High. Educ. **42**(6), 914–929 (2017). https://doi.org/10.1080/02602938.2016.1209733
6. Tai, J., Adachi, C.: Peer assessment and professional behaviours: what should we be assessing, how, and why? Med. Educ. **51**, 342–350 (2017). https://doi.org/10.1111/medu.13254
7. Sluijsmans, D., Brand-Gruwel, S., Merriënboer, J.J.G.V.: Peer assessment training in teacher education: effects on performance and perceptions. Assess. Eval. High. Educ. **27**(5), 443–454 (2002). https://doi.org/10.1080/0260293022000009311
8. Kiliç, D.: An examination of using self-, peer-, and teacher-assessment in higher education: a case study in teacher education. High. Educ. Stud. **6**(1), 136–144 (2016). https://doi.org/10.5539/hes.v6n1p

The Use of Graphical Communication in Academic Documents

Antonio M. Carretero Diaz[1](✉), M. Luisa Mtz Muneta[1], David Díaz-Gutiérrez[2], Rodrigo Pérez-Fernández[2], Jessica Díaz Fernández[3], Javier García-Martin[3], and M. del Mar de la Fuente García-Soto[1]

[1] ETS Ingenieros Industriales, Universidad Politécnica de Madrid, Madrid, Spain
a.carretero@upm.es
[2] ETS Ingenieros Navales, Universidad Politécnica de Madrid, Madrid, Spain
[3] ETS Ingenieros de Sistemas Informáticos, Universidad Politécnica de Madrid, Madrid, Spain

Abstract. The integration of the Spanish universities into the EHEA is marked by the entry into force of the Spanish Law RD 1393/2007, which begins the development and subsequent implementation of new study programs, which establishes the unavoidable requirement of the acquisition by students of the so-called generic competences. According to a recent study carried out to know and analyze how the training and evaluation of these generic competences is being addressed in Spanish technological universities that offer undergraduate studies in the branch of Engineering and Architecture. Among the 32 Tuning competences it is the oral and written communication competence in the mother language that appears collected most frequently, and prominently, in the curricula of the 40 universities analyzed. Written language is an important part of communication. This article proposes different recommendations for the preparation of academic documents, in order to support the communication of information making it clearer, faster and more effective.

Keywords: Generic competences · Written communication · Graphic communication · Engineering · Architecture · Graphics

1 Introduction

The entry of the Spanish law RD 1393/2007 [1] marked the beginning of the development and subsequent implementation of new study plans in which the unavoidable requirement of the acquisition by students of the so-called generic competences [2] (hereinafter GCC) was established. It was considered that these skills should facilitate the free movement of university students and graduates in Europe and access to an increasingly globalized labor market. During the 2018–2019 academic year, the Universidad Politécnica of Madrid, through its Educational Innovation Service, developed the initiative called "Analysis on how the generic competences are trained and evaluated in the Spanish universities" [3]. It faces the need to know to what extent they have been launched and developed the approaches set out in said RD 1393/2007 on the organization of university education,

F. Cavas Martínez et al. (Eds.): INGEGRAF 2021, LNME, pp. 404–414, 2022.
https://doi.org/10.1007/978-3-030-92426-3_47

and its subsequent amendments (RD 861/2010, RD 96/2014, RD 43/2015) [4, 5] in Spanish technological universities that offer degree studies in the branch of Engineering and Architecture.

The main objective of this study was to describe and analyze how these universities have incorporated the management of GCC into the curriculum of their degrees. The study was based on the following sources of information:

a) Identification and selection of 77 universities with degrees in the field of engineering and architecture.
b) Identification and selection of 21 works (among 119 primary studies on the implementation of GCC in the Spanish university), in which the implementation of the GCC is addressed, either at the university, center or degree level.
c) Systematic and exhaustive review of the websites of the 40 universities with the highest number of degrees in their offer, the highest number of students enrolled and/or more accreditations.
d) Surveys of Spanish universities on their process of implementation and treatment of GCC in their degrees (16 responses received).

Among the conclusions most supported by all the sources consulted, it can be highlighted that the GC of "Oral and written communication in one's mother language"

- It is among the 4 most cited GCC in the websites analyzed.
- It is the one with the highest presence in Spanish universities because it is included in 39 of the 40 websites of universities analyzed, so it is the GC that is developed in more universities. This means that an importance index of 97.5%, which is much higher than those that follow in importance that are the competencies of "Teamwork", "Problem solving", or "Organization and planning", to which 65% would correspond. On the other hand, it should be noted that this competence is also included in the most prestigious accreditations among universities and technological degrees:
- ABET accreditation [6], considered the most important worldwide in the fields of applied sciences, computer science, engineering, and technology, and in which since 2019. In it, 7 competences are evaluated and define competence #3 as "An ability to communicate effectively with a range of audiences".
- EUR-ACE accreditation [7], which incorporates the views and perspectives of key stakeholders (students, higher education institutions, employers, professional organizations, and accreditation agencies), identifies in its guideline 7.1 that the "Ability to use diverse methods to communicate clearly and unambiguously their conclusions, and the knowledge and rationale underpinning these, to specialist and non-audience specialists in national and international contexts" is valued.

2 Application Framework

It is the importance of this competence of "Oral and written communication in one's mother language" that motivates the presentation of this work, which seeks to help improve it by guiding students in the identification and use of the various resources usually used in the approach and writing of their academic documents, for which they will often be evaluated. It can also help teachers to evaluate some key performance of the assessment associated with the written communication of student work. The activities for which a student can write a document are varied but concrete, and in all of them a document will be generated classifiable into one of the following types of academic texts, also known as *genres* [7]:

- Written exams
- Reports of practices carried out in situ in the laboratory.
- Follow-up reports, for example, on the activity of external practices in companies.
- Reports of practices or experimental projects
- Academic works, which usually consist of the development and conceptual and argumentative analysis of a topic after the critical reading of bibliographic sources.
- Final degree/master project and doctoral theses.

From the point of view of writing, the purpose of writing a document is to transmit information in a clear, coherent, and effective manner between the sender, or the person who drafts the document, and the receiver, or the person who reads the document, as summarized in Fig. 1. For this it is necessary to clarify the ideas and thus order the message that you want to transmit, using tools such as style, and taking into account the expectations of the reader, grammatical rules, and all the graphic resources and means necessary to effectively communicate the information. In the document it should be easy to identify the central message, and follow it effortlessly.

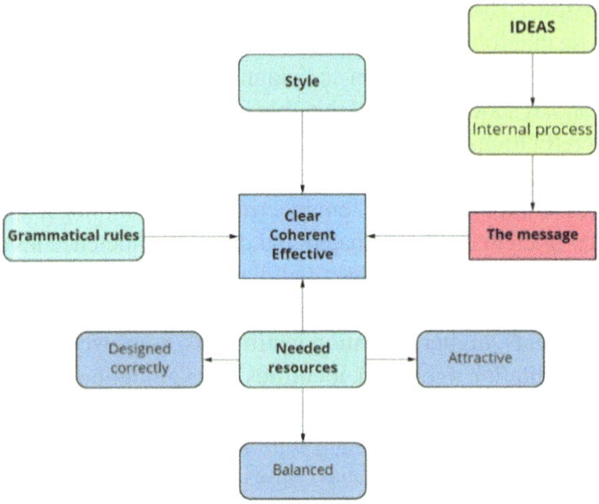

Fig. 1. Characteristics and components of a written document

Among the key performances that can be used to measure this communication competence are:

1. Organization of communication
2. Style

3. Required resources
4. Extension
5. Grammar rules

This communication focuses on key performance 3 "Required resources", identifying some of the resources usually used in academic documents, and showing indicative examples on the use and design of these resources according to different levels of performance that can be achieved and valued [9].

3 Developing Written Communication Competence: Graphic Resources

The use of the necessary means, including graphic resources, aims to support the communication of information by making it clearer, faster and more effective. But, they also aim to motivate reading and studying, by making it more pleasant and therefore effective. To do this, they must be properly designed and used in a balanced way.

The resources complement and reinforce writing, help synthesize and explain complex and difficult-to-understand concepts, and enable faster analysis of numerical results.

Table 1 shows what could be a rubric for the evaluation of the outcome "Use of the necessary resources", within the evaluation of the competence of "Written Communication". It describes four levels of performance as seen in its evaluation.

Table 1. Performance levels in the evaluation of the use of necessary resources

Required resources: It uses graphic resources and means to effectively communicate information			
Very Bad (1)	Bad (2)	Good (3)	Very good (4)
It does not use graphic resources or additional means of supporting communication	Use means of supporting communication that do not help understanding because they are poorly designed. Graphic resources do not help in the interpretation and analysis of information because they are excessive or insufficient or because they are poorly prepared	Use means of communication support that help understanding and maintain interest. The graphic resources used help in the interpretation and analysis of the information communicated	It stands out for an excellent and balanced design and use of means of support for communication and graphic resources. Properly employ text sizes, colors, bullets, image quality, etc. to convey ideas

The following tables show examples for each of the four levels of competence described in Table 1, on the design of some of the necessary resources that are commonly used in documents written with a specific use.

The examples have been extracted from works carried out by the students, individually or as a team, and have been organized according to specific types of resources used especially in academic documents with technical content (covers, indexes, tables, images (figures, graphs, diagrams, etc.), drawings, or texts) and by those levels of competence described.

Table 2 shows examples of covers corresponding to student works.

Table 2. Examples of covers according to the objectives that must be met

<table>
<tr><td rowspan="4">COVER</td><td>Level 1</td><td>Level 2</td></tr>
<tr><td>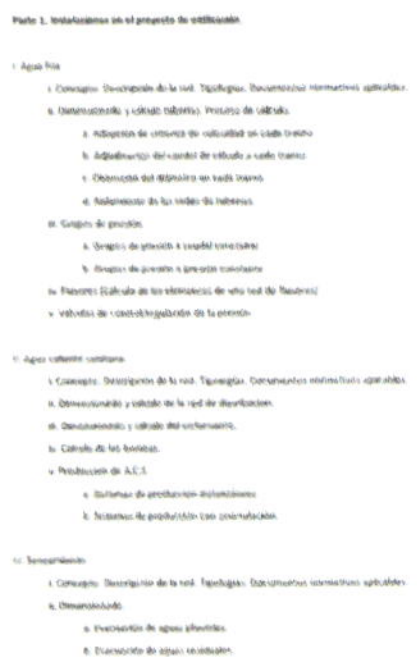</td><td>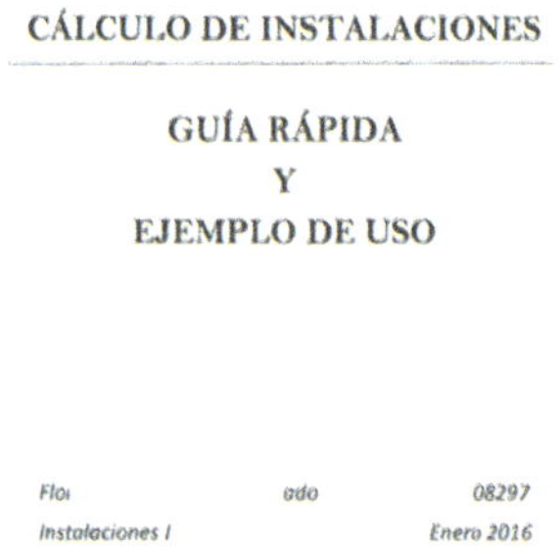
</td></tr>
<tr><td>Level 3</td><td>Level 4</td></tr>
<tr><td>
</td><td>
</td></tr>
</table>

The example for level 1 corresponds to the first page of a 48-page document presented as a course work. The content corresponds to the index of the work, so as a cover it lacks the most elementary information: author, title, date, and development environment.

The example for level 2 corresponds to the first page of a document presented as a work on a subject. The title does not clearly reflect the content and does not identify the development environment.

The example for level 3 contains the necessary information: author, title, date, development environment (subject, center).

The example for level 4 also shows an excellent and balanced design and use of communication support media and graphic resources, complementing the information with images that facilitate its identification and frame its content.

Table 3. Examples of document indexes according to the objectives to be met

	Level 1	Level 2
TABLE OD CONTENTS	No table of contents	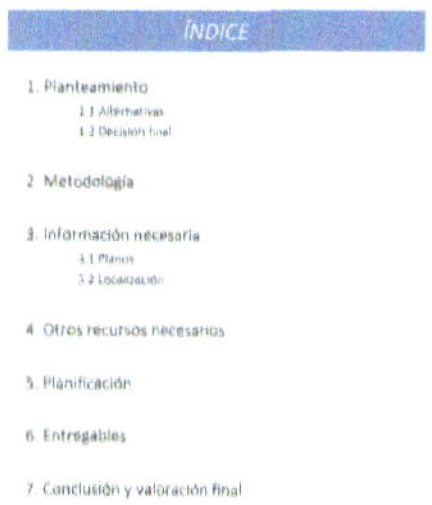
	Level 3	Level 4
		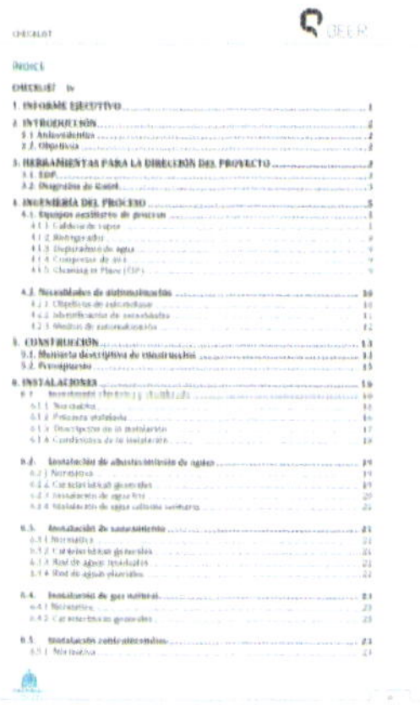

Table 3 shows examples of table of contents in documents. The table of contents provides very quickly information about the organization and hierarchy of the document. The example for level 1 corresponds to a document that simply lacks a table of contents, which we surprisingly know happens quite often.

The example for level 2 shows insufficient information about the content of the document, and that does not help to understand its organization, extension, or degree of development of the sections.

The example for level 3 presents an adequate and orderly content, with some formatting defects.

The example for level 4 shows adequate information, with a well-organized and formatted design, properly using text sizes, numbering, bullets, etc. to convey that information.

Table 4. Examples of data tables

DATA TABLES

Level 1

Planta alta	Vestuario mujeres	Ducha	0,2
Planta alta	Vestuario mujeres	Ducha	0,2
Planta alta	Vestuario mujeres	Ducha	0,2
Planta alta	Vestuario mujeres	Ducha	0,2
Planta alta	Vestuario mujeres	Ducha	0,2
Planta alta	Vestuario mujeres	Lavabo	0,1
Planta alta	Vestuario mujeres	Lavabo	0,1
Planta alta	Vestuario mujeres	Lavabo	0,1
Planta alta	Vestuario mujeres	Lavabo	0,1
		TOTAL	1,4

Level 2

CONSTRUCCIO N	I n	M	P	R	E	I
Aire	2	1	1	1	1	11
Suelo	8	4	4	3	3	41
Agua	4	2	1	1	1	18
Flora	2	4	2	2	3	20
Fauna	2	2	2	1	1	13
Medio perceptual	4	2	2	2	3	24
Población	2	1	2	2	2	15
Economía	4	2	2	2	1	20

EXPLOTACIO N	I n	M	P	R	E	I
Aire	2	1	1	1	1	11
Suelo	2	1	1	1	1	11
Agua	4	2	4	2	3	20
Flora	2	1	1	1	1	11
Fauna	2	1	1	1	1	11
Medio perceptual	2	1	4	3	3	20
Población	2	2	4	3	3	21
Economía	4	2	2	3	3	25

Level 3

CONSTRUCCIÓN	In	M	P	R	E	I
Aire	2	1	1	1	1	11
Suelo	8	4	4	3	3	41
Agua	4	2	1	1	1	19
Flora	2	4	2	2	3	20
Fauna	2	2	2	1	1	13
Medio perceptual	4	2	2	2	3	24
Población	2	1	2	2	2	15
Economía	4	2	2	2	1	20

EXPLOTACIÓN	In	M	P	R	E	I
Aire	2	1	1	1	1	11
Suelo	2	1	1	1	1	11
Agua	4	2	4	2	3	26
Flora	2	1	1	1	1	11
Fauna	2	1	1	1	1	11
Medio perceptual	2	1	4	3	3	20
Población	2	2	4	3	3	21
Economía	4	2	2	3	3	25

Level 4

Table 4 shows examples of tables with data collection. The use of tables requires additional effort to understand the essence of the data, and to organize, synthesize and present the information. With them, the understanding and interpretation by the reader is facilitated, and taking advantage of the multiple options of data format, the most relevant aspects can be highlighted.

The example for level 1 shows a table that lacks content identification, header in the various fields, or font. The example for level 2 presents the Information divided uncontrollably, which makes it difficult to understand.

The example for level 3 of the table takes advantage of the possibilities offered by computer tools to present the data in a way that facilitates the understanding and analysis of the information communicated. In the example for level 4 a homogeneous format is used throughout the table, the information is well organized and presented, which facilitates its understanding and highlights its objectives.

Tables 5 and 6 show examples of images included in the documents. The images have been valued for themselves, and for their illustrative function, although, like many of the resources, they must be valued in their context, in which they are usually accompanied by some explanatory text.

In Table 5, the example for level 1 corresponds to a case where the content is difficult to read and the content and source are not identified.

Table 5. Examples of the use of images in documents

	Level 1	Level 2
IMAGES: SIZE AND CLARITY	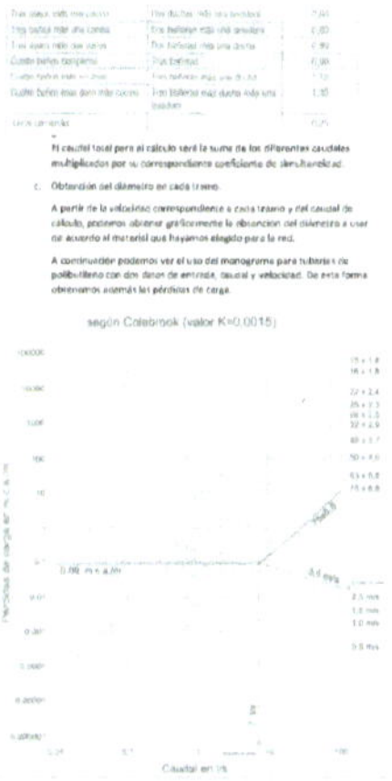	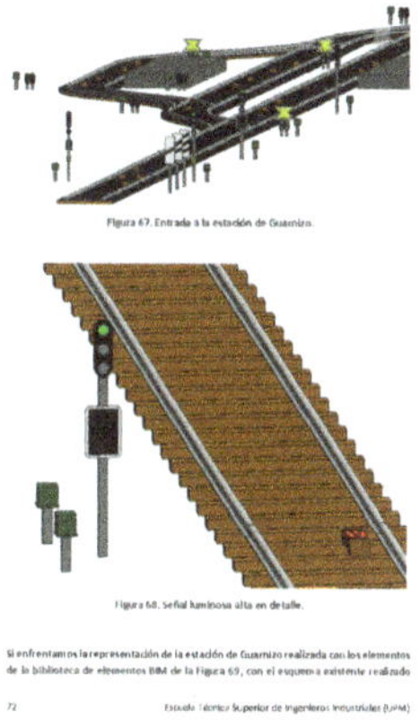
	Level 3	**Level 4**
	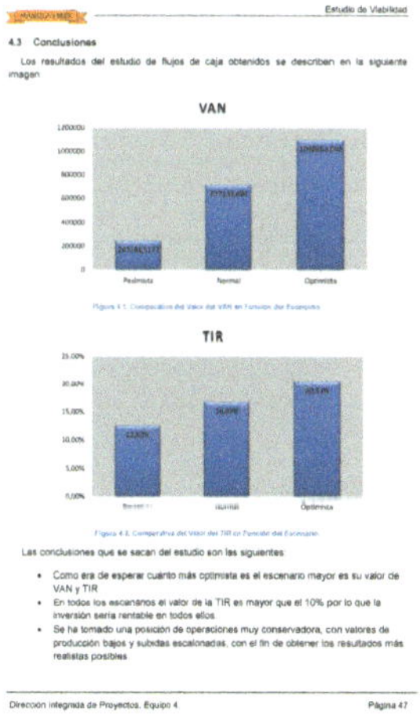	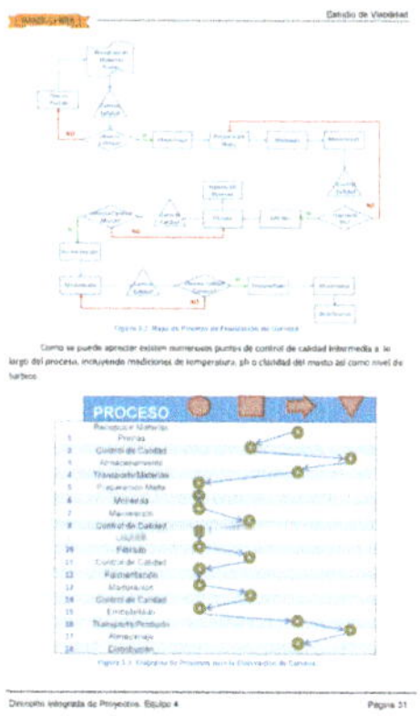

The example for level 2 corresponds to a case in which the graphic resources are excessive and poorly prepared, so they do not help in the interpretation and analysis of the information. The example for level 3 shows information of interest and well-explanatory, but with an excessive size of the images.

The example for level 4 shows information of interest and is well-organized to facilitate communication. It properly uses graphic resources, taking advantage of the size of the texts, the use of colors, vignettes, and simbology, as well as taking care of the quality of the images, to transmit the ideas.

Table 6. Examples of technical drawings according to the requirements and objectives that must be met

	Level 1	Level 2
TECHNICAL DRAWINGS		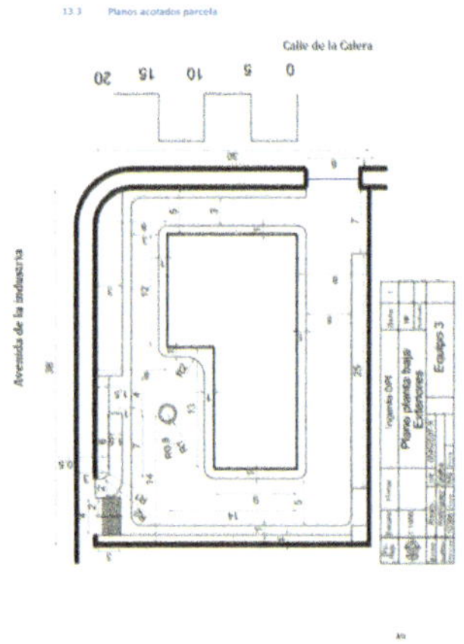
	Level 3	**Level 4**
	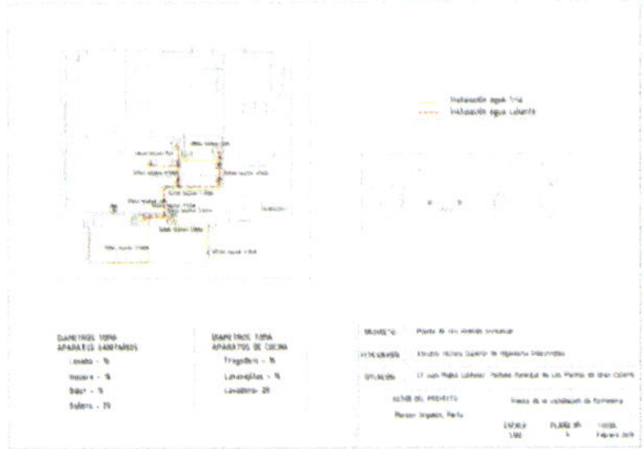	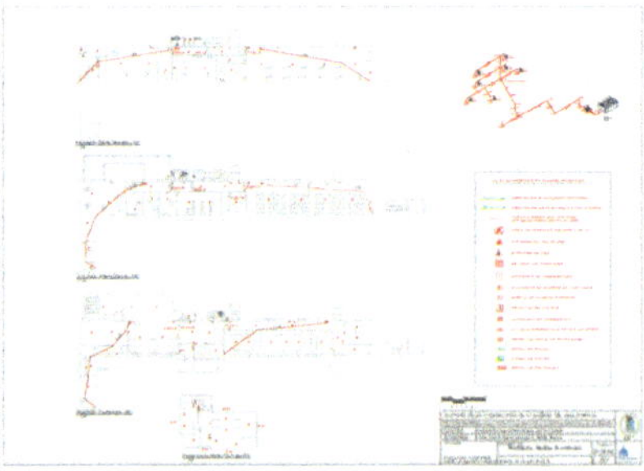

Specifically, Table 6 shows different additions of technical drawings in the documentation. The example for level 1 corresponds to information with graphic format that is presented as if it were a plane, but that does not meet some of the essential elements in a document of that type, such as the box.

The example for level 2 corresponds to a technical drawing with adequate information, but with some formal deficiencies.

The example for level 3 is a well-crafted, presented, and formally correct technical drawing. And the example for level 4 in addition to being a well-crafted and presented, and formally correct, technical drawing presents an excellent and balanced design and use of the means of supporting communication to convey ideas.

4 Conclusions

According to the study carried out by the UPM to know and analyze how the training and evaluation of these generic competences is being addressed in the Spanish technological universities that offer degrees in the branch of Engineering and Architecture. It is the oral and written communication competence in the mother language that appears collected most frequently, and prominently, in the curricula of the 40 universities analyzed among the 32 Tuning competences.

This recognition should raise the importance of resources especially graphics within university studies. A rubric has been proposed for the assessment of these additional resources to the text. Examples have been shown with concrete cases drawn from student work to illustrate the different levels of performance in learning outcomes that help measure written communication competence. While representative examples of these levels of performance can be found in the same document, often the care given to all aspects that influence communication is similar, and is also related to the overall quality of the work presented. One might ask whether this competence reflects not only a specific aptitude, but also an attitude towards work in general.

Acknowledgements. This work is based on the results of the project "Analysis on how the generic compe- tences are trained and evaluated in the Spanish universities" (GIE1819-Comp) pro- moted by the Educational Innovation Service of the Polytechnic University of Madrid. Also our most sincere thanks to the universities that contributed to its elaboration by providing their information in detail (Universidad Politécnica de Valencia, Universidad Miguel Hernández, Universidad de La Laguna, Universidad de Zaragoza, Universidad de Jaén, Universidad de Navarra, Universidad Pública de Navarra, Universidad San Pablo-CEU, Universidad Europea de Madrid, Universidad de Málaga, Universidad Complutense de Madrid, Universidad a Distancia de Madrid, Universidad de Vigo, Universidad Pompeu Fabra, Universidad Loyola Andalucía, Universidad de A Coruña, Universidad Antonio de Nebrija).

References

1. Real Decreto 1393/2007, de 29 de octubre, por el que se establece la ordenación de las enseñanzas universitarias oficiales. Boletín Oficial del Estado, nº 260 de 30 de octubre de 2007 (2007). https://www.boe.es/buscar/act.php?id=BOE-A-2007-18770

2. Tuning Project. Página (2002). www.let.rug.nl/TuningProject or www.relint.deusto.ers/Tuning Project/
3. Análisis sobre cómo se está abordando la formación y evaluación de las competencias transversales en las universidades nacionales. Iniciativa GIEs. Universidad Politécnica de Madrid (2019)
4. Real Decreto 861/2010, de 2 de julio, por el que se modifica el Real Decreto 1393/2007, de 29 de octubre, por el que se establece la ordenación de las enseñanzas universitarias oficiales. Boletín Oficial del Estado, nº 161, de 3 de julio de 2010 (2010)
5. Real Decreto 43/2015, de 2 de febrero, por el que se modifica el Real Decreto 1393/2007, de 29 de octubre, por el que se establece la ordenación de las enseñanzas universitarias oficiales, y el Real Decreto 99/2011, de 28 de enero, por el que se regulan las enseñanzas oficiales de doctorado. Boletín Oficial del Estado, nº 29, de 3 de febrero de 2015 (2015)
6. ABET-UPM. (s.f.). Escuela Técnica Superior de Ingeniería Industrial. Obtenido de UPM. http://competencias.industriales.upm.es/index.php/plan-de-mejora/
7. EUR-ACE. https://www.enaee.eu/eur-ace-system/
8. Guía de utilización de la rúbrica de Comunicación Escrita: Catalogo de ejemplos. Obtenido de. http://competencias.industriales.upm.es/wp-content/uploads/2017/02/Guia-de-utilización-de-la-rúbrica-de-Comunicación-Escrita_definitivo.pdf
9. Marco de Desarrollo de Competencias en la ETS Ingenieros Industriales de Madrid. Obtenido de. http://competencias.industriales.upm.es/wp-content/uploads/2017/02/Folleto_Competencias_nov_2013.pdf

Applying Visual Thinking Techniques in Engineering Education

Alfonso Martín Erro[1](✉), María Luisa Martínez Muneta[2], and Ángel Antonio Rodríguez Sevillano[3]

[1] Escuela Técnica Superior de Ingeniería y Diseño Industrial, Universidad Politécnica de Madrid, Madrid, Spain
alfonso.martin.erro@alumnos.upm.es
[2] Escuela Técnica Superior de Ingenieros Industriales, Universidad Politécnica de Madrid, Madrid, Spain
[3] Escuela Técnica Superior de Ingeniería Aeronáutica y del Espacio, Universidad Politécnica de Madrid, Madrid, Spain

Abstract. Visual Thinking tools and formats, such as visual maps or visual notes, have been proven to be a valuable resource in communicating and recording ideas, solving problems, as well as a means of supporting creativity. These techniques currently have a wide diffusion and acceptance in the areas of management and secondary education. Since visual thinking is basic in engineering, it is understood that the application of these techniques brings interesting benefits in the development of engineering activities. This paper shows the teaching experience with students of the degree of Organizational Engineering of the ETSII-UPM, in the subject of "Industrial Drawing" implementing Visual Thinking techniques. After a training session, two in-group assignments were carried out around the use of visual maps. The favourable acceptance of the students encourages to consider this set of techniques as a teaching subject.

Keywords: Engineering education · Communication · Creativity · Visual thinking · Visual maps

1 Introduction

By Visual Thinking we mean a set of graphical techniques and methodologies used for learning, recording information, generating ideas, and solving problems. These techniques consist of the use of simple graphic representations, following the principle that everything can be represented using simple graphic elements, such as points, lines, curves, rectangles, or triangles [1]. Garbiñe Larralde presents *Visual Thinking* as "a work methodology that seeks to apprehend and represent reality through simple drawings and simple texts" [2]. On the other hand, Fernandez Fontecha [3] describes it as a methodology that uses linguistic elements, basic visual forms, and graphics with the purpose of clarifying complex ideas or in support of problem-solving tasks. For Sanz Segura [4], the Visual Thinking technique, or *Visual Thinking*, consists of the graphic representation of any problem by simplifying it into easily recognisable elements.

F. Cavas Martínez et al. (Eds.): INGEGRAF 2021, LNME, pp. 415–422, 2022.
https://doi.org/10.1007/978-3-030-92426-3_48

Sanz Segura [4] points to graphic representation as the basis of these methodologies, referring to the principle put forward by Betty Edwards [5] that everything can be explained using simple and universal drawings. Dan Roam [6] develops his approach to *Visual Thinking* techniques based on establishing connections between the visual and verbal modes of the mind. It is about bringing together the capacities of both brain hemispheres, integrating the analytical with the creative [2].

From a review of the most relevant literature [1, 6–8] as well as other sources consulted [2, 9] and academic works [3, 4] it has been possible to group the elements used in these techniques into three categories:

1. **Concrete and abstract pictorial representations.** Drawing of human figures and objects. Pictograms or images of real objects. Ideographs and visual metaphors. Representation of ideas or concepts.
2. **Graphic elements to organise information.** They are used to establish differences, links as well as relationships in space and time: bullets, speech bubbles, arrows, connectors, dividers, frames, lines, signs, containers, or posters.
3. **Text** that is integrated into the graphical environment with different fonts and sizes. The typography used (typeface and font size, as well as its thickness), among other issues (Fig. 1).

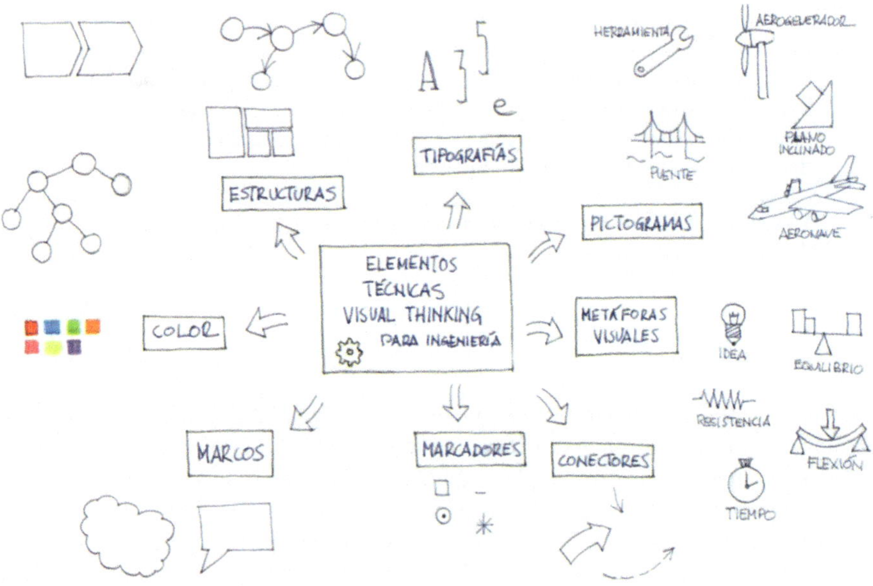

Fig. 1. Elements of *Visual Thinking* applicable in engineering (Illustration by the author)

Within these *Visual Thinking* methodologies, different techniques or formats are used:

- Visual maps [2]. Medium and large format.
- Visual notes - Sketchnoting [10]. Small format.
- Graphic Recording [8]. Large format
- *Storyboards* [4]. Medium format.
- Visual Notes [8]. Small format.

In recent years, there has been a growing interest in visual thinking techniques in professional fields such as marketing, business, and journalism, with several manuals promoting the practice of visual thinking [3]. Likewise, different *Visual Thinking* techniques and formats are applied in group dynamics, meetings or brainstorming sessions [8]. They are also widely used in group activities, in which a Graphic Facilitator intervenes, who applies the Graphic Recording technique to bring together and order the ideas put forward by everyone. The use of *Visual Thinking* techniques is also aimed at problem solving [3] and creative support as a technique for generating ideas [4].

Another of the applications studied revolves around teaching. Although *Visual Thinking* methodology can be applied to all types of educational levels [2], these techniques have mainly been disseminated in the secondary education environment, such as the initiative of Spain´s Ministry of Education's course for secondary school trainers, as well as the Ministry of Education's Aprende INTEF module [11].

Within the engineering environment, Visual Thinking techniques have been shown to be applied in design projects [4]. Specifically in engineering education, these techniques have been included in educational models [12, 13]. Likewise, Inelmen [14] implemented these techniques in an orientation course for first-year students, while Paepke-Hielmes [15] studied the implementation of the *Skecthnoting* methodology [10] with first-year engineering students, to foster their learning and retention skills, as well as their communication skills. The Design Thinking methodology [16] with visual formats is applied by Fernando de Pablo and Miren Lasa [8] with students at the University of Navarra.

2 Teaching Experience

In the second four-month period of 2018, a study was carried out on the implementation of *Visual Thinking* techniques in the subject "Industrial Drawing", taught in the first year of the degree in Organisational Engineering at the ETSII of the UPM. This subject incorporates creativity development components in its syllabus, specifically one hour of theory and practical application exercises. The aim of incorporating new teaching content on visual thinking was to add new elements for the development of visual communication skills.

31 students participated in the study, 11 male and 20 female. They received a training session in which the fundamentals of Visual Thinking techniques and methodologies were presented. The session was theoretical and practical, in such a way that the thematic presentations were interspersed with exercises on the application of the principles

presented. After a presentation justifying the application of these techniques in engineering, the fundamentals, elements, and applications of Visual Thinking techniques were shown:

- What is Visual Thinking.
- Applications of Visual Thinking. Generation of ideas. Learning. Communication. Problem solving.
- Examples of visual thinkers in science and engineering
- Visual Thinking and Engineering.
- Visual Thinking Methodologies. Visual Design Thinking. InfoDoodle [7]. Storytelling. SCVID [6]. Idea sketching. Mind maps. Empathy map
- Visual Thinking Process.
- Visual Language. Elements of Visual language. Ideograms and Visual metaphors. Visual Library. Visual Alphabet. Typographies.
- Formats: Visual Maps. Graphic Recording. Visual Notes.

After the training session, an assignment was proposed, consisting of generating a visual map in a group, based on a specific topic (the engineering profession). Following guidelines on how to make visual maps, eight groups of four members each were organised. Figure 2 shows three examples of the visual maps produced by the students.

Fig. 2. Examples of visual mapping exercises.

In the middle of the course, a new visual mapping activity was assigned. In this case, students were enrolled in 13 groups. Each one had to present a specific topic of their choice. The production of these visual maps, some of which are shown in Fig. 3, shows clear progress in relation to those produced at the beginning of the course.

These visual maps show an increased use of colour, as well as an improvement in the structuring of graphical and verbal information. There is also a better balance between text and graphics. The concepts have been expressed in a very visual and clear way, and the logical relationships established between the different types of information to be able to draw them are very coherent.

Fig. 3. Examples of the second Visual Maps assignment.

After the teaching experience, the students participants were asked in the year 2020/2021 about the usefulness of the techniques taught, once they have advanced their studies, as well as their perception of the usefulness of these techniques for their career. To this purpose, a questionnaire was distributed, consisting of a set of questions that could be assessed using a Likert scale, ranging from 1 to 5.

The students rated mind maps as the most useful technique, both during their studies and for their future professional practice. On a scale from 1 (not at all) to 5 (very much), the average value of mind maps was 4.344 in the studies, and with a similar value of 4.323 in the future professional practice. On the other hand, visual maps had a very similar rating: an average of 4.281 in the studies and 4.290 in the future professional practice (Table 1).

Table 1. Usefulness of *Visual Thinking* techniques (1: a little, 5: a lot)

Technique/format	In studies		In professional practice	
	Average	Std. dev.	Average	Std. dev.
Visual Maps	4.281	0.683	4.290	0.824
Sketchnoting	3.594	1.214	4.258	0.729
Mind Maps	4.344	0.865	4.323	0.748
Visual Design Thinking	4.031	0.647	3.871	0.670
InfoDoodle	2.742	0.773	2.967	0.490
Storytelling	3.188	0.998	3.548	1.060
SCVID	2.871	0.718	3.258	0.965

Among the various applications of Visual Thinking techniques that were shown to students in the training session, they recognised that they were mainly applied in activities similar to those carried out in the teaching experience, such as presentations of class work or exercises (average 4.375) versus the approach/definition of a problem or problem ideation/solution tasks (Table 2) that were not approached with practical activities.

Table 2. Application of *Visual Thinking* techniques (1 a little: 5: a lot)

Activity	Average	Standard deviation
Define/state a problem	3.968	1.062
Search for ideas/solutions to a problem	3.562	0.840
Presentations of work or exercises in class	4.375	0.659

Finally, the evaluations of the Visual Thinking techniques by the participants in the study were very positive. They found the organisation of information, communication of complex concepts and understanding them as the most important benefits of these techniques (Table 3).

Table 3. Benefits of using *Visual Thinking* techniques

Visual thinking techniques have helped me to: (1 nothing: 5 a lot)	Average	Standard deviation
Organising information better	4.406	0.560
Improving my learning	3.719	0.813
Developing my creativity	3.938	0.840
Understanding complex concepts	4.219	0.832
Improving the performance of group activities	3.313	0.471
Expressing complex ideas and concepts to others	4.344	0.545
Improve my communication skills	3.781	0.706
Understanding a problem from other perspectives	3.625	1.008
Valuing the point of view of others on a problem	3.156	0.677

3 Conclusions

This research paper is an attempt to bring Visual Thinking techniques into the engineering classroom. The aim is to provide students with a resource to enhance teamwork, as well as encourage learning and retention by means of visual resources.

The results obtained, both from the analysis of the visual maps made, as well as the students' own assessments, indicate that Visual Thinking techniques are positively accepted by engineering students.

They actively participated in learning and practising *Visual Thinking* techniques, in the interpretation and use of colour, typography, visual metaphors, pictograms and connectors.

Participants in the study found *Visual Thinking* techniques to be a valuable resource in working together as a group, enabling them to broaden the view on a particular topic by effectively integrating the contributions of their members.

References

1. Gray, D., Brown, S., Macanufo, J.: Gamestorming: a playbook for innovators, rulebreakers, and changemakers. O'Reilly Media, Inc (2010)
2. Larralde, G.: Qué es el Visual Thinking (2016). Accessed abr. 16, 2021. [En línea]. https://issuu.com/garbine/docs/librillo_visual_thinking_garbine
3. Fernández-Fontecha, A., O'Halloran, K.L., Tan, S., Wignell, P.: A multimodal approach to visual thinking: the scientific sketchnote. Vis. Commun. **18**(1), 5–29 (2019)
4. Sanz-Segura, R., Romero-Piqueras, C.: Visual Thinking, el empleo de recursos gráficos para optimizar nuestros proyectos. In: Jornada INGEGRAF Ingegraf+D+i 2016 Contribución de la ingeniería gráfica a la sociedad (2016)
5. Edwards, B.: Drawing on the Right Side of the Brain. Jeremy P. Tarcher (2012)
6. Roam, D.: The Back of the Napkin: Solving Problems and Selling Ideas with Pictures. Portfolio (2008)

7. Brown, S.: The Doodle revolution: Unlock the Power to Think Differently. Portfolio (2015)
8. de Pablo, F., Lasa, Mi.: ¡Dibújalo! LID Editorial Empresarial (2015)
9. Gende, I.M.: Visual Thinking: dibujando el aprendizaje | UNIR, Feb. 19, 2018. https://www.unir.net/educacion/revista/noticias/visual-thinking-dibujando-el-aprendizaje/549203599112/. Accessed 21 Jul 2019
10. Rohde, M.: Sketching: The Visual Thinking Power Tool, ene. 25 (2011). http://alistapart.com/article/sketching-the-visual-thinking-power-tool. Accessed 15 Jan 2019
11. INTEF. Curso "Visual Thinking en educación" #VisualMooc, (Mar 29, 2017). [Video online]. https://www.youtube.com/watch?v=Bp85Qi7tN1M. Accessed 21 July 2019
12. Martin-Erro, A., Dominguez, M., Espinosa, M.M.: Creativity and engineering design: analysis into the educational environment. In: EDULEARN15 Proceedings, Barcelona, Spain, July 2015, pp. 1304–1310 (2015)
13. Martin-Erro, A., Nuere, S.: Searching the ways to creativity in engineering education through engineering graphics. In: INTED2021 Proceedings, Mar 2021, pp. 2849–2858 (2021). https://doi.org/10.21125/inted.2021.0608.
14. Inelmen, E.: Implementing "visual thinking" in the engineering orientation course. Eur. J. Eng. Educ. **26**(3), 291–299 (2001). https://doi.org/10.1080/03043790110054418
15. Paepcke-Hjeltness, V., Mina, M., Cyamani, A.: Sketchnoting: a new approach to developing visual communication ability, improving critical thinking and creative confidence for engineering and design students. In: 2017 IEEE Frontiers in Education Conference (FIE), pp. 1–5 (2017)
16. Brown, T.: Design thinking. Harvard Bus. Rev. **86**(6), 84

Methodology for the Virtualisation of Engineering Drawing Exercises for Use Through Extended Reality

Paula Triviño-Tarradas[1(✉)], Alejandro Mohedo Gatón[2], Pilar Carranza Cañadas[1], Enrique Burgos-Ladrón de Guevara[1], Francisco Javier Mesas-Carrascosa[1], and Rafael Enrique Hidalgo Fernandez[1]

[1] Departamento de Ingeniería Gráfica y Geomática, Universidad de Córdoba, Campus de Rabanales, 14071 Córdoba, Spain
{ig2trtap,ir1carr,ir1lague,ig2mecaf,rhidalgo}@uco.es

[2] Delegación de Educación de La Junta de Andalucía en Córdoba, Córdoba, Spain

Abstract. The present study is framed within the experiences of educational innovation related to the techniques of graphic representation. This article provides a new approach on the teaching-learning process that implements techniques of extended reality (XR), also known as mixed reality (MR), constituted by the set of solutions of extended reality (XR), Augmented Reality (AR), Virtual Reality (VR) and Mixed and Diminished Realities in teaching. A selection of exercises of Descriptive Geometry in 2D can be transformed into 3D formats, which can be visualised in the three-dimensional space through new and different devices through virtualisation. To this end, a specific methodology is followed, which adapts them to a new way of visualising them. This document explains the methodology carried out to attain this visualisation. The exercises are subjected to different checks to allow the students to delve into the contents of the subject in a completely immersive and friendly manner. The implementation of these improvements in the teaching-learning process shows that one of the most innovative ways of observation is through virtual-reality headphones and goggles (Oculus or equivalent), which implies a step forward in non-immersive 3D visualisation through digital screens in the field of graphic engineering.

Keywords: Extended reality · Virtual reality · Augmented reality · Graphic expression · Virtual-reality goggles

1 Introduction

The rapid and unexpected change in on-line teaching in the last two years has forced faculty members of graphic expression to adapt the procedures, incorporating new technologies. The large amount of technological innovations of current smart glasses, mobile devices, videogame engines and web applications is of great importance. In this context, there are different realities that comprise the broad spectrum between reality and virtuality (i.e., "virtuality continuum") [1]. Terms such as AR, VR, MR, XR and DR, as

F. Cavas Martínez et al. (Eds.): INGEGRAF 2021, LNME, pp. 423–432, 2022.
https://doi.org/10.1007/978-3-030-92426-3_49

well as other future realities, require an adequate definition to understand the future we can expect from their application in teaching [2].

These realities are integrating into current university education, specifically in the degrees of Agronomic Science and Rural Development Engineering and Forest Engineering of the University of Cordoba (Spain). One of the advantages of teaching the contents in the classroom is the feeling of surprise when the students integrate the exercises into their spatial vision and understand how a three-dimensional space is represented in two dimensions. These future professionals gradually develop their spatial vision as we advance in the complexity of the exercises.

However, courses such as this and the previous one, where the education system reevaluated the way of teaching and the classrooms became empty, this type of technologies have posed an essential element to bring the contents of the subject matter in Engineering Drawing in the abovementioned degrees to the homes of the students. In this context, it was necessary to create and apply a new way of sharing the exercises, in order to allow the students to delve into the subject with improvements regarding the experience of spatial visualisation. Remote learning is a work method that removes the limitations of the geographic location of the people involved [3].

MR has a marked multidisciplinary character that notably connects complementary scopes, such as the education sector and engineering [4]. There are currently numerous applications that coexist with some type of technology linked to MR. The aim of the present study is to show the importance of applying these technological innovations in university education, identifying the methodology used for the virtualisation of engineering drawing exercises.

1.1 The Different Realities

MR (or XR) is a combination of what we know as VR and AR. Nowadays, it is possible to create three-dimensional models of high geometric and visual quality through 3D reconstruction and photogrammetry, obtaining very good results, which makes it a fundamental tool for heritage documentation, analysis and dissemination [5]. The numerous applications require a rigorous multidisciplinary procedure (with architects, engineers, documentalists, pedagogists, historians...) that guarantees the quality of the results.

XR can be applied in very different fields [6]:

- History and documentation of the space or object of the intervention.
- Data collection for archiving with historical rigour.
- Processing of photogrammetric models
- Editing and advanced comparative analysis in photogrammetry.
- Optimisation and correction of digital models for 2D/3D visualisation.
- Presentation of the virtual model.

Zhou, Lou and Yang [7] describe AR as an overlapping of digital graphics in a visor that allows seeing the real world. AR overlaps alterations on an existing environment, keeping the user connected to the physical reality observed; on the other hand, in VR, the user plunges into an artificial virtual world that aims to replace the outer reality.

That is, although VR and AR share hardware technologies, the trend seems to indicate a constant growth in AR.

The use of VR goggles, mobile phones or tablets for AR, which allows adapting the learning pace, provides numerous benefits in this regard. The user can move freely throughout the content to obtain a better understanding and assimilation of the concepts. Thus, the use of these resources increases the spatial vision capacity of the user.

2 Methodology

The digital means employed by techniques such as WebXR [8] provide the necessary tools for remote experimentation. This practice allows not only to continue with the teaching task but also to improve the experience of the user and motivate his/her learning.

2.1 Exercises in CAD 3D

From a complete archive of engineering drawing exercises of the Department of Graphic Engineering and Geomatics at the University of Cordoba (Spain), some exercises of both the dihedral and axonometric systems were selected for visualisation (Figs. 1, 2, 3, and 4).

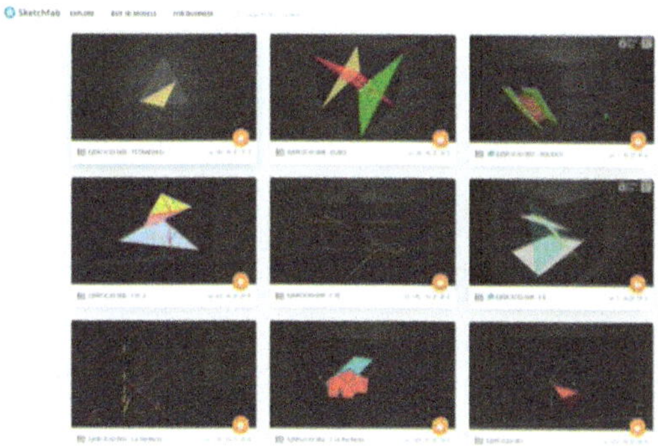

Fig. 1. An example from the online library of the engineering drawing exercises of the Department of Graphic Engineering and Geomatics (UCO).

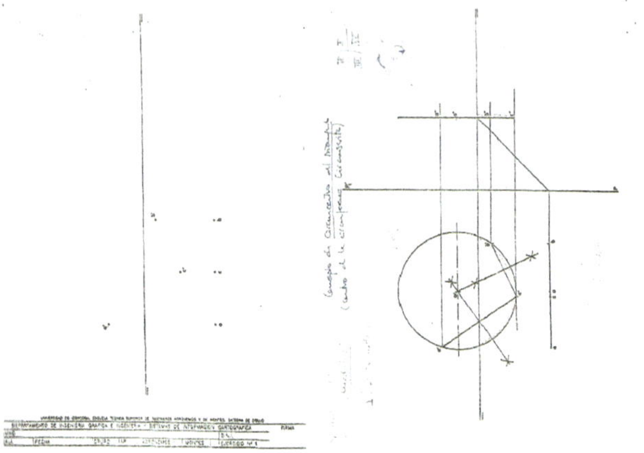

Fig. 2. Image of an exercise of the dihedral exercise (graphic formulation and solution) – 2D paper version.

Fig. 3. Schematic view of the representation of an exercise of the axonometric system.

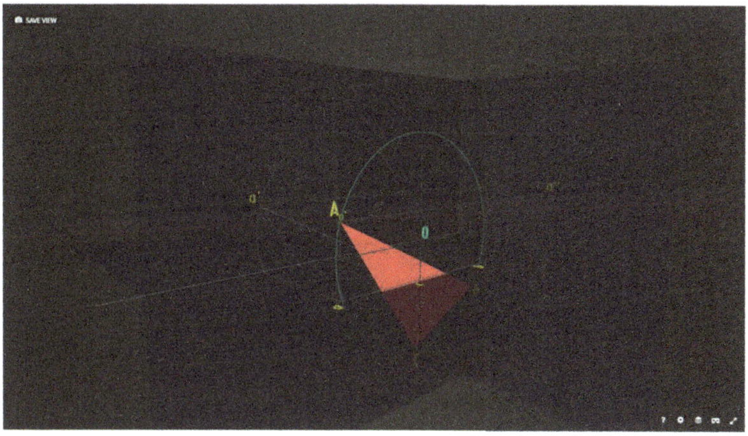

Fig. 4. Screenshot of what the user sees in the screen of the HMD visor.

2.2 Procedure Used in the Virtualisation

Sketchfab provides an online platform that is easy to use and allows converting these resources to their digital twins, which makes it a valuable tool for teaching. Students with poorer spatial vision are involved with greater interest in the exercises, as they visualise them from their own computers, tablets or mobile phones, through a web interface. The steps to follow for the virtualisation of the exercises are summarised as follows:

Step 01
Preparation of the exercise creating its 3D digital twin in a CAD programme. Three-dimensional geometry is employed in a modelling software, paying special attention to the concepts of scale and welding (unions) of the vertices.

Step 02
Selection of the thickness of the lines of the CAD figure to improve its spatial

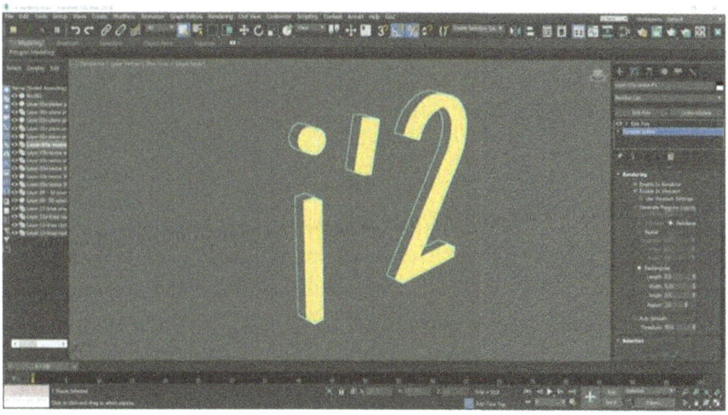

Fig. 5. Transformations applied through modifiers of the original geometry.

visualisation. To this end, there are two types: cylindrical and rectangular (Fig.5).

Step 03
Application of colour codes, to indicate greater or lower importance to each layer, with the aim of obtaining a visually more attractive space that improves the experience of the user (Fig. 6).

Fig. 6. Colour codes selected for an optimised narrative.

Step 04
Assignation of material to each layer of the model.

Step 05
Exportation in the format created by Wavefront Technologies in 1992 (object files: *.OBJ). To integrate the file in a web browser, it is necessary to export it in the OBJ format. This is an open-format file of geometric definition that has been adopted by most providers of 3D graphic applications.

Step 06
Online work in a professional WebXR visor, e.g., Sketchfab, in which the geometry is imported and a series of modifications are performed, such as: orientation of the geometry, lighting control, colour assignation, scale adjustments in VR, and some adjustments of movement sensations. The advanced lighting controls and the representation of materials itself allow recreating the virtual space with a good detail level. The annotation tool allows setting several predefined views, which enables the students to go over the exercise in a tutored and autonomous manner.

3 Results

The procedure allowed developing a uniform methodology to virtualise engineering drawing exercises in 3D, which can serve as a guide to other faculty members of similar subjects. This approach allows the specialised faculty to prepare a catalogue of practical exercises and share them in a website with their students. Figures 7 and 8 show how the platform employed in this study allows, through a university account, accessing advanced functions, such as space customisation and visualisation limit control.

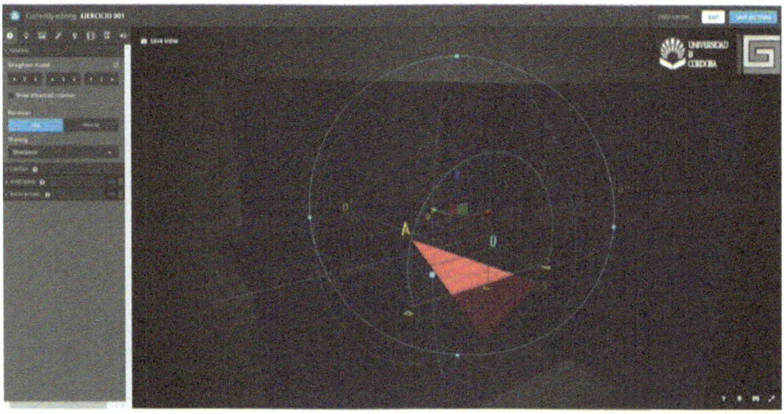

Fig. 7. Online editing and setting module through the Sketchfab platform.

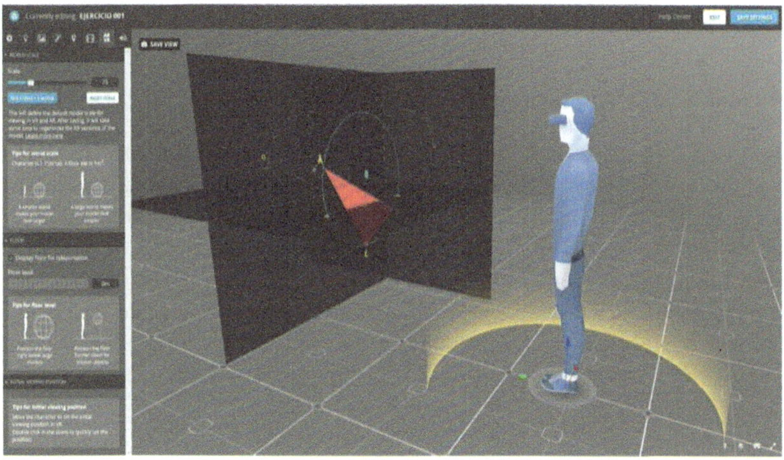

Fig. 8. Screenshot of online VR setting through the Sketchfab platform.

With the abovementioned controls, it is possible to configure the exercises to enable the students to experiment through the use of VR goggles (Fig. 9).

During the experience, it was observed that not all students used a computer with high performance, and the Internet connection was not always at high speed. Therefore, it is very important to optimise the exercise, in order to ensure that the graphic engine (mobile, tablet, computer) has enough resources to load the data fluently and quickly [9].

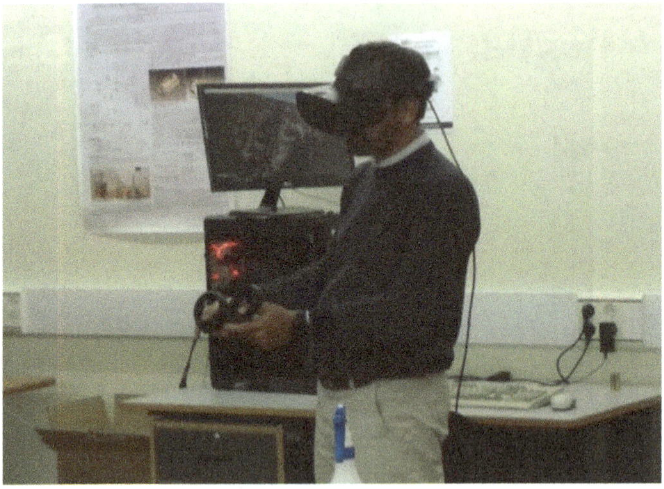

Fig. 9. User with Head-mounted virtual goggles visualising an exercise.

3.1 Virtual Reality Application: Metrics

The approval of an innovation project within the Innovation Plan and Good Teaching Practices of the University of Cordoba allowed obtaining the needed material to carry out and verify the suitability of its application in the classroom. A seminar on Mixted Reality: 'A tool in Covid-19 times' was held to evaluate the impact on teaching using this kind of innovative technologies.

Figure 10 shows the results extracted from the survey conducted to 17 people of different educative contexts, who were given the opportunity to visualize the exercises with their devices remotely. They were asked about their ICT knowledge.

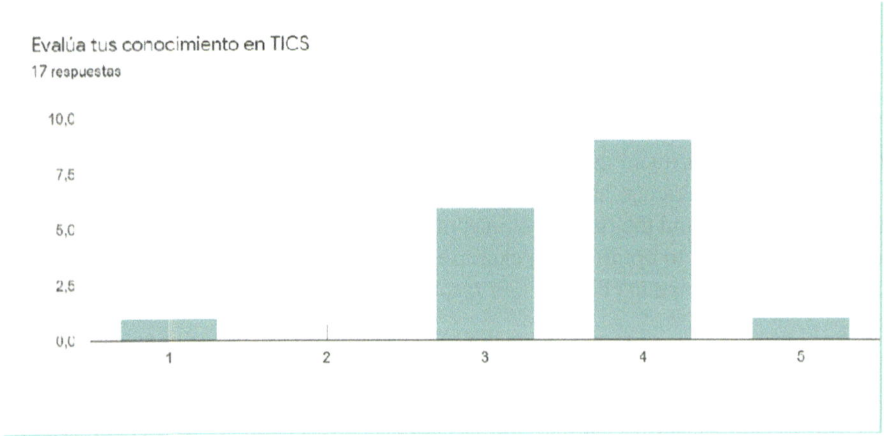

Fig. 10. Users ICTs knowledge.

Figure 11 tends to evaluate the improvement on the spatial vision and have a better understanding of the exercises by using this technology of VR.

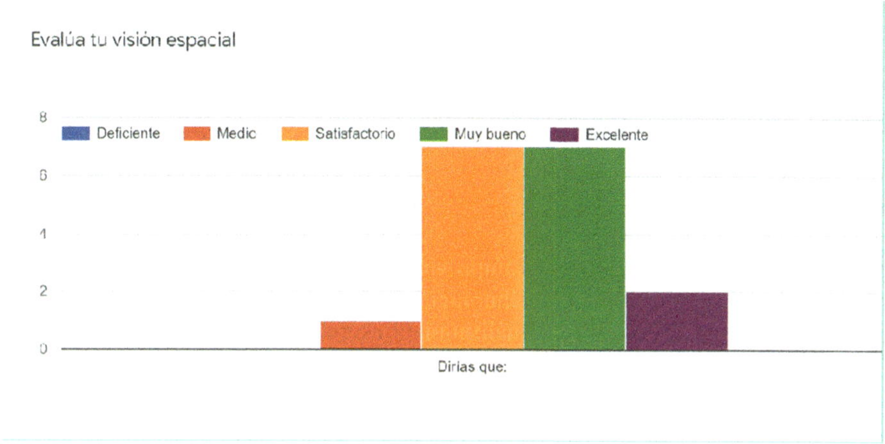

Fig. 11. Assessment of the user's spatial vision.

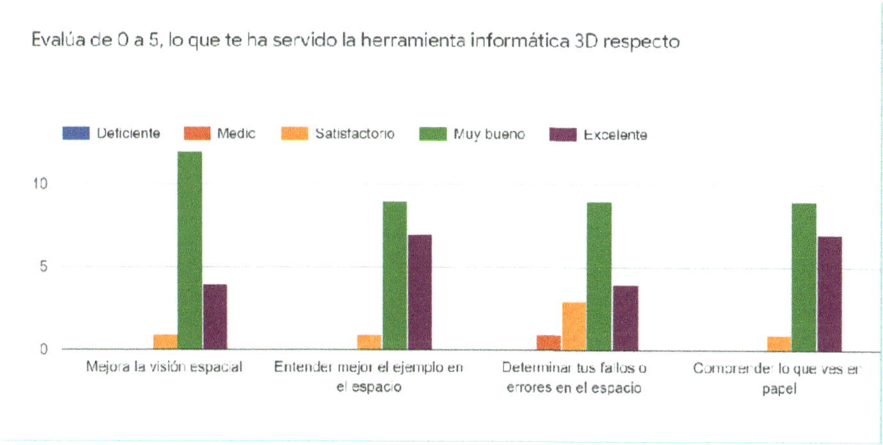

Fig. 12. Satisfaction level of using 3D tools on (1) spatial vison improvement; (2) having a better understanding of the exercise in the space; (3) identification of the mistakes in the space and (4) understand what is represented on the paper in 2D.

Finally, the level of acceptance of the visualization device (tablet, mobile, laptop or VR goggles) was analised and whether this new resource was considered motivating among other issues.

From our results, we can confirm that 90–100% of the surveyed people much improved their spatial vision in an excellent way. 80–90% people preferred the 3D system since they better understood the exercises spatially. 75–85% better understood their mistakes in the space. And, finally, around 90% better understood the exercises overview than with the 2D representation (Fig. 12).

4 Conclusions

With the aim of further adapting and improving the teaching-learning process, with the increasing implementation of new technologies in the field of education, it was possible to establish a method for the 3D virtualisation of exercises. Such virtualisation method allowed students to visualise the three-dimensional models of the exercises with any device from wherever they connected to the Internet, thus demonstrating its usefulness.

The procedure presented in this study shows the value of the physical archive of the engineering drawing exercises of the Department of Graphic Engineering and Geomatics at the University of Cordoba (Spain). The use of a digitalised resource allows guaranteeing its future application with the new technologies discussed in this article.

Acknowledgements. The authors would like to thank all the collaborators who participated in the experiences where this new procedure for exercise visualisation was implemented.

References

1. Milgram, P., Kishino, F.: A Taxonomy of Mixed Reality Visual Displays (1994)
2. Future Today Institute (2021)
3. Kadner, N.: The Virtual Production Field Guide, vol. II. Unreal Engine (2021)
4. Alvarez-Marin, A., Castillo-Vergara, M., Pizarro-Guerrero, J.: Augmented Reality as a Support to the Formation of Industrial Engineers (2017)
5. Yastikli, N.: Documentation of cultural heritage using digital photogrammetry and laser scanning. J. Cult. Herit. **8**(4), 423–427 (2007)
6. Azuma, R.T.: A survey of augmented reality. Presence: Teleoper. Virt. Environ. **6**(4), 355–385 (1997)
7. Zhou, Y., Lou, H., Yang, Y.: Automation in Construction (2017)
8. WebXR (2021). https://immersive-web.github.io/webxr/
9. Hughes, C.E., Stapleton, C.B., Hughes, D.E., Smith, E.M.: Mixed reality in education, entertainment, and training. IEEE Comp. Graph. Appl. **25**(6), 24–30 (2005)

Coordination of Subjects of the Product Design Mention

B. Micó-Vicent[1], J. Jordán-Nuñez[1(✉)], J. Gisbert Paya[2], and A. Molina-Picó[3]

[1] Departamento de Ingeniería Gráfica, Universitat Politècnica de València, Plaza Ferrándiz y Carbonell s/n, Alcoi, Spain
{barmivi,jorjornu}@upv.es

[2] Departamento de Ingeniería Textil y Papelera, Universitat Politècnica de València, Plaza Ferrandiz y Carbonell s/n, 03801 Alcoi, Spain
jaigispa@txp.upv.es

[3] Departamento de Informàtica de Sistemas y Computadores, Universitat Politècnica de València, Plaza Ferrándiz y Carbonell s/n, Alcoi, Spain
antoniomolina@disca.upv.es

Abstract. The last three subjects of the product design mention are: Foresight and Design, Design of products and environments for habitat, Product design for equipment. This year 2020/2021 in the degree in industrial design engineering and product development, has been coordinated through the faculty the development of a project for each subject with a common theme for the three subjects. It is the design and development of a kindergarten and some of its internal elements. In addition, in order to simulate the type of work in a company and to accustom students to teamwork with colleagues who do not choose, work teams have been selected in alphabetical order for the three subjects, taking into account the casuistry of the difference in the number of students enrolled in each subject. This methodology is intended to go a step further than the one applied in the previous course, where the teams were only applied in random order in only one subject. The response of the students during the first weeks was not very good due to the lack of habit of collaborating with colleagues who did not have feeling, but now they show to be in agreement with this organization.

Keywords: Coordination · Product mention · Teaching · Methodology · Team

1 Introduction

Within the degree in Industrial Design and Product Development Engineering taught at the Universitat Politècnica de València, is the mention of Product Design, which includes the three subjects involved in this work.

The subjects are:

- Foresight and Design (6 credits) [1]
- Product Design for Equipment (6 credits) [2]
- Design of Products and Environments for Habitat (6 credits) [3]

F. Cavas Martínez et al. (Eds.): INGEGRAF 2021, LNME, pp. 433–438, 2022.
https://doi.org/10.1007/978-3-030-92426-3_50

All of them are taught during the fourth year of the degree in the second semester, where in addition to the subjects, students usually devote their time to the development of the final degree project and internships.

During the 2020/2021 academic year, the professors of these three subjects have coordinated the subjects to allow students to work on a common theme and to organize them in a way that is most similar to what they will encounter in the future in the working world.

For this reason, the main objective of this teaching communication is to disseminate the coordination approach and the results achieved with this work methodology.

2 Subject Coordination

The coordination of subjects between the subjects of the Product Design mention is not the first time it has been done, but in the 2020/2021 edition, a further step is taken with the organization of the students. The methodology used is detailed below.

First of all, a common theme was decided to be developed in the three subjects in which the contents taught in each of them would be applied. The topic selected by consensus was the kindergarten and children's furniture, since this subject is quite broad and allows the development of all the contents.

From the course of Product and Environment Design for Habitat, a project of interior design of a daycare center is developed. In particular, we facilitate the location of a site of the University that must be reconditioned to convert the current use of Tatami (Gymnasium) to kindergarten. This is, of course, an academic exercise and does not involve practice. The kindergarten must also have all the measures contemplated in the current work for the prevention of COVID19.

The Product Design for Equipment course has worked as a project the creation and development of a piece of furniture for the interior of the kindergarten. The project consisted of making the furniture plans, the explosion, assembly plans, 3D rendering, calculation of efforts, and inserting this furniture in the kindergarten designed in the other subject.

In the Prospective and Design course, they work on the 25-year prospective project for furniture developed in the Equipment course. In such a way that they analyze and propose how should be a piece of furniture that performs these functions in the not immediate future for a day care center.

Working on these three projects with a common theme allows students to increase their knowledge of the subject, since most of the rules that they consult and restrictions are in agreement, and generates students to focus their skills. Otherwise, if we work with different topics for the three subjects, there is an endless number of rules to consult on different topics, which confuses the students during the semester.

The other part of the coordination of the subjects revolves around the organization of the students for the realization of the projects. In this 2020/2021 edition, a strategy has been applied to the three subjects that had been practiced in recent years only in the Prospective and Design subject. The idea is for students to carry out projects in groups in a way that is as similar as possible to what they would find working in a company. Based on the hypothesis that a worker does not decide his or her coworkers in a company, but must adapt and collaborate with the coworkers assigned to him or her by the employer, this concept has been applied to the subjects [4, 5].

For the realization of the projects, 5 groups of 5 students each have been organized, being the 5 groups the same for the three subjects. The selection of students for each group has been made by alphabetical order, because although it is not a random selection, it does not grant any selection based on any preference of talent or cronyism. The result is groups of students who almost certainly would not have been chosen in this way on their own, and the fact of working with another classmate of their choice is just a coincidence of the alphabet.

The difficulty encountered in this organization of people has been that the circumstances of each student to reach this last semester of the degree are very different, and therefore, not all students are enrolled in the three subjects. In particular, 22 of the students are enrolled in the three subjects, and with them 2 groups of 5 people and 3 groups of 4 people have been prepared. Subsequently, the groups of 4 persons have been completed with students coming from Erasmus or who only had one of the subjects of the mention.

3 Results

As expected, the students showed a willingness to work with a common theme in all three subjects, but there was poor acceptance of the imposition of groups. The reason for this is that any student prefers the comfort of grouping with people with whom they are more like-minded or with whom they have worked successfully before.

During the course of the semester, weekly tutorials have been held with all the groups to check that the required tasks are being fulfilled and that the internal organization is adequate. In all the groups, the attendance to the tutorials has been the same most of the weeks, except for a few occasions in which the workload was higher due to other subjects or personal circumstances. In these tutorials what was observed was that normally 2 or 3 people in the group were the ones who usually took the floor to ask questions or comment on aspects of the work. In all the groups, 1 or 2 people were more likely to be listeners. It was also observed that the people who interacted less were those who had only one of the three subjects or who were Erasmus students. This fact reflects that the methodology of coordinating the three subjects is only effective for students who have all three subjects, and for those who are not in this situation it does not seem to affect them in the same way, since for them it is like another subject of the degree.

After a final meeting with the students, they all agreed that despite the differences they have had during the projects, they all agree that this organization has prepared them to work with teams that they do not choose themselves, as they expect to happen when they arrive at the company.

The results obtained in the projects in terms of designs have been very satisfactory, as can be seen in Figs. 1, 2, 3, and 4.

Fig. 1. Separator and playground furniture for kindergarten (Equipment subject).

Fig. 2. Render 3d kindergarten project (Habitat subject).

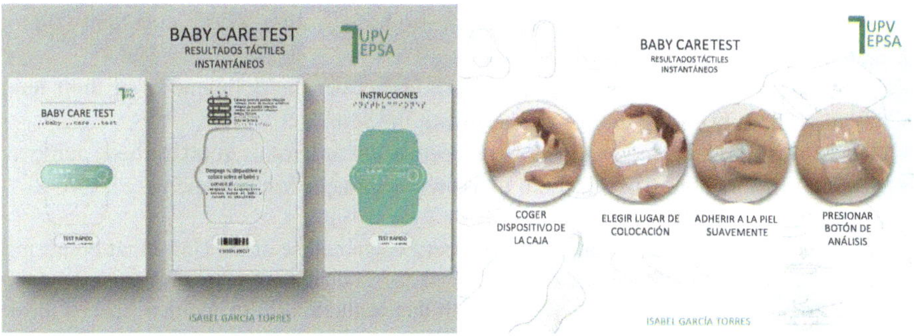

Fig. 3. Infant health monitoring device (Prospective subject).

Fig. 4. APP temperature control and alarm device (Prospective subject).

4 Conclusions

The most important conclusions reached in this work are that following this methodology of coordination and organization of people generates a work environment very similar to that of a company, and the students understand that this practice is good for them before finishing their degree. It is also true that the students believe that they only see it as convenient for final year subjects because throughout a semester important differences can arise between group members and there is no boss or tutor who is constantly on top of the team to calm situations.

The groups are able to develop teamwork and cover the deficiencies of their team members without bringing them to light during the semester.

The results of the projects in the design section are at least as good as those achieved in previous editions, but they all seem to show no signs of improvement.

Acknowledgments. The professors of the Product Design course of the Univertitat Politècnica de València at the Alcoy Campus would like to thank the students of the 2020/2021 fiscal year for their cooperation and predisposition to work with this methodology. In addition, special thanks to Group 2 formed by María Gandía Ferri, Isabel García Torres, Clara Gómez Sanchís, Alicia González Sánchez, and Claudia Olivares del Amor, for having developed the best projects that have been shown in this work.

References

1. Guía docente de la asignatura Prospectiva y Diseño. https://gdocu.upv.es/alfresco/service/api/internal/shared/node/content/Ylz8OSb2Qui23yC96lKOcg?a=true.pdf. última consulta 07 June 2021
2. Guía docente de la asignatura Diseño del Producto para Equipamiento. https://gdocu.upv.es/alfresco/service/api/internal/shared/node/content/aLdl6E6-QgaANHiIm6Zbng?a=true.pdf. última consulta 07 June 2021
3. Guía docente de la asignatura Diseño de Productos y Ambientes para Hábitat. https://gdocu.upv.es/alfresco/service/api/internal/shared/node/content/K3_UBt2EQ2qzPQ8gpcZlWA?a=true.pdf. última consulta 07 June 2021
4. Rojas, M., Jaimes, L., Valencia, M.: Efectividad, eficacia y eficiencia en equipos de trabajo. Revista espacios **39**(6) (2018)
5. Mañas, M.A., Díaz-Fúnez, P., Pecino, V., López-Liria, R., Padilla, D., Aguilar-Parra, J.M.: Consequences of team job demands: Role ambiguity climate, affective engagement, and extra-role performance. Front. Psychol. **8**, 2292 (2018)

Motivational Activity for Presentation Techniques

J. Jordán-Nuñez[1(✉)], B. Micó-Vicent[1], A. Molina-Picó[2], and M. Moncho-Santonja[3]

[1] Departamento de Ingeniería Gráfica, Universitat Politècnica de València, Plaza Ferrándiz y Carbonell s/n, Alcoi, Spain
{jorjornu,barmivi}@upv.es

[2] Departamento de Informàtica de Sistemas y Computadores, Universitat Politècnica de València, Plaza Ferrándiz y Carbonell s/n, Alcoi, Spain
antoniomolina@disca.upv.es

[3] Departamento de Ingeniería Gráfica, Centro de Investigación en Tecnologías Gráficas (CITG), Universitat Politècnica de València (UPV), Valencia, Spain
monsan@upv.es

Abstract. Traditionally the subject of "Presentation Techniques" taught in the Degree in Industrial Design and Product Development Engineering, has been approached from the point of view of the master class with sessions in which traditional techniques are explained and exhibitions or demonstrations of these same techniques. In the current 2020/2021 fiscal year, a new activity has been prepared that reinforces above all the motivation of the students in front of a presentation. The presentations that we usually ask the students almost always lack personal motivation for the subject or the work, since they are imposed by the teachers. The proposed activity consists of the adaptation and redesign of a board game to the student's taste, in this way the student works on a game that he/she likes and usually uses with friends or family. Of course there is a valuation of the game, but the most important and where the subject wins a lot is in the presentation of the game as the student acquires a personal motivation on the work, and strives to sell it in the best possible way.

Keywords: Game · Motivation · Redesign · Methodology

1 Introduction

One of the subjects of the Degree in Industrial Design and Product Development Engineering called "Product Presentation Techniques" [1] is taught at the Alcoy campus of the Universitat Politècnica de València. This same subject at the Vera campus is taught under the name "Graphic analysis and presentation of the industrial product".

The content of both subjects is practically identical and the credit loads and distribution of hours are the same, which is appropriate for the reaccreditation of the degree.

Within the contents of the course there are topics related to the presentation of products as the title of the course suggests, and a major problem detected during the

F. Cavas Martínez et al. (Eds.): INGEGRAF 2021, LNME, pp. 439–444, 2022.
https://doi.org/10.1007/978-3-030-92426-3_51

previous editions of the course is the lack of motivation of the students for the presentation of products. Given that in previous courses we have tried to replace the products to be presented by others more related to the student's profile, allowing them to choose a product from a thematic area and the results are still lack of motivation, it has been decided to try a new strategy to improve this aspect of the course.

Motivation turns out to be a key aspect in the person who presents anything, and it has been observed that people with a higher motivation are those who have developed the product because they feel a bond with the product [2]. It could be misnamed as a parental attachment to the product.

This link to the product makes the person have a greater interest in leaving the product in a good place and that his effort is greater. In addition, this extra effort is usually compensated by personal satisfaction and the student has no objection to devote the necessary hours. Obviously, the student also expects the reward in terms of appreciation and recognition from teachers and other colleagues to be greater due to all the factors mentioned above, and normally he will almost certainly be able to achieve it [3].

The main objective of this teaching communication is to show the results obtained with the incorporation of a new activity in project format, and the results of the students' opinion survey about the activity so that other teachers can consider the incorporation of this activity or similar ones that provoke that impulse in the motivation of the students.

2 Description of the Activity

The activity proposed for this subject is the creation or redesign of a board game or card game in the style of the students' choice. This activity will allow them to put into practice many of the skills and knowledge acquired during the course, and the application of new knowledge acquired in this subject such as editing documents with text and images in Adobe Indesign for the instructions of the game.

The development of the project has been entrusted to be carried out in groups of 2 or 3 people and at least one playable prototype must be built, containing all the usual elements of the games, such as: board, tokens, cards, instructions, packaging, ...

In addition, taking advantage of new technologies and social networks, they have been asked to document the entire creation process from the design to the manual fabrication of the prototype with photographs and publications on social networks such as Instagram.

This type of documentation gives them a nice reminder of the process, of working with their team at the university, and these are things that they are very used to practicing in their daily lives. In addition, the recognition they receive from their contacts on social networks in the form of "likes" or "I love it" motivates them a lot. Also seeing the progress of their classmates with the development of the game and seeing the fantastic visual results motivates them to improve their project to be the best.

The whole game project is planned so that the students can carry it out within the class hours using time from the practices, and therefore this edition of the course may be enough time.

The disadvantage of this project is that the manufacture of the prototype can generate a monetary cost that not all students are happy with. It must be said that this has not been a problem for the great majority and that many groups have decided to make a prototype for each group member in order to keep a souvenir of this activity.

The purpose of the activity is that when they make the presentation of the product they have that extra motivation that comes from selling your own product so that the presentation is much more powerful.

3 Project Results

The choice of games to redesign was based mostly on classic board and card games, the most selected being Hasbro's Cluedo. There have been other choices such as Monopoly, Who's Who, Trivial, Pictureka, El Lobo, and decks of cards. In all cases of redesign, students have renamed the game with names that hint at the theme of the redesign. In other cases, there are groups that have decided to develop a new game in which they have even had to create the game mechanics. This aspect of the game mechanics is clearly not related to the subject and degree, but it makes more evident the motivation that the students have to carry out this type of project. As the game cards would say, +1 in Motivation.

One of the most striking examples of redesign was the "Fourteen districts" based on the original game "Seven Wonders". In this work, the two members of the group made an adaptation to the theme of the movie "The Hunger Games", choosing to rewatch the entire saga and even stopping scene after scene to capture the relevant elements that they wanted to appear in the game as shown in Fig. 1. They also made more than 100 different action cards, 14 boards (the original game has 7), tokens, packaging and instructions. All this in digital format and physical format with the manual manufacturing of the prototype of the playable game as illustrated in Fig. 2.

MINUTO	NOMBRE	DESCRIPCIÓ N ESCENA	IMAGEN
00:11:37	Cosecha	Niños apilados en la plaza	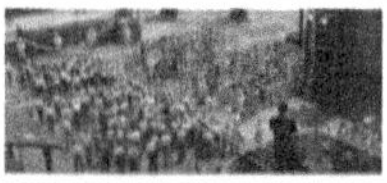
00:11:22	Pruebas genéticas	Niños en la plaza dejando una huella de sangre para análisis	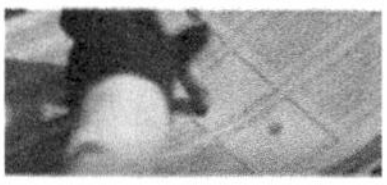
00:14:31		Bola papeletas selección tributos	
00:15:56		Katniss es llevada escoltada	

Fig. 1. Example of a table of scenes in Fourteen districts.

Fig. 2. Packaging and manufacturing of the Fourteen districts set.

As expected, when it was time for the presentation of the game in class the motivation was total, and the students conveyed a confidence and mastery of the product that reminded the great directors of the technology industries when they present their great novelties. It must be said that it could have been more emotional if it had been a face-to-face presentation, because due to the pandemic, the presentation had to be done by Teams for all the groups.

Another highlight was the "Clown fiesta", which was not a redesign, but a new game created by the student. The result in the presentation was so spectacular that I even suggested to the student to pitch resumes to companies in the board game industry because the prototype was a version that could look like it was sold in a store and her exposure was what they expect in companies from a marketing leader (Figs. 3 and 4).

Fig. 3. Manufacturing process of the Clown Party game.

Fig. 4. Final prototype of the game Clown party.

On the other hand, after seeing the results of the projects and presentations, a survey was made to the students to see if they had liked the activity and if they had noticed differences between the presentation made in this subject and others. The survey was prepared from the Google Forms application and was sent by e-mail to all students with the only restriction of only one response per student. The results observed were as follows:

- No. of responses received: 68
- Male: 25 (36.8%) Female: 43 (63.2%)
- Satisfied with the game project:
  Yes 62 (91.2%) No 6 (8.8%)
- Have you spent more hours than expected?:
  Yes 68 (100%) No 0 (0%)
- Do you think your involvement in the project has increased your motivation?:
  Yes 62 (91.2%) No 6 (8.8%)
- Was your presentation better, the same or worse than other subjects?:
  Better 60 (88.2%) Equal 8 (11.8%) Worse 0 (0%)
- What do you think has influenced the improvement of the presentation?:
  Program 5 (7.4%) Proyect 3 (4.4%) Both 58 (85.3%) Neither 2 (2.9%)

The most relevant answers of the survey highlight that the students have been satisfied with the project, that the project has considerably increased their involvement and thus their motivation, that their perception is that they have improved their presentations with respect to those made in other subjects, and that the vast majority believe that the main reason for their improvement in presentations is due to what they have learned in the subject and their involvement in the game project.

Therefore, it can be affirmed that the game activity has proved to be motivating and has helped them to improve their presentation techniques, and it is believed that this type of activity in which students become more personally involved in the development of a project can improve other aspects that they are expected to acquire in other subjects.

4 Exploitation of the Works

At the request of the students, it was decided to disseminate the work so that the university community could see what they are capable of doing and serve as an example for other subjects where they can do similar activities. The communication department of the Universitat Politècnica de València was contacted to see if it was possible to publish a news item on the university website, and after knowing the content of the news item they agreed immediately.

The news and images can be viewed at the following links:

- http://www.epsa.upv.es/noticia_detalle.php?lang=es&idnot=2464
- https://www.flickr.com/photos/152475265@N05/albums/72157718323098168

An exhibition of the prototypes will also be held at the Alcoy campus of the Universitat Politècnica de València during the "Employment Forum" event. The idea is to call companies from the toy sector to participate in the Employment Forum and thus make it easier for them to see the work developed by the students.

Acknowledgments. This work has been possible thanks to the collaboration of the students of the Degree in Industrial Design and Product Development Engineering enrolled in the subject product presentation techniques in the course 2020/2021 at the UPV. It has also been thanks to the collaboration of many professors of the Alcoy campus of the UPV by guiding students in some manufacturing processes, and the Universitat Politècnica de València for making available workshops and machinery not previously foreseen for the subject.

References

1. Guía docente de la asignatura Técnicas de presentación de productos. https://gdocu.upv.es/alfresco/service/api/internal/shared/node/content/Zy4MEYN-TAmem1NWJlZkww?a=true.pdf. última consulta 04 June 2021
2. Deci, E.L., Vallerand, R.J., Pelletier, L.G., Ryan, R.M.: Motivation and education: the self-determination perspective. Educ. Psychol. **26**(3–4), 325–346 (1991)
3. Ames, C.: Classrooms: goals, structures, and student motivation. J. Educ. Psychol. **84**(3), 261 (1992)

Author Index

F. Cavas Martínez et al. (Eds.): INGEGRAF 2021, LNME, pp. 445–447, 2022.
https://doi.org/10.1007/978-3-030-92426-3

The manufacturer's authorised representative in the EU is Springer Nature Customer Service Centre GmbH, Europaplatz 3, 69115 Heidelberg, Germany. If you have any concerns regarding our products, please contact ProductSafety@springernature.com

Printed and bound by CPI Group (UK) Ltd, Croydon, CR0 4YY
30/09/2026
02225291-0017